OGIC SERIES

Department of Computer Science, King's College, London,

t

tment of Computer Science, King's College, London, U.K.

SERIES

n an increasingly wide variety of disciplines, from the traditional subjects
mathematics to the more recent disciplines of cognitive science, compu-
al intelligence, and linguistics, leading to new vigor in this ancient subject.
s Applied Logic Series, seeks to provide a home for outstanding books and
phs in applied logic, and in doing so demonstrates the underlying unity and
gic.

in this series are listed at the end of this volume.

Foundations of Bayesianism

APPLIED L

VOLUME 24

Managing Editor

Dov M. Gabbay,
U.K.

Co-Editor

Jon Barwise†

Editorial Assista

Jane Spurr, *Depa*

SCOPE OF THE
Logic is applied i
of philosophy an
ter science, artific
Kluwer, through i
research monogra
applicability of lo

The titles publish

Foundations of Bayesianism

Edited by

DAVID CORFIELD
Department of Philosophy,
King's College London, U.K.

and

JON WILLIAMSON
Department of Philosophy,
King's College London, U.K.

KLUWER ACADEMIC PUBLISHERS
DORDRECHT / BOSTON / LONDON

A C.I.P. Catalogue record for this book is available from the Library of Congress.

ISBN 1-4020-0223-8

Published by Kluwer Academic Publishers,
P.O. Box 17, 3300 AA Dordrecht, The Netherlands.

Sold and distributed in North, Central and South America
by Kluwer Academic Publishers,
101 Philip Drive, Norwell, MA 02061, U.S.A.

In all other countries, sold and distributed
by Kluwer Academic Publishers,
P.O. Box 322, 3300 AH Dordrecht, The Netherlands.

Printed on acid-free paper

Printed in the Netherlands.

CONTENTS

Editorial Foreword vii

Editorial Preface xiii

Introduction: Bayesianism into the 21st Century 1
Jon Williamson and David Corfield

Bayesianism, Causality and Networks

Bayesianism and Causality, or, Why I am only a Half-Bayesian 19
Judea Pearl

Causal Inference without Counterfactuals 37
Philip Dawid

Foundations for Bayesian Networks 75
Jon Williamson

Probabilistic Learning Models 117
Peter Williams

Logic, Mathematics and Bayesianism

The Logic of Bayesian Probability 137
Colin Howson

Subjectivism, Objectivism and Objectivity in Bruno de Finetti's Bayesianism 161
Maria Carla Galavotti

Bayesianism in Mathematics 175
David Corfield

Common Sense and Stochastic Independence 203
Jeff Paris and Alena Vencovská

Integrating Probabilistic and Logical Reasoning 241
James Cussens

Bayesianism and Decision Theory

Ramsey and the Measurement of Belief 263
Richard Bradley

Bayesianism and Independence 291
Edward F. McClennen

The Paradox of the Bayesian Experts 309
Philippe Mongin

Criticisms of Bayesianism

Bayesian Learning and Expectations Formation: Anything Goes 341
Max Albert

Bayesianism and the Fixity of the Theoretical Framework 363
Donald Gillies

Principles of Inference and their Consequences 381
Deborah Mayo and Michael Kruse

Index 405

FOREWORD

COMBINING PROBABILISTIC AND LABELLED REASONING

I welcome this volume, on Bayesianism, to our Applied Logic Series. This is an important thematic volume containing papers in the interface area between probabilistic networks and ordinary logic. In fact, several of the papers in the volume address directly the problem of combining these two types of reasoning. I believe the next evolutionary step in the historical development of (formal methods of) practical reasoning must include theories of integration of ordinary logic (and its numerous varieties) with probabilistic reasoning and with neural reasoning and models.

Originally, the plan was that I include a contribution to the volume; a paper on probabilistic networks and labelled deductive systems. However, time being short and the task complex, I have only the initial ideas at this point in time. The paper was also going to serve as a background source on combining logics for the reader. This task I propose to do in this foreword.

Let us start with a case study which has meaning both from the probabilistic and from the pure logic point of view.

Consider a language with $\rightarrow$ only and some atomic statements $a, b, c, d \ldots$. let us read $d \rightarrow p$ as an insurance policy. A commitment that if at any point in time damage is done then payment will be made. First assume that the policy premium is paid by direct debit and so the policy is practically open ended.

Given d, we can do modus ponens and get p. However, the damage must be done at a time after the policy was taken, not before. We symbolise this restriction by writing the data as a sequence, indicating the temporal order of becoming true. Thus we have

$$(d \rightarrow p, d) \vdash p$$

but

$$(d, d \rightarrow p) \nvdash p.$$

When we think about this kind of model we find that many such examples arise in practice. Here are some

1. submit thesis $\rightarrow$ (pass viva $\rightarrow$ get PhD)‘

2. approve project $\rightarrow$ (spend expenses $\rightarrow$ get reimbursed)

The order of $x \to (y \to z)$ is first x and then y gives z.

Thus

$$(x \to (y \to z), x, y) \vdash z$$

but

$$(x \to (y \to z), y, x) \nvdash z.$$

In example (1) the student will probably not be allowed to take the viva before submitting his thesis, but in (2) it is quite possible that spending occurs a little before (and probably in anticipation of) the project being awarded.

Let us now describe another logic, a slight variation of the above. Suppose $x \to y$ has a limited validity. It is not open ended. Let us assume that time runs in days and for simplicity $x \to y$ is valid the next day only. Thus $x \to (y \to z)$ means that if next day we have x and the day after we have y then we can get z (on the same day as y).

Let us write $x \to_1 y$ for this kind of implication. Now we can have a logic with the implications $\to_1$ and $\to$. How are we going to integrate them? We need to see what kind of problems to expect.

Let us start with two implications $\Rightarrow_1$ and $\Rightarrow_2$ satisfying modus ponens and the deduction theorem. We want to put them together. If we just do that, unfortunately they collapse. Here is a proof.

To show

$$x \Rightarrow_1 y \vdash x \Rightarrow_2 y$$

we need to show

$$(x \Rightarrow_1 y, x) \vdash y$$

which holds by modus ponens.

Obviously we need to be careful. Let us look at the second line of the proof more closely. We have a database $\Delta = \{x \Rightarrow_1 y\}$ and we want to prove a wff of language 2 namely $x \Rightarrow_2 y$. So we are in a language 2 *mode*. We add x to Δ as a language 2 item of data and want to show y, using a language 2 proof. Why should x be accessible to modus ponens with $x \Rightarrow_1 y$?

This is the modus ponens of language 1 proofs.

If x, (which has been added during a language 2 proof) is accessible, then we can get y.

If we take another look at our case study with $\to$ and $\to_1$, then

$$x \to y \vdash ?x \to_1 y$$

does not hold because if we put x in the data it should be accessible to $x \to y$, since $\to_1$ insists on x being true on the next day, while $\to$ does not care about next or later days. The other way round does not work, i.e.

$$x \to_1 y \nvdash x \to y.$$

If we add x to the data, x may not be accessible to $\to_1$ because $\to$ does not insist on x being true the next day.

Now let us go back to

$$x \to y \vdash x \to_1 y.$$

We agreed that x is accessible to $x \to y$. We thus get y. But we got y in the logic of $\to$. We need to show that y is accessible in the logic of $\to_1$. Is y now accessible to the logic of $\to_1$?

The answer is yes, because both logics yield y in the same day as x.

To highlight that there can be a difference, assume that the insurance policy $x \to_2 y$ pays (gives y) the day after x while $x \to_1 y$ delivers y on the same day as x.

So let's try again

$$x \to_2 y \vdash ?x \to_1 y.$$

Since we assumed $\to_2$ accepts x at any time, x is accessible to $\to_2$ and we get y. However, y is available not on the same day as x but a day later. So we cannot 'export' y to the $\to_1$ proof procedures because $\to_2$ expects y on the same day as x.

The moral of the above is that when we put two languages $1, 2$ together and try to combine their proof procedures, we need two *fibring functions*, which we denote by $\mathbb{F}_{1,2}$ and $\mathbb{F}_{2,1}$. Such functions $\mathbb{F}$ take any database Δ of one logic into another database $\mathbb{F}(\Delta)$ of another logic. So given a database Δ in the mixed language and assume that we are in proof mode 1, we can switch to proof mode 2, provided we consult $\mathbb{F}_{1,2}$ to tell us what database is available for us to use at the moment of the switch, namely, we can use $\mathbb{F}_{1,2}(\Delta)$. We can now get, using language 2 proof rules a new Δ', and then if we want to switch back to the proof rules of language 1, we need to consult $\mathbb{F}_{2,1}(\Delta')$.

Obviously a discipline of labelling of data and proofs needs to be installed to allow us to define the function $\mathbb{F}$ in both directions, since what can be used will depend on how the data was historically proved, therefore the use of labelling.

Having considered the example above, let us see what we need to figure out to put together probabilistic networks and say implication. Networks use atoms only and connect them in an acyclic graph. Let us take a look at the following figure 1.

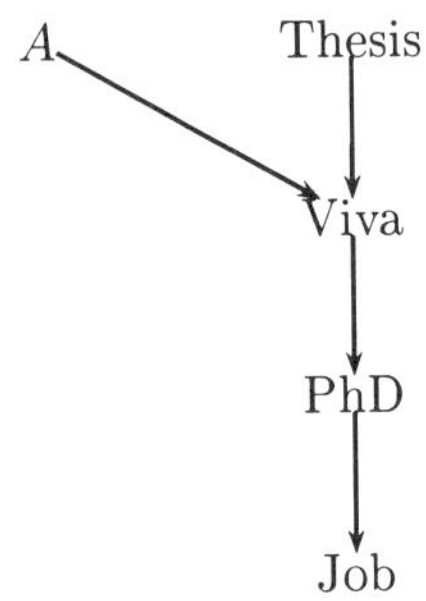

Figure 1.

We assume that A is some factor which can affect the outcome of the Viva.

Suppose now we turn this network into a combined logic and probability network by letting $A = (\text{Viva} \rightarrow \text{Job})$.[1]

In this case A certainly can influence the outcome of the Viva, especially for borderline candidates.

The question now is that suppose we add to the database the additional item Viva $= \top$, how do we construct the new probability network? How do we reason with it? What probability do we get for the event Job?

We have a mixed database Δ. A logical expression Viva $\rightarrow$ Job embedded in a node and a logical data item Viva $= \top$ as an additional member of Δ.

Our first restriction is to allow into the network $A = X \rightarrow Y$ only X, Y s.t. in the acyclic graph Y is a descendant of X. In other words, we are only short-circuiting the existing causal chain.

Our first step for coherence is to add to the network a direct link from X to Y and add a new conditional probability distribution. Our network looks as follows (figure 2).

Having made the new connection, A is now considered atomic. The effect of the implication has been fibred into the network by the new link and the new conditional probability function for the new link. This function must satisfy some coherence conditions.

(i) If the probability of A is identical to 0 we do not make the link.

[1]What we are doing is substituting an expression Viva $\rightarrow$ Job in the language of $\rightarrow$ for an atomic A of the language of the networks. If this $\rightarrow$ is the causal language of the insurance discussed above, then it is not a probabilistic connection but an absolute one. In terms of the insurance policy $d \rightarrow p$, an absolute interpretation means that the insurance company pays for sure when damage incurs. In probabilistic terms the policy just increases the chances of recovering some compensation for the damage, but it is not certain the company will pay, as policies have so many exclusions and insurance companies always look for excuses not to pay.

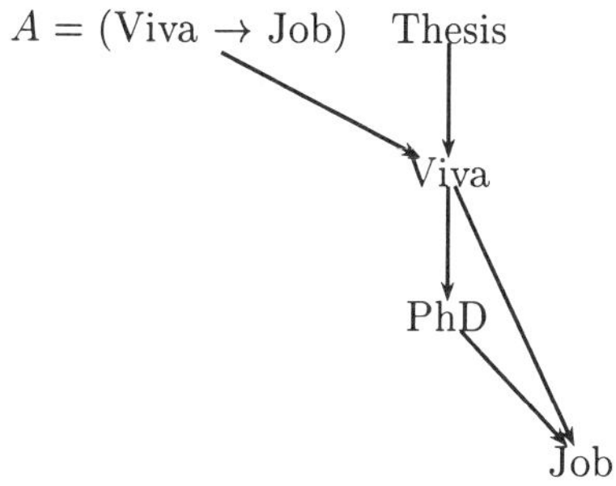

Figure 2.

(ii) If the probability of A is identical to 1 we must rearrange the conditional probabilities in the new network to yield *probability*(Job|PhD, Viva $= \top$) to be 1.

(iii) Otherwise some formula can be worked out for the general case.[2]

Now to reason from Viva $= \top$, we calculate as usual in the new network, regarding A as atomic.

Let us now consider putting networks into logic. Let us take the most typical logical deduction — modus ponens.

$$A \to B, A \vdash B.$$

Instead of $A \to B, A$ we substitute the networks of figure 3.

This looks like, in network terms, the case where we get $A = \top$ in the $A \to B$ network, except that here we set the probability of A as the new one of the A network. Some other combination may also be reasonable.

It may be more complicated to provide a formula for new probabilities for the case where A is inside a network, as in figure 4.

I hope the above discussion gave the reader a taste of the kind of problems we encounter in integrating probability and logic.

I have taken advantage of the relative freedom allowed in an editorial to present the case before all the details have been worked out.

---

[2] Jon Williamson and I will have a paper on this topic.

Note that if we read $A = X \to Y$ probabilistically, as remarked in the previous footnote, then X raises the probability of Y, conditional on Y's other parents, i.e.

$$\textit{probability}\ (Y|X \wedge \pi) > \textit{probability}\ (Y|\neg X \wedge \pi)$$

for each state π of Y's other parents.

Going back to our specific example of $A =$ Viva $\to$ Job, condition (ii) would need to be weakened. We only need to ensure that Job and Viva are probabilistically dependent on PhD.

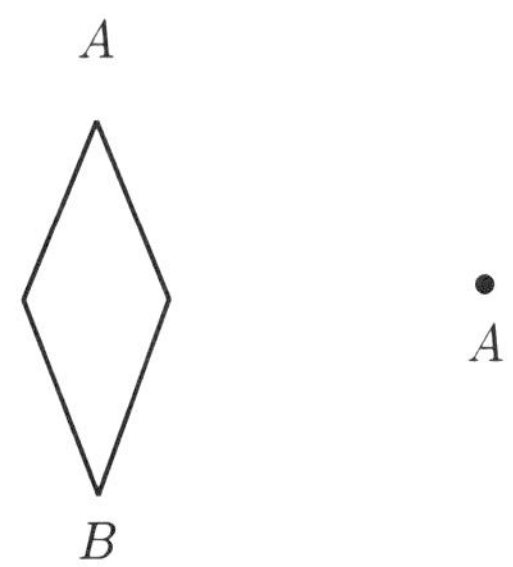

Figure 3.

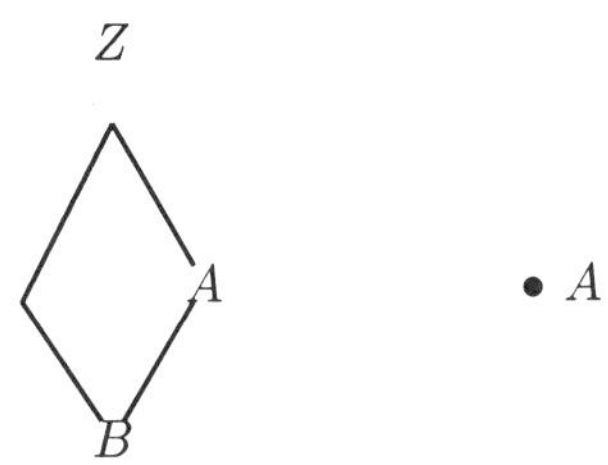

Figure 4.

BIBLIOGRAPHY

[Gabbay, 1998] D. M. Gabbay. *Fibring Logics*. Clarendon Press, Oxford, 1998.

[Gabbay, 1996] D. M. Gabbay. *Labelled Deductive Systems 1*. Clarendon Press, Oxford, 1996.

Dov M. Gabbay
London

PREFACE

Several chapters in this collection were presented at the conference 'Bayesianism 2000', held at King's College London on the 11th and 12th May 2000. We would like to thank the Centre for Philosophical Studies at King's College London and all the speakers and participants for helping to make it a great success. 'Causal inference without counterfactuals' appeared in the Journal of American Statistical Association 95 (June 2000), pages 407-427, and 'The paradox of the Bayesian experts and state-dependent utility theory', appeared in the Journal of Mathematical Economics 29 (no. 3, 1998), pages 331-361 — thanks to the American Statistical Association and to Elsevier Science respectively for allowing the reprinting of these papers. Thanks also to Oxford University Press for allowing Colin Howson to reproduce passages from his book *Hume's Problem: Induction and the Justification of Belief*.

We are also very grateful to Donald Gillies and Juliana Cardinale for valuable editorial advice, to Jane Spurr and Dov Gabbay for their publication assistance, and to the Leverhulme Trust and the UK Arts and Humanities Research Board for supporting this project financially.

Jon Williamson and David Corfield
London

JON WILLIAMSON AND DAVID CORFIELD

INTRODUCTION: BAYESIANISM INTO THE 21ST CENTURY

1 BAYESIAN BELIEFS

Bayesian theory now incorporates a vast body of mathematical, statistical and computational techniques that are widely applied in a panoply of disciplines, from artificial intelligence to zoology. Yet Bayesians rarely agree on the basics, even on the question of what Bayesianism actually is. This book is about the basics — about the opportunities, questions and problems that face Bayesianism today.

So what is Bayesianism, roughly? Most Bayesians maintain that an individual's degrees of belief ought to obey the axioms of the probability calculus. If, for example, you believe to degree 0.4 that you will be rained on tomorrow, then you should also believe that you will not be rained on tomorrow to degree 0.6. Most Bayesians also maintain that an individual's degrees of belief should take prior knowledge and beliefs into account. According to the *Bayesian conditionalisation* principle, if you come to learn that you will be in Manchester tomorrow (m) then your degree of belief in being rained on tomorrow (r) should be your previous conditional belief on r given m: $p^{t+1}(r) = p^t(r|m)$. By *Bayes' theorem* this can be rewritten $p^t(m|r)p^t(r)/p^t(m)$.[1]

Although Bayesianism was founded in the eighteenth century by Thomas Bayes[2] and developed in the nineteenth century by Laplace,[3] it was not until well into the twentieth century that Frank Ramsey[4] and Bruno de Finetti[5] provided credible justifications for the degree of belief interpretation of probability, in the shape of their *Dutch book* arguments. A Dutch book argument aims to show that if an agent bets according to her degrees of belief and these degrees are not probabilities, then the agent can be made to lose money whatever the outcome of the events on which she is betting. Already by this stage we see disagreement as to the nature of Bayesianism, centring on the issue of *objectivity*. De Finetti was a *strict subjectivist*: he believed that probabilities only represent degrees of rational belief, and that an agent's belief function is rational just when it is a probability function — no further constraints need to be satisfied.[6] Ramsey, on the other hand, was a *pluralist* in that he also accepted objective frequencies. Further, he advocated a kind of calibration between degrees of belief and frequencies:

[1][Howson & Urbach, 1989; Earman, 1992] and [Gillies, 2000] are good introductions to Bayesian thought.
[2][Bayes, 1764].
[3][Laplace, 1814].
[4][Ramsey, 1926].
[5][de Finetti, 1937].
[6]See Galavotti's paper in this volume.

D. Corfield and J. Williamson (eds.), Foundations of Bayesianism, 1–16.

> Thus given a single opinion, we can only praise or blame it on the ground of truth or falsity: given a habit of a certain form, we can praise or blame it accordingly as the degree of belief it produces is near or far from the actual proportion in which the habit leads to truth. We can then praise or blame opinions derivatively from our praise or blame of the habits that produce them.[7]

Such a view may be called *empirical Bayesianism*: degrees of belief should be calibrated with objective frequencies, where they are known.[8] Ramsey was cautious of too close a connection because of the reference class problem: Bayesian probabilities are single-case, defined over sentences or events, whereas frequencies are general-case, defined over classes of outcomes, and there may be no way of ascertaining which frequency is to be calibrated with a given degree of belief. The Principal Principle of [Lewis, 1980] aims to circumvent this problem by offering an explicit connection between degrees of belief and objective *single-case* probabilities. De Finetti shows that in certain circumstances, if degrees of belief are *exchangeable* then they will automatically calibrate to frequencies as Bayesian conditionalisation takes place.[9]

John Maynard Keynes advocated *logical Bayesianism*: a probability $p(b|a)$ is the degree to which a partially entails b, and also the degree to which a rational agent should believe b, if she knows a.[10] Thus for Keynes probability is truly objective — there is no room for two agents with the same knowledge to hold different belief functions yet remain perfectly rational. Moreover probability is fixed not by empirical frequencies but by logical constraints like the *principle of indifference*, which says that if there is no known reason for asserting one out of a number of alternatives, then all the alternatives must be given equal probability. There are problems with the principle of indifference which crop up when there is more than one way of choosing a suitable set of alternatives, but the *maximum entropy principle*, ardently advocated by Edwin Jaynes,[11] has been proposed as a generalisation of the principle of indifference which is more coherently applicable.

Empirical and logical Bayesianism may be grouped together under the banner of *objective Bayesianism*. Objective Bayesians may adopt a mixed approach: for example Rudolf Carnap had a position which incorporated both empirical and logical constraits on rational belief.[12] Objective Bayesians disagree with a strict subjectivist like de Finetti, since they claim that it is not sufficient that a belief function satisfies the axioms of probability — it must satisfy further constraints before it can be called rational. But objective Bayesianism harbours many views and proponents often disagree as to which extra constraints must be applied. Also, unlike Keynes many objective Bayesians accept

[7] [Ramsey, 1926] 51.

[8] See [Dawid, 1982].

[9] [de Finetti, 1937]. See also [Gaifman & Snir, 1982].

[10] [Keynes, 1921].

[11] [Jaynes, 1998].

[12] [Carnap, 1950], [Carnap & Jeffrey, 1971].

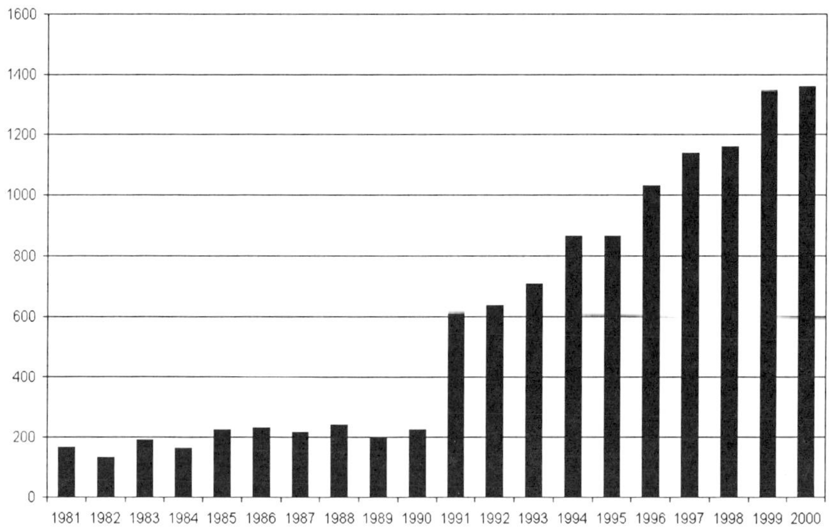

Figure 1. Number of Bayesian articles by year.

that in some situations there may be more than one rational probability function — two rational agents may have the same background knowledge but different belief functions.[13]

The question of objectivity remains an important issue for Bayesians today, and one that will crop up in several papers in this book.

2 BAYESIANISM TODAY

The last decade of the twentieth century has witnessed a dramatic shift in the profile of Bayesianism. Bayesianism has emerged from being thought of as a somewhat radical methodology — for enthusiasts rather than research scientists — into a widely applied, practical discipline well-integrated into many of the sciences. A search of the Web of Science database for articles whose subject contains the word or prefix 'Bayes' shows a dramatic upturn in the number of Bayesian papers in the 1990s — see Figure 1. A search for Bayesian books on the British library catalogue tells a similar story, as do other searches,[14] and the rise in the number of Bayesian meetings and the success of new organisations like the International Society for Bayesian Analysis[15] provide further evidence.

[13]See [Williamson, 1999] and the paper of Paris and Vencovská in this volume.

[14][Berger, 2000] §2.1.

[15]ISBA was established in 1992. See www.bayesian.org.

This renaissance has occurred largely thanks to computational and sociological considerations. The calculation of the posterior probability of a hypothesis given data can require, via Bayes theorem, determining the values of integrals. These integrals may often have to be solved using numerical approximation techniques, and it is only recently that computers have become powerful enough, and the algorithms efficient enough, to perform the integrations. The sociological changes have been on two main fronts. First, scientific researchers, who are usually taught to present their work as objectively as possible, were often discouraged from applying Bayesian statistics because of the perceived irreducible subjectivity of Bayesianism. This has changed as objective Bayesian techniques have become more popular. Second, Bayesian statistics has to a certain extent unified and absorbed classical techniques. Any religion worth its salt absorbs the gods of its competitors, and 'Bayesianity' is no different:[16] the diverse and seemingly unrelated techniques of classical statistics have been viewed as special-case approximations to Bayesian techniques, and Bayesianism has been invoked to shed light on the successes as well as the failures of classical statistics.[17] Present-day statistics is often a half-way house between the classical and Bayesian churches: increasingly one finds that Bayesian techniques are used to select an appropriate statistical model, while the probabilities within the model are tacitly treated as being objective.

In the field of artificial intelligence (AI) Bayesianism has been hugely influential in the last decade. Expert systems have moved from a logical rule-based methodology to probabilistic techniques, largely involving the use of Bayesian networks.[18] Statistical learning theory has helped integrate machine learning techniques into a probabilistic framework,[19] and Bayesian methods are often now used to ascertain the parameters of machine learning models, and to determine the error between model and data.[20] Applications in industry have followed quickly: Bayesian networks are behind several recent expert systems including the print trouble-shooter of Microsoft's Windows '95 (and, alas, the paperclip of Office '97);[21] Bayesian reasoning is widely implemented using neural networks, forming the core of Autonomy's software for dealing with unstructured information (which made Autonomy's director, Mike Lynch, Britain's first dollar-billionaire);[22] other graphical models also form the basis of applications of Bayesian statistics to med-

---

[16]The almost religious fervour with which Bayesians pursue the cause of Reverend Bayes, and with which non-Bayesians undergo the conversion to Bayesianism, has occasionally been noted. Jaynes appears to have coined the term 'Bayesianity'.

[17][Jaynes, 1998].

[18]See [Pearl, 1988] and the website of the Association for Uncertainty in AI at www.auai.org.

[19][Vapnik, 1995].

[20]See for example [Bishop, 1995], [Jordan, 1998] and Williams' paper in this volume.

[21]See research.microsoft.com/dtas/ and [Horvitz *et al.*, 1998].

[22]See the technology white paper at www.autonomy.com. Peter Williams reported at the conference Bayesianism 2000 that neural network based Bayesian reasoning also proved successful (and lucrative!) when applied to gold prospecting.

ical expert systems[23] and health technology assessment.[24]

These developments in AI and other sciences have stimulated work on more traditional philosophical issues. Bayesian networks integrate causality and probability in a particular way, and the question naturally arises as to how exactly Bayesian probability is related to causality, and whether techniques for learning Bayesian networks from data can be applied to the problem of discovering causal structure.[25] Probability logics and their AI implementations have prompted renewed investigations into the relationship between Bayesian probability and logic.[26] Objective Bayesian methods, often involving the use of the maximum entropy principle, have been successfully applied in physics,[27] and this has led to debate about the validity of objective Bayesianism[28] and further applications of maximum entropy.[29] Probabilistic decision-theoretic techniques have now been widely adopted in economics, and this has stimulated research in the foundations of Bayesian decision theory.[30] On the other hand, the application of Bayesianism to scientific methodology may lead to a corresponding application to mathematical methodology.[31]

In the context of this recent Bayesian upswell, it is all the more important to avoid complacency: criticisms of Bayesianism must be given due attention,[32] and the key messages of the early proponents of Bayesianism must be better understood.[33]

3 PROSPECTS FOR BAYESIANISM

Judging by the papers in this book, the future of Bayesianism will depend on progress on the following foundational questions.

- Is Bayesianism to be preferred over classical statistics?
- If so, what type of Bayesianism should one adopt — strict subjectivism, empirical objectivism or logical objectivism?
- How does Bayesian reasoning cohere with causal, logical, scientific, mathematical and decision-theoretic reasoning?

---

[23] [Spiegelhalter *et al.*, 1993].

[24] [Spiegelhalter *et al.*, 2000].

[25] See [Spirtes *et al.*, 1993], [McKim & Turner, 1997], [Hausman & Woodward, 1999], [Hausman, 1999], [Glymour & Cooper, 1999], [Pearl, 2000] and Pearl's, Dawid's and Williamson's papers in this volume.

[26] See [Williamson, 2000] and the papers of Cussens, Gabbay, Howson, and Paris and Vencovská in this volume.

[27] [Jaynes, 1998].

[28] See Howson's and Paris and Vencovská's papers.

[29] See Williamson's paper.

[30] See the papers of Mongin, McClennen, Bradley and Albert in this volume.

[31] See Corfield's paper in this volume.

[32] See the papers of Mayo and Kruse, Albert and Gillies.

[33] See Galavotti's paper.

These questions are, of course, intricately linked. The first two are well-worn but extremely important: much progress has been made, but it would be foolhardy to expect any conclusive answers in the near future. The last question is particularly pressing, given the recent applications of Bayesian methods to AI. AI is now faced with a confusing plethora of formalisms for automated reasoning, and unification is high on the agenda. If Bayesianism can provide a framework into which AI techniques slot then its future is guaranteed.

4 THIS VOLUME

The fifteen chapters of this book have been arranged in four parts. The first of these parts is entitled 'Bayesianism, Causality and Networks' and consists of four chapters. What unites the authors of the first three contributions is an eagerness to clarify the relationship between causal and probabilistic reasoning, two of them by way of the use of directed acyclic graphs. The author of the fourth chapter, on the other hand, reports on research on a different category of network — neural networks. In the opening chapter, Pearl proceeds from the fundamental idea of Bayesianism that we should integrate our background knowledge with observational data when we reason. He then argues that our everyday and scientific knowledge is largely couched in causal, rather than statistical, terms, and that as such it is not readily expressible in probabilistic terms. Now, clearly it would preferable to be able to feed background knowledge directly into our reasoning calculus, and so, if possible, we should devise a new mathematical language in which we can represent causal information and reason about it. The article advertises Pearl's exciting new research programme, detailed in his book 'Causality', whose central aim is the mathematisation of causality via directed graphs.[34] The key questions to be addressed then concern the benefits of adopting such a radically new language and the safety of the reasoning it warrants. Pearl himself says that is possible to cast his causal models in terms of probabilities using hypothetical variables, but then argues that the only purpose in doing so is to avoid confrontation with the consensus position in the statistics community, which sees no limitations to the expressiveness of probability theory. Indeed, for Pearl, there is a definite disadvantage in a choice of language which gives counterfactual propositions precedence over more readily comprehensible causal ones.

So Pearl's idea is that the previous failure to construct a mathematical system capable of integrating background causal knowledge has led to much of this most important way of encoding our beliefs about the world being overlooked. As such he has located a novel way in which we may take the Bayesian to be failing to act in as rational as possible a manner. On the other hand, a long-standing complaint of irrationality made against Bayesianism, one which will recur through the chapters of this volume, alleges that a Bayesian's tenets do not force her to test whether her degrees of belief are, in some sense or other, optimal. These two themes intertwine

[34] [Pearl, 2000].

in Dawid's article. Dawid is well known for his 'Popperian' Bayesianism which aims to assess an agent's degrees of belief by a process of calibration, where, for example, weather reporters are to be congratulated if it rains on roughly 30 percent of the occasions they give 0.3 as the probability that it will rain. This concern with testability recurs in Dawid's contribution to the volume. While he agrees with Pearl that statisticians have largely ignored causality and have been wrong to do so, still he finds some elements of Pearl's new thinking problematic. What is at stake here is the Popperian belief that anything worthy of scientific consideration is directly testable. For Dawid some of the counterfactual reasoning warranted by Pearl's calculus (and by statisticians adopting other schemes, such as Rubin's potential-outcome approach) just is untestable. An example illustrating this key difference between Dawid and Pearl is their respective treatments of counterfactual questions such as whether my last headache would have gone had I not taken an aspirin, given that I did take one and it did go. How should knowledge of the effects of aspirin on other headache incidents of mine bear on this question? Pearl says that, without evidence to the contrary, we should presume that such knowledge does have a bearing on the counterfactual statement. By contrast, Dawid claims that singular counterfactual statements are untestable and therefore should not be accepted by the scientifically minded.[35] For Dawid what may be justifiably said about counterfactuals does not involve their essential use.

As Dawid is a self-professed Popperian, a comparison that comes to mind is to think of Pearl as a Lakatosian. While Popper's philosophy allowed that metaphysical principles might guide the generation of novel scientific theories, thereby restoring some worth to them after the Logical Positivists had dismissed them as 'meaningless', still they accrued no further value even when those theories passed severe tests. Where Lakatos went further than Popper was to allow metaphysics to be an integral part of a research programme, which was to be assessed by its theoretical and empirical success as a whole. Similarly, we could say that Pearl has devised a research programme with a powerful heuristic and a new mathematical language. There is a metaphysical belief on Pearl's part in the regularity of a world governed by causal mechanisms which is integrated into this programme, hence his turn to structural equation models. Dawid, meanwhile, views the presuppositions behind the use of these models as unwarranted — the world for him is not so easily tamed.

In the third chapter Williamson questions the validity of the causal Markov condition, an assumption which links probability to causality and on which the theory of Bayesian networks and Pearl's recent account of causality depends. He argues that the causal Markov condition does not hold for an empirical account of probability, or for a strict subjectivist Bayesian interpretation, but does hold for an objective Bayesian interpretation, i.e., one using maximum entropy methods. If it can be established that the causal Markov condition does not hold with respect

[35] See the comments and rejoinder to Dawid's paper in the Journal of American Statistical Association 95 (June 2000), pages 424-448.

to a notion of empirical probability, this means that causal networks must be restructured if they are to be calibrated with frequency data. This leads Williamson to propose a two-stage methodology for using Bayesian networks: first build the causal network out of expert knowledge and then restructure it to fit observational data more closely. The validity of stage 1 of this methodology depends on the validity of a maximum-entropy based objective Bayesian interpretation of probability and so would not appeal to subjectivists like de Finetti or Howson (see below), while the validity of stage 2 depends on acceptance of the idea that one ought to calibrate Bayesian beliefs with empirical data.

Williams rounds out Part 1 of the book by offering us an overview of research carried out by the neural network community to provide a principled way of using data to fashion an accurate network. All forms of machine learning must find a way to reconcile the demands of accuracy and the risks of overfitting data. This relates to a long-standing debate in the philosophy of science about the desirability of choosing as simple as possible a model to represent empirical data. Now, some Bayesians, including those working in the tradition of Harold Jeffreys, claim to have found a principled way to effect this reconciliation by according a higher prior probability to a model with fewer free parameters. The potential for increased accuracy provided by an extra parameter will then be balanced by a lower prior probability for the more complicated model. Neural network researchers are now invoking these Bayesian notions to arrive at optimal network configurations and settings of connection strengths.

A frequently encountered point of disagreement between the different approaches to artificial intelligence concerns the need to represent data and inference in propositionally encoded form. Neural networks come in for criticism for acting like black boxes. They may work well in many situations, the thought is, but we do not really understand why. Thus, unlike in the case of Bayesian networks, they offer no insight to the expert hoping to use them to support decision-making. Of course, one might respond to this criticism by making the point that accuracy, not transparency, is the most important quality of a decision-making process, especially in critical situations such as medical diagnosis. In the context of Williams' chapter the lack of transparency relates to the fact that the space of weight configurations of a network bears no straightforward relation to an expert's qualitative understanding of a domain. Thus background knowledge cannot be encoded directly into a prior distribution over possible networks, but only through the mediation of real or simulated data. Perhaps this difficulty is the reason that we find such a great range of techniques employed by the neural network community, even though, in the case of the ones described by Williams at least, Bayesian principles are guiding them.

We turn next to the second part — Logic, Mathematics and Bayesianism. Here the five authors wish to investigate the relationship between Bayesian probabilistic reasoning and deductive logic. In two chapters (Howson, Paris & Vencovská) we find probability theory presented as an extension of deductive logic, while in two others (Galavotti, Corfield) it appears in the guise of realistic personalist degrees of belief. Finally, Cussens discusses his use of stochastic logic programming, an artificial intelligence technique, to encode probabilistic reasoning.

Howson views Bayesianism as an extension of deductive logic in the sense that, just as the use of deductive logic provides rules to ensure a consistent set of truth values for the statements of language, so the probability theory axioms ensure consistent degrees of belief. In doing so he rules out three widely held, yet disputed, aspects of Bayesian reasoning: its inextricable link to utility theory; the principle of indifference, along with any other notion of objective priors; and, conditionalisation. Justification of this logical core of Bayesianism is provided by the idea of probability as expected truth value, using the device of the indicator function of a proposition, where a proposition is taken *à la* Carnap as the set of structures in which a sentence is true. Howson's belief that probability theory is a form of logic sets him against decision theorists, such as Herman Rubin, who believe that 'you cannot separate probability from utility'.[36] Thus he aims to provide a justification for the probability axioms foregoing the use of Dutch Book arguments, thereby avoiding reliance on the notion of the desirability of acquiring money.

Howson also rejects the strain of Bayesianism which hopes to arrive at some values to enter into the probability calculus through the use of the principle of indifference or of maximum entropy. More radically still, he continues by arguing that conditionalisation has no place in a Bayesian logic, since it is a rule relating truth values held at different times. He illustrates this thesis in parallel deductive terms: if you held 'A implies B' to be true yesterday, then find out today that A is true, you are not now forced to accept B, since you may no longer believe that A implies B. Similarly, if yesterday you have $p(A|B) = x$, and today $p'(B) = 1$, this does not mean you need have $p'(A) = x$. Here the reader might wonder about the status of the commonly held notion that, unless you have good reason for this change of heart, you should stick to your original beliefs. Is it just an extra-logical rule of thumb that $p'(A|B) = p(A|B)$ unless there is good cause to change one's mind?

Galavotti has provided a largely historical piece on the Bayesianism of Bruno de Finetti. De Finetti is famous for his assertion that 'probability does not exist', preferring to see probabilities as subjective degrees of belief, rather than something inherent in the universe. But while he was keen to stress his disapproval of an objectivism which sees probabilities as simply out there in the world, this did not entail a disregard for objectivity. Empirical frequency data might be integrated into one's degrees of belief by the subjective judgement of the exchangeability of the data sequence. Moreover, and this may be a surprise for readers who share the commonly held impression that de Finetti was the arch-subjectivist Bayesian, he had a considerable interest in scoring rules used to judge the success of one's personal probability assignments. Comparisons of the accuracy of one's own previous probability judgements with those of others were to be integrated into one's current personal degrees of belief.

[36][Rubin, 1987].

Corfield bases his paper on the ideas of the Hungarian mathematician George Pólya, who in his description of plausible mathematical reasoning, which he interpreted by means of probabilistic degrees of belief, discerned what he took to be the common patterns of everyday reasoning. Corfield argues that no attempt to construe mathematical reasoning in Bayesian terms can assume logical omniscience — the requirement that rational agents accord the same degree of belief to any two logically equivalent statements. In the absence of this principle, logical and mathematical learning become thinkable in Bayesian terms. The idea that Bayesians should put logical and empirical learning on an equal footing goes back at least as far as de Finetti, and would seem to set Corfield against Howson who frames his Bayesian logic in such a way that logical omniscience comes already built in.

One could argue that a Bayesian reconstrual of mathematical reasoning as it occurs in practice is likely to be a largely empty exercise. Certainly, Bayesian reconstructions of scientific reasoning have come in for this kind of criticism. One may be able to explain why observing a white tennis shoe provides no support for the law 'all ravens are black', despite being an instance of the logically equivalent 'all non-black things are not ravens', these critics say, but it offers very little by way of insight into the rationality of decision making in science. However, one might reply that it has led Corfield to consider the rationality of certain overlooked styles of mathematical reasoning: use of analogy, choice of proof strategy, large scale induction. Regarding the latter, for instance, to date very little attention has been paid by philosophers of mathematics to the rationality of mathematicians raising their degrees of belief in conjectures due to confirmations. For example, should the computer calculation which shows that the first 1.5 billion nontrivial zeros of the Riemann zeta function have real part equal to $\frac{1}{2}$ be thought to lend support to the Riemann hypothesis, which claims that all of the infinitely many zeros lie on this line in the complex plane?

Paris and Vencovská share Howson's vision of probability theory as a logic, but unlike him they seek to isolate and justify principles which will allow the agent to select her priors rationally. In an earlier paper[37] they showed that the probability function which maximises entropy is the only choice if certain intuitively plausible constraints on objective Bayesian reasoning are to be respected. This was a significant result, but with one drawback: in their framework background knowledge is assumed to be encapsulated in a set of linear constraints. This rules out knowledge of, say, independencies amongst variables. In this chapter Paris and Vencovská extend their result to deal with non-linear constraints in the agent's background knowledge. There is now some room for subjectivity since there may be more than one most rational (i.e., maximum entropy) probability function. A point to note is that the framework adopted here is in the propositional calculus. This may be adequate for many AI applications, but it is not clear how it could be extended to the predicate calculus. If different reasoning principles are required for predicate reasoning, how does the resulting formalisation cohere with the propositional

[37] [Paris & Vencovská, 1990].

approach given here?

As uncertainty is now treated probabilistically by the majority of AI practitioners, those adopting a logic based approach who wish to discuss uncertain reasoning are faced with the thorny problem of integrating logic and probability. Philosophers have worked hard on this problem for many years with no consensus emerging. The line of thought that takes Bayesianism to be an extension of deductive logic would suggest that this should not be very problematic for a degree of belief interpretation of probability. However, this has not turned out to be the case — a very large number of disparate techniques have proposed by the AI community. Cussens bases his attempt to integrate probability theory and logic on what are called 'stochastic logic programs'. Stochastic logic programs (SLPs) originated in the inductive logic programming (ILP) paradigm of machine learning.[38] When presented with data, an ILP program will attempt to generate a logic program (essentially, a set of Horn clauses) which includes as successful goals as many positive examples as possible, while excluding as many negative examples. In cases where only positive examples are available, a common situation in science, to prevent overfitting, it was found necessary to generate a distribution over all possible ground instances. Muggleton did this by labelling the clauses of a proposed logic program with probabilities generated from the data. Elsewhere, Cussens has extended this idea to apply it to natural language processing, where a successful parsing of a sentence will be accorded a probability depending on the ways it may be generated by the grammar encoded by the logic program. In the present article, he takes SLPs to be capable of representing a very wide range of AI techniques, in particular showing how Bayesian networks may be encoded in its terms. He then compares his SLP approach to other techniques.

The reader might be interested to know of two other approaches to the integration of logic and Bayesian networks. Williamson develops 'logical Bayesian networks' (as opposed to causal Bayesian networks) whose nodes are sentences (rather than causes and effects) and whose arrows correspond to the logical implication relation (rather than the causal relation).[39] Meanwhile, Dov Gabbay is working on a way of representing Bayesian networks in the framework of his labelled deductive systems. His results were not ready in time for this volume, but they will appear in the near future.

Turning now to the third part we find the contributions of three Bayesian decision theorists. Probabilistic decision theory has a long heritage, stretching back to Pascal's Wager, but there still rage many disputes over its fundamental principles. Here, two of the contributors, Mongin and McClennen, scrutinise the acceptability of particular axioms, while in the first chapter of this part Bradley discusses the problem of the measurement of belief.

Bradley's claim is that the resources for resolving the issue of how to assess the strengths of beliefs and desires of an agent are to found in the writings of Ramsey

[38] [Muggleton & de Raedt, 1994].

[39] [Williamson, 2001]. See also [Williamson, 2000] where these logical networks form the basis of a proof theory of a probabilistic logic.

from the 1920s. While decision theorists have followed the lead of Savage, Ramsey has largely been overlooked. However, as Bradley points out, Savage relied on the assumption of state-independent utility, where the desirability of an outcome is independent of the state of the world in which it occurs. This assumption has come in for a great deal of criticism, which has given rise to highly complex theories of state-dependent utility. Bradley argues that if we revive Ramsey's notion of 'ethically neutral events', ones to whose outcome the agent is indifferent, we gain a means to access the strength of an agent's beliefs and desires without the need to invoke these complex theories.

The Independence Principle, and the closely related Sure-Thing Principle, are central to Bayesian decision theory. The independence principle states that if an agent shows no preference between two gambles, P and P', then, for any $0 < \alpha \leq 1$ and any gamble Q, she will also show no preference between the composite gambles $R = \alpha P + (1 - \alpha)Q$ and $R' = \alpha P' + (1 - \alpha)Q$. While it appears to be a highly plausible principle, McClennen investigates various arguments put forward to support it, both directly and via the Sure-Thing Principle, and finds them all wanting. One might have supposed that the independence principle holds, since these composite gambles are disjunctive in the sense that in the case of R the final outcome will either be the outcome of P or the outcome of Q but not both. Still, McClennen argues, there may be an interactive effect making the agent prefer R to R'. This may occur, for instance, if Q more closely resembles P than P' and the agent has a preference for a less varied composite gamble.

In the final chapter of Part 3, Mongin takes on the task of examining the difficulties created by the simultaneous assumption of Bayesian and Paretian principles. The latter refer to assumptions about preferences of outcomes in the light of group consensus about preferences. For example, for a group of experts working with different utilities or different probabilities in the framework of state-independent utility theory, there will not in general be a way to select a utility function and probabilities such that one outcome is preferred over another whenever all the experts agree to this ordering. This result lends support to the move towards state-dependent utility theory mentioned above. However, a pure form of this theory entails the undesirable consequence that subjective probabilities are not in general uniquely determined. Mongin then proceeds to scrutinise a form of state-dependence which entails unique probabilities. He shows, however, that the assumptions of this theory still conflict with Paretian principles.

The fourth and final part consists of three contributors' criticisms of Bayesianism. As the chapters up to this point amply demonstrate, Bayesians disagree amongst themselves about all manner of issues: the extent of rationality constraints, the link to utility theory, the role of conditionalisation, etc. This being so, the critic's task is made harder. Whatever principle she attacks, some Bayesian may claim not to hold to it.

Albert's criticism is aimed at the use of Bayesian principles by decision theorists. Adopting a line reminiscent of Popper's critical attitude towards psychoanalysis, he claims that for the Bayesian 'there is no such thing as irrational behavior'

— any set of actions can be construed as satisfying the constraints of Bayesianism. Albert argues for this conclusion by discussing a situation involving a chaotic clock which outputs a sequence of 0s and 1s. The agent must judge the likelihood of these digits occurring, based on the sequence to date, to help him win as much money as possible. What Albert shows is that, even with the agent's utility function given, whatever he does one can reconstruct it as rational according to some choice of prior distribution over the hypothesis space. Now, in response one may argue that the chaotic clock situation does not resemble the everyday conditions met with in economic life, but Albert argues that his example is sufficiently generic in this sense. Objective Bayesians may also claim that knowledge of the chaotic clock set up provides the rational agent with an obvious unique choice of prior. On the other hand, subjectivists might question whether the ways of recording the agent's behaviour are sufficient to pick up fully his belief structure. For example, the chaotic clock situation would not allow you to discover the incoherence of an agent who is certain that 0 will appear next, but also certain that 1 will appear next.

In his chapter, Gillies wants to argue that Bayesianism is appropriate only in a restricted range of situations, indeed, 'only if we are in a situation in which there is a fixed and known theoretical framework which it is reasonable to suppose will not be altered in the course of the investigation'. As soon as the reasoner departs from the current theoretic framework, Bayesianism is of no assistance. In saying this, Gillies appears to be aligning himself with one side of an argument heard before in the philosophy of science about the room for Bayesian principles to operate when a new theory is proposed. Earman, for instance, who is very sympathetic to Bayesianism, argues that changes of conceptual framework require the resetting of priors, which occurs in a non-algorithmic fashion by plausibility arguments.[40] He also indicates that these exogenous redistributions of prior probabilities will occur frequently, perhaps within the course of an everyday conversation. Now, Howson's Bayesianism might have no problem with this — after all it is not a diachronic theory — but Bayesians of a different stripe, wishing to salvage a substantial role for conditionalisation, might prefer to concede that the advent of novel ideas will have their effects not through conditionalisation, but still look to 'normal' science to see Bayes' theorem at work. However, the two examples put forward by Gillies could hardly be considered revolutionary changes. Rather they appear to involve the kind of reasoning that any statistician will have to perform in the course of their work, i.e., the assessment of the validity of the current model. The indication is that where the error statistician is always eager to challenge the current model and put it to severe tests, the Bayesian has neither the means nor the incentive to look beyond the current framework. It is true that Dutch Book arguments by themselves do not require an agent to take the slightest trouble to make an observation or to challenge a modelling assumption that would be beneficial to a bet they are making. However, under reasonable assumptions, one can show that it is always worth seeking cost-free information before making a decision. It is therefore not

[40][Earman, 1992].

surprising to find Bayesian statisticians engaging in what Box and Tiao call 'model criticism'.[41] Readers may care to see how a Bayesian statistician works on a very similar problem to Gillies' second example in [Gelman *et al.*, 1995], 170-171.

In their article, Mayo and Kruse take on Bayesian statistics on the issue of stopping rules. For the error statistician the conditions stipulated before the start of an experiment as to when it will be deemed to have ended will usually be relevant to the significance of the test. For instance, when testing for a proportion in some population, even if the data turns out the same, it makes a difference whether it has been generated by deciding to stop after a fixed number of positive cases have been observed or whether it has been generated by deciding to stop after a fixed number of trials. For most Bayesians (see [Box & Tiao, 1973], 44-46 for an exception), on the other hand, acceptance of the likelihood principle entails that such considerations should play no part in the calculation of their posterior distributions. This marks a very significant difference between the schools. *Pace* Gillies, as Mayo has said elsewhere, 'one cannot be just a little bit Bayesian'. Most Bayesians would agree.

The Bayesian position has a considerable plausibility to it. Should we condemn an experimenter who intends to test 100 cases, finds half way through that the proportion of positive cases is low, decides then to wait for 10 such cases to occur, which duly happens after precisely 100 trials, and then writes up the experiment as originally planned? Are all experiments called off early because of funding problems worthless? There appears to be something magical occurring in that the 'real' intentions of the experimenter make a difference. On the other hand, as critics of the Bayesian indifference to stopping rules, Mayo and Kruse need only point out the problematic consequences of operating with a single rule of their choosing. What then if an experimenter in a binomial situation with, say, $p = 0.2$ is to continue testing until this value of p becomes unlikely to a specified degree, putting some upper limit on the number of trials to make it a proper rule? Here at least the Bayesian can provide bounds for the likelihood of a test achieving this end. But Mayo and Kruse go on to discuss an experimental situation where due to the stopping rule the Bayesian statistician will necessarily reason to a foregone conclusion on the basis of the likelihood principle. This type of situation arises when an improper prior, failing to satisfy countable additivity, is employed. One could, of course, maintain that countable additivity be enforced, but improper priors are commonplace in the Bayesian literature. Mayo and Kruse present several quotations revealing that some Bayesian statisticians are struggling to come to terms with this apparent paradox. Readers who wish to read more on this topic may well enjoy the discussion of this phenomenon by the Bayesians Kadane, Schervish and Seidenfeld.[42] Not wishing to forego the use of improper priors, these authors consider that further work is required to ascertain when their use is admissible.

King's College, London.

[41] [Box & Tiao, 1973].

[42] [Kadane *et al.*, 1999] §§3.7 & 3.8.

BIBLIOGRAPHY

[Bayes, 1764] Thomas Bayes. An essay towards solving a problem in the doctrine of chances. *Philosophical Transactions of the Royal Society of London*, **53**, 370–418, 1764.

[Berger, 2000] James O. Berger. Bayesian analysis: a look at today and thoughts of tomorrow. *Journal of the American Statistical Association*, **95** (December), 2000.

[Bishop, 1995] Christopher M. Bishop. *Neural networks for pattern recognition*. Oxford University Press, 1995.

[Box & Tiao, 1973] G. Box & G. Tiao. *Bayesian Inference in Statistical Analysis*. Addison-Wesley, 1973.

[Carnap, 1950] Rudolf Carnap. *Logical foundations of probability*. Routledge & Kegan Paul Ltd, 1950. Second edition 1962.

[Carnap & Jeffrey, 1971] Rudolf Carnap & Richard C. Jeffrey(eds.). *Studies in inductive logic and probability*, Volume 1, University of California Press, 1971.

[Dawid, 1982] A.P. Dawid. The well-calibrated Bayesian. With discussion, *Journal of the American Statistical Association*, **77**, 604–613, 1982.

[Earman, 1992] John Earman. *Bayes or bust?*. M.I.T. Press, 1992.

[de Finetti, 1937] Bruno de Finetti. Foresight. Its logical laws, its subjective sources. In [Kyburg & Smokler, 1964], 53–118, 1937.

[Gaifman & Snir, 1982] H. Gaifman & M. Snir. Probabilities over rich languages. *Journal of Symbolic Logic*, **47**, 495–548, 1982.

[Gelman *et al.*, 1995] Andrew B. Gelman, John S. Carlin, Hal S. Stern & Donald B. Rubin. *Bayesian data analysis*. Chapman & Hall/CRC, 1995.

[Gillies, 2000] Donald Gillies. *Philosophical theries of probability*. Routledge, 2000.

[Glymour & Cooper, 1999] Clark Glymour & Gregory F. Cooper(eds.). *Computation, causation, and discovery*. M.I.T. Press, 1999.

[Hausman, 1999] Daniel M. Hausman. The mathematical theory of causation. Review of [McKim & Turner, 1997], *British Journal for the Philosophy of Science*, **50**, 151–162, 1999.

[Hausman & Woodward, 1999] Daniel M. Hausman & James Woodward. Independence, invariance and the causal Markov condition. *British Journal for the Philosophy of Science*, **50**, 521–583, 1999.

[Horvitz *et al.*, 1998] Eric Horvitz, Jack Breese, David Heckerman, David Hovel & Koos Rommelse. The Lumière Project: Bayesian user modeling for inferring the goals and needs of software users. In *Proceedings of the Fourteenth Conference on Uncertainty in Artificial Intelligence*, Morgan Kaufmann, pages 256–265, 1998.

[Howson & Urbach, 1989] Colin Howson & Peter Urbach. *Scientific reasoning: the Bayesian approach*. Open Court, 1989. Second edition, 1993.

[Jaynes, 1998] E.T. Jaynes. *Probability theory: the logic of science*. http://bayes.wustl.edu/etj/prob.html.

[Jordan, 1998] Michael I. Jordan(ed.). *Learning in Graphical Models*. MIT Press, Cambridge, MA, 1998.

[Kadane *et al.*, 1999] J. Kadane, M. Schervish & T. Seidenfeld(eds.). *Rethinking the Foundations of Statistics*. Cambridge University Press, 1999.

[Keynes, 1921] John Maynard Keynes. *A treatise on probability*. Macmillan, 1948.

[Kyburg & Smokler, 1964] H.E. Kyburg & H.E. Smokler(eds.). *Studies in subjective probability*. 1964. Second edition, Robert E. Krieger Publishing Company, 1980.

[Laplace, 1814] Pierre Simon - Marquis de Laplace. *A philosophical essay on probabilities*. Dover, 1951.

[Lewis, 1980] David K. Lewis. A subjectivist's guide to objective chance. In [Lewis, 1986], 83–132, 1980.

[Lewis, 1986] David K. Lewis. *Philosophical papers II*. Oxford University Press, 1986.

[McKim & Turner, 1997] Vaughn R. McKim & Stephen Turner. *Causality in crisis? Statistical methods and the search for causal knowledge in the social sciences*. University of Notre Dame Press, 1997.

[Muggleton & de Raedt, 1994] Stephen Muggleton & Luc de Raedt. Inductive logic programming: theory and methods. In *Journal of Logic Programming*, **19, 20**, 629–679, 1994.

[Paris & Vencovská, 1990] J.B. Paris & A. Vencovská. A note on the inevitability of maximum entropy. *International Journal of Approximate Reasoning*, **4**, 181–223, 1990.

[Pearl, 1988] Judea Pearl. *Probabilistic reasoning in intelligent systems: networks of plausible inference.* Morgan Kaufmann, 1988.

[Pearl, 2000] Judea Pearl. *Causality: models, reasoning, and inference.* Cambridge University Press, 2000.

[Ramsey, 1926] Frank Plumpton Ramsey. Truth and probability. In [Kyburg & Smokler, 1964], 23–52, 1926.

[Rubin, 1987] H. Rubin. A Weak System of Axioms for "Rational" Behavior and the Nonseparability of Utility from Prior. *Statistics and Decisions*, **5**, 47–58, 1987.

[Spiegelhalter *et al.*, 1993] David J. Spiegelhalter, A. Philip Dawid, Steffen L. Lauritzen & Robert G. Cowell. Bayesian analysis in expert systems. *Statistical Science*, **8**(3), 219–283, with discussion, 1993.

[Spiegelhalter *et al.*, 2000] D.J. Spiegelhalter, J.P. Myles, D.R. Jones & K.R. Abrams. Bayesian methods in health technology assessment: a review. *Health Technology Assessment*, **4**(38), 2000.

[Spirtes *et al.*, 1993] Peter Spirtes, Clark Glymour & Richard Scheines. Causation, Prediction, and Search. *Lecture Notes in Statistics*, **81**, Springer-Verlag, 1993.

[Vapnik, 1995] Vladimir N. Vapnik. *The nature of statistical learning theory.* Springer-Verlag, 1995. Second edition 2000.

[Williamson, 1999] Jon Williamson. Countable additivity and subjective probability. *British Journal for the Philosophy of Science*, **50**(3), 401–416, 1999.

[Williamson, 2000] Jon Williamson. Probability logic. In Dov Gabbay, Ralph Johnson, Hans Juergen Ohlbach & John Woods (eds.), *Handbook of the Logic of Inference and Argument: The Turn Toward the Practical*, Elsevier, 393–419, 2000.

[Williamson, 2001] Jon Williamson. Bayesian networks for logical reasoning. *Proceedings of the 8th Workshop on Automated Reasoning*, A. Voronkov (ed.), 55–56, 2001.

PART I

BAYESIANISM, CAUSALITY AND NETWORKS

JUDEA PEARL

BAYESIANISM AND CAUSALITY, OR, WHY I AM ONLY A HALF-BAYESIAN

1 INTRODUCTION

I turned Bayesian in 1971, as soon as I began reading Savage's monograph *The Foundations of Statistical Inference* [Savage, 1962]. The arguments were unassailable: (i) It is plain silly to ignore what we know, (ii) It is natural and useful to cast what we know in the language of probabilities, and (iii) If our subjective probabilities are erroneous, their impact will get washed out in due time, as the number of observations increases.

Thirty years later, I am still a devout Bayesian in the sense of (i), but I now doubt the wisdom of (ii) and I know that, in general, (iii) is false. Like most Bayesians, I believe that the knowledge we carry in our skulls, be its origin experience, schooling or hearsay, is an invaluable resource in all human activity, and that combining this knowledge with empirical data is the key to scientific enquiry and intelligent behavior. Thus, in this broad sense, I am a still Bayesian. However, in order to be combined with data, our knowledge must first be cast in some formal language, and what I have come to realize in the past ten years is that the language of probability is not suitable for the task; the bulk of human knowledge is organized around causal, not probabilistic relationships, and the grammar of probability calculus is insufficient for capturing those relationships. Specifically, the building blocks of our scientific and everyday knowledge are elementary facts such as "mud does not cause rain" and "symptoms do not cause disease" and those facts, strangely enough, cannot be expressed in the vocabulary of probability calculus. It is for this reason that I consider myself only a half-Bayesian.

In the rest of the paper, I plan to review the dichotomy between causal and statistical knowledge, to show the limitation of probability calculus in handling the former, to explain the impact that this limitation has had on various scientific disciplines and, finally, I will express my vision for future development in Bayesian philosophy: the enrichment of personal probabilities with causal vocabulary and causal calculus, so as to bring mathematical analysis closer to where knowledge resides.

2 STATISTICS AND CAUSALITY: A BRIEF SUMMARY

The aim of standard statistical analysis, typified by regression and other estimation techniques, is to infer parameters of a distribution from samples drawn of that population. With the help of such parameters, one can infer associations among variables, estimate the likelihood of past and future events, as well as update the

D. Corfield and J. Williamson (eds.), Foundations of Bayesianism, 19–36.

likelihood of events in light of new evidence or new measurements. These tasks are managed well by statistical analysis so long as experimental conditions remain the same. Causal analysis goes one step further; its aim is to infer aspects of the data generation process. With the help of such aspects, one can deduce not only the likelihood of events under static conditions, but also the dynamics of events under *changing conditions*. This capability includes predicting the effect of actions (e.g., treatments or policy decisions), identifying causes of reported events, and assessing responsibility and attribution (e.g., whether event x was necessary (or sufficient) for the occurrence of event y).

Almost by definition, causal and statistical concepts do not mix. Statistics deals with behavior under uncertain, yet static conditions, while causal analysis deals with changing conditions. There is nothing in the joint distribution of symptoms and diseases to tell us that curing the former would not cure the latter. In general, there is nothing in a distribution function that would tell us how that distribution would differ if external conditions were to change—say from observational to experimental setup—every conceivable difference in the distribution would be perfectly compatible with the laws of probability theory, no matter how slight the change in conditions.[1]

Drawing analogy to visual perception, the information contained in a probability function is analogous to a precise description of a three-dimensional object; it is sufficient for predicting how that object will be viewed from any angle outside the object, but it is insufficient for predicting how the object will be viewed if manipulated and squeezed by external forces. The additional properties needed for making such predictions (e.g., the object's resilience or elasticity) is analogous to the information that causal models provide using the vocabulary of directed graphs and/or structural equations. The role of this information is to identify those aspects of the world that remain invariant when external conditions change, say due to an action.

These considerations imply that the slogan "correlation does not imply causation" can be translated into a useful principle: one cannot substantiate causal claims from associations alone, even at the population level—behind every causal conclusion there must lie some causal assumption that is not testable in observational studies. Nancy Cartwright [1989] expressed this principle as "no causes in, no causes out", meaning we cannot convert statistical knowledge into causal knowledge.

The demarcation line between causal and statistical concepts is thus clear and crisp. A statistical concept is any concept that can be defined in terms of a distribution (be it personal or frequency-based) of observed variables, and a causal con-

[1]Even the theory of stochastic processes, which provides probabilistic characterization of certain dynamic phenomena, assumes a fixed density function over time-indexed variables. There is nothing in such a function to tell us how it would be altered if external conditions were to change. If a parametric family of distributions is used, we can represent some changes by selecting a different set of parameters. But we are still unable to represent changes that do not correspond to parameter selection; for example, restricting a variable to a certain value, or forcing one variable to equal another.

cept is any concept concerning changes in variables that cannot be defined from the distribution alone. Examples of statistical concepts are: correlation, regression, dependence, conditional independence, association, likelihood, collapsibility, risk ratio, odd ratio, and so on.[2] Examples of causal concepts are: randomization, influence, effect, confounding, disturbance, spurious correlation, instrumental variables, intervention, explanation, attribution, and so on. The purpose of this demarcation line is not to exclude causal concepts from the province of statistical analysis but, rather, to make it easy for investigators and philosophers to trace the assumptions that are needed for substantiating various types of scientific claims. Every claim invoking causal concepts must be traced to some premises that invoke such concepts; it cannot be derived or inferred from statistical claims alone.

This principle may sound obvious, almost tautological, yet it has some far reaching consequences. It implies, for example, that any systematic approach to causal analysis must acquire new mathematical notation for expressing causal assumptions and causal claims. The vocabulary of probability calculus, with its powerful operators of conditionalization and marginalization, is simply insufficient for expressing causal information. To illustrate, the syntax of probability calculus does not permit us to express the simple fact that "symptoms do not cause diseases", let alone draw mathematical conclusions from such facts. All we can say is that two events are dependent—meaning that if we find one, we can expect to encounter the other, but we cannot distinguish statistical dependence, quantified by the conditional probability $P(disease|symptom)$ from causal dependence, for which we have no expression in standard probability calculus.[3] Scientists seeking to express causal relationships must therefore supplement the language of probability with a vocabulary for causality, one in which the symbolic representation for the relation "symptoms cause disease" is distinct from the symbolic representation of "symptoms are associated with disease." Only after achieving such a distinction can we label the former sentence "false," and the latter "true."

The preceding two requirements: (1) to commence causal analysis with untested,[4] judgmentally based assumptions, and (2) to extend the syntax of probability calculus, constitute, in my experience, the two main obstacles to the acceptance of causal analysis among statisticians, philosophers and professionals with traditional training in statistics. We shall now explore in more detail the nature of these two barriers, and why they have been so tough to cross.

---

[2]The term 'risk ratio' and 'risk factors' have been used ambivalently in the literature; some authors insist on a risk factor having causal influence on the outcome, and some embrace factors that are merely associated with the outcome.

[3]Attempts to define causal dependence by conditioning on the entire past (e.g., Suppes, 1970) violate the statistical requirement of limiting the analysis to "observed variables", and encounter other insurmountable difficulties (see Eells [1991], Pearl [2000a], pp. 249-257).

[4]By "untested" I mean untested using frequency data in nonexperimental studies.

2.1 *The Barrier of Untested Assumptions*

All statistical studies are based on some untested assumptions. For examples, we often assume that variables are multivariate normal, that the density function has certain smoothness properties, or that a certain parameter falls in a given range. The question thus arises why innocent causal assumptions, say, that symptoms do not cause disease or that mud does not cause rain, invite mistrust and resistance among statisticians, especially of the Bayesian school.

There are three fundamental differences between statistical and causal assumptions. First, statistical assumptions, even untested, are testable in principle, given sufficiently large sample and sufficiently fine measurements. Causal assumptions, in contrast, cannot be verified even in principle, unless one resorts to experimental control. This difference is especially accentuated in Bayesian analysis. Though the priors that Bayesians commonly assign to statistical parameters are untested quantities, the sensitivity to these priors tends to diminish with increasing sample size. In contrast, sensitivity to priors of causal parameters, say those measuring the effect of smoking on lung cancer, remains non-zero regardless of (nonexperimental) sample size.

Second, statistical assumptions can be expressed in the familiar language of probability calculus, and thus assume an aura of scholarship and scientific respectability. Causal assumptions, as we have seen before, are deprived of that honor, and thus become immediate suspect of informal, anecdotal or metaphysical thinking. Again, this difference becomes illuminated among Bayesians, who are accustomed to accepting untested, judgmental assumptions, and should therefore invite causal assumptions with open arms—they don't. A Bayesian is prepared to accept an expert's judgment, however esoteric and untestable, so long as the judgment is wrapped in the safety blanket of a probability expression. Bayesians turn extremely suspicious when that same judgment is cast in plain English, as in "mud does not cause rain." A typical example can be seen in Lindley and Novick's [1981] treatment of Simpson's paradox.

Lindley and Novick showed that decisions on whether to use conditional or marginal contingency tables should depend on the story behind the tables, that is, on one's assumption about how the tables were generated. For example, to decide whether a treatment $X = x$ is beneficial ($Y = y$) in a population, one should compare $\Sigma_z P(y|x, z)$ to $\Sigma_z P(y|x', z)$ if Z stands for the gender of patients. In contrast, if Z stands for a factor that is affected by the treatment (say blood pressure), one should compare the marginal probabilities, $P(y|x)$ *vis-à-vis* $P(y|x')$, and refrain from conditioning on Z (see [Pearl, 2000a; pp. 174-182] for details). Remarkably, instead of attributing this difference to the causal relationships in the story, Lindley and Novick wrote: "We have not chosen to do this; nor to discuss causation, because the concept, although widely used, does not seem to be well-defined" (p. 51). Thus, instead of discussing causation, they attribute the change in strategy to another untestable relationship in the story—exchangeability [DeFinetti, 1974] which is cognitively formidable yet, at least formally, can be cast in a probability expression. In Section 4.2, we will return to discuss this trend among Bayesians of equating "definability" with expressibility in probabilistic language.

The third resistance to causal (*vis-à-vis* statistical) assumptions stems from their intimidating clarity. Assumptions about abstract properties of density functions or about conditional independencies among variables are, cognitively speaking, rather opaque, hence they tend to be forgiven, rather than debated. In contrast, assumptions about how variables cause one another are shockingly transparent, and tend therefore to invite counter-arguments and counter-hypotheses. A co-reviewer on a paper I have read recently offered the following objection to the causal model postulated by the author:

> "A thoughtful and knowledgeable epidemiologist could write down two or more equally plausible models that leads to different conclusions regarding confounding."

Indeed, since the bulk of scientific knowledge is organized in causal schema, scientists are incredibly creative in constructing competing alternatives to any causal hypothesis, however plausible. Statistical hypotheses in contrast, having been several levels removed from our store of knowledge, are relatively protected from such challenges.

I conclude this subsection with a suggestion that statisticians' suspicion of causal assumptions, *vis-à-vis* probabilistic assumptions, is unjustified. Considering the organization of scientific knowledge, it makes prefect sense that we permit scientists to articulate what they know in plain causal expressions, and not force them to compromise reliability by converting to the "higher level" language of prior probabilities, conditional independence and other cognitively unfriendly terminology.[5]

2.2 *The Barrier of New Notation*

If reluctance to making causal assumptions has been a hindrance to causal analysis, finding a mathematical way of expressing such assumptions encountered a formidable mental block. The need to adopt a new notation, foreign to the province of probability theory, has been traumatic to most persons trained in statistics; partly because the adaptation of a new language is difficult in general, and partly because statisticians have been accustomed to assuming that all phenomena, processes, thoughts, and modes of inference can be captured in the powerful language of probability theory.[6]

---

[5]Similar observations were expressed by J. Heckman [2001].

[6]Commenting on my $set(x)$ notation [Pearl, 1995a, b], a leading statistician wrote: "Is this a concept in some new theory of probability or expectation? If so, please provide it. Otherwise, 'metaphysics' may remain the leading explanation." Another statistician, commenting on the $do(x)$ notation used in *Causality* [Pearl, 2000a], insisted: "...the calculus of probability is the calculus of causality."

Not surprisingly, in the bulk of the statistical literature, causal claims never appear in the mathematics. They surface only in the verbal interpretation that investigators occasionally attach to certain associations, and in the verbal description with which investigators justify assumptions. For example, the assumption that a covariate is not affected by a treatment, a necessary assumption for the control of confounding [Cox, 1958], is expressed in plain English, not in a mathematical equation.

In some applications (e.g., epidemiology), the absence of notational distinction between causal and statistical dependencies seemed unnecessary, because investigators were able to keep such distinctions implicitly in their heads, and managed to confine the mathematics to conventional probability expressions. In others, as in economics and the social sciences, investigators rebelled against this notational tyranny by leaving mainstream statistics and constructing their own mathematical machinery (called Structural Equations Models). Unfortunately, this machinery has remained a mystery to outsiders, and eventually became a mystery to insiders as well.[7]

But such tensions could not remain dormant forever. "Every science is only so far exact as it knows how to express one thing by one sign," wrote Augustus de Morgan in 1858 — the harsh consequences of not having the signs for expressing causality surfaced in the 1980-90's. Problems such as the control of confounding, the estimation of treatment effects, the distinction between direct and indirect effects, the estimation of probability of causation, and the combination of experimental and nonexperimental data became a source of endless disputes among the users of statistics, and statisticians could not come to the rescue. [Pearl, 2000a] describes several such disputes, and why they could not be resolved by conventional statistical methodology.

3 LANGUAGES FOR CAUSAL ANALYSIS

3.1 The language of diagrams and structural equations

How can one express mathematically the common understanding that symptoms do not cause diseases? The earliest attempt to formulate such relationship mathematically was made in the 1920's by the geneticist Sewall Wright [1921]. Wright used a combination of equations and graphs to communicate causal relationships. For example, if X stands for a disease variable and Y stands for a certain symptom of the disease, Wright would write a linear equation:

(1) $y = ax + u$

supplemented with the diagram $X \longrightarrow Y$, where x stands for the level (or severity) of the disease, y stands for the level (or severity) of the symptom, and u stands

[7] Most econometric texts in the last decade have refrained from defining what an economic model is, and those that attempted a definition, erroneously view structural equations models as compact representations of probability density functions (see [Pearl, 2000a, pp. 135-138]).

for all factors, other than the disease in question, that could possibly affect Y (U is called "exogenous", "background", or "disturbance".) The diagram encodes the possible existence of (direct) causal influence of X on Y, and the absence of causal influence of Y on X, while the equation encodes the quantitative relationships among the variables involved, to be determined from the data. The parameter a in the equation is called a "path coefficient" and it quantifies the (direct) causal effect of X on Y; given the numerical value of a, the equation claims that, ceteras paribus, a unit increase in X would result in an a-unit increase of Y. If correlation between X and U is presumed possible, it is customary to add a double arrow between X and Y.

The asymmetry induced by the diagram renders the equality sign in Eq. (1) different from algebraic equality, resembling instead the assignment symbol (:=) in programming languages. Indeed, the distinctive characteristic of structural equations, setting them apart from algebraic equations, is that they stand for a value-assignment process — an autonomous mechanism by which the value of Y (not X) is determined. In this assignment process, Y is committed to track changes in X, while X is not subject to such commitment.[8]

Wright's major contribution to causal analysis, aside from introducing the language of path diagrams, has been the development of graphical rules for writing down (by inspection) the covariance of any pair of observed variables in terms of path coefficients and of covariances among disturbances. Under certain causal assumptions, (e.g. if $Cov(U, X) = 0$), the resulting equations may allow one to solve for the path coefficients in terms of observed covariance terms only, and this amounts to inferring the magnitude of (direct) causal effects from observed, non-experimental associations, assuming of course that one is prepared to defend the causal assumptions encoded in the diagram.

The causal assumptions embodied in the diagram (e.g, the absence of arrow from Y to X, or $Cov(U, X) = 0$) are not generally testable from nonexperimental data. However, the fact that each causal assumption in isolation cannot be tested does not mean that the sum total of all causal assumptions in a model does not have testable implications. The chain model $X \longrightarrow Y \longrightarrow Z$ for example, encodes seven causal assumptions, each corresponding to a missing arrow or a missing double-arrow between a pair of variables. None of those assumptions is testable in isolation, yet the totality of all those assumptions implies that Z is unassociated with X, conditioned on Y. Such testable implications can be read off the diagrams (see [Pearl 2000a, pp. 16–19]), and these constitute the only opening through which the assumption embodies in structural equation models can be tested in observational studies. Every conceivable statistical test that can be applied to the model is entailed by those implications.

---

[8]Clearly, if we intervene on X, Y would continue to track changes in X. Not so when we intervene on Y, X will reman unchanged. Such intervention (on Y) would alter the assignment mechanism for Y and, naturally, would cause the equality in Eq. (1) to be violated.

3.2 *From path-diagrams to do-calculus*

Structural equation modeling (SEM) has been the main vehicle for causal analysis in economics, and the behavioral and social sciences [Goldberger 1972; Duncan 1975]. However, the bulk of SEM methodology was developed for linear analysis and, until recently, no comparable methodology has been devised to extend its capabilities to models involving discrete variables, nonlinear dependencies, or situations in which the functional form of the equations is unknown. A central requirement for any such extension is to detach the notion of "effect" from its algebraic representation as a coefficient in an equation, and redefine "effect" as a general capacity to transmit *changes* among variables. One such extension, based on simulating hypothetical interventions in the model, is presented in Pearl [1995a, 2000a]

The central idea is to exploit the invariant characteristics of structural equations without committing to a specific functional form. For example, the non-parametric interpretation of the chain model $Z \longrightarrow X \longrightarrow Y$ corresponds to a set of three functions, each corresponding to one of the variables:

$$\begin{aligned} z &= f_Z(w) \\ x &= f_X(z, v) \\ y &= f_Y(x, u) \end{aligned} \tag{2}$$

together with the assumption that the background variables W,V, U (not shown in the chain) are jointly independent but, otherwise, arbitrarily distributed. Each of these functions represents a causal process (or mechanism) that determines the value of the left variable (output) from those on the right variables (input). The absence of a variable from the right hand side of an equation encodes the assumption that it has no direct effect on the left variable. For example, the absence of variable Z from the arguments of f_Y indicates that variations in Z will leave Y unchanged, as long as variables U and X remain constant. A system of such functions are said to be *structural* (or *modular*) if they are assumed to be autonomous, that is, each function is invariant to possible changes in the form of the other functions [Simon 1953; Koopmans 1953].

This feature of invariance permits us to use structural equations as a basis for modeling actions and counterfactuals. This is done through a mathematical operator called $do(x)$ which simulates physical interventions by deleting certain functions from the model, replacing them by constants, while keeping the rest of the model unchanged. For example, to represent an intervention that sets the value of X to x_0 the model for Eq. (2) would become

$$\begin{aligned} z &= f_Z(w) \\ x &= x_0 \\ y &= f_Y(x, u) \end{aligned} \tag{3}$$

The distribution of Y and Z calculated from this modified model characterizes the effect of the action $do(X = x_0)$ and is denoted as $P(y, z|do(x_0))$. It is not hard to show that, as expected, the model of Eq. (2) yields $P(y|do(x_0)) = P(y|x_0)$ and $P(z|do(x_0)) = P(z)$ regardless of the functions f_X, f_Y and f_Z. The general rule is simply to remove from the factorized distribution $P(x, y, z) = P(z)P(x|z)P(y|x)$ the factor that corresponds to the manipulated variable (X in our example) and to substitute the new value of that variable (x_0 in our example) into the truncated expression — the resulting expression then gives the post-intervention distribution of the remaining variables [Pearl, 2000a; section 3.2]. Additional features of this transformation are discussed in the Appendix; see [Pearl, 2000a; chapter 7] for full details.

The main task of causal analysis is to infer causal quantities from two sources of information: (i) the assumptions embodied in the model, and (ii) the observed distribution $P(x, y, z)$, or from samples of that distribution. Such analysis requires mathematical means of transforming causal quantities, represented by expressions such as $P(y|do(x))$, into *do*-free expressions derivable from $P(z, x, y)$, since only *do*-free expressions are estimable from non-experimental data. When such a transformation is feasible, we say that the causal quantity is *identifiable*. A calculus for performing such transformations, called *do*-calculus, was developed in [Pearl, 1995a]. Remarkably, the rules governing this calculus depend merely on the topology of the diagram; it takes no notice of the functional form of the equations, nor of the distribution of the disturbance terms. This calculus permits the investigator to inspect the causal diagram and

1. Decide whether the assumptions embodied in the model are sufficient to obtain consistent estimates of the target quantity;

2. Derive (if the answer to item 1 is affirmative) a closed-form expression for the target quantity in terms of distributions of observed quantities; and

3. Suggest (if the answer to item 1 is negative) a set of observations and experiments that, if performed, would render a consistent estimate feasible.

4 ON THE DEFINITION OF CAUSALITY

In this section, I return to discuss concerns expressed by some Bayesians that causality is an undefined concept and that, although the *do*-calculus can be an effective mathematical tool in certain tasks, it does not bring us closer to the deep and ultimate understanding of causality, one that is based solely on classical probability theory.

4.1 Is causality reducible to probabilities?

Unfortunately, aspirations for reducing causality to probability are both untenable and unwarranted. Philosophers have given up such aspirations twenty years ago, and were forced to admit extra-probabilistic primitives (such as "counterfactuals" or "causal relevance") into the analysis of causation (see Eells [1991] and Pearl [2000a, Section 7.5]). The basic reason was alluded to in Section 2: probability theory deals with beliefs about an uncertain, yet static world, while causality deals with changes that occur in the world itself, (or in one's theory of such changes). More specifically, causality deals with how probability functions change in response to influences (e.g., new conditions or interventions) that originate from outside the probability space, while probability theory, even when given a fully specified joint density function on all (temporally-indexed) variables in the space, cannot tell us how that function would change under such external influences. Thus, "doing" is not reducible to "seeing", and there is no point trying to fuse the two together.

Many philosophers have aspired to show that the calculus of probabilities, endowed with a time dynamic, would be sufficient for causation [Suppes, 1970]. A well known demonstration of the impossibility of such reduction (following Otte [1981]) goes as follows. Consider a switch X that turns on two lights, Y and Z, and assume that, due to differences in location, Z turns on a split second before Y. Consider now a variant of this example where the switch X activates Z, and Z, in turns, activates Y. This case is probabilistically identical to the previous one, because all functional and temporal relationships are identical. Yet few people would perceive the causal relationships to be the same in the two situations; the latter represents cascaded process, $X \longrightarrow Z \longrightarrow Y$, while the former represents a branching process, $Y \longleftarrow X \longrightarrow Z$. The difference shows, of course, when we consider interventions; intervening on Z would affect Y in the cascaded case, but not in the branching case.

The preceding example illustrates the essential role of *mechanisms* in defining causation. In the branching case, although all three variables are symmetrically constrained by the functional relationships: $X = Y$, $X = Z$, $Z = Y$, these relationships in themselves do not reveal the information that the three equalities are sustained by only two mechanisms, $Y = X$ and $Z = X$, and that the first equality would still be sustained when the second is violated. A set of mechanisms, each represented by an equation, is not equivalent to the set of algebraic equations that are implied by those mechanisms. Mathematically, the latter is defined as *one* set of n equations, whereas the former is defined as n separate sets, each containing one equation. These are two distinct mathematical objects that admit two distinct types of solution-preserving operations. The calculus of causality deals with the dynamics of such modular systems of equations, where the addition and deletion of equations represent interventions (see Appendix).

4.2 *Is causality well-defined?*

From a mathematical perspective, it is a mistake to say that causality is undefined. The *do*-calculus, for example, is based on two well-defined mathematical objects: a probability function P and a directed acyclic graph (DAG) D; the first is standard in statistical analysis while the second is a newcomer that tells us (in a qualitative, yet formal language) which mechanisms would remain invariant to a given intervention. Given these two mathematical objects, the definition of "cause" is clear and crisp; variable X is a *probabilistic-cause* of variable Y if $P(y|do(x)) \neq P(y)$ for some values x and y. Since each of $P(y|do(x))$ and $P(y)$ is well-defined in terms of the pair (P, D), the relation "probabilistic cause" is, likewise, well-defined. Similar definitions can be constructed for other nuances of causal discourse, for example, "causal effect", "direct cause", "indirect cause", "event-to-event cause", "scenario-specific cause", "necessary cause", "sufficient cause", "likely cause" and "actual cause" (see [Pearl, 2000a, pp. 222–3, 286–7, 319]; some of these definitions invoke functional models).

Not all statisticians/philosophers are satisfied with these mathematical definitions. Some suspect definitions that are based on unfamiliar non-algebraic objects (i.e., the DAG) and some mistrust abstract definitions that are based on unverifiable models. Indeed, no mathematical machinery can ever verify whether a given DAG really represents the causal mechanisms that generate the data — such verification is left either to human judgment or to experimental studies that invoke interventions. I submit, however, that neither suspicion nor mistrust are justified in the case at hand; DAGs are no less formal than mathematical equations, and questions of model verification need be kept apart from those of conceptual definition.

Consider, for example, the concept of a distribution *mean*. Even non-Bayesians perceive this notion to be well-defined, for it can be computed from any given (non-pathological) distribution function, even before ensuring that we can estimate that distribution from the data. We would certainly not declare the mean "ill-defined" if, for any reason, we find it hard to estimate the distribution from the available data. Quite the contrary; by defining the mean in the abstract, as a functional of *any* hypothetical distribution, we can often prove that the defining distribution need not be estimated at all, and that the mean can be estimated (consistently) directly from the data. Remarkably, by taking seriously the abstract (and untestable) notion of a distribution, we obtain a license to ignore it. An analogous logic applies to causation. Causal quantities are first defined in the abstract, using the pair (P, D), and this abstract definition then provides a theoretical framework for deciding, given the type of data available, which of the assumptions embodied in the DAG are ignorable, and which are absolutely necessary for establishing the target causal quantity from the data.[9]

The separation between concept definition and model verification is even more pronounced in the Bayesian framework, where purely judgmental concepts, such as the prior distribution of the mean, are perfectly acceptable, as long as they can be assessed reliably from one's experience or knowledge. Dennis Lindley has remarked recently (personal communication) that "causal mechanisms may be easier

[9] I have used a similar logic in defense of counterfactuals [Pearl, 2000a], which Dawid [2000] deemed dangerous on account of being untestable. (See, also Dawid [2001], this volume.) Had Bernoulli been constrained by Dawid's precautions, the notion of a "distribution" would have had to wait for another "dangerous" scientist, of Bernoulli's equal, to be created.

to come by than one might initially think". Indeed, from a Bayesian perspective, the newcomer concept of a DAG is not an alien at all — it is at least as legitimate as the probability assessments that a Bayesian decision-maker pronounces in constructing a decision tree. In such construction, the probabilities that are assigned to branches emanating from a decision variable X correspond to assessments of $P(y|do(x))$ and those assigned to branches emanating from a chance variable X correspond to assessments of $P(y|x)$. If a Bayesian decision-maker is free to assess $P(y|x)$ and $P(y|do(x))$ in any way, as separate evaluations, the Bayesian should also be permitted to express his/her conception of the mechanisms that entail those evaluations. It is only by envisioning these mechanisms that a decision maker can generate a coherent list of such a vast number of $P(y|do(x))$ type assessments.[10] The structure of the DAG can certainly be recovered from judgments of the form $P(y|do(x))$ and, conversely, the DAG combined with a probability function P dictates all judgments of the form $P(y|do(x))$. Accordingly the structure of the DAG can be viewed as a qualitative parsimonious scheme of encoding and maintaining coherence among those assessments. And there is no need to translate the DAG into the language of probabilities to render the analysis legitimate. Adding probabilistic veneer to the mechanisms portrayed in the DAG may make the *do* calculus appear more traditional, but would not change the fact that the objects of assessment are still causal mechanisms, and that these objects have their own special grammar of generating predictions about the effect of actions. In summary, recalling the ultimate Bayesian mission of fusing judgment with data, it is not the language in which we cast judgments that legitimizes the analysis, but whether those judgments can reliably be assessed from our store of knowledge and from the peculiar form in which this knowledge is organized.

If it were not for this concern to maintain reliability (of judgment), one could easily translate the information conveyed in a DAG into purely probabilistic formulae, using hypothetical variables. (Translation rules are provided in [Pearl, 2000a, p. 232]). Indeed, this is how the potential-outcome approach of Neyman [1923] and Rubin [1974] has achieved statistical legitimacy: judgments about causal relationships among observables are expressed as statements about probability functions that involve mixtures of observable and counterfactual variables. The difficulty with this approach, and the main reason for its slow acceptance in statistics, is that judgments about counterfactuals are much harder to assess than judgments about causal mechanisms. For instance, to communicate the simple assumption that symptoms do not cause diseases, we would have to use a rather roundabout expression and say that the probability of the counterfactual event "disease had symptoms been absent" is equal to the probability of "disease had symptoms been present". Judgments of conditional independencies among such counterfactual events are even harder for researchers to comprehend or to evaluate.

---

[10]Coherence requires, for example, that for any x, y, and z, the inequality $P(y|do(x), do(z)) \geq P(y, x|do(z))$ be satisfied. This follows from the property of composition (see Appendix, Eq. (6), or [Pearl, 2000a; pp. 229]

5 SUMMARY

This paper calls attention to a basic conflict between mission and practice in Bayesian methodology. The mission is to express prior knowledge mathematically and reliably so as to assist the interpretation of data, hence the acquisition of new knowledge. The practice has been to express prior knowledge as prior probabilities — too crude a vocabulary, given the grand mission. Considerations of reliability (of judgment) call for enriching the language of probabilities with causal vocabulary and for admitting causal judgments into the Bayesian repertoire. The mathematics for interpreting causal judgments has matured, and tools for using such judgments in the acquisition of new knowledge have been developed. The grounds are now ready for mission-oriented Bayesianism.

APPENDIX

CAUSAL MODELS, ACTIONS AND COUNTERFACTUALS

This appendix presents a brief summary of the structural-equation semantics of causation and counterfactuals as defined in Balke and Pearl [1995], Galles and Pearl [1997, 1998], and Halpern [1998]. For detailed exposition of the structural account and its applications see [Pearl, 2000a].

Causal models are generalizations of the structural equations used in engineering, biology, economics and social science.[11] World knowledge is represented as a modular collection of stable and autonomous relationships called "mechanisms", each represented as a function, and changes due to interventions or unmodelled eventualities are treated as local modifications of these functions.

A causal model is a mathematical object that assigns truth values to sentences involving causal relationships, actions, and counterfactuals. We will first define causal models, then discuss how causal sentences are evaluated in such models. We will restrict our discussion to recursive (or feedback-free) models; extensions to non-recursive models can be found in Galles and Pearl [1997, 1998] and Halpern [1998].

DEFINITION 1 (Causal model).
A *causal model* is a triple

$$M - \langle U, V, F \rangle$$

where

(i) U is a set of variables, called *exogenous*. (These variables will represent background conditions, that is, variables whose values are determined outside the model.)

---

[11] Similar models, called "neuron diagrams" [Lewis, 1986, p. 200; Hall, 1998] are used informally by philosophers to illustrate chains of causal processes.

(ii) V is an ordered set $\{V_1, V_2, \ldots, V_n\}$ of variables, called *endogenous*. (These represent variables that are determined in the model, namely, by variables in $U \cup V$.)

(iii) F is a set of functions $\{f_1, f_2, \ldots, f_n\}$ where each f_i is a mapping from $U \times (V_1 \times \ldots \times V_{i-1})$ to V_i. In other words, each f_i tells us the value of V_i given the values of U and all predecessors of V_i. Symbolically, the set of equations F can be represented by writing[12]

$$v_i = f_i(pa_i, u_i) \quad i = 1, \ldots, n$$

where pa_i is any realization of the unique minimal set of variables PA_i in V (connoting *parents*) sufficient for representing f_i.[13] Likewise, $U_i \subseteq U$ stands for the unique minimal set of variables in U that is sufficient for representing f_i.

Every causal model M can be associated with a directed graph, $G(M)$, in which each node corresponds to a variable in V and the directed edges point from members of PA_i toward V_i (by convention, the exogenous variables are usually not shown explicitly in the graph). We call such a graph the *causal graph* associated with M. This graph merely identifies the endogenous variables PA_i that have direct influence on each V_i but it does not specify the functional form of f_i.

For any causal model, we can define an *action* operator, $do(x)$, which, from a conceptual viewpoint, simulates the effect of external action that sets the value of X to x and, from a formal viewpoint, transforms the model into a *submodel*, that is, a causal model containing fewer functions.

DEFINITION 2 (Submodel).
Let M be a causal model, X be a set of variables in V, and x be a particular assignment of values to the variables in X. A submodel M_x of M is the causal model

$$M_x = \langle U, V, F_x \rangle$$

where

$$F_x = \{f_i : V_i \notin X\} \cup \{X = x\} \tag{4}$$

In words, F_x is formed by deleting from F all functions f_i corresponding to members of set X and replacing them with the set of constant functions $X = x$.

If we interpret each function f_i in F as an independent physical mechanism and define the action $do(X = x)$ as the minimal change in M required to make

[12] We use capital letters (e.g., X, Y) as names of variables and sets of variables, and lower-case letters (e.g., x, y) for specific values (called realizations) of the corresponding variables.

[13] A set of variables X is *sufficient* for representing a given function $y = f(x, z)$ if f is trivial in Z—that is, if for every x, z, z' we have $f(x, z) = f(x, z')$.

$X = x$ hold true under any u, then M_x represents the model that results from such a minimal change, since it differs from M by only those mechanisms that directly determine the variables in X. The transformation from M to M_x modifies the algebraic content of F, which is the reason for the name *modifiable structural equations* used in [Galles and Pearl, 1998].[14]

DEFINITION 3 (Effect of action).
Let M be a causal model, X be a set of variables in V, and x be a particular realization of X. The *effect of action* $do(X = x)$ on M is given by the submodel M_x.

DEFINITION 4 (Potential response).
Let Y be a variable in V, let X be a subset of V, and let u be a particular value of U. The *potential response* of Y to action $do(X = x)$ in situation u, denoted $Y_x(u)$, is the (unique) solution for Y of the set of equations F_x.

We will confine our attention to actions in the form of $do(X = x)$. Conditional actions, of the form "$do(X = x)$ if $Z = z$" can be formalized using the replacement of equations by functions of Z, rather than by constants [Pearl, 1994]. We will not consider disjunctive actions, of the form "$do(X = x$ or $X = x')$", since these complicate the probabilistic treatment of counterfactuals.

DEFINITION 5 (Counterfactual).
Let Y be a variable in V, and let X be a subset of V. The counterfactual expression "The value that Y would have obtained, had X been x" is interpreted as denoting the potential response $Y_x(u)$.

Definition 5 thus interprets the counterfactual phrase "had X been x" in terms of a hypothetical external action that modifies the actual course of history and imposes the condition "$X = x$" with minimal change of mechanisms. This is a crucial step in the semantics of counterfactuals [Balke and Pearl, 1994], as it permits x to differ from the actual value $X(u)$ of X without creating logical contradiction; it also suppresses abductive inferences (or backtracking) from the counterfactual antecedent $X = x$.[15]

It can be shown [Galles and Pearl, 1997] that the counterfactual relationship just defined, $Y_x(u)$, satisfies the following two properties:
Effectiveness:
For any two disjoint sets of variables, Y and W, we have

(5) $Y_{yw}(u) = y.$

[14] Structural modifications date back to Marschak [1950] and Simon [1953]. An explicit translation of interventions into "wiping out" equations from the model was first proposed by Strotz and Wold [1960] and later used in Fisher [1970], Sobel [1990], Spirtes et al. [1993], and Pearl [1995]. A similar notion of sub-model is introduced in Fine [1985], though not specifically for representing actions and counterfactuals.

[15] Simon and Rescher [1966, p. 339] did not include this step in their account of counterfactuals and noted that backward inferences triggered by the antecedents can lead to ambiguous interpretations.

In words, setting the variables in W to w has no effect on Y, once we set the value of Y to y.

Composition:

For any two disjoint sets of variables X and W, and any set of variables Y,

$$(6) \quad W_x(u) = w \Longrightarrow Y_{xw}(u) = Y_x(u).$$

In words, once we set X to x, setting the variables in W to the same values, w, that they would attain (under x) should have no effect on Y. Furthermore, effectiveness and composition are *complete* whenever M is recursive (i.e., $G(M)$ is acyclic) [Galles and Pearl, 1998; Halpern, 1998], that is, every property of counterfactuals that follows from the structural model semantics can be derived by repeated application of effectiveness and composition.

A corollary of composition is a property called *consistency* by [Robins, 1987]:

$$(7) \quad (X(u) = x) \Longrightarrow (Y_x(u) = Y(u))$$

Consistency states that, if in a certain context u we find variable X at value x, and we intervene and set X to that same value, x, we should not expect any change in the response variable Y. Composition and consistency are used in several derivations of Section 3.

The structural formulation generalizes naturally to probabilistic systems, as is seen below.

DEFINITION 6 (Probabilistic causal model).
A probabilistic causal model is a pair

$$\langle M, P(u) \rangle$$

where M is a causal model and $P(u)$ is a probability function defined over the domain of U.

$P(u)$, together with the fact that each endogenous variable is a function of U, defines a probability distribution over the endogenous variables. That is, for every set of variables $Y \subseteq V$, we have

$$(8) \quad P(y) \triangleq P(Y = y) = \sum_{\{u \mid Y(u)=y\}} P(u)$$

The probability of counterfactual statements is defined in the same manner, through the function $Y_x(u)$ induced by the submodel M_x. For example, the *causal effect* of X on Y is defined as:

$$(9) \quad P(Y_x = y) = \sum_{\{u \mid Y_x(u)=y\}} P(u)$$

Likewise, a probabilistic causal model defines a joint distribution on counterfactual statements, i.e., $P(Y_x = y, Z_w = z)$ is defined for any sets of variables Y, X, Z, W, not necessarily disjoint. In particular, $P(Y_x = y, X = x')$ and $P(Y_x = y, Y_{x'} = y')$ are well defined for $x \neq x'$, and are given by

$$(10) \quad P(Y_x = y, X = x') = \sum_{\{u | Y_x(u)=y \ \& \ X(u)=x'\}} P(u)$$

and

$$(11) \quad P(Y_x = y, Y_{x'} = y') = \sum_{\{u \mid Y_x(u)=y \ \& \ Y_{x'}(u)=y'\}} P(u).$$

When x and x' are incompatible, Y_x and $Y_{x'}$ cannot be measured simultaneously, and it may seem meaningless to attribute probability to the joint statement "Y would be y if $X = x$ *and* Y would be y' if $X = x'$." Such concerns have been a source of recent objections to treating counterfactuals as jointly distributed random variables [Dawid, 2000]. The definition of Y_x and $Y_{x'}$ in terms of two distinct submodels, driven by a standard probability space over U, demonstrates that joint probabilities of counterfactuals have solid mathematical and conceptual underpinning and, moreover, these probabilities can be encoded rather parsimoniously using $P(u)$ and F.

Computer Science Department, University of California, USA.

BIBLIOGRAPHY

[Balke and Pearl, 1994] A. Balke and J. Pearl. Probabilistic evaluation of counterfactual queries. In *Proceedings of the Twelfth National Conference on Artificial Intelligence*, volume I, pages 230–237. MIT Press, Menlo Park, CA, 1994.

[Balke and Pearl, 1995] A. Balke and J. Pearl. Counterfactuals and policy analysis in structural models. In P. Besnard and S. Hanks, editors, *Uncertainty in Artificial Intelligence 11*, pages 11–18. Morgan Kaufmann, San Francisco, 1995.

[Cartwright, 1989] N. Cartwright. *Nature's Capacities and Their Measurement*. Clarendon Press, Oxford, 1989.

[Cox, 1958] D.R. Cox. *The Planning of Experiments*. John Wiley and Sons, NY, 1958.

[Dawid, 2000] A.P. Dawid. Causal inference without counterfactuals (with comments and rejoinder). *Journal of the American Statistical Association*, 95(450):407–448, June 2000.

[DeFinetti, 1974] B. DeFinetti. *Theory of Probability: A Critical Introductory Treatment,* 2 volumes (Translated by A. Machi and A. Smith). Wiley, London, 1974.

[Duncan, 1975] O.D. Duncan. *Introduction to Structural Equation Models*. Academic Press, New York, 1975.

[Eells, 1991] E. Eells. *Probabilistic Causality*. Cambridge University Press, Cambridge, MA, 1991.

[Fine, 1985] K. Fine. *Reasoning with Arbitrary Objects*. B. Blackwell, New York, 1985.

[Fisher, 1970] F.M. Fisher. A correspondence principle for simultaneous equations models. *Econometrica*, 38(1):73–92, January 1970.

[Galles and Pearl, 1997] D. Galles and J. Pearl. Axioms of causal relevance. *Artificial Intelligence*, 97(1-2):9–43, 1997.

[Galles and Pearl, 1998] D. Galles and J. Pearl. An axiomatic characterization of causal counterfactuals. *Foundation of Science*, 3(1):151–182, 1998.

[Goldberger, 1972] A.S. Goldberger. Structural equation models in the social sciences. *Econometrica: Journal of the Econometric Society*, 40:979–1001, 1972.

[Hall, 1998] N. Hall. Two concepts of causation, 1998. In press.

[Halpern, 1998] J.Y. Halpern. Axiomatizing causal reasoning. In G.F. Cooper and S. Moral, editors, *Uncertainty in Artificial Intelligence*, pages 202–210. Morgan Kaufmann, San Francisco, CA, 1998.

[Heckman, 2001] J.J. Heckman. Econometrics and empirical economics. *Journal of Econometrics*, 100(1):1–5, 2001.

[Koopmans, 1953] T.C. Koopmans. Identification problems in econometric model construction. In W.C. Hood and T.C. Koopmans, editors, *Studies in Econometric Method*, pages 27–48. Wiley, New York, 1953.

[Lewis, 1986] D. Lewis. *Philosophical Papers*. Oxford University Press, New York, 1986.

[Lindley and Novick, 1981] D.V. Lindley and M.R. Novick. The role of exchangeability in inference. *The Annals of Statistics*, 9(1):45–58, 1981.

[Marschak, 1950] J. Marschak. Statistical inference in economics. In T. Koopmans, editor, *Statistical Inference in Dynamic Economic Models*, pages 1–50. Wiley, New York, 1950. Cowles Commission for Research in Economics, Monograph 10.

[Neyman, 1923] J. Neyman. On the application of probability theory to agricultural experiments. Essay on principles. Section 9. *Statistical Science*, 5(4):465–480, 1990. [Translation]

[Otte, 1981] R. Otte. A critique of Suppes' theory of probabilistic causality. *Synthese*, 48:167–189, 1981.

[Pearl, 1994] J. Pearl. A probabilistic calculus of actions. In R. Lopez de Mantaras and D. Poole, editors, *Uncertainty in Artificial Intelligence 10*, pages 454–462. Morgan Kaufmann, San Mateo, CA, 1994.

[Pearl, 1995a] J. Pearl. Causal diagrams for empirical research. *Biometrika*, 82(4):669–710, December 1995.

[Pearl, 1995b] J. Pearl. Causal inference from indirect experiments. *Artificial Intelligence in Medicine*, 7(6):561–582, 1995.

[Pearl, 2000a] J. Pearl. *Causality: Models, Reasoning, and Inference*. Cambridge University Press, New York, 2000.

[Pearl, 2000b] J. Pearl. Comment on A.P. Dawid's, Causal inference without counterfactuals. *Journal of the American Statistical Association*, 95(450):428–431, June 2000.

[Robins, 1987] J.M. Robins. A graphical approach to the identification and estimation of causal parameters in mortality studies with sustained exposure periods. *Journal of Chronic Diseases*, 40(Suppl 2):139S–161S, 1987.

[Rubin, 1974] D.B. Rubin. Estimating causal effects of treatments in randomized and nonrandomized studies. *Journal of Educational Psychology*, 66:688–701, 1974.

[Savage, 1962] L. J. Savage. *The Foundations of Statistical Inference*. Methuen and Co. Ltd., London, 1962.

[Simon and Rescher, 1966] H.A. Simon and N. Rescher. Cause and counterfactual. *Philosophy and Science*, 33:323–340, 1966.

[Simon, 1953] H.A. Simon. Causal ordering and identifiability. In Wm. C. Hood and T.C. Koopmans, editors, *Studies in Econometric Method*, pages 49–74. Wiley and Sons, Inc., 1953.

[Sobel, 1990] M.E. Sobel. Effect analysis and causation in linear structural equation models. *Psychometrika*, 55(3):495–515, 1990.

[Spirtes *et al.*, 1993] P. Spirtes, C. Glymour, and R. Scheines. *Causation, Prediction, and Search*. Springer-Verlag, New York, 1993.

[Strotz and Wold, 1960] R.H. Strotz and H.O.A. Wold. Recursive versus nonrecursive systems: An attempt at synthesis. *Econometrica*, 28:417–427, 1960.

[Suppes, 1970] P. Suppes. *A Probabilistic Theory of Causality*. North-Holland Publishing Co., Amsterdam, 1970.

[Wright, 1921] S. Wright. Correlation and causation. *Journal of Agricultural Research*, 20:557–585, 1921.

A. PHILIP DAWID

CAUSAL INFERENCE WITHOUT COUNTERFACTUALS

PART I: INTRODUCTION

1 CAUSAL MODELLING

Association is not causation. Many have held that Statistics, while well suited to investigate the former, strays into treacherous waters when it makes claims to say anything meaningful about the latter. Yet others have proceeded as if inference about the causes of observed phenomena were indeed a valid object of statistical enquiry; and it is certainly a great temptation for statisticians to attempt such 'causal inference'. Among those who have taken the logic of causal statistical inference seriously I mention in particular Rubin (1974, 1978), Holland (1986), Robins (1986, 1987), Pearl (1995a) and Shafer (1996). This paper represents my own attempt to contribute to the debate as to what are the appropriate statistical models and methods to use for causal inference, and what causal conclusions can be justified by statistical analysis.

There are many philosophical and statistical approaches to understanding and uncovering causation, and I shall not here attempt to attack the problem on a broad front. Attention will be confined to a simple decision-based understanding of causation, wherein an external agent can make interventions in, and observe various properties of, some system. Rubin (1978) and Heckerman and Shachter (1995), among others, have emphasized the importance of a clear decision-theoretic description of a causal problem. Understanding of the 'causal effects' of intervention will come through the building, testing and application of *causal models*, relating interventions, responses and other variables.

In my own view, the enterprise of causal statistical modelling is not essentially different from any other kind of statistical modelling, and is most satisfactorily understood from a Popperian hypothetico-deductive viewpoint. A model is not a straightforward reflection of external reality, and to propose a model is not to assert or to believe that Nature behaves in a particular way (Nature is surely utterly indifferent to our attempts to ensnare her in our theories). Rather, a model is a construct within the mental universe, by means of which we attempt somehow to describe certain, more or less restricted, aspects of the empirical universe. In order to do this we need to have a clear understanding of the semantics of such a description. This involves setting up a clear correspondence between the very different features of these two universes. In particular, we require very clear (if possibly implicit) understandings of:

D. Corfield and J. Williamson (eds.), Foundations of Bayesianism, 37–74.

- What the system modelled is (and so, in particular, how to distinguish a valid from an invalid instance of the system).
- What real world quantities are represented by variables appearing in the model.
- What an intervention involves. 'Setting' a patient's treatment to 'none' by (a) withholding it from him, (b) wiring his jaw shut, or (c) killing him are all very different interventions, with different effects, and must be modelled as such. We must also be clear as to what variables are affected by the intervention, directly or indirectly, and how.
- What is meant by replication (in time, space, ...).

In addition, it is vital that we have clearly defined methods for understanding, assessing and measuring the empirical success of any such attempt at description of the real world by a mathematical model[1].

So long as a model appears to describe the relevant aspects of the world satisfactorily, we may continue, cautiously, to use it; when it fails to do so, we need to search for a better one. In particular, any causal understandings we may feel we have attained must always be treated as tentative, and subject to revision should further observation of the world require it.

To be fully general we should consider models for complex problems, such as those discussed by Robins (1986), Pearl (1995a), wherein interventions, of various kinds, are possible at various points in a system, with effects that can cascade through a collection of variables. While such problems can be modelled and analysed (using structures such as influence diagrams) within the general philosophical and methodological framework of this paper, that would involve additional theoretical development. To keep things simple, we restrict attention here to systems on which it is possible to make a single external intervention, which we refer to as *treatment*, and observe a single eventual *response*. We also suppose, with no further real loss of generality, that there are just two treatments available. Another restriction, that could again be relaxed at the cost of further elaboration, is that we shall not here address the important and challenging problems arising from non-ignorable treatment assignment or observational studies (e.g. Rubin, 1974, 1978). See, however, §8.1 for some related analysis.

2 COUNTERFACTUALS

Much recent analysis of causal inference is grounded in the manipulation of *counterfactuals*. Philosophically, a counterfactual statement is an assertion of the form "If X had been the case, then Y would have happened", made when it is known to

[1]One approach to such understanding and assessment in the case of ordinary probability modelling, based on the concept of probability calibration, may be found in Dawid (1985).

be false that X is the case. In a famous historical counterfactual, Pascal (Pensées, 1669, §162), opines:

> Le nez de Cléopâtre: s'il eût été plus court, toute la face de la terre aurait changé.

More recent, an intriguing, seemingly self-referring, assertion is that of Shafer (1996, *p.*108):

> Were counterfactuals to have objective meaning, we might take them as basic, and define probability and causality in terms of them.

It is one of the aims of this paper to persuade the reader of the genuinely counterfactual nature of this claim.

An archetype of the use of counterfactuals in a causal statistical context is the assertion: "If only I had taken aspirin, my headache would have gone by now". It is implicit that I did not take aspirin, and I still have the headache. Such an assertion, if true, could be regarded as justifying an inference that not taking aspirin has 'caused' my headache to persist this long; and that, if I had taken aspirin, that would have 'caused' my headache to disappear by now. The assignment of cause is thus based on a comparison of the real and the counterfactual outcome.

If Y_A denotes the duration of my headache when I take aspirin, and $Y_{\bar{A}}$ its duration when I don't, the above assertion is of the form "$Y_{\bar{A}} > y$, $Y_A < y$", and relates jointly to the pair of values for $(Y_A, Y_{\bar{A}})$. An important question, which motivates much of the development in this paper, is to what extent such assertions can be validated or refuted by empirical observation. My approach is grounded in a Popperian philosophy, in which the meaningfulness of a purportedly scientific theory, proposition, quantity or concept is related to the implications it has for what is or could be observed, and, in particular, to the extent to which it is possible to conceive of data that would be affected by the truth of the proposition, or the value of the quantity. When this is the case, assertions are empirically refutable, and considered 'scientific'. When not so, they may be branded 'metaphysical'. I shall argue that counterfactual theories are essentially metaphysical. This in itself might not be automatic grounds for rejection of such a theory, if the causal inferences it led to were unaffected by the metaphysical assumptions embodied in it. Unfortunately, this is not so, and the answers which the approach delivers to its inferential questions are seen, on closer analysis, to be dependent on the validity of assumptions that are entirely untestable, even in principle. This can lead to distorted understandings and undesirable practical consequences.

3 TWO PROBLEMS

There are several different problems of causal inference, which are often conflated. In particular, I consider it important to distinguish between causal queries of the following two different types (Holland, 1986):

I "I have a headache. Will it help if I take aspirin?"

II "My headache has gone. Is it because I took aspirin?"

Query I requires inference about the *effects of causes*, i.e. comparisons among the expected consequences of various possible interventions in a system. Such queries have long been the focus of the bulk of the standard statistical theory of Experimental Design (which, it is worth remarking, has in general displayed little eagerness for counterfactual analyses). Query II, by contrast, relates to *causes of effects*: we seek to understand the causal relationship between an already observed outcome and an earlier intervention. Queries of this second kind might arise in legal inquiries, for example into whether responsibility for a particular claimant's leukæmia can be attributed to the fact that her father worked in a nuclear power station for 23 years. The distinction between queries I and II is closely related to that sometimes made between problems of *general* and of *singular* causation (Hitchcock, 1997), although, in our formulation, both queries relate to singular circumstances.

I consider both types of query valid and important, but they are different, and, as we shall see, require different, though related, treatments. Evidence, e.g. findings from epidemiological surveys, that is directly relevant to query I, is often used, inappropriately, to address query II, without careful attention to the difference between the queries.

4 PREVIEW

I first consider, in Part II, the problem of 'effects of causes'. Section 5 introduces the essential ingredients of the problem, and distinguishes two varieties of model: a *metaphysical model*, which allows direct formulation of counterfactual quantities and queries, and a *physical model*, which does not. By means of a simple running example I illustrate how certain inferences based on a metaphysical model are not completely determined by the data, however extensive, but remain sensitive to untestable additional assumptions. I also delimit the extent of the resulting arbitrariness. In Section 6, I describe an entirely different approach, based on physical modelling and decision analysis, and show how it delivers an unambiguous conclusion, avoiding the above problems. Section 7 questions the rôle of an implicit attitude of 'fatalism' in some counterfactual causal models and methods. Section 8 extends the discussion to cases in which additional covariate information is available on individual systems. In Section 9, I investigate whether certain analyses stemming from a counterfactual approach might nevertheless be acceptable for 'physical' purposes; examples are given of both possible answers. Section 10 asks whether it might ever be strictly advantageous to base physical analyses on a metaphysical structure. This appears to be sometimes the case for causal *modelling*, but arguably not so for causal *inference*.

In Part III, I turn to address the distinct problem of 'causes of effects'. For this, purely physical modelling appears inadequate, and the arbitrariness already identified in metaphysical modelling becomes a much more serious problem. I show in Section 11 how this arbitrariness can be reduced by taking account of concomitant variables. Section 12 introduces a convention of conditional independence across alternative universes, which helps to clarify the counterfactual inference, and possibly reduce the intrinsic ambiguity. Section 13 considers the possibility of using underlying deterministic relations to clarify causal questions and inferences. I argue that, to be useful, these must involve genuine concomitant variables. A contrast is drawn with 'pseudo-deterministic models', which are always available in the counterfactual framework. These do have a deterministic mathematical structure, but need not involve true concomitants. Such a purely formal structure, I argue, is not enough to support meaningful inferences about the causes of effects. In Section 14, I discuss more deeply the meaning of concomitance, and argue that this is partly a matter of convention, relative to a specific causal inquiry, rather than a property of the physical world.

The general message of this paper is that inferences based on counterfactual assumptions and models are generally unhelpful and frequently plain misleading. Alternative approaches can avoid these problems, while continuing to address meaningful causal questions. For inference about the effects of causes, a straightforward 'black-box' decision-analytic approach, based on models and quantities which are empirically testable and discoverable, is perfectly adequate. For inference about the causes of effects, we need to suit our causal models to the questions addressed as well as to the empirical world, and to seek understanding of the relationships between observed variables and possibly unobserved, but empirically meaningful, concomitant variables. The causal inferences which are justified by empirical findings will still, in general, retain a degree of arbitrariness and convention, which should be fully admitted.

PART II: EFFECTS OF CAUSES

5 COMPARISON OF TREATMENTS: COUNTERFACTUAL APPROACH

As a simple and familiar setting to discuss and contrast different approaches to inference about the effects of causes, I investigate the problem of making comparisons between two treatments, t and c, for example, aspirin and placebo control, on the basis of an experiment. In this section, I shall consider counterfactual approaches to this problem, and show how they can produce ambiguous answers, unless arbitrary and unverifiable assumptions are imposed.

Consider a large homogeneous population $\mathcal{U}$ of clearly distinguishable *individuals*, or *systems*, or (as we shall generally call them) *units*, u, to each of which we can choose to apply any one treatment, i, out of the treatment set $\mathcal{T} = \{t, c\}$, and observe the resulting response, Y. Once one treatment has been applied, the other treatment can no longer be applied. This property can be ensured by appropriate definition of experimental unit u (e.g. headache episode rather than patient) and treatment (combinations of treatments, if available, being redefined as new treatments).

Experimentation consists in selecting disjoint sets of units $\mathcal{U}_i \subseteq \mathcal{U}$ $(i = t, c)$, applying treatment i to each unit in $\mathcal{U}_i$, and observing the ensuing responses (e.g., time for the headache to disappear). The experimental units might be selected for treatment by some form of randomisation, but this is inessential to our argument. For further clarification of the argument, I shall assume that the treatment groups are sufficiently large that all inferential problems associated with finite sampling can be ignored.

Homogeneity of the population is an intuitive concept, which can be formalised in a number of ways. From a classical standpoint, the individuals might be regarded as drawn randomly and independently from some large population; a Bayesian might regard them as exchangeable. In this context, homogeneity is also taken to imply that no specific information is available on the units that might serve to distinguish one from another (this constraint will be relaxed in §8). In particular, the experimenter is unable to take any such information into account, either deliberately or inadvertently, in deciding which treatment a particular unit is to receive. To render this scenario more realistic and versatile, suppose that we did in fact have additional measured *covariate* information on each unit, determined by (but not uniquely identifying) that unit. Then we could confine attention to a subpopulation having certain fixed covariate values, and this subpopulation might then be reasonably regarded as homogeneous. That is, our discussion should be understood as applying at the level of the residual variation, after all relevant observed covariates have been allowed for. (We can then also allow treatment assignment to take these observed covariates into account.)

Counterfactual framework. The counterfactual approach to causal analysis for this problem focuses on the collection of *potential responses* $\mathcal{Y} := (Y_i(u) : i \in \mathcal{T}, u \in \mathcal{U})$, where $Y_i(u)$ is intended to denote "the response that would be observed if treatment i were assigned to unit u". One can consider $\mathcal{Y}$ as arranged in a two-way layout of treatments by units, with $Y_i(u)$ occupying the cell for row i and column u. Note that many of the variables in $\mathcal{Y}$ are (to borrow a term from Quantum Physics) *complementary*, in that they are not simultaneously observable. Specifically, for any unit u, one can observe $Y_i(u)$ for at most one treatment i. The assignment of treatments to units will determine just which (if any) of these complementary variables are to be observed, yielding a collection $\mathcal{X}$ of responses that I call a *physical array* — in contrast with the *metaphysical array* $\mathcal{Y}$. Although the full collection $\mathcal{Y}$ is intrinsically unobservable, counterfactual analyses are based on consideration of all the $(Y_i(u))$ simultaneously. Current interest in the counterfactual approach was instigated by Rubin (1974, 1978), although it can be traced back at least to Neyman (1935; see also Neyman, 1923).

5.1 Metaphysical model

What kind of models is it reasonable to entertain for the metaphysical array $\mathcal{Y}$? The assumption of homogeneity essentially requires us to model the various pairs $(Y_t(u), Y_c(u))$, for $u \in \mathcal{U}$, as independent and identically distributed, given their (typically unknown) bivariate distribution P. I shall denote the implied marginal distributions for Y_t and Y_c by P_t, P_c respectively. It is important to note that the full bivariate distribution P is not completely specified by these marginals, without further specification of the dependence between Y_t and Y_c.

Although the major points of our discussion will apply to a general model of the above form, for definiteness I shall concentrate on the following specific bivariate normal model.

EXAMPLE 1.

> The pairs $\{(Y_t(u), Y_c(u)) : u \in \mathcal{U}\}$ are modelled as independent and identically distributed, each having the bivariate normal distribution with means (θ_t, θ_c), common variance ϕ_Y, and correlation ρ.

When $\rho \geq 0$, which seems a reasonable judgment[2] one can also represent this structure by means of the mixed model:

$$(1) \quad Y_i(u) = \theta_i + \beta(u) + \gamma_i(u),$$

where all the $(\beta(u))$ and $(\gamma_i(u))$ are mutually independent normal random variables, with mean 0 and respective variances $\phi_\beta := \rho\phi_Y$ and $\phi_\gamma := (1-\rho)\phi_Y$. Inversely, one could start with (1) as our model, in which case we have

$$(2) \quad \phi_Y = \phi_\beta + \phi_\gamma,$$

$$(3) \quad \rho = \frac{\phi_\beta}{\phi_\beta + \phi_\gamma}.$$

In the usual parlance of the Analysis of Variance, Model (1) expresses $Y_i(u)$ as composed of: a fixed *treatment effect* θ_i associated with the applied treatment i, common to all units; a random *unit effect* $\beta(u)$, unique to unit u, but common to both treatments; and a random *unit-treatment interaction*, $\gamma_i(u)$, varying from one treatment application to another, even on the same unit. (This last term could also be interpreted as incorporating intrinsic random variation, which can not be distinguished from interaction since replicate observations on $Y_i(u)$ are impossible.)

5.2 Causal effect

The counterfactual approach typically takes as the fundamental object of causal inference the *individual causal effect*: a suitable numerical comparison, for a given

[2]See §12. One can also regard (1) as a (fictitious) representation of the bivariate normal model even when $\rho < 0$, in which case we must have $-\phi_Y \leq \phi_\beta \leq 0$ and $0 \leq \phi_\gamma \leq 2\phi_Y$. Then the calculations below, though based on this fictitious representation, are still valid.

unit, between the various potential responses it would exhibit, under the various treatments that might be applied. Note that such a quantity is meaningless unless one regards the several potential responses, complementary though they are, as having simultaneous existence.

The individual causal effect (ICE) for unit u will here be identified with the difference

(4) $\tau(u) := Y_t(u) - Y_c(u).$

Alternative possibilities might be $\log Y_t(u) - \log Y_c(u)$, or $Y_t(u)/Y_c(u)$. There seems no obvious theoretical reason, within this framework, to prefer any one such comparison to any other, the choice perhaps being made according to one's understanding of the applied context and the type of inferential conclusion desired. But, however defined, an ICE involves direct comparison of complementary quantities, and is thus intrinsically unobservable.

In most studies, the specific units used in the experiment are of no special interest in themselves, but merely provide a basis for inference about generic properties of units under the influence of the various treatments. For this purpose, it is helpful to conceive of an entirely new *test unit*, u_0, from the same population, that has not yet been treated; and to regard the purpose of the experiment as to assist us in making the decision as to which treatment to apply to it. If we decide on treatment t, we shall obtain response $Y_t(u_0)$; if c, we shall obtain $Y_c(u_0)$. We thus need to make inference about these two quantities, and compare them somehow. Note that, although $Y_t(u_0)$ and $Y_c(u_0)$ are complementary, neither is (as yet) counterfactual.

The counterfactual approach might focus on the ICE $\tau(u_0) = Y_t(u_0) - Y_c(u_0)$, or a suitable variation thereon. Under model (1), we have

(5) $\tau(u) = \tau + \lambda(u),$

with $\tau := \theta_t - \theta_c$, the *average causal effect* (ACE), and $\lambda(u) := \gamma_t(u) - \gamma_c(u)$, the *residual causal effect*, having distribution

(6) $\lambda(u) \sim N(0, 2\phi_\gamma).$

Thus

(7) $\tau(u) \sim N(\tau, 2\phi_\gamma).$

This model holds, in particular, for the inferential target $\tau(u_0)$. Since $\tau(u_0)$ is probabilistically independent of any data on the units in the experiment, inference about $\tau(u_0)$ essentially reduces to inference about the pair (τ, ϕ_γ).

5.3 *Physical model*

Suppose a particular experimental assignment has been specified. Label, arbitrarily, the units receiving treatment i as $u_{i1}, u_{i2} \cdots, u_{in_i}$. Then the observed response on unit u_{ij} will be $X_{ij} := Y_i(u_{ij})$. The collection $(X_{ij} : i = t, c; j = 1, \ldots, n_i)$

constitutes the physical array $\mathcal{X}$. The mean response on all units receiving treatment i is $\bar{X}_i := \frac{1}{n_i}\sum_{j=1}^{n_i} X_{ij}$.

It follows trivially from the model assumptions of Example 1 that the joint distribution over $\mathcal{X}$ is described by:

$$(8)\quad X_{ij} \sim N(\theta_i, \phi_Y)$$

independently for all (i, j). Equivalently, from (1),

$$(9)\quad X_{ij} = \theta_i + \epsilon_{ij},$$

with $\epsilon_{ij} := \beta(u_{ij}) + \gamma_i(u_{ij}) \sim N(0, \phi_Y)$, independently for all (i, j).

Now to the extent that the model (1) says anything about the empirical world, this has to be fully captured in the implied models (8) (one such for each possible physical array). Clearly, from extensive data having the structure (8), one can identify θ_t, θ_c and ϕ_Y; but the individual components ϕ_β and ϕ_γ in (2) — or, equivalently, the correlation ρ satisfying (3) — are not identifiable: we have *intrinsic aliasing* (McCullagh and Nelder, 1989, §3.5) of unit effect and unit-treatment interaction. So far as the desired inference about $\tau(u_0)$ is concerned, we can identify its mean, τ = ACE, in (7). However its variance, $2\phi_\gamma$, is not identifiable from the data, beyond the requirement $\phi_\gamma \leq \phi_Y$ (if we restrict to $\rho \geq 0$, or $\phi_\gamma \leq 2\phi_Y$ for ρ unrestricted).

5.4 A quandary

Here we have an inferential quandary. Consider two statisticians, both of whom believe in model (1). However, statistician S1 further assumes that $\phi_\beta = 0$ ($\rho = 0$), and statistician S2 that $\phi_\gamma = 0$ ($\rho = 1$). Both S1 and S2 accept model (8) for the physical array, with no further constraints on its parameters. Extensive data, assumed to be fully consistent with model (8) for the physical array, lead to essentially exact estimates of θ_t, θ_c and ϕ_Y. However, S1 infers $\phi_\beta = 0, \phi_\gamma = \phi_Y$, while S2 has $\phi_\beta = \phi_Y, \phi_\gamma = 0$. When they come to inference about $\tau(u_0)$, from (7), they will agree on its mean, τ, but differ about its variance, $2\phi_\gamma$. A third statistician, making different assumptions (e.g. $\phi_\beta = \phi_\gamma$, equivalent to $\rho = \frac{1}{2}$) will come to yet another distinct conclusion. Is it not worrying that models that are intrinsically indistinguishable, on the basis of any data that could ever be observed, can lead to such different inferences? How can we possibly choose between these inferences?

The above state of affairs is clearly in violation of what, in another context (Dawid, 1984, §5.2), I have called *Jeffreys's Law*: the requirement that mathematically distinct models that cannot be distinguished on the basis of empirical observation should lead to indistinguishable inferences. This property can be demonstrated mathematically in cases where those inferences concern future observables; and I consider that it has just as much intuitive force in the present context of causal inference.

There is one important, but very special, case where the above ambiguity vanishes: when ϕ_Y is essentially zero, and hence so are both ϕ_β and ϕ_γ. In this case, the units are not merely homogeneous, but *uniform*, in that, for each i, $Y_i(u)$ is the same for all units u. The property $\phi_Y \doteq 0$ can, of course, be investigated empirically, and might be regarded as a distinguishing feature of at least some problems in the 'hard' sciences. When it holds, one can, in effect, observe both $Y_t(u)$ and $Y_c(u)$ simultaneously, by employing distinct units, thus enabling direct measurement of causal effects. I shall further consider this case of uniformity, and its extensions, in §13 below.

5.5 *Additional constraints*

How should we proceed if we do not have uniformity? It is common in studies based on counterfactual models to impose additional constraints. In the present context, a common additional constraint is that of *treatment-unit additivity* (TUA), which asserts that $\tau(u)$ in (4) is the same for all $u \in \mathcal{U}$. In terms of (1), this is equivalent to $\phi_\gamma = 0$ ($\rho = 1$), and leads to a simple inference: $\tau(u_0) = \tau$, with no further uncertainty (τ having been identified, from a large experiment, as $\bar{X}_t - \bar{X}_c$). However, as pointed out above, there is simply no way that TUA can be tested on the basis of any empirically observable data in the context of model (1); and it is intuitively clear that the same holds for any other models that might be considered: when, for each pair $(Y_t(u), Y_c(u))$, it is never possible to observe both components, how can one ever assess empirically the assertion that $Y_t(u) - Y_c(u)$ (unobservable for each u) is the same for all u?[3]

A similar untestable assumption commonly made in the case of binary responses (Imbens and Angrist, 1994) is *monotonicity*, which requires that $P(Y_c = 1, Y_t = 0) = 0$ (where the response 1 represents a successful, and 0 an unsuccessful, outcome).

5.6 *What can we say?*

If we are to restrict our inferences to those that *are* justified by the data, without the imposition of untestable additional constraints, the most that can be said about $\tau(u_0)$ (assuming model (1)) is:

(10) $\tau(u_0) \sim N(\tau, 2\phi_\gamma)$,

with τ estimated precisely, but ϕ_γ subject only to the inequality $0 \leq \phi_\gamma \leq \phi_Y$,[4] whose right-hand-side only is estimated precisely. Only if we are fortunate enough

[3]If we were to have used a more general model in Example 1, whereby we allowed the variance to be different for two responses, say ϕ_t and ϕ_c, then TUA does have the testable implication $\phi_t = \phi_c$, and so could be rejected on the basis of data casting doubt on this property. But such data would still not distinguish between TUA and any of the other models considered above, all of which would likewise be rejected. We have assumed thoughout that the data are consistent with the physical model (8), so that this issue does not arise.

[4]Or $0 \leq \phi_\gamma \leq 2\phi_Y$ if we allow $\rho < 0$.

to find that ϕ_Y is negligible (the situation of uniformity) can we obtain an unambiguous inference for $\tau(u_0)$.

A very similar analysis can be conducted for other metaphysical models. Although the physical model only allows us to identify the marginal distributions P_t and P_c of the joint distribution P, the distribution of an individual causal effect (however defined) will depend further on the dependence structure of P.[5] Consequently, even when we have conducted very large experiments, we cannot make unambiguous inferences about such causal effects without making further untestable assumptions, such as TUA or monotonicity.

There are two contrasting morals that may be drawn from the above analysis, both grounded in the principle that we should be careful not to make 'metaphysical inferences', sensitive to assumptions that can not be put to empirical test. Moral 1 is that inference about individual causal effects should be carefully circumscribed, as below (10). Alternatively, we might draw the more revolutionary Moral 2: that if we can not get a sensible answer to our question, perhaps the question itself, with its focus on inference for $\tau(u_0)$, is not well posed. In the next section I shall reformulate the question in an entirely different manner, that does allow a clear and unambiguous answer.

6 DECISION-ANALYTIC APPROACH

As we have seen in the above example, the principal difficulty in the counterfactual approach is that the desired inference depends on the joint probability structure of the complementary variables $(Y_t(u), Y_c(u))$, whereas one is only ever able to observe (at most) one of these for each u. We can, however, consistently estimate both marginal distributions P_t, P_c. Can we put these separate marginal distributions to good use?

I take a straightforward Bayesian decision-analytic approach (see e.g. Raiffa, 1968). We have to decide whether to apply treatment t or treatment c to a new unit u_0. The marginal distributions P_t and P_c of Y_t and of Y_c having been identified, from our extensive experimental data on each separate treatment group, these now express the appropriate predictive uncertainty about the response on u_0, conditional on its being given t or c. The consequence (loss) of our decision may be measured by some function $L(\cdot)$ of the eventual yield Y. The decision tree for this problem is given in Figure 1.

At node ν_t, $Y \sim P_t$, and the (negative) value of being at ν_t is measured by the expected loss $E_{P_t}\{L(Y)\}$. Similarly, ν_c has value $E_{P_c}\{L(Y)\}$. The principles of Bayesian decision analysis now require that, at the decision node ν_0, we choose that treatment i leading to the smaller expected loss.

Note that, whatever loss function is used, this solution only involves the two identifiable marginal distributions, P_c and P_t. In particular, our statisticians S1

[5]There is a large literature on properties, and inequalities for, joint distributions with known marginals: see e.g. Rüschendorf *et al.* (1996).

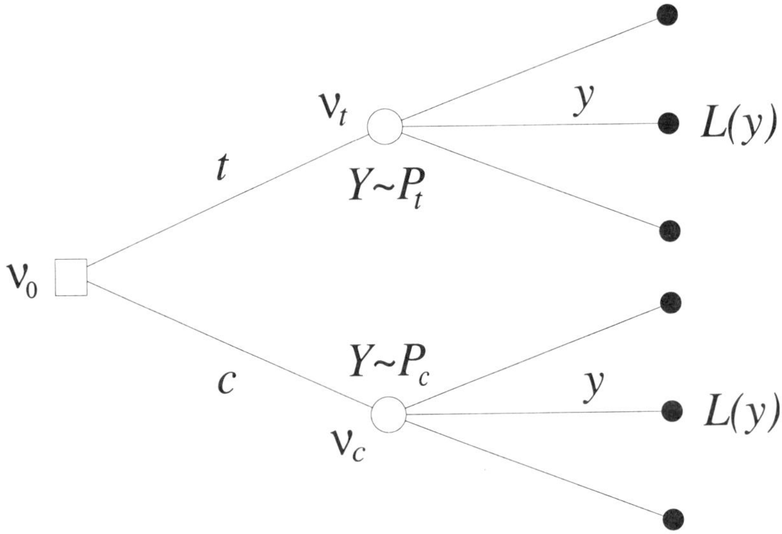

Figure 1. Decision tree

and S2 of §5.4, who agree on model (1) and obtain common estimates of θ_t, θ_c and ϕ_Y, while disagreeing about ρ, will be led to the identical decision. It simply does not matter that S2 believes that the time for a headache to disappear if aspirin is taken will be *exactly* 10 minutes less than if it is not taken, while S1 regards the difference of these times as uncertain, although again with expectation 10 minutes; there is no way in which such differences in beliefs can affect the decision problem.

It is only for simplicity of the argument that I have assumed the experiment large enough to allow full identification of P_t and P_c. With a more limited experiment, we could either replace these with suitable estimates, or, for a wholeheartedly Bayesian approach, use the appropriate predictive distributions for the response on u_0 (under either hypothetical treatment application, separately), given the experimental data.

Our analysis extends readily to the case where we consider a number of future units, and want to decide how to apply treatments to them. In a quality control setting, our loss might be a combination of the sample mean and variance of all the responses, for example.

One can also consider models for more complex problems, involving nonhomogeneous populations. For example, Dawid (1988) used symmetry arguments to justify the construction of certain random-effect type models for complex experimental layouts, generalising models such as those of (1) for the metaphysical array or (9) for the physical array. In the general case, we again need to use the data of the experiment to make appropriate predictive inferences for test units, under varying hypothetical treatment assignments; but these predictive inferences will now be more complex, and will also depend upon the relationship assumed between the test units and the experimental units. For example, if our experiment involved planting different varieties of cereal on plots (units) nested within blocks nested within fields, and recording their yields, one might wish to consider predictions for the yield, of each variety, if planted on: a new plot in an old (i.e. experimental) block in an old field; a plot in a new block in an old field; or (more usefully) a plot in a new field. So long as our models relate the responses of the new and the old units (under arbitrary treatment assignments), and so support the required predictive inferences, one can conduct whatever decision-analytic analysis appears most relevant to one's purpose, eschewing counterfactuals entirely.

7 FATALISM

Many counterfactual analyses are based, explicitly or implicitly, on an attitude that I term *fatalism*. This conceives of the various potential responses $Y_i(u)$, when treatment i is applied to unit u, as pre-determined attributes of unit u, waiting only to be uncovered by suitable experimentation. (It is implicit that that the unit u, and its properties and propensities, exist independently of, and are unaffected by, any treatment that may be applied.) Note that, since each unit label u is regarded as individual and unrepeatable, there is never any possibility of empirically testing this assumption of fatalism, which can thus be categorised as metaphysical.

The fatalistic world-view runs very much counter to the philosophy underlying statistical modelling and inference in almost every other setting. For example, it leaves no scope for introducing realistic stochastic effects of external influences acting between the times of application of treatment and of the response. Any account of causation that requires us to jettison all our familiar statistical framework and machinery should be treated with the utmost suspicion, unless and until it has shown itself completely indispensable for its purpose.

7.1 Some fatalistic concepts

I do not wish to give the impression that all counterfactual analyses must be fatalistic; there are notable exceptions, for example Robins and Greenland (1989). However, it is a very natural bed-fellow of counterfactual inference, much of which can not proceed without it. For example, only if one takes a fatalistic attitude does it make sense even to talk of such properties as 'treatment-unit additivity' or 'monotonicity' (§8).

A fundamental use of fatalism underlies certain counterfactual analyses of treatment non-compliance (see e.g. Imbens and Rubin, 1997), where each patient is supposed categorisable as: a *complier* (who would take the treatment if prescribed, and not take it if not prescribed); a *defier* (not take it if prescribed, take it if not prescribed); an *always taker* (take it, whether or not prescribed); or a *never taker* (not take it, whether or not prescribed). Some causal inferences are based on consideration of the responses to treatment of, say, the group of compliers. However, it is only under the unrealistic assumption of fatalism that this group has any meaningful identity, and thus only in this case could such inferences even begin to have any useful content.

SUTVA

An assumption that has often been considered essential to useful causal inferences is the *stable unit-treatment value assumption* (SUTVA) (Rubin, 1980, 1986). To describe this, we have to start from a more general metaphysical model of the effect of experimentation on responses, wherein the response $Y_\xi(u)$ of unit u could in principle depend on the full treatment assignment ξ over all units, not just the specific treatment i applied to u. Then SUTVA requires that, in fact, this potential complicating feature is absent, so that we can replace $Y_\xi(u)$ by $Y_i(u)$, thus returning us to the situation already considered. But again, without the fatalistic assumption of pre-existing values of the $(Y_\xi(u))$, for any assignment ξ, it is not possible to make sense of SUTVA (but see §10.1 for a non-fatalistic reinterpretation of SUTVA).

Decision analysis and fatalism

By contrast, the decision-analytic approach requires no commitment to (or, for that matter, against) fatalism. There is no conceptual or mathematical difficulty in regarding the probability distributions of the response (i.e. P_t and P_c in Example 1) as incorporating further uncontrollable influences, over and above effects attributable directly to treatment. So far as SUTVA is concerned, the decision analyst has no need of it. In the context of Example 1, SUTVA can be replaced by the much weaker assumption that the application of treatments does not destroy the homogeneity of the units, beyond the obvious difference that some will now have one treatment, some another. Then we will still have complete homogeneity of the responses for all units (experimental or future) receiving the same treatment, and can thus use the experimental data to identify the distribution, P_i, of response within treatment group i, which also expresses our uncertainty about the response $Y_i(u_0)$ of a new unit u_0, if it were given treatment i. Hence we are still in a position to set up, and solve, the basic decision problem for u_0.

8 USE OF ADDITIONAL INFORMATION

Now suppose that it is possible to gather, or at least to conceive of gathering, additional information about individual units, which might be used to refine our uncertainties about their responses to treatments. Any such information can be described in terms of a generic variable K, determined by a measurement protocol which, when applied to unit u, leads to a measurement $K(u)$. For present purposes we restrict our attention to generic variables that are *covariates*, i.e. features of units that can be observed prior to experimentation. Nevertheless, before it is observed each $K(u)$ must be treated as a random variable.

There are several cases to consider, according as whether or not the covariates are observed on the experimental units, and/or on test units.

1. **Covariates on experimental and test units.** Suppose that a covariate K is measured on all experimental units, and also that, for a test unit u_0, $K(u_0)$ will be measured before the treatment decision has to be made.

   If K takes values in a finite set, we can simply restrict attention to the subset (assumed large) of the experimental units for which $K(u) = K(u_0)$. Then we essentially recover the homogeneous population problem that has already been analysed.

   Otherwise, or if the above restricted subset is not sufficiently large, we can conduct appropriate statistical modelling. In a counterfactual treatment, we would need to model a joint conditional distribution of (Y_c, Y_t) given K; for the decision-analytic treatment, we only need to use the data to assess and compare the associated predictive distributions of $Y(u_0)$ given $K(u_0)$, for each treatment. Again, the decision-analytic approach, in contrast to the counterfactual approach, is essentially insensitive to any further assumptions about, or modelling of, the joint distribution of potential responses.

2. **Covariates on experimental units only.** In this case, it is appropriate to ignore altogether the covariate information on the experimental units — except that, when the experiment is not large, modelling of this more detailed information might enhance the accuracy of estimation of the required marginal predictive distributions of $Y(u_0)$ for each treatment.

3. **Covariate on test unit only.** This is more problematic, since even for the less demanding decision-analytic approach the experiment gives no direct information about the required predictive distributions of response given covariate and treatment. Whichever approach one takes, there is no escape from the fact that the solution will be highly dependent on untested (though in principle testable) assumptions about these distributions. One possibility would be to ignore $K(u_0)$ altogether, but this is itself tantamount to an empirically untested assumption of independence between K and Y for each treatment. In any event, however one proceeds, there is no advantage to be gained from the introduction of counterfactuals. Similar comments apply when information of differing extents is available on the experimental and test units.

8.1 Alternatives to additivity

One argument that can be made for the need for a metaphysical assumption such as treatment-unit additivity (§5.5) is the following. An experiment (e.g. a clinical trial) will often have very specific *inclusion criteria* that render the experimental units non-representative of the population to which it is intended to generalise the findings. Then, although we may still have homogeneity of units within the experiment, it might no longer be reasonable to regard the test unit u_0 as exchangeable with the experimental units. But if we can assume TUA, so that $Y_t(u) - Y_c(u) \equiv \tau$ for all units, experimental and test, then our estimate of the treatment effect τ from the experiment will still be applicable to u_0. Thus counterfactual analysis based on TUA appears unaffected by this modification to the framework. For the decision-analytic approach, however, the required separate predictive inferences about the response $Y(u_0)$, given either treatment, for a test unit u_0, would be simultaneously more complicated and less reliable when the experimental units cannot be regarded as representative of the test units.

However, there is an alternative way of proceeding that avoids metaphysical assumptions. For each unit u let $Q(u)$ be a variable, taking values $0, t, c$, generated by the experimenter as part of the process of designing his experiment. He intends to include u in the experiment, and apply treatment t to it, if $Q(u) = t$; to include u in the experiment, and apply treatment c to it, if $Q(u) = c$; and to exclude u from the experiment if $Q(u) = 0$. These intentions do not, however, preclude us from considering other possibilities; we can, for example, meaningfully assess probabilistic uncertainty about $Y(u)$, given that the assignment $Q(u) = t$ has been made, on the hypothesis that u will receive treatment c.

We assume that, for some covariate K, the distribution of $Q(u)$ given $K(u)$ is the same for all units u. Thus K is the information that the experimenter takes account of in generating Q, and so embodies the inclusion and treatment criteria. The distribution of Q given K is supposed unaffected by further conditioning on the applied treatment i and the eventual response Y: using the notation and properties of conditional independence (Dawid, 1979), $Q \perp\!\!\!\perp (i, Y) | K$, whence

(11) $Y \perp\!\!\!\perp Q | K, i.$

Consider now the following model assumption:

(12) $\mathrm{E}(Y|K, i) = \theta_i + \gamma(K) \qquad (i = t, c),$

for some unknown parameters θ_t, θ_c and parametric function $\gamma(\cdot)$. If this holds, define $\tau = \theta_t - \theta_c$.

Note that, by (11), the left-hand side of (12) is unaffected by further conditioning on Q. In particular, (12) implies: $\mathrm{E}\{Y|K, i, Q = i\} = \theta_i + \gamma(K)$ $(i = t, c)$, so that, for any k,

(13) $\mathrm{E}\{Y|K = k, t, Q = t\} - \mathrm{E}\{Y|K = k, c, Q = c\} \equiv \tau.$

Conversely, (13), with (11), implies (12). But we can estimate $\mathrm{E}\{Y|K = k, i, Q = i\}$ straightforwardly from the measurements of covariate K and outcome Y on the set of experimental units to which treatment i has been applied. Consequently property (12) is testable from the experimental data, and, if it can be assumed to hold, the parameter τ is estimable (a simple unbiased estimator of τ is given by the difference of the mean responses for the two treated groups).

Also we can compare hypothetical treatment applications on a test unit u_0, with observed $K(u_0) = k$ and, by construction, $Q(u_0) = 0$, as follows:

$$\begin{aligned} &\mathrm{E}\{Y(u_0)|K(u_0) = k, t\} - \mathrm{E}\{Y(u_0)|K(u_0) = k, c\} \\ &= \mathrm{E}\{Y|K = k, t, Q = 0\} - \mathrm{E}\{Y|K = k, c, Q = 0\} \\ &= \mathrm{E}\{Y|K = k, t\} - \mathrm{E}\{Y|K = k, c\}, \end{aligned}$$

on again using (12). But this is just τ, as identified from the experiment. (If $K(u_0)$ is not observed, we have to take a further expectation over K, but this clearly has no effect.)

The above approach, based on the testable assumption (12) rather than the metaphysical assumption of TUA, thus allows us to generalise readily from the experiment to the target population, even in the face of differential selection and treatment criteria.

It has been assumed in the above that it is appropriate to focus directly on the expected response. In the general framework of §6, with a loss function L, we could replace $\mathrm{E}(Y)$ by $\mathrm{E}\{L(Y)\}$ throughout (a counterfactual analysis would similarly require that TUA be modified to: $L\{Y_t(u)\} - L\{Y_c(u)\} \equiv \tau$, all u).

9 SHEEP AND GOATS

I have argued that any elements of a theory that have no observable or testable consequences (for example, TUA) are to be regarded as metaphysical, and, in accordance with Jeffreys's Law, should not be permitted to have any inferential consequences, either. Causal analyses can be classified into *sheep* (those obeying this dictum) and *goats* (the rest). We have seen that the decision-analytic approach is a sheep.

What of the counterfactual approach? It certainly has the potential to generate goats. In particular, any inference that is dependent on assumptions requiring the acceptance of fatalism (for example TUA, or monotonicity, or assertions about the group of compliers in clinical trial) must be a goat. However, specific inferential uses of counterfactual models may turn out to be sheep. Here is one such use.

9.1 ACE

Suppose, in the counterfactual approach, one were to define the ICE for unit u as $f\{Y_t(u)\} - f\{Y_c(u)\}$, for some function f. For example, one might use the linear form $Y_t(u) - Y_c(u)$, or the logarithmic form $\log\{Y_t(u)/Y_c(u)\}$. If $\mathcal{U}$ is effectively infinite, the ACE (population average of ICE(u)) is $E_P\{f(Y_t) - f(Y_c)\}$. But this is just $E_{P_t}\{f(Y)\} - E_{P_c}\{f(Y)\}$, and thus depends only on the marginal distributions P_c and P_t (and is exactly the criterion determining the solution of the decision problem having $L \equiv f$). Hence this particular use of counterfactual analysis, focusing on an infinite-population ACE, is consistent with the decision-analytic approach, and involves only terms subject to empirical scrutiny. It is fortunate that many of the superficially counterfactual analyses in the literature, from Rubin (1978) onwards, have in fact confined attention to ACE, and thus lead to acceptable conclusions.

However, seemingly minor variations of the above form for ICE, such as $Y_t(u)/Y_c(u)$, can not be handled in this way: $E_P(Y_t/Y_c)$ is not determined by the marginals P_t and P_c alone (although these can be used to set bounds: Rachev, 1985). So any form of inference focusing on such causal effects, at either the individual or the population average level, would be a metaphysical goat, dependent on untestable ingredients of the metaphysical model, and hence likely to be misleading.

9.2 Neyman and Fisher

Here is a variation on ACE, even using the simple definition (4), that is nevertheless a goat. It is the basis of the approach introduced by Neyman (1935), and followed through by Wilk and Kempthorne (1955, 1956, 1957).

Let $\mathcal{U}^* := \mathcal{U}_t \cup \mathcal{U}_c$ be the set of experimental units[6], say N in total. Neyman expressed the null hypothesis of 'no treatment effect' as asserting that $Y_t^* = Y_c^*$, where $Y_i^* := N^{-1} \sum_{u \in \mathcal{U}^*} Y_i(u)$ is the average response that would have been observed in the experiment, had all units been given treatment i (thus both Y_t^* and Y_c^* are genuinely counterfactual quantities). Wilk and Kempthorne (1955) considered averages over a larger, but still finite, population $\mathcal{U}$ from which $\mathcal{U}^*$ was drawn. In these approaches, inference is based on the distribution generated by random treatment assignment (and, where appropriate, random sampling of the levels used for the experiment), under assumed values for the metaphysical array of all potential responses $(Y_i(u))$, these values playing the rôle of parameters in the randomisation model. Such an approach (even when extended by introducing random errors of observation) is clearly based on a fatalistic world view.

Neyman showed that, for the Latin square, the usual t-test was an unbiased test of his null hypothesis only if TUA could be assumed; similarly, the analyses of Wilk and Kempthorne give different answers, according as whether or not one assumes TUA. These workers concluded that one needs to think very carefully, in each particular context, about the validity of the TUA assumption, and tailor one's inferences accordingly. However, since there are no conceivable data that could shed any light on this validity, it is not clear how to act on this advice. Two statisticians with observationally equivalent models could arrive at discrepant conclusions. This suggests very strongly that Neyman's approach is not a helpful one, and that his metaphysical null hypothesis is misguided.

[6]In the literature, the units are not completely homogeneous, but are classified in an experimental layout, e.g. a row-column structure with treatments imposed to form a Latin Square. However, this does not affect the essential logic.

Fisher, in the rapporteur's account of his comments on Neyman's paper, rejected this approach, arguing instead that the appropriate null hypothesis was

$$H_0 : \tau = 0,$$

for which the standard t-test *is* valid.

Fisher's null hypothesis is often taken to have been

$$H_0^* : \tau(u) \equiv 0,$$

i.e. $\tau = 0$ and $\phi_\gamma = 0$, implying $Y_t(u) = Y_c(u)$ for all u. This, too, is a metaphysical hypothesis. However, it is not certain that this was Fisher's intention. In any case, so far as the observable structure (8) is concerned, these two hypotheses are indistinguishable, as are the resulting tests. This identity extends to more complex layouts: Dawid (1988) shows how the standard tests may be justified purely on the basis of a hypothesis of invariance of the joint distribution of responses under suitable relabelling of units, which is very much weaker than H_0^*: see also Cox (1958). The broader hypothesis H_0 is equivalent to $P_t = P_c$, which is all that is needed for indifference in the decision problem — and is, of course, a sheep, being testable from the data.

10 INSTRUMENTAL USE OF COUNTERFACTUALS

Even if one accepts that the output of a causal analysis should not involve any direct assertions about counterfactuals, the example of Section 9.1 demonstrates that it is at least possible to use counterfactual models for acceptable purposes. However, that example also shows no obvious advantage to doing so, and the use of counterfactual models always lays one open to the danger of producing 'goat-like' inferences, without signalling when that is the case (as for the variant forms of ACE considered at the end of Section 9.1).

It nevertheless remains conceivable that purely mathematical use of the richer structure inherent in the modelling of the metaphysical array might actually simplify some derivations and analyses of acceptable 'sheep-like' inferences. An analogy might be the fruitfulness of coupling arguments in probability theory, or of complex analysis in number theory.

In my view, there may be a limited place for such instrumental use of counterfactuals in the context of causal *model-building*. However, I remain to be persuaded of the usefulness of counterfactuals, even in a purely instrumental rôle, for causal *inference*.

10.1 Counterfactuals for modelling

The model (9) for the physical array was derived by marginalising the metaphysical model of Example 1, so as to focus on the subcollection of variables picked out by the experimental design. This may be regarded as an instrumental use of counterfactuals for the purposes of modelling. However, in this simple example this looks like overkill: model (9) is itself a very natural structure to impose on the physical array directly.

In more complicated problems, there may be some genuine advantage to modelling at the metaphysical level. Thus, suppose the experimental units are laid out in a row-column structure. One way to build appropriate models for outcomes is to apply the ideas of symmetry modelling (Dawid, 1988). If we associate with each plot the full vector of (complementary) potential responses it would exhibit under the various different possible treatment applications, then it might be reasonable to regard the joint distribution for all these vectors as invariant under separate relabellings of rows and columns. If (less compellingly, and purely for simplicity of exposition) we also impose invariance under relabellings of the treatments, symmetry arguments imply that we can represent the probability structure of the metaphysical array $\mathcal{Y} = (Y_{irc})$ (where i labels treatments, i rows and j columns) by the random effects model:

$$(14)\quad Y_{irc} = \mu + \alpha_i + \beta_r + \gamma_c + (\alpha\beta)_{ir} + (\alpha\gamma)_{ic} + (\beta\gamma)_{rc} + (\alpha\beta\gamma)_{irc},$$

with all the terms uncorrelated, $\mathrm{var}(\alpha_i) = \sigma^2_\alpha$, *etc.*

If we consider the implications of this model for the marginal joint distribution of some physical array $\mathcal{X} = (X_{rc})$, in which a specified treatment $i = i(rc)$ is applied to the unit in row r and column c, we find a similar representation, but with the last two terms intrinsically confounded (just as the separate terms $\beta(u)$ and $\gamma_i(u)$ in (1) are confounded in the term ϵ_{ij} of (9)). If we further confine attention to Latin Square designs, so that no treatment appears more than once in any row or column, there is additional (extrinsic) confounding, resulting in the model:

$$(15)\quad X_{rc} = \mu + \alpha_i + \beta_r + \gamma_c + \epsilon_{rc},$$

where, with $i = i(r, c)$,

$$(16)\quad \epsilon_{rc} = (\alpha\beta)_{ir} + (\alpha\gamma)_{ic} + (\beta\gamma)_{rc} + (\alpha\beta\gamma)_{irc}.$$

This is of course the (random-effects version of) the usual model for the observables in the Latin Square design[7].

On the other hand, we could have initially restricted attention to the physical array $\mathcal{X}$, and considered the group of symmetries that preserve its structure. Such a symmetry is represented by the combination of a row-permutation and a column-permutation having the additional property that any two units receiving identical treatments before permutation also receive identical treatments after permutation.

[7]The extrinsic confounding between the $(\alpha\beta)$, $(\alpha\gamma)$ and $(\beta\gamma) + (\alpha\beta\gamma)$ terms in (16) will, however, make predictive inferences, which depend on these terms individually, especially sensitive to assumptions that cannot be tested with such a design.

This group will depend very specifically on the way in which treatments are assigned to units, and can have highly variable structure for different Latin Square layouts (Bailey, 1991, Example 4). Because of these additional restrictions on the symmetry of the physical array $\mathcal{X}$, the implied symmetry model constructed directly for $\mathcal{X}$ can be considerably more complex than that expressed by (15). In such a case, modelling the metaphysical array directly, for the purely instrumental use of deriving an appropriate model for the physical array, appears the more fruitful approach.

Another example of the usefulness (or at least convenience), for constructing models of the physical domain, of direct modelling of the metaphysical domain (using 'pseudo-structural nested distribution models') is given by Robins and Wasserman (1997).

Compatibility

If we do take the approach of modelling each possible physical array by marginalising from a single joint model for the metaphysical array, the resulting collection of physical models will have a property I term *compatibility*: for two different experimental layouts that both result in unit u receiving treatment i, the marginal models for the associated response on that unit are identical. This identity extends to the joint model for the responses of a collection of units that happen to be treated in the same way in both experiments. This property can be regarded as a non-counterfactual counterpart of the counterfactual Stable Unit-Treatment Value Assumption (SUTVA) (see §7).

I further distinguish two forms, *strong* and *weak*, of compatibility for a collection of physical models under varying treatment assignments. Weak compatibility (which seems the more natural, and makes no reference whatsoever to counterfactuals) simply requires the above-stated property of identity of common marginal models. Strong compatibility requires the existence of a single joint model for the metaphysical array that can be used to generate, by appropriate marginalisation, the various different physical models. To extend the analogy with Quantum Theory, strong compatibility requires the existence of 'hidden variables', underlying all observations that might be made. Although strong compatibility always implies weak compatibility, in full generality the converse need not hold. Consider, for example, variables (Y_1, Y_2, Y_3), where Y_i is either 1 or -1, and where we can observe any of the pairs (Y_1, Y_2), (Y_2, Y_3), (Y_3, Y_1), but cannot observe all three variables simultaneously. The corresponding bivariate distributions are specified by: $Y_1 = Y_2$, $Y_2 = Y_3$, $Y_3 = -Y_1$, with Y_i either 1 or -1, each with probability $1/2$. Then these distributions are weakly, but not strongly, compatible (I am grateful to Steffen Lauritzen for this example). Although the structure of this example is not quite the same as that of our current problem, it is conceivable that in this, too, we could have weak compatibility without strong compatibility. This opens up the possibility of a still deeper analogy with Quantum Theory, where observable behaviour cannot be explained by means of a 'hidden variable' theory.

In the decision-analytic approach, the property of compatibility, while it may be very useful in streamlining the modelling, has no fundamental role to play. All we need do is construct appropriate models relating the outcomes on the experimental units, according to the treatment assignments actually made, with those on as yet untreated units, under various assumptions about how those new units might be treated. Then we can use these to make predictive inferences under the varying assumptions, and so to assess the relative value of future interventions.

10.2 Counterfactuals for inference?

There are many problems where workers who have grown familiar and comfortable with counterfactual modelling and analysis evidently consider that it forms the only satisfactory basis for *causal inference*. However, I have not as yet come any use of counterfactual models for inference about the effects of causes which is not either (a) a goat, delivering misleading inferences of no empirical content; or (b) interpretable, or readily reinterpretable, in non-counterfactual terms. I have already given examples of (a), and also, in §9.1, of (b). Here are some more cases of (b).

Robins initially (Robins, 1986) developed causal inferential methods on the basis of a counterfactual model. However in recent work (Robins and Wasserman, 1997) both the underlying model and the associated methods are re-expressed in non-counterfactual terms.

Conversely, Pearl, in introducing a semantics for graphicalmodels of causal structures (Pearl, 1993), did so in a way that avoided counterfactuals. Later (Pearl, 1995a) he translated this into a counterfactual language, based on functional models, but to no obvious advantage: his specific analyses (for example, in the Appendix to Pearl, 1995a) make no necessary use of this additional structure.

An interesting problem that did initially appear to require a counterfactual model is the development of inequalities for (sheep-like) causal effects in clinical trials with imperfect treatment compliance (Balke and Pearl, 1994b); however, I have been able to derive the identical inequalities without the additional baggage of functional models or counterfactuals (indeed, an example of just such a derivation was given by Pearl, 1995b)

Another interesting recent example of (b) is contained in Greenland *et al.* (1999), which purports to define confounding in terms of counterfactuals, but which explicitly introduces an alternative interpretation based on exchangeability. Most of its analyses make no essential use of counterfactuals. Two appendices, considering carefully the interpretation of counterfactual assertions in a number of cases, represent to me convincing demonstrations of their meaninglessness and pointlessness (although the authors themselves stop short of this conclusion).

PART III: CAUSES OF EFFECTS

11 INFERENCE ABOUT CAUSES OF EFFECTS

I now turn to address the problem of inference about the causes of effects. As we shall see, this is still more problematic than inference about the effects of causes, and it may be impossible to avoid a degree of ambiguity in the resulting inferences.

The major new ingredient is that, as well as having the experimental data, we now have a further unit u_0, of individual interest, to which treatment t has already been applied, and the response $Y_t(u_0) = y_0$ observed. (We may also have further relevant information about u_0 or its environment, perhaps even gathered between the application of treatment and observation of response. This possibility will be considered below — for the moment we assume that this is not so.) We are interested in whether, for the specific unit u_0, the application of t 'caused' the observed response. It appears that, to address this question, we have no alternative but to compare, somehow, the observed valued y_0 with the counterfactual quantity $Y_c(u_0)$, the response that would have resulted from application of c to u_0. Equivalently, we require inference about the individual causal effect $\tau(u_0) = y_0 - Y_c(u_0)$. However, the fact that such an inference may be desirable does not, of itself, render it possible. Let us see what can be justified scientifically from data.

EXAMPLE 2. Consider again the bivariate normal counterfactual model of Example 1. Suppose that there is no possibility of ever measuring any other relevant information on any unit, beyond its response to treatment.

The conditional distribution of $\tau(u_0) \equiv Y_t(u_0) - Y_c(u_0)$, given $Y_t(u_0) = y_0$, is normal, with mean and variance

$$\lambda := E\{\tau(u_0)|Y_t(u_0) = y_0\} = y_0 - \theta_c - \rho(y_0 - \theta_t), \tag{17}$$

$$\delta^2 := \text{var}\{\tau(u_0)|Y_t(u_0) = y_0\} = (1 - \rho^2)\phi_Y. \tag{18}$$

Now, as already emphasised, from the extensive experimental data (even when extended with the additional observation $Y_t(u_0) = y_0$) we can learn only θ_t, θ_c, and ϕ_Y. We can not identify the correlation ρ. Hence, even with extensive data, we have residual arbitrariness. When $\rho = 0$ ($\phi_\beta = 0$, or independence of Y_t and Y_c), we get $\lambda = y_0 - \theta_c$, $\delta^2 = \phi_Y$. The value $\rho = 1$ ($\phi_\gamma = 0$, or TUA) yields $\lambda = \theta_t - \theta_c$, $\delta^2 = 0$ (or, at the other extreme, if we allow $\rho = -1$, we have $\lambda = 2y_0 - \theta_t - \theta_c$, and $\delta^2 = 0$ again). Assuming $\rho \geq 0$, we can only infer the inequalities

$$\lambda \text{ lies between } \theta_t - \theta_c \text{ and } y_0 - \theta_c,$$

and

$$\delta^2 \leq \phi_Y.$$

Thus only when y_0 is sufficiently close to θ_t will we get an unambiguous conclusion about λ, insensitive to empirically untestable assumptions about ρ; and only when ϕ_Y is sufficiently small will we be able to say anything empirically supportable and unambiguous about δ^2. If we take $\rho = 1$, equivalent to TUA, we obtain a seemingly deterministic inference, $\tau(u_0) = \theta_t - \theta_c$; but this is of little real value when the data give us no reason to choose any particular value of ρ over any other.[8]

Note that, if we do assume TUA, but not otherwise, the retrospective inference about $\tau(u_0)$ is not affected by the additional information $Y_t(u_0) = y_0$ on the new unit, and thus is the same as for the case of arguing about effects of causes. Because the TUA assumption is so prevalent in the literature, the essential distinction between inference about the effects of causes and inference about the causes of effects has not usually been remarked.

The above sensitivity to assumptions extends to e.g. Bayesian inference, which would require integration of the distribution defined by (17) and (18) over the posterior distribution of all the parameters. In this posterior, θ_t, θ_c and ϕ_Y will be essentially degenerate at their sample estimates, so that we can substitute these in (17) and (18), and just integrate over the conditional distribution of the non-identified parameter ρ, given $(\theta_t, \theta_c, \phi_Y)$. However, this will be exactly the same in the posterior as in the prior, and thus the inference will remain sensitive to the assumed form of the prior.

No amount of wishful thinking, clever analysis or arbitrary untestable assumptions can license unambiguous inference about causes of effects, even when the model is simple and the data are extensive (unless we are lucky enough to discover uniformity among units).

11.1 Concomitants

It appears from the above that there is an inherent ambiguity in inference about the causes of effects. However, some progress towards reducing this may be possible if we are able to probe more deeply into the hidden workings of our units, by observing suitable additional variables. This is the basis and purpose of scientific investigation. As we have seen in §§ 6 and 8, such deeper scientific understanding is not essential for assessing 'effects of causes', which can proceed by an essentially 'black box' approach, simply modelling dependence of the response on whatever covariate information happens to be observed for the test unit. However, it is vital for any study of inference about 'causes of effects', which has to take into account what has been learned, from experiments, about the inner workings of the black box.

Thus suppose that it is possible to measure *concomitant variables* associated with a unit. These might be covariates, as already considered. However, we can

[8]The inequalities developed above rely on the assumption, itself untestable, of joint normality. Even though the data may support marginal normality for each of Y_t and Y_c, any further aspects of the joint distribution must remain unknowable, and, in principle, the distribution of Y_c, given the observed value $Y_t = y$, could be anything (so long as $\phi_Y > 0$). Thus a complete sceptic could hold that inference about the causes of effects, on the basis of empirical evidence, is impossible.

also allow other quantities, so long as they can be assumed to be unaffected[9] by the treatment applied. An example might be the weather between the times of planting and of harvesting a crop. Typically the variation in the response conditional on concomitants will be smaller than unconditionally.

EXAMPLE 3. Suppose that, in the context of Example 1, detailed experiments have measured a concomitant K, and have found that, conditional on $K(u) = k$ and the application of treatment i, the response $Y(u)$ is normally distributed with residual variance ψ_K, say, and mean $\theta_i + k$. From these experiments the values of ψ_K and the θ's have been calculated.

Define $\phi_K := \text{var}(K)$, and $\psi_0 := \phi_Y - \phi_K + \psi_K$. Then $\text{cov}(K, Y_c) = \text{cov}(K, Y_t) = \phi_K$. Combining these with the covariance structure for the complementary pair (Y_c, Y_t) implied by (1), the full dispersion matrix of (K, Y_c, Y_t) is seen to be

$$\begin{pmatrix} \phi_K & \phi_K & \phi_K \\ \phi_K & \phi_Y & \rho\phi_Y \\ \phi_K & \rho\phi_Y & \phi_Y \end{pmatrix}.$$

Thus the conditional correlation between Y_c and Y_t, given K, is

$$(19)\quad \rho_{ct\cdot K} := \frac{\rho\phi_Y - \phi_K}{\phi_Y - \phi_K} = 1 - (1-\rho)\frac{\psi_0}{\psi_K}.$$

In parallel to Example 2, we can not identify the arbitrary parameter $\rho_{ct\cdot K} \in [-1, 1]$ from these more refined experiments (although it might be reasonable to take $\rho_{ct\cdot K} \geq 0$).

We now consider inference about 'causes of effects' on a test unit u_0. We again distinguish between the cases that concomitant information is, or is not, available for u_0.

1. If we were to observe $K(u_0) = k$, say, we could conduct an analysis very similar to that of Example 2. In particular, (17) would be replaced by $E\{\tau(u_0) | Y_t(u_0) = y, K(u_0) = k\} = (y - \theta_c - k) - \rho_{ct\cdot K}(y - \theta_t - k)$, which, since the final term in parentheses is now of order $\sqrt{\psi_K}$, rather than $\sqrt{\psi_0}$ previously, should be less sensitive to the arbitrariness in the correlation, now $\rho_{ct\cdot K}$. Similarly, (18) would be replaced by $\text{var}\{\tau(u_0) | Y_t(u_0) = y, K(u_0) = k\} = (1 - \rho_{ct\cdot K}^2)\psi_K$, now bounded above by $\psi_K < \psi_0$, rather than by $\phi_Y = \psi_0$. Clearly these improvements are the more substantial the smaller, relatively, is the residual variance ψ_K of Y given K.

2. Now suppose that we do not observe $K(u_0)$, or any other concomitant variable, on u_0. In this case — in contrast to Case 2 of §8 for effects of causes — our analysis *is* affected by the more detailed findings in the experiments performed. Define $\gamma_K := \phi_K/\phi_Y = 1 - \psi_K/\psi_0$. By (19) we have (assuming $\rho_{ct\cdot K} \geq 0$)

[9] Use of this adjective itself begs many causal and counterfactual questions: see §14.

(20) $\gamma_K \leq \rho \leq 1.$[10]

Consequently the experimental identification of K, even though it can not be observed on u_0, has reduced the 'interval of ambiguity' for ρ from $[0, 1]$ to $[\gamma_K, 1]$[11], and thus yields tighter limits on λ and δ^2 in (17) and (18).

From this perspective, the ultimate aim of scientific research may be seen as discovery of a concomitant variable, K^* say, that yields the smallest achievable residual variance $\psi^* := \psi_{K^*}$, and thus, with $\gamma^* := \gamma_{K^*} = 1 - \psi^*/\psi_0$, the shortest possible interval of ambiguity, $[\gamma^*, 1]$, for ρ.[12] I term such a variable a *sufficient concomitant*. (The collection of all concomitants is always sufficient in this sense, but one would hope to be able to reduce it without explanatory loss.) However, unless $\psi^* = 0$, and rarely even then, it will not usually be possible to know whether we have attained this goal.

Nonetheless, using (20) with (17) and (18), we can still make scientifically sound (though imprecise) inferences on the basis of whatever current level of understanding, in terms of discovered explanatory concomitant variables K, we have attained. This will take into account that there is a non-statistical component of uncertainty or arbitrariness in our inferences, expressed by interval bounds on our quantitative causal conclusions.

We have assumed that the experiments performed have been sufficiently large that purely statistical uncertainty can be ignored. In practice this will rarely be the case. However, we do not as yet have any appropriate methodology for combining such statistical uncertainty with the intrinsic ambiguity that still remains in the limit. Techniques for dealing with this problem are urgently needed.

12 CONDITIONAL INDEPENDENCE

Suppose K^* is a sufficient concomitant. Assuming $\rho_{ct.K^*} \geq 0$, we have, from (19), the ultimate residual variance $\psi^* \geq (1 - \rho)\psi_0$. In particular, $\rho < 1$ implies $\psi^* > 0$. If $\psi^* = 0$ (and thus $\rho = 1$), the value of K^* determines both potential responses Y_t and Y_c, without error, and so, once we have identified K^*, the ambiguity in our inferences entirely disappears. We call such a situation *deterministic*, and consider it further in §13 below.

However, for reasons to be discussed in §14, we regard determinism as exceptional, rather than routine. In this section we consider further the non-deterministic case, having $\psi^* > 0$, and, by (19), ρ constrained only to the interval of ambiguity $[\gamma^*, 1]$ (as $\rho_{ct.K^*}$ ranges from 0 to 1), with $\rho^* = 1 - \psi^*/\psi_0$.

[10] Or, for $\rho_{ct.K}$ unrestricted, $2\gamma_K - 1 \leq \rho \leq 1$.

[11] Or, for $\rho_{ct.K}$ unrestricted, from $[-1, 1]$ to $[2\gamma_K - 1, 1]$.

[12] We are here assuming, for simplicity, that the model of Example 3 applies, for any concomitant K that might be considered. Although the mathematics is more complicated if this assumption is dropped, the essential logic continues to apply.

So far as any empirical evidence is concerned, there is no constraint whatsoever on $\rho_{ct.K^*}$. However, it would seem odd to hypothesise, for example, $\rho_{ct.K^*} = 1$, since this would imply $\rho = 1$, complete dependence between real and counterfactual responses, at the same time as asserting non-determinism, in the sense that there is no concomitant information we could gather that would allow us to predict the response perfectly. Likewise, to hypothesise any other value of $\rho_{ct.K^*} > 0$ would appear to leave open the possibility of finding a more powerful set of predictors that would explain away this residual dependence, thus further reducing the residual variance.

In order to limit the arbitrariness in the value of ρ, one could attempt to give ρ further meaning by requiring that $\rho_{ct.K^*} = 0$: the totally inexplicable components of variation of the response, in the real and in the counterfactual universes, should be independent. Extending this, we might require that all variables be treated as conditionally independent across complementary universes, given all the concomitants (which are of course constant across universes). Under this assumption, the interval of ambiguity for ρ shrinks to the point $\gamma^* = 1 - \psi^*/\psi_0$.

The above conditional independence assumption is best regarded as a *convention*, providing an interpretation of just what one intends by a counterfactual query. It leads to a factor-analysis type decomposition of the joint probabilistic structure of complementary variables, into (a) a part fully explained by the concomitants, and common to all the complementary universes, and (b) residual 'purely random' errors, modelled as independent (for any given unit) across universes. In this way, we can at last give a clear structure and meaning (albeit partly conventional) to a metaphysical probability model for the collection of all potential responses. Note that, if we accept this conditional independence convention, we obtain, on using (19), $\rho = \gamma_{K^*} \geq 0$ — providing some justification for imposing this condition[13].

If we have identified a sufficient concomitant K^* (leaving aside, for the moment, the question of how one could know this), the conditional independence convention renders counterfactual inference, in principle, straightforward and unambiguous. In the context of Example 3, we can take $\rho = \gamma^* = \psi^*/\psi_0$, thus eliminating the ambiguity. More generally, from detailed experiments on treated and untreated units, we can discover the joint distribution of K^* and Y_t, and of K^* and Y_c. For a new unit u_0 on which no concomitants are observed, on observing $Y_t(u_0) = y$ we can condition (e.g. using Bayes' theorem) in the joint distribution of (K^*, Y_t) to find the revised distribution of K^*; and then combine this with the conditional distribution of Y_c given K^* to obtain the appropriate distribution of the counterfactual Y_c. This two-stage procedure is valid if and only if one accepts the conditional independence property. Alternatively (and equivalently), we can use this property to combine the two experimentally determined distributions into a single joint distribution for (K^*, Y_t, Y_c), and marginalise to obtain that of (Y_t, Y_c); finally we condition on $Y_t(u_0) = y$ in this bivariate distribution. Minor

[13]Without the convention, and with no constraints on $\rho_{ct.K^*}$, we can only assert $\rho \geq 2\gamma_{K^*} - 1$.

variations will handle the case where we have, in addition, observed the value of some concomitant variables on u_0.

EXAMPLE 4. *(with acknowledgments to V. G. Vovk).* A certain company regularly needs to send some of its workers into the jungle. It knows that the probability that a typical worker will die (D) if sent to the jungle (J) is $\text{prob}(D|J) = \frac{3}{4}$, compared with $\text{prob}(D|\bar{J}) = \frac{1}{4}$ if retained at Head Office. Joe is sent to the jungle, and dies. What is the probability that Joe would have died if he had been kept at Head Office?

1. Suppose first all workers are equally robust, and that the risk of dying is governed purely by the unspecified dangers of the two locations. One might then regard the complementary outcomes as independent, so that the answer to the question is $\frac{1}{4}$.

2. Now suppose that, in addition to external dangers, the fate of a worker depends in part on his natural strength. With probability $\frac{1}{2}$ each, a worker is either strong (S) or weak ($\bar{S}$). A strong worker has probability of dying in the jungle $\text{prob}(D|J, S) = \frac{1}{2}$, and at Head Office $\text{prob}(D|\bar{J}, S) = 0$. A weak worker has respective probabilities $\text{prob}(D|J, \bar{S}) = 1$, and $\text{prob}(D|\bar{J}, \bar{S}) = \frac{1}{2}$. (These values are consistent with the earlier probabilities assigned to $\text{prob}(D|J)$ and $\text{prob}(D|\bar{J})$.) Given that Joe died in the jungle, the posterior probability that he was strong is $\frac{1}{3}$. If one assumes conditional independence, given strength, between the complementary outcomes, the updated probability that he would have died if kept at Head Office now becomes $\frac{1}{3} \times 0 + \frac{2}{3} \times \frac{1}{2} = \frac{1}{3}$.

3. In fact Joe was replaced at Head Office by Jim, who took his desk. Jim died when his filing cabinet fell on him. This gives additional information about the dangers Joe might have faced had he stayed behind. How should we take it into account? There is no right answer. If we regard the toppling of the filing cabinet, killing whoever is at the desk, as unaffected by who that occupant may be, and include it as a concomitant, then the answer becomes 1. Or, we could elaborate, allowing the probability that the occupant is killed by the falling cabinet to depend on whether he is strong or weak. But it would be equally reasonable to consider that, had Joe stayed behind, the dangers he would have met would have been different from those facing Jim. In this case the previous arguments and answers (according as whether or not we account for strength) could still be reasonable.

As should be clear from the above example, even with the conditional independence convention the answer to a query about 'causes of effects' must depend, in part, on what variables it is considered reasonable to regard as concomitants. We consider this issue further in §14.

12.1 Undiscovered sufficient concomitants

What if, as will usually be the case, we have measured concomitants K in our experiments, but have not as yet identified a sufficient concomitant K^*? In Example 3, we could then only assert $\psi^* \leq \psi_K$, and thus, using the conditional independence property $\rho = \gamma^*$, $\rho \geq \gamma_K$. Hence the convention of conditional independence at the level of the sufficient concomitant has not, in this case, resulted in any reduction in the interval of ambiguity for ρ.

We can, nevertheless, think, in the light of current knowledge, and having regard to the potentially available concomitants (see §14 below), about plausible values of the ultimate residual variance ψ_{K^*}, and use this in setting reasonable limits, or distributions, for $\rho = 1 - \psi_{K^*}/\psi_0$. This still leaves our inference dependent on (as yet) experimentally unverified assumptions, but it might at least be possible to present reasoned arguments for the assumptions made. This approach based on conditional independence also obviates the need for new methods of statistical inference, combining ambiguity and uncertainty.

13 DETERMINISM

In certain problems of the 'hard' sciences it can happen that, by taking account of enough concomitant variables, the residual variation in the response, for any treatment, can be made to disappear completely (at least for all practical purposes), thus inducing, at this more refined level, the situation of uniformity considered in §5.4 above, when all problems of causal inference and prediction disappear. In Example 3, this would occur if we found $\psi_K = 0$, which would imply $\rho = 1$, and so eliminate all ambiguity. Such problems may be termed *deterministic*, since the response is then given as a function $Y = f(i, D)$ of the appropriate *determining concomitant* D (which is then necessarily sufficient) and the treatment i, without any further variability. This property is, in principle, testable when D is given (if it is rejected, it may be possible to reinstate it, at a deeper level, by refining the definition of D). However, even when such underlying determinism does exist, discovering that this is the case, and identifying the determining concomitant D and the form of f, may be practically difficult or impossible, requiring a large-scale, detailed and expensive scientific investigation, and sophisticated statistical analyses.

If we had a deterministic model, we could use it to *define* potential responses: $Y_i(u) = f(i, D(u))$.[14] We could determine the value of any potential response on unit u by measuring $D(u)$. Thus in this special case we can indeed consider the complementary variables $(Y_i(u)) \equiv (f(i, D(u)))$, for fixed unit u but varying treatment i, as having real, rather than merely metaphysical, simultaneous existence.

[14]We need here the property that D, being a concomitant, is unaffected by treatment. However, since D need not be a covariate, this model is not necessarily fatalistic.

Note especially that, even in this rare case where we can give empirical meaning to counterfactuals, we are not basing our causal modelling on the primitive notion of counterfactual; rather, it is the counterfactuals that are grounded in, and take their meaning from, the model. (In the same way, I consider that Lewis's (1973) interpretation of counterfactuals in terms of 'closest possible worlds' is question-begging, since closeness can not be sensibly defined except in terms of an assumed causal model).

A deterministic model, when available, can also be used to make sense of non-manipulative accounts of causation. Given D the potential responses, for various real or hypothetical values of the variable 'treatment', are determined, and can be compared directly, however the specification of treatment may be effected.

For inference about the causes of effects, assume that we have observed $Y_t(u_0) = y_0$, but not $D(u_0)$, and wish to assess our uncertainty about $Y_c(u_0)$. In the context of Example 3, we have $\rho = 1$, eliminating all ambiguity, and (in this rare case) justifying TUA and the inference $\tau(u_0) = \theta_t - \theta_c$. More generally, suppose that detailed experimentation has identified a deterministic model $Y_i(u) = f(i, D(u))$. Although we have not observed $D(u_0)$, we can assess a distribution for it. This should reflect both typical natural variation of D across units (as discovered from experiments), and any additional concomitant information we may have on u_0. From this distribution we can derive the induced joint distribution over the collection $(f(i, D(u_0)))$ of complementary potential responses. Then we can condition the distribution of $D(u_0)$ on the observation $f(t, D(u_0)) = y_0$, and thus arrive at appropriate posterior uncertainty about a genuine counterfactual such as $Y_c(u_0) \equiv f(c, D(u_0))$. In this way, a fully deterministic model (if known) allows an unambiguous solution to the problem of assessing the 'causes of effects'. The essential step is the generation of the joint distribution over the set of complementary responses (together with any observed concomitants), this being fully grounded in an understanding of their dependence on determining concomitants, and a realistic probabilistic assessment of the uncertainty about those determining concomitants.

The above procedure is merely a special case of that described in §12, but not now dependent on the convention of conditional independence of residual variation across parallel universes — because in this case there is no residual variation.

EXAMPLE 5. Suppose that a major scientific investigation has demonstrated the validity of the model (1), but now reinterpreted as a deterministic model, with all the β's and γ's identified as concomitant variables that can, with suitable instruments, be measured for any unit, and have been so measured in the experimental studies. Further, from these studies, the previously specified independent normal distributions for these quantities have been verified, and all the parameters $(\theta_t, \theta_c, \phi_\beta, \phi_\gamma)$ have been identified.

We now examine a new unit u_0, which has been given treatment t, and observe the associated response $Y_t(u_0) = y$. The individual causal effect $\tau(u_0)$ is $\gamma_t(u_0) - \gamma_c(u_0)$, which is now, in principle, measurable. In practice measurement of the

β's and γ's for unit u_0 may not be possible. Then (in the absence of any further relevant information) we might describe our uncertainty about their values using their known joint population distribution. The appropriate uncertainty about $\tau(u_0)$ is then expressed by the normal distribution with mean λ and variance δ^2 given by (17) and (18); however, since the value of $\rho = \phi_\beta/(\phi_\beta + \phi_\gamma)$ is now available from the scientific study, the ambiguity in this inference has been eliminated.

Note that it is vital for the above analysis that the quantities $\gamma_t(u)$ and $\gamma_c(u)$ be *simultaneously* measurable, with the specified independent distributions. It is not enough only to identify $\beta(u)$, and define the γ's as error terms: $\gamma_i(u) = Y_i(u) - \theta_i - \beta(u)$; in that case, since we can not simultaneously observe both $Y_t(u)$ and $Y_c(u)$, we could not verify the required assumption of independence between $\gamma_t(u)$ and $\gamma_c(u)$.

13.1 Undiscovered determinism

If we believe that our problem is deterministic, but have not, as yet, completely identified the determining concomitant D or the function f, we can propose parametric forms for f and the distribution of D, and attempt to estimate these (or integrate out over the posterior distribution of their parameters), using the available data. In principle, sufficiently detailed experimentation would render such assumptions empirically testable, and identify the parameters. In practice, however, this may be far from the case. Thus consider Example 2, in which we have not been able to measure any concomitants. We could propose an underlying deterministic model of the form:

$$Y = \theta_i + D_i \qquad (i = t, c)$$

with D_t, D_c determining concomitants, supposedly measurable on any unit by further, more refined, experiments. In the current state of knowledge, however, we can say no more than $D_i \sim N(0, \phi_Y)$. Further, we have no information on the correlation ρ between D_t and D_c. It is clear that, until we are able to conduct the more detailed experiments, merely positing such an underlying deterministic structure makes no progress towards removing current ambiguities, and our inferences remain highly sensitive to our assumptions. In such a case there seems to be no obvious advantage in assuming determinism; we might just as well conduct analyses such as that of Example 3, basing them only on experimentally observed quantities, and deriving suitably qualified inferences encompassing the remaining ambiguity — which should not be artificially eliminated by imposing unverified constraints on the model. (It may nevertheless be, as suggested in §12.1, that thinking about the possibilities for what we might discover in further experiments could help us towards reasonable and defensible resolution — subject to later empirical confirmation or refutation — of some of the ambiguities).

13.2 Pseudo-determinism

It seems to me that behind the popularity of counterfactual models lies an implicit view that all problems of causal inference can be cast in the deterministic paradigm (which in my view is only rarely appropriate), for a suitable (generally unobserved) determining concomitant D. If so, this would serve to justify the assumption of simultaneous existence of complementary potential responses. Heckerman and Shachter (1995), for example, take a lead in this from Savage (1954), who based his axiomatic account of Bayesian decision theory on the supposed existence of a 'state of Nature', entirely unaffected by any decisions taken, which, together with those decisions, determines all variables. Shafer (1986) has pointed up some of the weaknesses of this conception.

The functional graphical model framework of Pearl (1995a) posits that, underlying observed distributional stabilities of observed variables there are functional relationships, involving the treatments and further latent variables. When such a deterministic structure can be taken seriously, with all its variables in principle observable, it leads to the possibility, at least, of well-defined counterfactual inferences, as described above. These will again, quite reasonably, be sensitive to the exact form of the functional relationships involved, over and above any purely distributional properties of the manifest variables; but these functional relationships are, in principle, discoverable. Balke (1995), Balke and Pearl (1994a) investigate the dependence of causal inferences on the functional assumptions.

However, often the 'latent variables' involved in such models are not genuine concomitants (measurable variables, unaffected by treatment). Then there is no way, even in principle, of verifying the assumptions made — which will, nevertheless, affect the ensuing inferences, in defiance of Jeffreys's law. I term such functional models *pseudo-deterministic*, and regard it as misleading to base analyses on them. In particular, I regard it as unscientific to impose intrinsically unverifiable assumed forms for functional relationships, in a misguided attempt to eliminate the essential ambiguity in our inferences.

Within the counterfactual framework it is always possible to construct, mathematically, a pseudo-deterministic model: simply *define* $D(u)$ to be the complementary collection of all potential outcomes on unit u. In Example 1 we would thus take $D = (Y_t, Y_c)$. We then have the trivial deterministic functional relationship $Y = f(i, D)$, where f has the *canonical form*: $f(i, (y_t, y_c)) = y_i$ $(i = t, c)$. If we were now to assign a joint distribution to (Y_t, Y_c), the analysis presented above for inferring 'causes of effects' in deterministic models could be formally applied.

This is not a true deterministic model: D is not a true concomitant, since it is not, even in principle, observable. Construction of such a pseudo-deterministic model makes absolutely no headway towards addressing the non-uniqueness problems exposed in §§5.4 and 11: it remains the case that no amount of scientific investigation will suffice to justify any assumed dependence structure for (Y_t, Y_c), or eliminate the sensitivity to this of our inferences about causes of effects. This can only be done by taking into account genuine concomitants.

14 CONTEXT

In basing inference about the causes of effects on concomitant variables (as in §11.1), it appears that I am departing from my insistence that metaphysical assumptions should not be allowed to affect inferences. For to say that a variable is a concomitant involves an assertion that it is unaffected by treatment, and hence would take the same value, both in the real universe and in parallel counterfactual universes in which different treatments were applied. Such an assumption is, clearly, not empirically testable. Nevertheless, our causal inferences will depend on the assumptions we make as to which variables are to be treated as concomitants. This arbitrariness is over and above the essential inferential ambiguity we have already identified, which remains even after the specification of concomitants has been made.

My attitude is that there is indeed an arbitrariness in the models we can reasonably use to make inferences about causes of effects, and hence in the conclusions that are justified. But I would regard this as relating, at least in part, to differences in the nature of the questions being addressed. The essence of a specific causal inquiry is captured in the largely conventional specification of what we may term the *context* of the inference, namely the collection of variables that it is considered appropriate to regard as concomitants: see Example 4. Appropriate specification of context, relevant to the specific purpose at hand, is vital to render causal questions and answers meaningful. It may be regarded as providing necessary clarification of the *ceteris paribus* ('other things being equal') clause that is often invoked in attempts to explicate the idea of cause. Differing purposes will demand differing specifications, requiring differing scientific and statistical approaches, and yielding differing answers. In particular, whether it is reasonable to employ a deterministic model must depend on the context of the problem at hand, since this will determine whether it is appropriate to regard a putative determining variable D as a genuine concomitant, unaffected by treatment. For varying contexts we might have varying models, some deterministic (involving varying definitions of D), some non-deterministic.

EXAMPLE 6. Consider an experiment in which the treatments are varieties of corn, and the units are field plots. Suppose we have planted variety 1 on a particular field-plot, and measured its yield. We might ask "What would the yield have been on this plot if we had planted variety 2?". Before we can address this question, we need to make it more precise; and this can be done in various ways, depending on our meaning and purpose.

First, the answer must in part depend on the treatment protocol. For example, this might lay down the weight, or alternatively the number, of seeds to be planted. In the former case our counterfactual universe would be one in which the weight of variety 2 to be planted would the same as the weight of variety 1 actually planted; in the latter case, we would need to change 'weight' to 'number' in the above, so specifying different counterfactual conditions, and leading us to expect a different answer. (In either case the actual and counterfactual responses will depend in part on the particular seeds chosen, introducing an irreducibly random element into each universe.) We might choose to link the treatments in the two universes in further ways: for example, if we had happened to choose larger than average seeds of variety 1, we might want to consider a counterfactual universe in which we also chose larger than average seeds of variety 2. This would correspond to a fictitious protocol in which the treatment conditions were still more closely defined.

The same counterfactual question might be asked by a farmer who had planted variety 1 in non-experimental conditions. In this case there was no treatment protocol specified, and there is correspondingly still more freedom to specify the fictitious protocol linking the real and the counterfactual universe. But only when we have clearly specified our hypothetical protocol can we begin to address the counterfactual query.

This done, we must decide what further variables we are to regard as concomitants, unaffected by treatment. It might well be reasonable to include among these certain physical properties of the field-plot at the time of planting, and perhaps also the weather in its neighbourhood, subsequent to planting.

We might also want to take into account the effect of insect infestation on yield. It would probably not be reasonable to treat this as a concomitant, since different crops are differentially attractive to insects. Instead, we might use some specification of the abundance and whereabouts of the insects prior to planting. However, it would be simply unreasonable to expect this specification to be in any sense complete. Would we really want to consider the exact initial whereabouts and physical and mental states of all insects as identical in both the real and the counterfactual universe, and so link (though still far from perfectly) the insect infestations suffered in the two universes? If we did, we would need a practically unattainable understanding of insect behaviour before we could formulate and interpret, let alone answer, our counterfactual query. Furthermore, to insist (perhaps in an attempt to justify a deterministic model) on fixing the common properties of the two universes at an extremely fine level of detail risks embroiling us in unfathomable arguments about determinism and free will (would we really have been at liberty to apply a different treatment in such a closely determined alternative universe?). To go down such a path seems to me to embark upon a quest entirely inappropriate to any realistic interpretation of our query. Instead, we could imagine a counterfactual universe agreeing with our own at a much less refined level of detail (in which initial insect positions are perhaps left unspecified). This corresponds to a broader view of the relevant context, with fewer variables considered constant across universes. It is up to the person asking the counterfactual query, or attempting causal inference, to be clear about the appropriate specification, explicit or implicit, of the relevant context.

The conditional independence convention further allows us to tailor our counterfactual inferences to the appropriate context, as in Example 4, without embarking on fruitless searches for 'ultimate causes'. In Example 6, we may wish to omit from our specification of context any information about, or relevant to, the population and behaviour of the insects. We could then take the amounts of insect infestation, in the real and the counterfactual universes, as independent, conditionally on whatever concomitants are regarded as determining our context. This choice may be regarded as making explicit our decision to exclude insect information from the context, rather than as saying anything meaningful about the behaviour of the world. With this understanding, we see that the very meaning (and, hence, the unknown value) of the correlation ρ between Y_t and Y_c (or of any other measure of the dependence between such complementary quantities) will involve, in part, our own specification of the context we consider appropriate to counterfactual questions.

The relation between the partly conventional specification of context and general scientific understanding is a subtle one. Certainly the latter should inform the former, even when it does not determine it: general scientific or intuitive understandings of meteorological processes must underlie any identification of the weather as a concomitant, unaffected by treatment. Moreover, it is always possible that further scientific understanding might lead to a refinement of what is regarded as the appropriate context: thus the discovery of genetics has allowed us to identify previously unrecognised invariant features of an individual, and thus to discard previously adequate, but now superseded, causal theories. Causal inference is, even more than other forms of inductive inference, only tentative; causal models and inferences need to be revised, not only when theories and assumptions on which they are based cease to be tenable in the light of empirical data, but also when the specification of the relevant context has to be reformulated — be this due to changing scientific understanding, or to changing requirements of the problem at hand.

15 CONCLUSION

I have argued that the counterfactual approach to causal inference is essentially metaphysical, and full of temptations to make 'inferences' that can not be justified on the basis of empirical data, and are thus unscientific. An alternative approach based on decision analysis, naturally appealling and fully scientific, has been presented. This approach is completely satisfactory for addressing the problem of inference about the effects of causes, and the familiar 'black box' approach of experimental statistics is perfectly adequate for this purpose.

However, inference about the causes of effects poses greater difficulties. A completely unambiguous solution can only be obtained in those rare cases where it is possible to reach a sufficient scientific understanding of the system under investigation as to allow the identification of essentially deterministic causal mechanisms (relating responses to interventions and concomitants, appropriately defined). When this is not achievable (whether the difficulties in doing so be fundamental or merely pragmatic), the inferences justified even by extensive data are not uniquely determined, and we have to be satisfied with inequalities. However, these may be reduced by modelling the relevant context, and conducting experiments in which concomitants are measured. A major and detailed scientific study may be required to reduce the residual ambiguity to its minimal level (and, even then, there can be no prior guarantee that it will do so).

Thus, if we want to make meaningful and useful assertions about the causes of effects, we have to be very clear about the meaning and context of our queries. And then there is no magical statistical route that can by-pass the need to do real science, in order to attain the clearest possible understanding of the operation of relevant (typically non-deterministic) causal mechanisms.

ACKNOWLEDGMENTS

The ideas finally presented in this paper have been festering for many years, in the course of which I have had valuable discussions (and often heated arguments) with many people. I particularly wish to acknowledge the major contributions of Don Rubin, Judea Pearl, Glenn Shafer, Jamie Robins, Ross Shachter and Volodya Vovk.

Department of Statistical Science, University College London.

BIBLIOGRAPHY

[Bailey, 1991] Bailey, R. A. (1991). Strata for randomized experiments (with Discussion). *J. Roy. Statist. Soc.* B **53**, 27–78.

[Balke, 1995] Balke, A. A. (1995). *Probabilistic Counterfactuals: Semantics, Computation, and Applications*. PhD Dissertation, Department of Computer Science, University of California, Los Angeles.

[Balke and Pearl, 1994a] Balke, A. A. and Pearl, J. (1994a). Probabilistic evaluation of counterfactual queries. In *Proceedings of the Twelfth National Conference on Artificial Intelligence (AAAI-94)*, Seattle, WA. Vol. I, 230–237.

[Balke and Pearl, 1994b] Balke, A. A. and Pearl, J. (1994b). Counterfactual probabilities: Computational methods, bounds and applications. *Proc. 10th Conf. UAI*, 46–54.

[Cox, 1958] Cox, D, R. (1958). The interpretation of the effects of non-additivity in the Latin Square. *Biometrika* **45**, 69–73.

[Dawid, 1979] Dawid, A. P. (1979). Conditional independence in statistical theory (with Discussion). *J. Roy. Statist. Soc.* B **41**, 1–31.

[Dawid, 1984] Dawid, A. P. (1984). Present position and potential developments: some personal views. Statistical theory. The prequential approach (with Discussion). *J. Roy. Statist. Soc.* A **147**, 278–292.

[Dawid, 1985] Dawid, A. P. (1985). Calibration-based empirical probability (with Discussion). *Ann. Statist.* **13**, 1251–1285.

[Dawid, 1988] Dawid, A. P. (1988). Symmetry models and hypotheses for structured data layouts (with Discussion). *J. Roy. Statist. Soc.* B **50**, 1–34.

[Greenland *et al.*, 1999] Greenland S., Robins J. M. and Pearl, J. (1999). Confounding and collapsibility in causal inference. *Stat. Sci.* **14**, 29–46.

[Heckerman and Shachter, 1995] Heckerman, D. and Shachter, R. (1995). Decision-theoretic foundations for causal reasoning. *J. Artificial Intell. Res.* **3**, 405–430.

[Hitchcock, 1997] Hitchcock, C. (1997). Causation, probabilistic. In *Stanford Encyclopedia of Philosophy*. Online at: `http://plato.stanford.edu/entries/causation-probabilistic`

[Holland, 1986] Holland, P. W. (1986). Statistics and causal inference (with Discussion). *J. Amer. Statist. Ass.* **81**, 945–970.

[Imbens and ngrist, 1994] Imbens, G. W. and Angrist, J. (1994). Identification and estimation of local average treatment effects. *Econometrica* **62**, 467–476.

[Imbens and Rubin, 1997] Imbens, G. W. and Rubin, D. B. (1997). Bayesian inference for causal effects in randomized experiments with noncompliance. *Ann. Statist.* **25**, 305–327.

[Lewis, 1973] Lewis, D. K. (1973). *Counterfactuals*. Oxford: Blackwell.

[McCullagh and Nelder, 1989] McCullagh, P. and Nelder, J. A. (1989). *Generalized Linear Models* (Second Edition). London: Chapman and Hall.

[Neyman, 1923] Neyman, J. (1923) On the application of probability theory to agricultural experiments. Essay on principles. *Roczniki Nauk Rolniczych* **X**, 1–51 (in Polish). English translation of Section 9 (D. M. Dabrowska and T. P. Speed): *Statistical Science* **9** (1990), 465–480.

[Neyman, 1935] Neyman, J. (1935). Statistical problems in agricultural experimentation (with Discussion). *Suppl. J. Roy. Statist. Soc.* **2**, 107–180.

[Pascal, 1669] Pascal, B. (1669) *Pensées sur la Réligion, et sur Quelques Autres Sujets*. Paris: Guillaume Desprez. (Edition Garnier Frères, 1964).

[Pearl, 1993] Pearl, J. (1993) Aspects of graphical models connected with causality. *Proc. 49th Session ISI*, 391–401.

[Pearl, 1995a] Pearl, J. (1995a). Causal diagrams for empirical research (with Discussion). *Biometrika* **82**, 669–710.

[Pearl, 1995b] Pearl, J. (1995b). Causal inference from indirect experiments. *Artificial Intelligence in Medicine* **7**, 561–582.

[Rachev, 1985] Rachev, S. T. (1985). The Monge-Kantorovich mass transference problem and its stochastic applications. *Th. Prob. Appl.* **29**, 647– 671.

[Raiffa, 1968] Raiffa, H. (1968). *Decision Analysis: Introductory Lectures on Choices under Uncertainty*. Reading, Mass.: Addison-Wesley.

[Robins, 1986] Robins, J. M. (1986). A new approach to causal inference in mortality studies with sustained exposure periods — application to control of the healthy worker survivor effect. *Math. Modelling* **7**, 1393–1512.

[Robins, 1987] Robins, J. M. (1987). Addendum to "A new approach to causal inference in mortality studies with sustained exposure periods — application to control of the healthy worker survivor effect." *Comp. Math. Appl.* **14**, 923–945.

[Robins and Greenland, 1989] Robins, J. M. and Greenland, S. (1989). The probability of causation under a stochastic model for individual risk. *Biometrics* **45**, 1125–1138.

[Robins and Wasserman, 1997] Robins, J. M. and Wasserman, L. A. (1997). Estimation of effects of sequential treatments by reparameterizing directed acyclic graphs. Technical Report 654, Department of Statistics, Carnegie Mellon University.

[Rubin, 1974] Rubin, D. B. (1974). Estimating causal effects of treatments in randomized and nonrandomized studies. *J. Educ. Psychol.* **66**, 688–701.

[Rubin, 1978] Rubin, D. B. (1978). Bayesian inference for causal effects: the role of randomization. *Ann. Statist.* **6**, 34–68.

[Rubin, 1980] Rubin, D. B. (1980). Comment on "Randomization analysis of experimental data: the Fisher randomization test", by D. Basu. *J. Amer. Statist. Ass.* **81**, 961–962.

[Rubin, 1986] Rubin, D. B. (1986). Which Ifs have causal answers? (Comment on "Statistics and causal inference", by P. W. Holland). *J. Amer. Statist. Ass.* **81**, 961–962.

[Rüschendorf *et al.*,] Rüschendorf, L., Schweizer, B. and Taylor, M. D. (Eds). *Distributions with Fixed Marginals and Related Topics*. Institute of Mathematical Statistics Lecture Notes — Monograph Series, Volume 28. Hayward, California: Institute of Mathematical Statistics.

[Savage, 1954] Savage, L. J. (1954). *The Foundations of Statistics*. New York: Wiley
[Shafer, 1986] Shafer, Glenn (1986). Savage revisited (with Discussion). *Statistical Science* **4**, 463–501.
[Shafer, 1996] Shafer, Glenn (1996). *The Art of Causal Conjecture.* Cambridge, Mass.: MIT Press.
[Wilk and Kempthorne, 1955] Wilk, M. B. and Kempthorne, O. (1955). Fixed, mixed and random models. *J. Amer. Statist. Ass.* **50**, 1144–1167.
[Wilk and Kempthorne, 1956] Wilk, M. B. and Kempthorne, O. (1956). Some aspects of the analysis of factorial experiments in a completely randomized design. *Ann. Math. Statist.* **27**, 950–985.
[Wilk and Kempthorne, 1957] Wilk, M. B. and Kempthorne, O. (1957). Non-additivities in a Latin square. *J. Amer. Statist. Ass.* **52**, 218–236.

JON WILLIAMSON

FOUNDATIONS FOR BAYESIAN NETWORKS

Bayesian networks are normally given one of two types of foundations: they are either treated purely formally as an abstract way of representing probability functions, or they are interpreted, with some causal interpretation given to the graph in a network and some standard interpretation of probability given to the probabilities specified in the network. In this chapter I argue that current foundations are problematic, and put forward new foundations which involve aspects of both the interpreted and the formal approaches.

One standard approach is to interpret a Bayesian network objectively: the graph in a Bayesian network represents causality in the world and the specified probabilities are objective, empirical probabilities. Such an interpretation founders when the Bayesian network independence assumption (often called the causal Markov condition) fails to hold. In §2 I catalogue the occasions when the independence assumption fails, and show that such failures are pervasive. Next, in §3, I show that even where the independence assumption does hold objectively, an agent's causal knowledge is unlikely to satisfy the assumption with respect to her subjective probabilities, and that slight differences between an agent's subjective Bayesian network and an objective Bayesian network can lead to large differences between probability distributions determined by these networks.

To overcome these difficulties I put forward logical Bayesian foundations in §5. I show that if the graph and probability specification in a Bayesian network are thought of as an agent's background knowledge, then the agent is most rational if she adopts the probability distribution determined by the Bayesian network as her belief function. Specifically, I argue that causal knowledge constrains rational belief via what I call the causal irrelevance condition, and I show that the distribution determined by the Bayesian network maximises entropy given the causal and probabilistic knowledge in the Bayesian network.

Now even though the distribution determined by the Bayesian network may be most rational from a logical point of view, it may not be close enough to objective probability for practical purposes. I show in §6 that by adding arrows to the Bayesian network according to a conditional mutual information arrow weighting, one can decrease the cross entropy distance between the Bayesian network distribution and the objective distribution. This can be done within the context of constraints on the Bayesian network which limit its size and the time taken to calculate probabilities from the network, in order to minimise computational complexity.

This leads to two-stage foundations for Bayesian networks:§4 first adopt the probability function determined by a Bayesian network (this, according to the logical Bayesian interpretation, is the best subjective probability function one can adopt given the knowledge encapsulated in the network), and secondly refine the

D. Corfield and J. Williamson (eds.), Foundations of Bayesianism, 75–115.

Bayesian network to better fit objective probability (this process of calibration is required by empirical Bayesianism).[1]

To start with I shall give an introduction to Bayesian networks and their foundations in §1, before proceeding to criticisms of the standard interpretations of Bayesian networks in §2 and §3. The remainder of the paper will be taken up with my suggestions for new foundations.

1 BAYESIAN NETWORKS

Suppose we have a domain of N variables, $C_1, \ldots, C_N$, each of which takes finitely many values, $v_i^1, \ldots, v_i^{K_i}, i = 1, \ldots, N$. A *literal* is an expression c_i of the form $C_i = v_i^j$ and a *state* is a conjunction of literals. A *Bayesian network* consists of a directed acyclic graph, or *dag*, G over the nodes $C_1, \ldots, C_N$ together with a set of specifying probability values $S = \{p(c_i|d_i) : c_i$ is a literal involving node C_i and d_i is a state of the parents of C_i in G, $i = 1, \ldots, N\}$.[2] Now, under an independence assumption,[3] namely that given its parents D_i, each node C_i is probabilistically independent of any set S of other nodes not containing the descendants of C_i, $p(c_i|d_i \wedge s) = p(c_i|d_i)$, a Bayesian network suffices to determine a joint probability distribution p over the nodes $C_1, \ldots, C_N$.[4] Furthermore, any probability distribution on $C_1, \ldots, C_N$ can be represented by some Bayesian network.

Bayesian networks are important in many areas where probabilistic inference must be performed efficiently, such as in expert systems for artificial intelligence. Diagnosis constitutes a typical problem area for expert systems: here one is presented with a state of symptoms s and, under the probabilistic approach to diagnosis, one must find $p(c_i|s)$ for a range of causal literals c_i.[5] Depending on the structure of the graph G, both the number of specifiers required to determine a probability distribution p and the computational time required to calculate $p(c_i|s)$ may be substantially lower for a Bayesian network under the independence assumption than for a representation of p which makes no assumptions. Thus Bayesian net-

[1] See the introduction to this volume for more on the distinction between logical and empirical Bayesianism. Such forms of Bayesianism are often referred to as 'objective' Bayesian positions, and confusion can arise because physical or empirical probability (frequency, propensity or chance) is often called 'objective' probability in order to distinguish it from Bayesian 'subjective' probability. In this chapter I will draw the latter distinction, using 'objective' to refer to empirical interpretations of causality and probability that are to do with objects external to an agent, and using 'subjective' to refer to interpretations of causality and probability that depend on the perspective of an agent subject.

[2] If C_i has no parents, $p(c_i|d_i)$ is just $p(c_i)$.

[3] The Bayesian network independence assumption is often called the Markov or causal Markov condition.

[4] The joint distribution p can be determined by the *direct method*: $p(c_1 \wedge \ldots \wedge c_N) = \prod_{i=1}^N p(c_i|d_i)$ where d_i is the state of the direct causes of C_i which is consistent with $c_1 \wedge \ldots \wedge c_N$. Alternatively p may be determined by potentially more efficient *propagation algorithms*. See [Pearl 1988] or [Neapolitan 1990] here and for more on the formal properties of Bayesian networks.

[5] See [Williamson 2000] for more on the probabilistic approach to diagnosis.

works can offer key pragmatic advantages over formalisms without an assumption like independence.

There are two main types of philosophical foundations given to Bayesian networks. One can treat Bayesian networks as *abstract structures*, and use machine learning techniques to learn from a database of past case data (for instance of the symptoms and diagnoses of past patients) a Bayesian network that represents, or represents an approximation to, a target probability distribution.[6] More commonly, Bayesian networks are *interpreted*. Here the graph is taken to represent a causal structure, either objective or subjective. In the former case the graph contains an arrow from C_i to C_j if C_i is a direct cause of C_j, but in the subjective case the graph represents the causal knowledge of an agent X, with an arrow from C_i to C_j if X believes, or knows, that C_i is a direct cause of C_j. The specified probabilities are also given an interpretation, either objective in terms of empirical frequencies, propensities or chances, or more often subjective in terms of degrees of rational belief. Finally the independence assumption is posited as a relation between the causal interpretation and the interpretation of probability.

In my view the most important limitation of the abstract approach is that there is often not enough initial data for it to get off the ground. The abstract approach requires a database of past case data, but there may simply not be enough such data to invoke a machine learning algorithm for generating a Bayesian network. Furthermore, new case data may trickle in slowly and it may take a while before the learning algorithm yields dependable results. Even if there is plenty of data, the data may not be reliable enough to generate a reliable network — in my experience this is a significant problem, since different people often measure or categorise variables in different ways even when collecting data for the same database. There is also a difficulty when certain variables are not measured at all: diagnostic data, for example, rarely includes the presence or absence of every possible symptom of a patient, but just the most significant symptoms, and the symptoms considered most significant are subject to biases of individual doctors. In sum, the abstract approach is not appropriate for applications which require an expert system operating right from the outset, but where the data is not available, is of poor quality, or is subject to mixtures of unknown biases. However the interpreted approach does not face this sort of problem: an expert can often from the outset provide qualitative causal knowledge, subjective degrees of belief and even estimates of objective probabilities, and this information can be used to construct a Bayesian network right away — no past case data is required.

On the other hand the interpreted approach also has its problems, largely to do with the status of the independence assumption.[7] In the next two sections I shall outline these problems with the independence assumption and then go on to develop a hybrid methodology incorporating aspects of both the interpreted and abstract accounts: the basic idea behind the hybrid methodology is to form an

[6]See [Jordan 1998].

[7]One problem that I will not consider here is the *knowledge elicitation problem*: the expert may find it hard to articulate her knowledge, and the elicitation process can be quite slow.

initial Bayesian network from expert knowledge, and to further refine this network in the light of new case data. First we shall tackle the problems with an objective interpretation, and then investigate the subjective approach in §3.

2 OBJECTIVE NETWORKS

Under an objective interpretation, the Bayesian network independence assumption makes a substantive claim about the relationship between objective causality and objective, empirical probability. I will show here that this claim is highly problematic, rendering an objective interpretation inadequate.

It will be useful to note that the *principle of the common cause* is a logical consequence of the independence assumption.[8] The principle of the common cause claims the following. Suppose two variables are probabilistically dependent and neither causes the other, then

- **existence**: they have one or more causes in common,[9] and
- **screening**: they are probabilistically independent conditional on those common causes.

We can exploit the link between independence and the common cause principle because when an objective interpretation is given to both principles one can find many counterexamples to the latter principle which thereby contradict the former. In effect we can translate doubts about probabilistic analyses of causality in the philosophical literature — such analyses often appeal to the objectively-interpreted principle of the common cause — into doubts about the objective interpretation of Bayesian networks. Many of the counterexamples are well-known and, when considered in isolation, thought to be so unusual as to be unimportant, or thought to be susceptible to particular rebuttals. I want to provide a taxonomy of the counterexamples in order to show that the problem is more widespread than often considered and so general that the rebuttals are either too particular or unappealing when generalised.[10]

[8]This principle is due to Reichenbach (see [Reichenbach 1956], §19, pages 157-167). It is also often assumed as a basis for statistical experimentation — [Fisher 1935]. One can see that the principle of the common cause is a consequence of the independence assumption by generalising the following example in the obvious way. Suppose we have a Bayesian network with graph $A \longrightarrow B, C \longrightarrow D$. Thus neither B nor D cause the other, nor do they have a common cause. B and D must then be unconditionally probabilistically independent since for literals b and d on B and D respectively, their joint probability $p(b \wedge d) = \sum_{a,c} p(b|a)p(a)p(d|c)p(c) = [\sum_a p(b|a)p(a)][\sum_c p(d|c)p(c)] = p(b)p(d)$, where the first equality follows from the direct decomposition of probability in a Bayesian network (see [Neapolitan 1990] theorem 5.1 for example).

[9]Existence of a common cause resembles Mill's Fifth Canon of Inductive Reasoning: 'Whatever phenomenon varies in any manner whenever another phenomenon varies in some particular manner, is either a cause or an effect of that phenomenon, or is connected with it through some fact of causation.' [Mill 1843], page 287.

[10]A large literature touches on the independence assumption in one way or another. Thus there are

I shall argue against the independence assumption by documenting two types of counterexample to the principle of the common cause: the causal variables C_i and C_j may be *accidentally correlated*, or there may be some *extra-causal constraint* which ensures that they are probabilistically correlated.[11] There may either be no suitable common cause to account for a correlation, contradicting the existence condition above, or if there are common causes, they will not account for all of the correlation, contradicting the screening condition.

2.1 Accidental Correlations

Christmas trees tend to be sold when most oranges ripen and are sold. Let C represent the number of Christmas trees sold on any day and O represent the number of oranges sold on any day (C and O are random variables). Then $p(C > x|O > y) > p(C > x)$ for some suitable constants x and y. Now it seems clear that sales of Christmas trees do not cause sales of oranges, nor vice versa. Hence, some common cause must be found to explain their probabilistic dependence if the independence assumption is to hold. If there is a common cause it would have to be something like the time of year or the season. However, intuitively one does not endow the time of the year with causal powers, and there are no obvious mechanisms at play underlying any such causation. Intuitively there is no common causal explanation for the correlation — it is accidental. If such intuitions are right, then the independence assumption must fail for this causal scenario.

In order to save the independence assumption one may well be tempted to maintain that the time of year really is the common cause here. I shall call this strategy *causal extension*. The idea is that one tries to extend the intuitive concept of cause by counting intuitively non-causal variables, like the time of the year, as causal. In the context of Bayesian networks, causal extension often takes the form of an assumption that there is a 'hidden', 'latent' or 'unmeasured' common cause whenever two variables are found to be correlated, even when there is no intuitively plausible common cause.[12] Unfortunately, there are a number of difficulties with the strategy of causal extension. Firstly, extending the concept of cause creates epistemic problems. Identifying causal variables and the causal relationships between them is a hard problem. Any extension of the concept of cause is likely to make the task harder. In particular, it may be very difficult for an expert to provide a causal graph under the causal extension approach: one is asking the expert to identify variables that render the independence assumption valid, rather than to identify the causes and effects that she is used to dealing with. Furthermore,

criticisms (for example [Humphreys & Freedman 1996], [Humphreys 1997], [Lemmer 1993], [Lemmer 1996], [Lad 1999]) and defences (for example [Spirtes et al. 1997], [Hausman 1999], [Pearl 2000] §2.9.1) of the independence assumption which I will not cover here. I will however cover the criticisms I believe most telling and the most viable reactions to these criticisms.

[11] 'Correlation' is occasionally used to denote some kind of *linear* dependence, but I shall just use it as a synonym for 'probabilistic dependence' here.

[12] See [Binder et al. 1997] and [Pearl 2000] for example.

if one increases the number of nodes and arrows that must be considered in the graph of a Bayesian network then one risks the network becoming too complex for practical use. The amount of space required to store a Bayesian network and the amount of time required to calculate probabilities from the network both increase exponentially with the number of nodes in the worst case. This worst case occurs when the graph is dense — that is, there are many arrows in the graph. Thus causal extension is a dangerous tactic from an epistemic and practical point of view.

The second major problem is that by extending the concept of cause we are liable to lose qualities that are important to causality. Genuine causal variables tend to have various characteristics in common: for example one can normally view them as spacio-temporally localised events, and causes and effects tend to be related by physical mechanisms. If we allow variables which do not have these qualities then we can no longer be said to be explicating the notion of cause — the extension is ad hoc and the word 'cause' loses meaning, just becoming a synonym for 'variable' if the process is pursued indefinitely. This is clearly undesirable if we require a genuinely causal interpretation of the graph in the Bayesian network, as opposed to more abstract foundations.

Elliott Sober produced the following counterexample to the principle of the common cause:

> Consider the fact that the sea level in Venice and the cost of bread in Britain have both been on the rise in the past two centuries. Both, let us suppose, have monotonically increased. Imagine that we put this data in the form of a chronological list; for each date, we list the Venetian sea level and the going price of British bread. Because both quantities have increased steadily in time, it is true that higher than average sea levels tend to be associated with higher than average bread prices. The two quantities are very strongly positively correlated.
>
> I take it that we do not feel driven to explain this correlation by postulating a common cause. Rather, we regard Venetian sea levels and British bread prices as both increasing for somewhat isolated endogenous reasons. Local conditions in Venice have increased the sea level and rather different local conditions in Britain have driven up the cost of bread. Here, postulating a common cause is simply not very plausible, given the rest of what we believe.[13]

Here Sober calls the existence of a common cause into question — there is a causal explanation of the correlation, but it is not an explanation involving *common causes*, so in a sense the correlation is accidental. Postulating a common cause conflicts with intuitions here. In particular there appears to be no common causal *mechanism*. We often appeal to non-probabilistic issues like mechanisms to help determine which correlations are causal and which are accidental. As Schlegel points out, 'we reject a correlation between sun spots and economic cycles as

[13] [Sober 1988] 215.

probably spurious, because we know of no relating process, but accept a correlation between sun spots and terrestrial magnetic storms because there is a plausible physical relationship.'[14]

Besides causal extension, there is a separate line of response one can make to such counterexamples, that of *restriction*, whereby one restricts the application of the independence assumption so that it does not apply to awkward cases like Sober's.[15] This response can take one of two forms, *correlation restriction* or *causal restriction*. Regarding the former, some, such as Papineau and Price, claim that British bread prices and the Venetian water level do not have the right type of correlation for the principle of the common cause to be applied since their correlation can be predicted from the co-variation within each time-series[16] or from determinism within each physical process.[17] They thus attempt to avoid the counterexample to the common cause principle by restricting the principle itself. However, it should be noted that they pursue this strategy in the context of a defence of a probabilistic analysis of causality. Whether or not this move is successful in that context, it is no help here when thought of in terms of the Bayesian network framework, for restricting the principle of the common cause restricts the independence assumption too, and the reduction of a probability function to a Bayesian network is not possible without full-blown independence. Hence correlation restriction is not a viable move when considering Bayesian networks.

The other variety of restriction, causal restriction, is more promising. Here the strategy is to argue that the variables themselves are not of the sort to which the independence assumption applies. One may claim that the correlated variables are not causal variables, although this is rather implausible when it comes to the examples above. Alternatively one may accept that they are causal, but have not been individuated correctly for the independence assumption to apply. For example, the variables may need to be indexed by time,[18] may need to be complete descriptions of their corresponding single-case events, or may need to be properties that can be repeatedly instantiated.

While it is possible that for any particular counterexample to independence there is another way of individuating the variables so that the dependency is removed, it is less clear that one rule of individuation will overcome all counterexamples. I have used examples which exhibit temporal correlation here because it is easy to see how such variables could be correlated, but any two events might exhibit accidental correlation, in which case alternative individuation will not help. The independence assumption rules out accidental correlation a priori, and such a restriction does not appear a priori to be any more plausible applied to one individuation than another. Thus an appeal to individuation is by no means guaranteed to overcome the problem of accidental correlation.

---

[14][Schlegel 1974] 10.

[15]Lakatos called this type of defence 'monster-barring'.

[16][Papineau 1992] 243.

[17][Price 1992] 264.

[18]See [Spirtes et al. 1993] page 63 for example.

Causal restriction also induces epistemic problems of its own. If individuation matters then one has to do a certain amount of analysis before tackling a problem, making the application of Bayesian networks harder. Furthermore, in a particular problem one may be interested in variables which must be individuated in a way for which independence does not hold, in which case the machinery of Bayesian networks cannot be applied at all.

I have illustrated the problem of accidental correlations and introduced strategies for defending the independence assumption, including causal extension and causal restriction. These strategies are somewhat less than effective at dealing with the problem, and if they can be made to work will only do so at an epistemic and intuitive cost. In §2.2 we will see how these strategies can be applied to other common types of counterexample. Our conclusions will be much the same. Yet these costs are not ones we have to reluctantly accept. In the foundations I propose later, we will stick with our intuitive notion of cause and the individuation of variables will not matter.

2.2 *Extra-Causal Constraints*

I shall now consider counterexamples to the principle of the common cause where probabilistic dependencies have an explanation that relates the dependent variables — thus the dependencies are not accidental — but where the explanation is not causal. There are a number of non-causal correlators: two causal variables can be correlated

- in virtue of their meaning,
- because they are logically related,
- because they are mathematically related,
- because they are related by (non-causal) physical laws, or
- because they are constrained by local laws or boundary conditions.

Let us look at each of these situations in turn.

First, the meanings of expressions can constrain their probabilities. 'Flu and orthomyxoviridae infection are probabilistically dependent, not because they have a common cause, but because 'flu is an example of orthomyxoviridae infection — the variables have overlapping meaning.

In response one can advocate a kind of causal restriction. One can argue that causes should be individuated so as to avoid overlapping meaning, and that one should remove a node from a Bayesian network if there is another with related meaning. But this is not always a sensible move for a number of reasons. One can lose valuable information from a Bayesian network by deleting a node, since both the original nodes may be important to the application of the network. Meaning might be related through vagueness rather than classification overlap, for example if one symptom is a patient's report of fever and another is a thermometer reading, and it may be useful to consider all such related nodes. In some cases one may even want to include synonyms in a Bayesian network, for example in a network for natural language reasoning. Furthermore, removing a node can invalidate the independence assumption if the removed node is a common cause of other nodes. Or one simply may not know that two nodes have related meaning: Yersin's discovery that the black death coincides with Pasteurella pestis was a genuine example of scientific inference, not the sort of thing one can do at one's desk while building an expert system.

Causal extension is no better a ploy here. One could suggest that a common cause variable called 'synonymy' or 'meaning overlap' should be introduced. But this will not in general screen off such dependencies, and as before we have epistemic cost in terms of identifying dependencies in virtue of meaning and the likely added complexity of incorporating new variables and arrows, as well as a commitment to a counterintuitive concept of cause.

Probabilistic correlations can also be explained by logical relations. For instance, logically equivalent sentences are necessarily perfectly correlated,[19] and if one sentence c logically implies sentence d, the probability of d must be greater than or equal to that of c. Thus one should be wary of Bayesian networks which involve logically complex variables. Suppose C causes complaints D, E and F, and that we have three clinical tests, one of which can determine whether or not a patient has both D and E, another tells us whether or not the patient has one of E and F, and the third tells us whether the patient has C. Thus there is no direct way of determining $p(d|c), p(e|c)$ or $p(f|c)$ for literals c, d, e and f of C, D, E, and F respectively, but one can find $p(d \wedge e|c)$ and $p(e \vee f|c)$. One might then be tempted (in the spirit of causal extension) to incorporate $C \longrightarrow (D \wedge E), C \longrightarrow (E \vee F)$ in one's causal graph, so that the probability specification of the corresponding Bayesian network can be determined objectively. In such a situation, however, C will not screen node $D \wedge E$ off from node $E \vee F$ and the independence assumption is not satisfied.

This problem seriously affects situations where causal relata are genuinely logically complex, as happens with context-specific causality. A may cause B *only if* the patient has genetic characteristic C: if the patient has any other genetic characteristic then there is no possible causal mechanism from A to B. Then the conjunction $A \wedge C$ is the cause of B, not A or C on their own. However, A may be able to cause D in everyone, so the causal graph would need to contain a node $A \wedge C$ and a second node A. One would not expect these two nodes to be screened off by any common causes.

Next we turn to mathematical relations as a probabilistic correlator. By way of example, consider the application of Bayesian network theory to colon endoscopy as documented in [Sucar et al. 1993] and [Kwoh & Gillies 1996]. The object is

[19] At least according to standard axiomatisations of probability.

to guide the endoscope inside the colon towards the lumen, avoiding the diverticulum. A Bayesian network was used to identify the lumen and diverticulum from the endoscope image. The presence of the lumen causes a large dark region to appear on the endoscope screen while the diverticulum causes a small dark region. The size of the region can be directly measured, but its darkness was measured by its mean intensity level together with its intensity variance in the region. A Bayesian network was constructed incorporating these variables and the independence assumption was tested and found to fail: the mean and variance variables were found to be correlated when, according to the causal graph under the independence assumption, they should not have been. The problem was that there is no obvious common cause for this correlation: mean and variance are related mathematically, not causally. We have that $VarX = EX^2 - (EX)^2$, where $VarX$ is the variance of random variable X, and E signifies expectation so that EX is the mean of X. To take the simplest example, if X is a Bernoulli random variable and $EX = x$ then $VarX = x(1-x)$, making the mean and variance perfectly correlated. In the endoscopy case, the light intensity will have a more complicated distribution, but the mean value will still constrain the variance, making the mean and variance probabilistically dependent. To try to resolve this failure of the independence assumption, at first one of the two correlated nodes was removed (causal restriction). This gave some improvement in performance but suffered from significant loss of information. Next (causal extension) [Kwoh & Gillies 1996] attempted to introduce an extra common cause to screen off the correlation, but while this move improved the success rate of the Bayesian network, it raised fundamental problems. Firstly it is not clear what the new node represents (it was just called a 'hidden node'), so a causal interpretation may no longer be appropriate for the graph. Secondly, the distribution specifying probabilities relating the new node to the other nodes had to be ascertained: this could only be done mathematically, by finding what the probabilities should be if the introduction of the new node allowed the unwanted correlation to be fully screened off, and could not be tested empirically or equated with any objective probability distribution. Therefore the Bayesian network lost both the objective causal and the objective probabilistic components of its interpretation. An objective interpretation is just not feasible, given extra-causal dependencies like this.

That extra-causal constraints include physical laws has been exemplified by Arntzenius:[20]

> Suppose that a particle decays into 2 parts, that conservation of total momentum obtains, and that it is not determined by the prior state of the particle what the momentum of each part will be after the decay. By conservation, the momentum of one part will be determined by the momentum of the other part. By indeterminism, the prior state of the particle will not determine what the momenta of each part will be after the decay. Thus there is no prior screener off.

[20] [Arntzenius 1992] pages 227-228, from [van Fraassen 1980] page 29.

The principle of the common cause fails here because there is nothing obvious that we can call a common cause — the existence component of the principle fails. But even if some weird and wonderful common cause could be found in such quantum situations, independence would still fail because screening condition would fail. Suppose we consider the spins B and C of two particles: B and C have values *up* or *down*. The two particles are fired such that one has spin up (represented by literal b) if and only if the other does (c). Suppose also that either one being spin up is as likely as not, $p(b) = p(c) = 1/2$, but that a common cause A is found which explains the spins, so $A \longrightarrow B, A \longrightarrow C$, and $p(b|a), p(c|a) = x > 1/2$. But since $p(b|c) = 1$, screening off is satisfied if and only if $1 = p(b|a \wedge c) = p(b|a)$, so the cause must be deterministic, a wildly inappropriate assumption in the quantum world. Thus we must conclude that there are quantum constraints on objective probability which are extra-causal.[21]

The philosophical literature also contains several examples of how local non-causal constraints and initial conditions can account for dependencies amongst causal variables. Cartwright, for instance, points out that

> independence is not always an appropriate assumption to make. ... A typical case occurs when a cause operates subject to constraint, so that its operation to produce one effect is not independent of its operation to produce another. For example, an individual has \$10 to spend on groceries, to be divided between meat and vegetables. The amount that he spends on meat may be a purely probabilistic consequence of his state on entering the supermarket; so too may be the amount spent on vegetables. But the two effects are not produced independently. The cause operates to produce an expenditure of n dollars on meat if and only if it operates to produce an expenditure of $10 - n$ dollars on vegetables. Other constraints may impose different degrees of correlation.[22]

Salmon[23] gives another counterexample to the screening condition. Pool balls are set up such that the black is pocketed (B) if and only if the white is (W), and a beginner is about to play who is just as likely as not to pot the black if she attempts the shot (S), and is very unlikely to pot the white otherwise. Thus if we let b, w and s be literals representing the occurrence of B, W and S respectively, $p(b \leftrightarrow w) = 1$ and $p(b|s) = 1/2$, so $1/2 = p(w|s) \neq p(w|s \wedge b) = 1$ and the cause S does not screen off its effects B and W from each other. As Salmon says:

---

[21] Note that [Butterfield 1992] looks at Bell's theorem and concludes (page 41) that, 'the violation of the Bell inequality teaches us a lesson, ... namely, some pairs of events are not screened off by their common past.' [Arntzenius 1992] has other examples and also argues on a different front against the principle of the common cause assuming determinism. See also [Healey 1991] and [Savitt 1996] pages 357-360 for a survey.

[22] [Cartwright 1989] 113-114.

[23] [Salmon 1980] pp. 150-151, [Salmon 1984] pp. 168-169.

> It may be objected, of course, that we are not entitled to infer ... that there is no event prior to B which does the screening. In fact, there is such an event — namely, the compound event which consists of the state of motion of the cue-ball shortly after they collide. The need to resort to such artificial compound events does suggest a weakness in the theory, however, for the causal relations among S, B and W seem to embody the salient features of the situation. An adequate theory of probabilistic causality should, it seems to me, be able to handle the situation in terms of the relations among these events, without having to appeal to such *ad hoc* constructions.[24]

I would echo this sentiment in the current context: in my view an adequate objective causal-probabilistic interpretation of Bayesian networks should not have to appeal to ad hoc constructions. Spirtes, Glymour and Scheines give a causal-restriction defence against Salmon's counterexample by arguing that the collision should be more specifically individuated (in particular the momentum of the cue ball should be described).[25] Again this is less than satisfactory in the absence of a general theory as to how causes should be individuated.

A further example: repeatedly pull one of two beads (a blue bead B and red bead R, otherwise identical) out of a bag. Then $p(b|r) = 0 < 1/2 = p(b)$. But rather than saying that pulling out the red bead is a preventative of pulling out the blue bead, the correlation is explained by the set-up of the repeatable experiment: only one bead is pulled out of the bag in any trial. Here the set-up constrains the probabilities and isn't the sort of thing that counts as a cause.

In response to the problem of extra-causal constraints, one might admit defeat in problems such as the diagnosis of apparatus for the investigation of quantum mechanical systems,[26] or troubleshooting pool players, but maintain that most applications of intelligent reasoning may be unaffected. But extra-causal constraints occur just about anywhere, including central diagnosis problems for example. When diagnosing circuit boards, one may be constrained by the fact that two components cannot fail simultaneously ($F_1 \wedge F_2$), for if one of them fails the circuit breaks and the other one cannot fail. Suppose there is a common cause C for the failures as in Figure 1. Then C fails to screen F_1 off from F_2 for $p(f_2|c \wedge f_1) = 0 \neq p(f_2|c)$. In medicine the opposite is the case: failure of one component in the human body increases the chances of failure of another, as resources are already weakened. In both these cases the constraints are very general and not the sort of thing one would want to call causes.

But why not pursue causal extension and include these extra-causal constraints in a Bayesian network? Besides the problem of a loss of the causal interpretation, we have further difficulties. Knowledge of extra-causal constraints is often in some sense superfluous to an intelligent agent's needs. An agent performing diag-

[24] [Salmon 1980] 151 (my notation).
[25] [Spirtes et al. 1993] 63.
[26] As [Spirtes et al. 1993] do, pages 63-64.

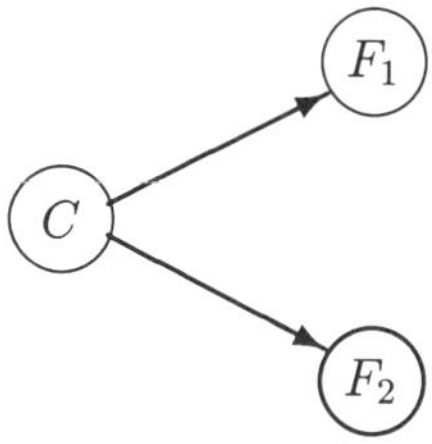

Figure 1. Failure of circuitry components.

Figure 2. Christmas tree sales, festivity, spending and orange sales

nosis, for instance, needs to know about causes and effects because she has to find the probabilities of various causes given some symptoms, but she is not directly concerned with facts about meaning, experimental set-ups or physical laws. Thus if there is a requirement to keep the agent's language and causal graph small, as in the Bayesian network formalism where computational complexity is an issue, extra-causal constraints are the things to leave out. Second, it may be much harder for domain experts to provide the relevant extra-causal information than the causal information. In particular, discovering all physical laws which have correlational consequences on a domain is no mean feat. Third, even if a general constraint is identified, it is often difficult to say exactly how it should be connected to the other variables in a causal graph. Should there be an arrow between the set-up of a pool table and each possible pot, or just some? Extra-causal constraints are generally symmetric while causal relations are not. Fourthly, these constraints often vary between cases in the way that causal laws don't. If the set-up of a pool table is included in a causal graph and we are interested in predicting the next pot then, since the set-up changes as play progresses, the causal graph will also have to vary radically from shot to shot. This obviously complicates the task.

Note finally that accidental and extra-causal correlations can combine to complicate matters. If two variables are accidentally correlated then a common cause is very unlikely to completely screen off that correlation. More plausibly, the common cause would account for part of the correlation, and there would be a surplus that we might call accidental. An inefficient English bakery might partly explain why the water level rises in Venice (through global warming) and also partly why bread prices rise in the UK, but the remaining bulk of the correlation might be completely accidental. Likewise direct causes of an effect may not fully screen it off from their causes. In response to our first example of accidental correlation, one might put forward some causal story: high Christmas tree sales (C) causes people to be festive (F) which causes people to spend more (S) which causes orange sales to rise (O), as in Figure 2. But even if this explains some of the correlation (and this is rather dubious), it will not explain it all, for $p(o|c) = 1$, but people spend money on many other occasions in the year and $p(o|s)$ is not much bigger than $p(o)$. So $p(o|c \wedge s) > p(o|s)$.

I hope to have shown that many types of dependency can be invoked to contest the validity of the objectively-interpreted independence assumption. Two strategies present themselves if we look for a defence against the counterexamples, causal restriction and causal extension. However each strategy is subject to epistemological, practical and intuitive difficulties, rendering an objective interpretation of Bayesian networks at worst impossible and at best undesirable.

3 SUBJECTIVE NETWORKS

We have seen how problems arise for an objective interpretation of the components of a Bayesian network. But there is a further reason why an objective interpretation is unattractive in practice: one may simply not know of all the causal variables or causal relations relevant to a domain of interest, and one may not be able to accurately estimate the corresponding objective probabilities required in the specification of a Bayesian network. In practice our knowledge is limited, and information in a Bayesian network will often be incomplete and inaccurate.

Thus it makes sense to relativise the Bayesian network to an agent's perspective. In this section we shall suppose that the Bayesian network expresses the knowledge of a particular agent, X say — that the graph G is interpreted as X's representation of causality, and that the probability specification S is interpreted as containing her degrees of belief in literals conditional on parent states. The independence assumption then links the agent's picture of causality to her belief function p: if it holds then her belief function is reducible to her Bayesian network.

Does the independence assumption hold here? There is little reason to suppose that it might. X's knowledge of causality may be very limited, and her degrees of belief may wildly differ from objective probability: according to strict-subjectivist Bayesian theory X may hold whatever beliefs she likes, as long as her belief function is formally a probability function. Yet the independence assumption is a very strong constraint, for it fixes X's belief function given her Bayesian network, thereby restricting X's subjectivity. If X's causal knowledge or the degrees of belief in her probability specification were to change slightly then her other degrees of belief would have to change correspondingly, leaving no room for subjectivity with regard to these other beliefs. Therefore a strong constraint like independence does not fit well with subjectivism, whose appeal is based on the freedom it allows causal knowledge and degrees of belief.

So how can a subjective interpretation of Bayesian networks be maintained? One line of reasoning goes something like this: if independence holds objectively, and the subjective network is similar to the objective network, then the subjective distribution determined by the subjective network will be close enough to objective probability to be put to practical use. Suppose we require an expert system for diagnosis of liver disease. We may think we have a fair idea of the causal picture relating this area, and may be able to obtain estimates of the objective probabilities for a probability specification, thereby forming a Bayesian network that is in some sense close to an objective version. If the independence assumption were to hold in the objective case then one might expect it to hold approximately in the subjective case. One might further suppose that if independence approximately held in the subjective case then the probability distribution determined by the subjective network might approximate objective probability, at least closely enough for the practical purposes of liver diagnosis.

It is such a position that I want to argue against in this section. There are two flaws in the above reasoning. First, as we saw in the last section, there is often reason to doubt the independence assumption as made of objective causality and probability. Secondly, even if independence were to hold objectively, small differences between a subjective network and the objective network can lead to significant differences in the probability distributions determined by these networks. It is this second claim that I want to argue for here.

For this argument it will be necessary to consider subjective and objective distributions and networks simultaneously, and so it will be worth spelling out the notation and concepts clearly in advance. The objective probability distribution is p^*. We also have an objective Bayesian network consisting of causal graph G^* and the associated probability specification S^*. Independence is assumed to hold of objective causality G^* with respect to objective probability p^*, and this has the repercussion that the objective network (G^*, S^*) determines p^*. Agent X has a subjective Bayesian network consisting of causal graph G and associated probability specification S. This subjective network (G, S) determines probability function p under the independence assumption. The question of whether independence holds subjectively and p matches X's full belief function is not of concern here. Instead, we are concerned with the above alternative justification of the subjective interpretation which claims that if the subjective network (G, S) closely resembles the objective network (G^*, S^*) then the function p will be close enough to objective probability p^* to be of practical use. I argue that differences between the objective and subjective networks that are likely to occur in practice will yield significant differences between resulting probability distributions.

It will be useful to distinguish two types of difference between the subjective and objective networks: differences between the causal graphs G and G^* and differences between the probability specifications S and S^*.

3.1 Causal Subjectivity

First I shall argue as follows. Even if we make the assumption that independence holds objectively, we assume that X's belief specification S consists of objective probabilities, and assume that her causal knowledge is correct (G is a subgraph

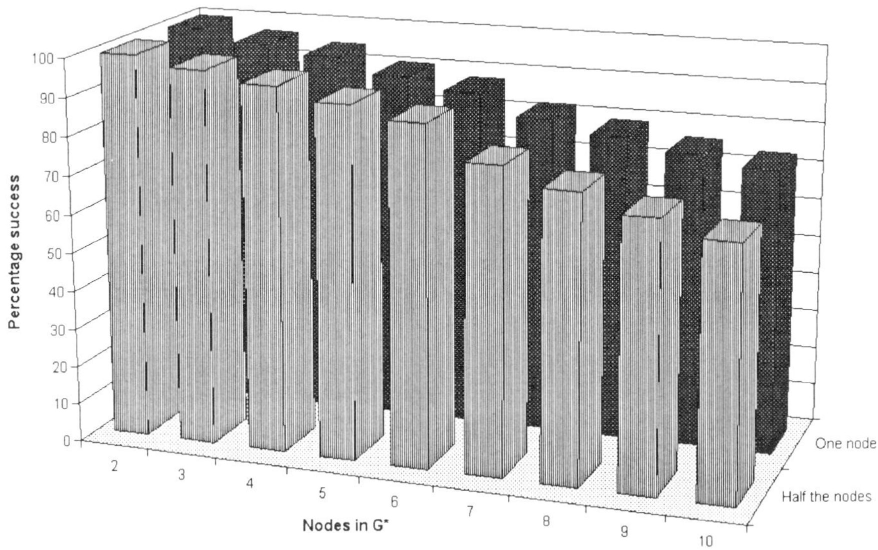

Figure 3. Nodes removed.

of G^*), then if, as one would expect, her causal knowledge is incomplete (a strict subgraph), p may be not be close enough to p^* for practical purposes.

There are two basic types of incompleteness. X may well not know about all the variables (G has fewer nodes than G^*) or even if she does, she may not know about all the causal relations between the variables (G has fewer arrows than G^*).

To deal with the first case, suppose G is just G^* minus one node C and the arrows connecting it to the rest of the graph. Even if G^* satisfies independence with respect to p^* then G can only be guaranteed (for all p^*) to satisfy independence if all the direct causes of C are direct causes of C's direct effects, each pair D, E of its direct effects have an arrow between them say from D to E, and the direct causes of each such D are direct causes of E.[27] Needless to say, such a state of affairs is rather unlikely and a failure of independence will have practical repercussions.

I ran a simulation to indicate just how close the subjectively-determined distribution p will be to the objective distribution p^*, the results of which form Figure 3. The bars in the background of the graph show the performance of Bayesian networks formed by removing a single node and its incident arrows from networks known to satisfy independence. For $N = 2, \ldots, 10$ I randomly generated Bayesian networks on N nodes, and for each net removed a random node, chose a random

[27] See [Pearl et al. 1990] 82.

state of nodes s and calculated $p(c|s)$ for each literal c not in s. The new networks were deemed successful if their values for $p(c|s)$ differed from the values determined by the original network by less than 0.05, that is, $|p(c|s) - p^*(c|s)| < 0.05$. For each N the percentage success was calculated over a number of trials[28] and each bar in the chart represents such a percentage. The bars in the foreground of the graph represent the percentage success where half the nodes[29] and their incident arrows were removed.

Such experiments are computationally time-consuming and only practical for small values of N. While one should be wary of reading too much into a small data set, the results do suggest a trend of decreasing success rate as the size of the networks increase. Thus it appears plausible that if one removes a node and its incident arrows from a large Bayesian network that satisfies independence, then the resulting network will not be useful, in the sense that the probability values it determines will not be sufficiently close to objective probability. Moreover, removing more nodes from a Bayesian net is likely to further reduce its probability of success, as the graph shows.

This trend may be surprising, in that if one removes a node from a large causal graph one is changing a smaller portion of it than if one removes a node from a small graph, so one might expect that removing a node changes the resulting distribution less as the original number of nodes N increases. But one must bear in mind that the independence assumption is non-local: removing a node can imply an independency between two nodes which are very far apart in the graph. Thus removing a node from a small graph is likely to change fewer implied independencies than removing a node from a large graph.

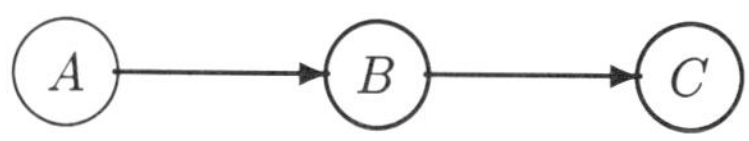

Figure 4. Objective causal graph G^*.

Figure 5. B and its incident arrows removed.

Figure 6. B removed but its incident arrows redirected.

[28]At least 2000 trials for each N, and more in cases where convergence was slow.
[29]In fact the nearest integer less than or equal to half the nodes was chosen.

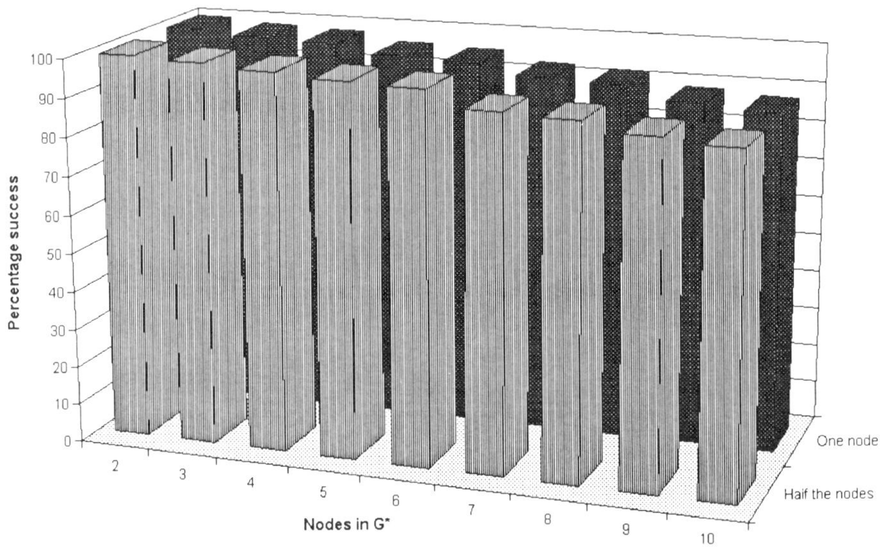

Figure 7. Nodes removed - arrows re-routed.

Of course one may complain that such a simulation is unrealistic in some way. For instance, if one doesn't know about some intermediary cause in an objective causal graph, one may yet know about the causal chain on which it exists. Thus if Figure 4 represents the objective causal graph and one doesn't know about B, one may know that A causes C, as in Figure 6 rather than Figure 5. In this case removing B's incident arrows introduces an independence assumption which is not implied by the original graph, whereas redirecting them does not. In simulations I found that while redirecting rather than removing arrows improved success (see Figure 7) the qualitative lesson remained: the general trend was still that success decreases as the number of nodes increases.

There is another way that the simulation may be unrealistic. Some types of cause may be more likely to be unknown than others, so perhaps one should not remove a node at random in the simulation. However, if we adjust for this factor we should not expect our conclusions to be undermined. To the extent that effects are more likely to be observable and causes to be unobservable, one will be more likely to know about nodes in the latter parts of causal chains than in the earlier parts. But while removing a leaf in a graph will not introduce any new independence constraints, removing common causes can do so. Thus if X is less likely to know about causes than effects, her subjective causal graph is even less likely to satisfy independence than one with nodes removed at random.

There may be other factors which render the simulations inappropriate, based on the way the networks are chosen at random. Here I made it as likely as not that two nodes have an arrow between them, and as likely as not that an arrow is in one direction as in another, while maintaining acyclicity. Thus the graphs are unlikely to be highly dense or highly sparse. I chose the specifying probabilities uniformly over machine reals in $[0, 1]$. Roughly half the nodes ($N/2$ nodes if N was even otherwise $(N-1)/2$ nodes) were chosen to be symptoms in s and the nodes and their values were selected uniformly. In the face of a lack of knowledge about the large-scale structure of the objective causal graph I suggest these explications of 'at random' are appropriate. In any case, the trend indicated by the simulation does not seem to be sensitive to changes in the way a network is chosen at random.

In sum then, for a G^* large enough to be an objective causal graph the removal of an arbitrary node is likely to change the independencies implied by the graph, and to change the resulting distribution determined by the Bayesian network. This much is arguably true whether or not the objective situation (G^*, p^*) satisfies independence itself, for if independence fails, removing arbitrary nodes is hardly likely to make it hold.

Having looked at what happens when agent X is ignorant of causal variables, we shall now turn to the case where she is ignorant of causal relations.

Suppose then that G is formed from G^* by deleting an arrow, say from node C_i to node C_j. Then G can not be guaranteed to satisfy independence with respect to p^*. For suppose $C_i, D_1, \ldots, D_k$ are the direct causes of C_j in G^*. Then the independence of G with respect to p^* requires that C_i be independent of C_j, conditional on $D_1, \ldots, D_k$, which is not implied by the independence of G^* with respect to p^*.

The situation is worse if the following condition holds, which I shall call the *dependence* principle.[30] This corresponds to the intuition that a cause will either increase the probability of an effect, or, if it is a preventative, make the effect less likely. More precisely,

- **dependence**: if $C_i, D_1, \ldots, D_k$ are the direct causes of C_j then C_i and C_j are probabilistically dependent conditional on $D_1, \ldots, D_k$: there are some literals c_i and c_j of C_i and C_j and some state d of $D_1, \ldots, D_k$ such that $p^*(c_j|c_i \wedge d) \neq p^*(c_j|d)$, as long as these probabilities are non-extreme (that is, neither 0 nor 1).

Now if G^* satisfies dependence with respect to p^*, the arrow between C_i and C_j is removed to give G as before, and the probabilities are non-extreme, the independence assumption will *definitely fail* for G with respect to p^*. This is simply because the independence of G with respect to p^* requires that C_i and C_j be independent conditional on $D_1, \ldots, D_k$ which contradicts the assumption that dependence holds for G^* with respect to p^*. Note that this result only depends on

[30]See [Williamson 1999] for a defence of this principle. Note that the dependence principle is a partial converse to the independence assumption.

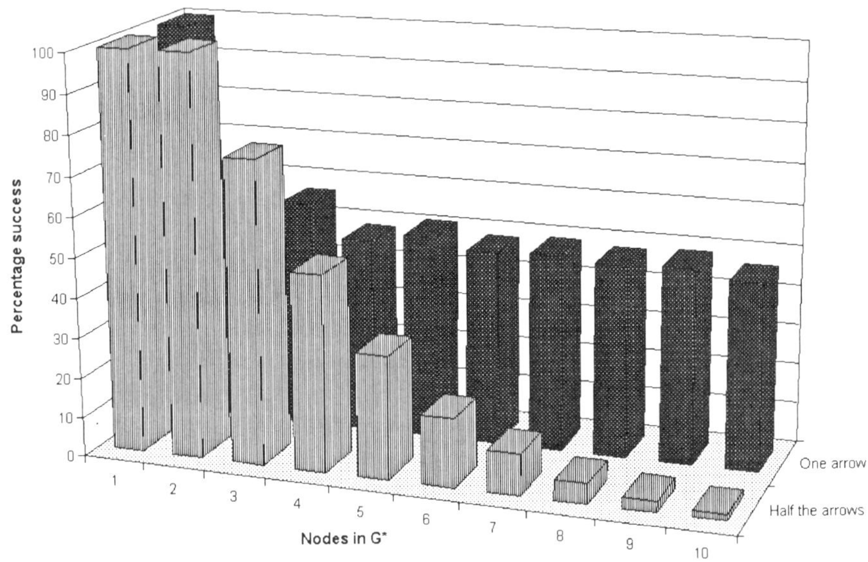

Figure 8. Arrows removed.

the local situation involving C_i, C_j and the other direct causes $D_1, \ldots, D_k$ of C_j, so that further changes elsewhere in the graph cannot rectify the situation.[31] Note also that this result does *not* require that objective causality G^* satisfy independence with respect to objective probability p^*. Thus if the dependence principle holds of causality in the world it is extremely unlikely that independence will hold of a subjective causal theory.

Of course, we are arguing against independence by appealing to an alternative principle here and the sceptical reader may not be convinced by this last argument. But we can perform simulations as before to indicate the general trends. The back row of Figure 8 represents the results of the same simulation as before (the dependence principle is not assumed to hold), except with a random arrow rather than a node removed. In this case there is no clear downward trend, but success rate is uniformly low. If more arrows are removed, then for all but small N the resulting network is less likely still to satisfy independence, as the front row of Figure 8 shows, and again we see a downward trend as the number of nodes in G^* increases.

[31] If one or more of the other direct causes or their arrows to C_j are also absent in G, then independence may be reinstated, although this would be a freak occurrence and the extra change may break a further independence relation elsewhere in the graph.

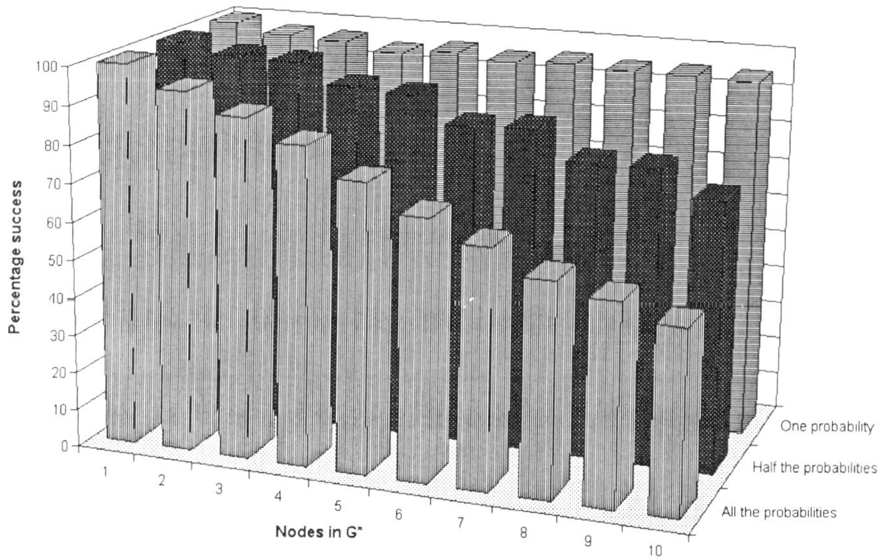

Figure 9. Node probabilities perturbed.

In sum, causal subjectivity can lead to a significant difference between the subjective and objective probability distributions.

3.2 *Probabilistic Subjectivity*

Turning now to X's degrees of belief, it is not hard to see how p can differ from p^*. We suppose that the objective situation satisfies independence, and that X's causal graph G matches the objective causal graph G^*. However, if her specification S differs from the objective specification then the probability function p determined by the subjective network (G, S) would not be expected to agree exactly with p^*. The back row of Figure 9 shows what happens if one of the nodes has its associated probability specifiers perturbed by 0.03, the middle row shows what happens if half the nodes' probabilities are perturbed by 0.03, and the front rows gives the case where all nodes have their probabilities perturbed.

In practice probabilistic and graphical subjectivity will occur together, making it even less likely that p is close enough to p^* for practical purposes. The back row of Figure 10 shows what happens if a node is removed (arrows re-routed), then an arrow is removed, and then one node's probabilities are perturbed by 0.03. The front row shows what happens if half the nodes then half the remaining arrows are removed, then half the remaining nodes are perturbed.

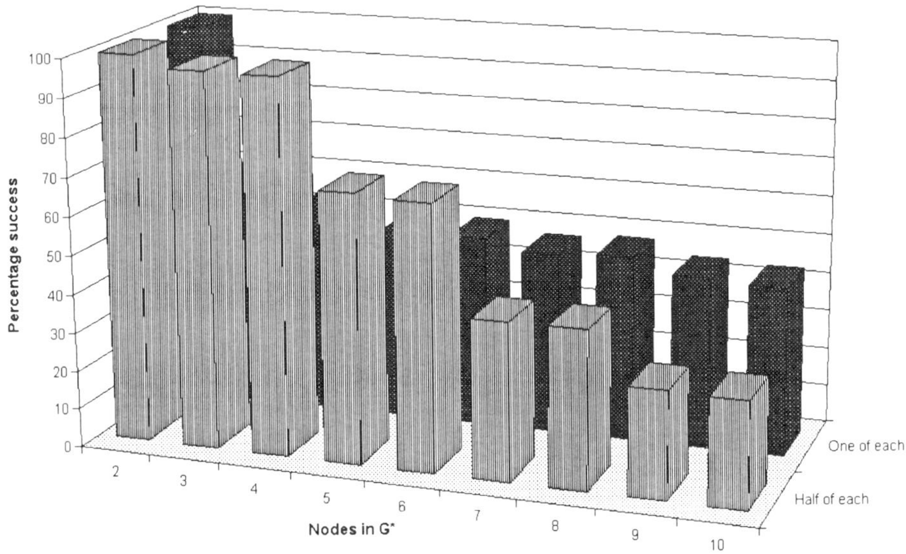

Figure 10. Nodes and arrows removed, node probabilities perturbed.

Thus subjectivity in a Bayesian network can lead, significantly often, to practical problems: the distribution determined by a subjective network may differ too much from the objective distribution to be of practical use.

4 TWO-STAGE BAYESIAN NETWORKS

We have seen some of the problems that face interpretations of Bayesian networks. The independence assumption can fail for an objective interpretation because correlations may be accidental or have non-causal explanations. Independence can hardly be expected to hold for a subjective interpretation — the agent's Bayesian network will generally give rise to a probability function p which differs from her true belief function — but more importantly p is also likely to differ from objective probability, which upsets the alternative justification of subjective networks.

I want to argue for another view of Bayesian networks, which I believe rests on firmer foundations. The view I put forward here initially adopts a subjective interpretation, where the graph in the Bayesian network is an agent's representation of causal structure and the probability specifiers are her degrees of rational belief. I acknowledge the fact that, according to the above arguments, the distribution specified by an agent's Bayesian network may not be close enough to the objective distribution to be of much practical use, but I argue that it is a good starting point, and can be refined to better approximate reality. This gives a *two-stage methodology* where stage one is the representation of X's belief function p by an initial Bayesian network and stage two is the further refinement of the network. In terms of foundations, stage one yields a subjective interpretation (but a different subjective interpretation to those given in §3), while stage two borrows techniques from the abstract approach in order to deliver a network whose distribution more closely approximates the objective distribution (and in the process of refinement the causal interpretation may be dropped as we shall see).

Two key questions require attention before we can be convinced of these two-stage foundations for Bayesian networks. Firstly, how can stage one be justified? I have argued against a strict subjective interpretation, and so must somehow demonstrate that some other kind of subjective interpretation of the Bayesian network is a good starting point. I shall do this in the rest of this section and the next section. Secondly, how can stage two be performed? I shall discuss the refinement of Bayesian networks in §6.

I shall interpret X's Bayesian network as her background knowledge: the causal graph G contains her knowledge of causal variables and their causal relations, and the probability specification S is her knowledge of conditional probabilities of causes given parent-states.[32] The independence assumption may then be used to determine X's degrees of belief from her background knowledge: her full belief function will be the probability function determined by the Bayesian network on G and S under the independence assumption.

Thus independence is no longer a substantive assumption linking the agent's causal graph with some pre-determined rational belief function, it is a *logic*, used to derive undetermined degrees of belief from those that are given in X's probability specification.

The central issue then is how we can justify the use of the independence assumption as a means of determining a rational belief function.

This issue of finding a single rational belief function given some background knowledge has received plenty of attention in the literature. Approaches range from Laplace's *principle of indifference* to Jaynes' *maximum entropy principle*. The former says that if X is indifferent as to which of J alternatives is true then she should believe each of them to degree $1/J$. The latter explicates and generalises the former as follows. A probability function over $C_1, \ldots, C_N$ may be fully specified by specifying values for each of the parameters $x^{k_1,\ldots,k_N} = p(C_1 = v_1^{k_1} \wedge \ldots \wedge C_N = v_N^{k_N})$, where $v_i^{k_i} \in \{v_i^1, \ldots, v_i^{K_N}\}$ for $i = 1, \ldots, N$. We have the constraints that each $x^{k_1,\ldots,k_N} \in [0, 1]$, and by additivity $\sum_{k_1,\ldots,k_N} x^{k_1,\ldots,k_N} = 1$, together with any constraints implied by background knowledge. The maximum entropy principle says that in the absence of any further information X should select a most rational belief function by choosing the $x^{k_1,\ldots,k_N}$ subject to these

[32] I shall leave it open as to whether these probabilities are taken to be estimates of objective probabilities or informed degrees of belief. It suffices that they count as knowledge and may be used to guide X's other beliefs.

constraints which maximises the entropy

$$H = - \sum_{k_1,\ldots,k_N} x^{k_1,\ldots,k_N} \log x^{k_1,\ldots,k_N}.$$

There are several convincing justifications for the maximum entropy principle. The most well-known involves Shannon's information-theoretic interpretation of entropy as a measure of uncertainty, in which case we maximise entropy subject to some background knowledge if we determine a probability function whose informativeness is as close as possible to that of just the background knowledge itself. A second justification is based on Boltzmann's work with entropy in physics, and a third involves Paris and Vencovská's demonstration that the maximum entropy solution is the only completion to satisfy various intuitively compelling desiderata, such as language invariance.[33] Grünwald gives a fourth, game-theoretic justification: maximum entropy is the (worst-case) optimal distribution for a game requiring the prediction of outcomes under a logarithmic loss function.[34]

Where does this leave independence and stage one of our two-stage methodology? Stage one is justified because the probability function determined by the independence assumption from the Bayesian network coincides with that determined by the maximum entropy principle, as we shall now see.

5 BAYESIAN NETWORKS HAVE MAXIMUM ENTROPY

The argument for the identity of the Bayesian network and maximum entropy functions requires first making the constraints imposed by the background knowledge explicit, and next showing that if we maximise entropy subject to these constraints then we get the same solution as that determined by the Bayesian network under the independence assumption.

5.1 Background Knowledge

Agent X's background knowledge consists of the components of a causally interpreted Bayesian network: a causal graph and the specified probabilities of literals conditional on states of their parents. We first need to formulate this knowledge in a way that can more formally be applied to the maximum entropy procedure. Regarding the probability specification, there is no problem. We can simply maximise entropy subject to the constraints that certain probabilities, namely those in the Bayesian network specification, are fixed from the outset. However, the causal graph does not provide obvious constraints — it is of qualitative form, free from notions like entropy or probability. Therefore we need some procedure for turning the causal information into a constraint on probability.

---

[33] See [Paris 1994], [Paris & Vencovská 1997], [Paris 1999] and [Paris & Vencovská 2001] for the details of these justifications.

[34] [Grünwald 2000].

I suggest that the causal interpretation imposes the following constraint. Suppose we are presented with the components of a Bayesian network involving variables $C_1, \ldots, C_N$ and then use these to determine a single rational belief function p_1, whether by independence, maximum entropy or some other means. Then we find out further causal information, namely that there are some new variables $D_1, \ldots, D_M$ to be added to the causal graph, and that these variables are not causes of the current C-variables $C_1, \ldots, C_N$. Intuitively, this new information should not affect our understanding of the original problem on the C-variables. More precisely, suppose the new information takes the form of an extension of the original causal graph where the D-variables do not cause C-variables, and an extension to the probability specification incorporating new conditional probabilities of the D-variables given their parents. If we use this new Bayesian network to determine a new rational belief function p_2 over the larger domain $C_1, \ldots, C_N, D_1, \ldots, D_M$, then the restriction of p_2 to the C-variables should agree with p_1, the function based just on the C-variables. I shall call this the principle of *causal irrelevance*: learning of the new variables should be irrelevant to degrees of belief on the previous domain.

This principle is based on an asymmetry of causation whereby information about causes can lead to information about their effects, but knowledge of effects does not provide useful information about causes. This is not to say that information about the *value* or *occurrence* of an effect is irrelevant to the question of what the value of its cause is (which is clearly wrong), but that information of the form that a variable *has an effect of unknown value* is irrelevant to its own value. The same need not be true of causes: if two variables thought to be causally unrelated are found to have a common cause, one may be wise to suppose that these variables are probabilistically dependent to a greater extent than previously thought.

Take a simple example: suppose L signifies lung cancer and B bronchitis. We know of no causal relations linking the two variables, and have the probabilities $p(l), p(b)$ for each literal l, b involving L, B respectively. We then use this information to determine a joint probability distribution p_1 over L and B. Suppose we later learn that S, smoking, is a cause of lung cancer and of bronchitis, and we find the probabilities $p(l|s), p(b|s), p(s)$ for each literal l, b, s involving L, B, S respectively. Then, because S is a common cause, we might be inclined to form a new belief function p_2 over L, B and S which renders L and B more dependent than they were under p_1: $p_2(l|b) > p_1(l|b)$ for some literals l and b. The motivation is that if we find out b, then we now know this may be because some literal s has caused b, in which case s may also have caused l, making it more likely than we would previously have thought.

Suppose next we learn that each of lung cancer and bronchitis cause chest pains C, as in Figure 11. If we find values for $p(c|l \wedge b)$ for each literal c, l and b, and form a new belief function p_3, the causal irrelevance condition requires that p_3 must not differ from p_2, over S, L and B. For example, $p_3(l|b) = p_2(l|b)$, for each l and b. The idea here is that if we learn b, then knowledge of the existence of the common effect C does not give us a new way l may occur and so our degree of

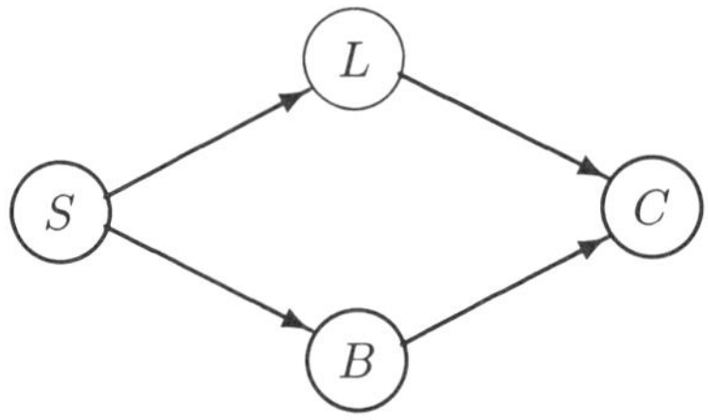

Figure 11. Smoking, lung cancer, bronchitis and chest pains.

belief in l should not change. C is irrelevant to S, L and B.

In sum, I shall assume that the process of determining a single rational belief function is constrained not only by the probability values in the specification of the Bayesian network, but also by the causal graph under the principle of causal irrelevance. The principle of causal irrelevance is strong enough to allow causal information to constrain rational belief, and thereby play a part in our new justification of the independence assumption, yet, unlike the independence assumption, weak enough to be uncontroversial in itself.

5.2 *Maximising Entropy*

The key proposition is this:

BAYESIAN NETWORKS MAXIMISE ENTROPY Given the probability specification and causal graph of a Bayesian network and the principle of causal irrelevance, the distribution which maximises entropy is just the distribution determined by the Bayesian network under the independence assumption.

Proof. The strategy of the proof will be to use Lagrange multipliers to derive conditions for entropy to be maximised, and then show that the Bayesian network distribution satisfies these conditions. This straightforward method is possible for the following reason. The constraints — which consist of the specified probabilities, certain probabilities fixed by the causal graph under causal irrelevance, and additivity constraints common to all probability distributions — are linear and restrict the domain of the entropy function to a compact convex set in $[0,1]^{K_1} \times \ldots \times [0,1]^{K_N}$,[35] and on that domain, entropy is a strictly concave function (as shown below). Thus the problem has a unique local maximum, the global maximum, and if the Bayesian network distribution satisfies the conditions for an optimal solution then it must be the unique global maximum.

We can see that entropy is strictly concave as follows. H is strictly concave if and only if, for any two distinct vectors a and b of the parameters $x^{k_1,\ldots,k_N}$ and

[35] See [Paris 1994], proposition 6.1, page 66.

$\lambda \in (0,1)$,

$$H(\lambda a + (1-\lambda)b) > \lambda H(a) + (1-\lambda)H(b) \Leftrightarrow$$

$$\lambda \sum a_i \log a_i + (1-\lambda) \sum b_i \log b_i - \sum(\lambda a_i + (1-\lambda)b_i)\log(\lambda a_i + (1-\lambda)b_i) > 0$$

$$\Leftrightarrow \lambda \sum a_i \log \frac{a_i}{\lambda a_i + (1-\lambda)b_i} + (1-\lambda)\sum b_i \log \frac{b_i}{\lambda a_i + (1-\lambda)b_i} > 0$$

$$\Leftrightarrow \lambda d(a, \lambda a + (1-\lambda)b) + (1-\lambda)d(b, \lambda a + (1-\lambda)b) > 0,$$

where d signifies *cross entropy*, a measure of distance of probability distributions, and a, b and $\lambda a + (1-\lambda)b$ are non-zero since $\sum a_i = 1 - \sum b_i, \lambda \in (0,1)$. d is well known to be non-negative and strictly positive if its arguments are distinct.[36] Thus $d(a, \lambda a + (1-\lambda)b)$ is strictly positive if $a \neq \lambda a + (1-\lambda)b$, which is true since a and b are distinct and $\lambda \in (0,1)$. Therefore H is strictly concave and the Lagrange multiplier approach will yield the global maximum.

The next thing to do is to reformulate the optimisation problem to make it suit the Bayesian network framework. This means finding more appropriate parameters than the standard $x^{k_1,\ldots,k_N}$ mentioned above. Without loss of generality we can suppose the nodes $C_1, \ldots, C_N$ are ordered ancestrally with respect to the causal graph G in the Bayesian network: that is, all the parents of C_i in G come before C_i in the ordering.[37] To make the proof clearer we shall also suppose that all the probabilities in the specification are positive — we shall see later that zeros do not affect the result. Let $c_i^{k_i}$ represent the literal $C_i = v_i^{k_i}$, for $k_i = 1, \ldots, K_i, i = 1, \ldots, N$. The new parameters are

$$y_{i,k_i}^{k_1,\ldots,k_{i-1}} = p(c_i^{k_i} | c_1^{k_1} \wedge \ldots \wedge c_{i-1}^{k_{i-1}}),$$

for $i = 1, \ldots, N$. The main thing to note about this parameterisation is that by the chain rule of probability,

$$x^{k_1,\ldots,k_N} = \prod_{i=1}^{N} y_{i,k_i}^{k_1,\ldots,k_{i-1}}.$$

Now we shall translate the entropy formula into this framework (in what follows we shall minimise negative entropy $-H$, which is equivalent to maximising entropy H):[38]

$$-H = \sum_{k_1,\ldots,k_N} x^{k_1,\ldots,k_N} \log x^{k_1,\ldots,k_N}$$

$$= \sum_{k_1,\ldots,k_N} \left[\prod_{j=1}^{N} y_{j,k_j}^{k_1,\ldots,k_{j-1}}\right] \sum_{i=1}^{N} \log y_{i,k_i}^{k_1,\ldots,k_{i-1}}$$

[36]See [Paris 1994] proposition 8.5 for example.

[37]Recall that such an ordering is always possible because of the dag structure of the causal graph.

[38]Note that the existence and uniqueness of a maximum is independent of parameterisation.

$$= \sum_{i=1}^{N} \sum_{k_1,\ldots,k_N} \left[\prod_{j=1}^{N} y_{j,k_j}^{k_1,\ldots,k_{j-1}} \right] \log y_{i,k_i}^{k_1,\ldots,k_{i-1}}$$

$$= \sum_{i=1}^{N} \sum_{k_1,\ldots,k_i} \left[\prod_{j=1}^{i} y_{j,k_j}^{k_1,\ldots,k_{j-1}} \right] \log y_{i,k_i}^{k_1,\ldots,k_{i-1}},$$

where we make this last step because for each i we can separate out

$$\sum_{k_{i+1},\ldots,k_N} \left[\prod_{j=i+1}^{N} y_{j,k_j}^{k_1,\ldots,k_{j-1}} \right],$$

and these terms cancel to 1 by additivity of probability.

We shall deal with three types of constraints. The specification constraints are determined by those values provided in the Bayesian network specification. Causal constraints are determined by the causal graph under the causal irrelevance condition. Finally additivity constraints are imposed by the axioms of probability. While one might suspect that all these constraints would lead to a complicated optimisation problem, we will see that by adopting an inductive approach we will be able to form a Lagrangian function which only incorporates relatively few specification and additivity constraints.

Within the new framework we can write the specification constraints as

$$p(c_i^{k_i} | c_{r_1}^{k_{r_1}} \wedge \ldots \wedge c_{r_L}^{k_{r_L}}) = a_{i,k_i}^{k_{r_1},\ldots,k_{r_L}},$$

where the $c_{r_1}, \ldots, c_{r_L}$ involve the parents of C_i, $r_1, \ldots, r_L < i$ (thanks to the ancestral order) and $i = 1, \ldots, N$.[39] We also have constraints imposed by additivity: $\sum_{k_i} y_{i,k_i}^{k_1,\ldots,k_{i-1}} = 1$ for each $k_1, \ldots, k_{i-1}, i = 1, \ldots, N$.

Decomposing the entropy as $H = \sum_{i=1}^{N} H_i$ where

$$H_i = \sum_{k_1,\ldots,k_i} \left[\prod_{j=1}^{i} y_{j,k_j}^{k_1,\ldots,k_{j-1}} \right] \log y_{i,k_i}^{k_1,\ldots,k_{i-1}},$$

we shall prove the proposition by induction on N. The case $N = 1$ is trivial since the constraints $p(c_1^{k_1}) = a_{1,k_1}$ completely determine the probability distribution over C_1: there is nothing to do to maximise entropy and so the Bayesian network distribution, which satisfies the constraints, maximises entropy. Suppose the induction hypothesis holds for $N-1$ and consider the case for N. It is here that we apply the principle of causal irrelevance to generate the causal constraints on the maximisation process from the causal graph. Since the variables are ordered ancestrally, the move from $N-1$ to N essentially involves incorporating a new variable

[39] Note that the $r_1, \ldots, r_L$ depend on i. I am inclined to avoid any further subscripting however.

C_N which is not a cause of any of the previous variables $C_1, \ldots, C_{N-1}$. Hence if we maximise entropy on this new domain and restrict the resulting probability function to $C_1, \ldots, C_{N-1}$ then by causal irrelevance we must have maximised entropy on this smaller domain. Applying the induction hypothesis on this smaller domain $\{C_1, \ldots, C_{N-1}\}$, we see that entropy is maximised if the distribution is determined by the Bayesian network on $C_1, \ldots, C_{N-1}$. Thus for $i = 1, \ldots, N-1$, the parameters $y_{i,k_i}^{k_1,\ldots,k_{i-1}}$ must be fixed to $a_{i,k_i}^{k_{r_1},\ldots,k_{r_L}}$. Now $H_1, \ldots, H_{N-1}$ involve only these fixed parameters, so in order to maximise H all that remains is to maximise H_N with respect to $y_{N,k_N}^{k_1,\ldots,k_{N-1}}$, subject to the specification constraints fixing the values $a_{N,k_N}^{k_{r_1},\ldots,k_{r_L}}$ and the additivity constraints $\sum_{k_N} y_{N,k_N}^{k_1,\ldots,k_{N-1}} = 1$ for each $k_1, \ldots, k_{N-1}$.

We shall now adapt the specification constraints.

Let $b^{k_{r_1},\ldots,k_{r_L}} = p(c_{r_1}^{k_{r_1}} \wedge \ldots \wedge c_{r_L}^{k_{r_L}})$ and $e^{k_1,\ldots,k_{N-1}} = \prod_{j<N} y_{j,k_j}^{k_1,\ldots,k_{j-1}}$ be constants, fixed by having maximised entropy on $C_1, \ldots, C_{N-1}$. Then

$$
\begin{aligned}
a_{N,k_N}^{k_{r_1},\ldots,k_{r_L}} b^{k_{r_1},\ldots,k_{r_L}} &= p(c_N^{k_N} \wedge c_{r_1}^{k_{r_1}} \wedge \ldots \wedge c_{r_L}^{k_{r_L}}) \\
&= \sum_{k_i, i \neq r_1,\ldots,r_L,N} p(c_1^{k_1} \wedge \ldots \wedge c_N^{k_N}) \\
&= \sum_{k_i, i \neq r_1,\ldots,r_L,N} \prod_{j \leq N} y_{j,k_j}^{k_1,\ldots,k_{j-1}}, \\
&= \sum_{k_i, i \neq r_1,\ldots,r_L,N} e^{k_1,\ldots,k_{N-1}} y_{N,k_N}^{k_1,\ldots,k_{N-1}}.
\end{aligned}
$$

We are now in a position to specify the Lagrangian function for the minimisation of $-H_N$:

$$
\begin{aligned}
\Lambda_N = \sum_{k_1,\ldots,k_N} & e^{k_1,\ldots,k_{N-1}} y_{N,k_N}^{k_1,\ldots,k_{N-1}} \log y_{N,k_N}^{k_1,\ldots,k_{N-1}} + \sum_{k_{r_1},\ldots,k_{r_L},k_N} \lambda_{k_N}^{k_{r_1},\ldots,k_{r_L}} \times \\
& \left[\sum_{k_i, i \neq r_1,\ldots,r_L,N} e^{k_1,\ldots,k_{N-1}} y_{N,k_N}^{k_1,\ldots,k_{N-1}} - a_{N,k_N}^{k_{r_1},\ldots,k_{r_L}} b^{k_{r_1},\ldots,k_{r_L}} \right] \\
& + \sum_{k_1,\ldots,k_{N-1}} \mu^{k_1,\ldots,k_{N-1}} \left[\sum_{k_N} y_{N,k_N}^{k_1,\ldots,k_{N-1}} - 1 \right] \\
= \sum_{k_1,\ldots,k_N} & \Big(e^{k_1,\ldots,k_{N-1}} y_{N,k_N}^{k_1,\ldots,k_{N-1}} \log y_{N,k_N}^{k_1,\ldots,k_{N-1}} + \\
& \lambda_{k_N}^{k_{r_1},\ldots,k_{r_L}} \left[e^{k_1,\ldots,k_{N-1}} y_{N,k_N}^{k_1,\ldots,k_{N-1}} - a_{N,k_N}^{k_{r_1},\ldots,k_{r_L}} b^{k_{r_1},\ldots,k_{r_L}} \right] +
\end{aligned}
$$

$$\mu^{k_1,\ldots,k_{N-1}}\left[y_{N,k_N}^{k_1,\ldots,k_{N-1}} - 1/K_N\right]\Big).$$

By Lagrange's theorem,[40] in order to find conditions for a minimum we must first check a constraint qualification. Enumerate the constraints $f_1, \ldots, f_J$. Form a matrix A by letting each row i consist of the partial derivatives

$$\frac{\partial f_i}{\partial y_{N,k_N}^{k_1,\ldots,k_{N-1}}}, 1 \le k_j \le K_j, j = 1, \ldots, N.$$

Finally check that the rank of A is J — this is easily done and I shall avoid the details here.

Entropy is maximised if the partial derivatives of the Lagrangian are zero,

$$\frac{\partial \Lambda_N}{\partial y_{N,k_N}^{k_1,\ldots,k_{N-1}}} = e^{k_1,\ldots,k_{N-1}}\left[1 + \log y_{N,k_N}^{k_1,\ldots,k_{N-1}} + \lambda_{k_N}^{k_{r_1},\ldots,k_{r_L}}\right] + \mu^{k_1,\ldots,k_{N-1}} = 0$$

Given any such equation we can eliminate the Lagrange multiplier $\mu^{k_1,\ldots,k_{N-1}}$ by finding another equation involving $k'_N \neq k_N$,

$$\frac{\partial \Lambda_N}{\partial y_{N,k'_N}^{k_1,\ldots,k_{N-1}}} = 0$$

(there will always be another such equation since C_N has at least two values), and substituting to give a new equation

$$\lambda_{k_N}^{k_{r_1},\ldots,k_{r_L}} - \lambda_{k'_N}^{k_{r_1},\ldots,k_{r_L}} = \log y_{N,k'_N}^{k_1,\ldots,k_{N-1}} - \log y_{N,k_N}^{k_1,\ldots,k_{N-1}}$$

We next eliminate the multiplier expression on the left-hand side by finding another such equation involving $k'_1, \ldots, k'_{N-1}$ such that $k'_{r_1} = k_{r_1}, \ldots, k'_{r_L} = k_{r_L}$. There will always be another such equation unless $L = N - 1$, in which case the constraints uniquely determine the Bayesian network distribution, and entropy is trivially maximised. This then gives

$$\log y_{N,k'_N}^{k_1,\ldots,k_{N-1}} - \log y_{N,k_N}^{k_1,\ldots,k_{N-1}} = \log y_{N,k'_N}^{k'_1,\ldots,k'_{N-1}} - \log y_{N,k_N}^{k'_1,\ldots,k'_{N-1}}.$$

Finally, all we need do is note that in the Bayesian network distribution the constraints are satisfied and the independence assumption implies that

$$y_{N,k_N}^{k_1,\ldots,k_{N-1}} = y_{N,k_N}^{k_{r_1},\ldots,k_{r_L}} = a_{N,k_N}^{k_{r_1},\ldots,k_{r_L}},$$

$$y_{N,k_N}^{k'_1,\ldots,k'_{N-1}} = y_{N,k_N}^{k'_{r_1},\ldots,k'_{r_L}} = y_{N,k_N}^{k_{r_1},\ldots,k_{r_L}} = a_{N,k_N}^{k_{r_1},\ldots,k_{r_L}},$$

[40] See for example [Sundaram 1996] §5.2.1

in which case we substitute into our condition:

$$a_{N,k'_N}^{k_{r_1},\ldots,k_{r_L}} - a_{N,k_N}^{k_{r_1},\ldots,k_{r_L}} = a_{N,k'_N}^{k_{r_1},\ldots,k_{r_L}} - a_{N,k_N}^{k_{r_1},\ldots,k_{r_L}},$$

and find that it clearly holds. Thus the Bayesian network distribution is the entropy maximiser, as required.

All that remains is to point out what happens when specifiers may be zero. There are two (compatible) scenarios: if some $a_{j,k_j}^{k_{r_1},\ldots,k_{r_L}} = 0$ for $j < N$ then the corresponding $e^{k_1,\ldots,k_{N-1}} = \prod_{j<N} y_{j,k_j}^{k_1,\ldots,k_{j-1}}$, which by the induction hypothesis is $\prod_{j<N} a_{j,k_j}^{k_{r_1},\ldots,k_{r_L}}$, vanishes. This eliminates entropy terms and constraints equally, leaving fewer partial derivative conditions. These conditions are satisfied as above. The second scenario is that some $a_{N,k_N}^{k_{r_1},\ldots,k_{r_L}} = 0$. In this case the Lagrangian and partial derivatives are as before, the constraints are satisfied as before, but when substituting zeros in the partial derivatives we make use of the convention, common when dealing with the cross entropy measure, that $0[\log 0 - \log 0] = 0 \log 0/0 = 0$. Thus the conditions are satisfied by null specifiers. ■

Thus we see that the independence assumption can be justified after all. The important thing to remember is that under the two-stage foundations, the independence assumption is neither a fact of causality nor even an assertion about an agent's knowledge. It is a mechanism that can be used to derive new probability statements from those in the agent's background knowledge. Independence is justified because as a logic it coincides with maximum entropy, which has well known justifications.

6 STAGE TWO

Given background knowledge consisting of a causal graph G and associated probability specification S, we can represent the rational (maximum entropy) belief function p by the Bayesian network on G and S. This is stage one of the two-stage methodology. However, while p is rational given background knowledge, it may not bear a close enough resemblance to objective probability to be put to practical use. If that is the case then we need to transform the Bayesian network into one which more closely approximates objective probability. This is stage two of the two-stage methodology. Bayesian networks may be applied to medical diagnosis for example, or fault-finding in aeroplanes. In such high risk scenarios it is not sufficient that any decisions are deemed reasonable given a lack of relevant information: it would be negligent not to collect enough relevant information to reliably model the objective situation.

Thus the next step is to refine the Bayesian network in the light of new information, in order to achieve greater reliability. Many of the algorithms from the extensive literature on learning Bayesian networks from data[41] can be applied here. In the rest of this section I will summarise my own ideas in this respect — these are simple techniques which I believe have a clear justification that coheres well with the entropy-based approach of the last section.[42] First I shall deal with the case where new causal information comes to be known. After this I shall address the following questions. What sort of information should one collect in order to best refine the network? How one can limit the complexity of the network?

6.1 Causal Information

Suppose our agent X finds out that C_i causes C_j. I suggest that she should just add an arrow from C_i to C_j to her initial causal graph (if there is no arrow there already), and she should ensure her specifying probabilities $p(c_j|d_j)$ take this new parent into account. There are two possible justifications of this *adding-arrows* strategy. One can apply the arguments of the last section. If X learns of the new causal link and the corresponding probabilities then her background knowledge now includes an extended causal graph and probability specification, in which case she should maximise entropy by adopting the new Bayesian network formed by adding the arrow and the specifiers.

The second possible justification relies on the dependence principle[43] as opposed to causal irrelevance, as follows. Suppose we start off with Bayesian network (G, S_G), where G is X's causal graph and S_G is her associated probability specification, whose entries we shall assume agree with the objective probabilities $p^*(c_i|d_i)$. Then we add an arrow from C_i to C_j and change the specified probabilities to give a new network (H, S_H). We measure the improvement of the new network over the old by how much closer its induced probability function p_H is to the objective probability function p^* than p_G, according to the usual measure of distance between probability functions, cross entropy. Then we have the following facts:

IMPROVEMENT OF ADDING ARROWS
(i) the new network is no worse a network than the initial network;
(ii) the new network is a better network if and only if C_j is probabilistically dependent on C_i, conditional on C_j's other parents D.

In particular, if the dependence principle holds then the fact that C_i is a cause of C_j entails that the two nodes are conditionally probabilistically dependent and thus that the probability distribution represented by the new network is closer to

[41] See [Jordan 1998] and [Buntine 1996] for good surveys.

[42] Some related work: the Kutató algorithm of [Herskovitz 1991] also has an entropy-based justification. However it involves minimising entropy and poses significant computational problems in the worst case. [Jitnah 1999] employs mutual information as I do, but as a technique for probabilistic inference given a Bayesian network rather than a means for deriving the network itself.

[43] Recall that the dependence principle says that a direct cause changes the probability of its effect conditional on the effect's other causes.

the target objective distribution than that of the old network: we are justified in adding an arrow from C_i to C_j.

Proof. For simplicity (but without loss of generality as we shall see shortly) we shall assume that p_G and p_H are strictly positive over the *atomic states* $c_1 \wedge \ldots \wedge c_N$.

For (i) we need to show that $d(p^*, p_H) - d(p^*, p_G) \leq 0$, where d is cross entropy distance. So,

$$d(p^*, p_H) - d(p^*, p_G) = \sum_s p^*(s) \ln \frac{p^*(s)}{p_H(s)} - \sum_s p^*(s) \ln \frac{p^*(s)}{p_G(s)}$$

$$= \sum_s p^*(s) \ln \frac{p_G(s)}{p_H(s)},$$

where the s are the atomic states, and bearing in mind that $p_H(s) > 0$. Now for real $x > 0, \ln(x) \leq x - 1$. By assumption $p_G(s)/p_H(s) > 0$, so

$$\sum_s p^*(s) \ln \frac{p_G(s)}{p_H(s)} \leq \sum_s p^*(s) \left[\frac{p_G(s)}{p_H(s)} - 1\right]$$

$$= \sum_s p^*(s) \frac{p_G(s)}{p_H(s)} - 1,$$

and thus we need to show that

$$\sum_s p^*(s) \frac{p_G(s)}{p_H(s)} \leq 1.$$

Now since we are dealing with Bayesian networks,

$$\frac{p_G(s)}{p_H(s)} = \frac{\prod p^*(c_k | d_k^G)}{\prod p^*(c_k | d_k^H)},$$

for each literal c_k consistent with s, where d_k^G is the state of the parents of C according to G which is consistent with s, and likewise for d_k^G. But H is just G but with an arrow from C_i to C_j, so the terms in each product are the same and cancel, except when it comes to literals c_j involving node C_j. Thus

$$\frac{p_G(s)}{p_H(s)} = \frac{p^*(c_j | d_j^G)}{p^*(c_j | d_j^H)} = \frac{p^*(c_j | d)}{p^*(c_j | c_i \wedge d)},$$

where we just let d be d_j^G and c_i the remaining literal in d_j^H. Substituting and simplifying,

$$\sum_s p^*(s) \frac{p_G(s)}{p_H(s)} = \sum p^*(c_i \wedge c_j \wedge d) \frac{p^*(c_j | d)}{p^*(c_j | c_i \wedge d)}$$

$$= \sum p^*(c_j|d)p^*(d|c_i)p^*(c_i).$$

Consider the new set of variables $\{C_i, C_j, D\}$ where C_i and C_j are as before and D takes as values the states of the parents of C_j according to G. Form a Bayesian network T incorporating the graph $C_i \longrightarrow D \longrightarrow C_j$ (with specifying probabilities determined as usual from the probability function p^*). Then since T is a Bayesian network, $\sum p^*(c_j|d)p^*(d|c_i)p^*(c_i) = \sum p_T(c_i \wedge c_j \wedge d) = 1$ by the additivity of probability. Thus $\sum_s p^*(s)p_G(s)/p_H(s) = 1$ so $d(p^*, p_H) - d(p^*, p_G) \leq 0$, as required.

Let us now turn to (ii). From the above reasoning we see that

$$d(p^*, p_H) - d(p^*, p_G) < 0 \Leftrightarrow \ln \frac{p_G(s)}{p_H(s)} < \frac{p_G(s)}{p_H(s)} - 1$$

for some atomic state s. But $\ln x < x - 1 \Leftrightarrow x \neq 1$, and

$$\frac{p_G(s)}{p_H(s)} \neq 1 \Leftrightarrow \frac{p^*(c_j|c_i)}{p^*(c_j|c_i \wedge d)} \neq 1 \Leftrightarrow p^*(c_j|c_i \wedge d) - p^*(c_j|d) \neq 0,$$

where the c_i, c_j, d are consistent with s. Therefore, $d(p^*, p_H) - d(p^*, p_G) < 0$ if and only if there is some c_i, c_j, d for which the conditional dependence holds.

The assumption that p_G and p_H are positive over atomic states is not essential. Suppose p_H is zero over some atomic states. Then in the above,

$$\sum_s p^*(s) \ln \frac{p_G(s)}{p_H(s)} =$$

$$\sum_{s:p_H(s)>0} p^*(s) \ln \frac{p_G(s)}{p_H(s)} + \sum_{s:p_H(s)=0} p^*(s) \ln \frac{p_G(s)}{p_H(s)}.$$

The first sum on the right hand side is ≤ 0 as above. The second sum is zero because each component is, as we shall see now. Suppose $p_H(s) = 0$. Then $\prod_{k=1}^N p^*(c_k|d_k^H) = 0$ so $p^*(c_k \wedge d_k^H) = 0$ for at least one such k, in which case $p(s) = 0$ since by the axioms of probability, $p(u) = 0 \Rightarrow p(u \wedge v) = 0$. Now in the sum read $p^*(s) \ln p_G(s)/p_H(s)$ to be $p^*(s) \ln p_G(s) - p^*(s) \ln p_H(s)$. In dealing with cross entropy by convention $0 \ln 0$ is taken to be 0. Therefore $p^*(s) \ln p_G(s)/p_H(s) = 0 \ln p_G(s) - 0 = 0$. The same reasoning applies if p_G is zero over some atomic states. Likewise, if $p^*(s)$ is zero then $p^*(s) \ln p_G(s)/p_H(s)$ is zero too. ∎

This justifies the adding-arrows approach if X learns of a new causal link amongst the current variables. If she learns of a new variable C_{N+1} that is causally related to one or more of the other variables, and she also learns the probabilities $p(c_{N+1}|d_{N+1})$, then we can apply the above argument (or equally the arguments of §5) to show that X's new network should be constructed from her old network by adding the new node and causal arrows to her graph and the new probabilities to her specification.

Finally note that the above argument only requires that the added arrow links conditionally probabilistically dependent nodes. As we have discussed in §2, nodes need not be causally related to be probabilistically dependent. Therefore, if our agent is presented with information to the effect that two nodes are conditionally dependent, she is justified in adding the corresponding arrow to her network, regardless of whether those nodes are causally related. But as a result of this generalisation, the graph in the agent's Bayesian network need no longer be causally interpreted: the Bayesian network becomes an abstract tool for representing a probability function.

6.2 *Mutual Information*

We now have a strategy for changing the network when causal information or other probabilistic dependencies are presented to the agent. But is there a strategy for *seeking out* a good arrow to add? By adding arrows we increase both the size of the specification required in the Bayesian network (the *space complexity*) and the time taken to calculate probabilities from the network (the *time complexity*) — is there a means of limiting these complexities to prevent the network from becoming impractical? I shall address both these questions in this section.

The key to limiting complexity consists in finding constraints $\mathcal{C}$ such that Bayesian networks satisfying $\mathcal{C}$ have acceptable complexity, and then ensuring that (i) the current network satisfies $\mathcal{C}$, and (ii) an arrow is only added to the current network if the resulting network continues to satisfy $\mathcal{C}$. Consider by way of example the following constraints.

- $\mathcal{C}_1$: no node has more than K parents, for some constant K. This bound on the number of parents serves to restrict the space complexity of a Bayesian network. For instance if $K = 0$ then the discrete network (no arrows) is the only available network, if $K = 1$ then all networks satisfying $\mathcal{C}_1$ have graphs that are forests, and if $K = N - 1$ there is no restriction at all on the networks. It is easy to see that if all variables are binary, the complexity of a network satisfying $\mathcal{C}_1$ is less than or equal to $(N - K + 1)2^K - 1$, a value that is linear in N.

- $\mathcal{C}_2$: the Bayesian network has space complexity of at most κ. Now if $\kappa = N$ the only network to satisfy $\mathcal{C}_2$ is the discrete network and if $\kappa = 2^N - 1$ any network satisfies the constraint. Depending on the problem in hand and available resources we will want to choose an appropriate value for κ or K which balances the range of networks available with their complexity.

- $\mathcal{C}_3$: the graph is singly-connected. Having a singly connected graph ensures that the Bayesian network can be used to calculate required probabilities efficiently (in time linear in the number of nodes N). Note however that a singly-connected network can have space complexity up to $2^{N-1}+N-1$ on binary-valued nodes, so in practice this constraint may best be used with another which limits space complexity.

In sum, if we fix some constraints $\mathcal{C}$ the goal then is to find a *constrained network* (a Bayesian network satisfying $\mathcal{C}$) which gives a good approximation to the target objective distribution p^* (using cross entropy as a measure of degree of approximation).

We shall associate a *weight* with each arrow in a Bayesian network as follows. In order to weigh the arrows going into a node C_i we enumerate the parents of C_i as $D^1, \ldots, D^k$. Then for $j = 1, \ldots, k$ we weigh the arrow from D^j to C_i by the *conditional mutual information*,

$$I(C_i, D^j|\{D^1, \ldots, D^{j-1}\}) = \sum_{c_i,d,d^j} p^*(c_i \wedge d \wedge d^j) \log \frac{p^*(c_i \wedge d^j|d)}{p^*(c_i|d)p^*(d^j|d)},$$

where d ranges over the states $d^1 \wedge \ldots \wedge d^{j-1}$. Then:

> MAX-WEIGHT APPROXIMATION
> The network subject to constraints $\mathcal{C}$ which affords the closest approximation to p^* (according to the cross entropy measure of distance) is the network satisfying $\mathcal{C}$ whose arrow weights are maximised.

Proof. The distance between the probability function p determined by X's Bayesian network and the target function p^* is

$$d(p^*, p) = \sum_s p^*(s) \log \frac{p^*(s)}{p(s)}$$

$$= \sum_s p^*(s) \log p^*(s) - \sum_s p^*(s) \log \prod_{i=1}^N p^*(c_i|d_i)$$

where the c_i and d_i are consistent with s,

$$= \sum_s p^*(s) \log p^*(s) - \sum_s p^*(s) \sum_{i=1}^N \log p^*(c_i|d_i)$$

$$= \sum_s p^*(s) \log p^*(s) - \sum_s p^*(s) \sum_{i=1}^N \log \frac{p^*(c_i \wedge d_i)}{p^*(c_i)p^*(d_i)} - \sum_s p^*(s) \sum_{i=1}^N \log p^*(c_i)$$

$$= -H(p^*) - \sum_{i=1}^N I(C_i, D_i) + \sum_{i=1}^N H(p^*|_{C_i})$$

where $H(p^*)$ is the entropy of function p^*, $I(C_i, D_i)$ is the mutual information between C_i and its parents and $H(p^*|_{C_i})$ is the entropy of p^* restricted to node

C_i. The entropies are independent of the choice of Bayesian network so the distance between the network and target distributions is minimised just when the total mutual information is maximised.[44]

Note that

$$I(A,B)+I(A,C|B)$$
$$=\sum_{a,b,c} p^*(a\wedge b\wedge c)\left[\log\frac{p^*(a\wedge b)}{p^*(a)p^*(b)}+\log\frac{p^*(a\wedge c|b)}{p^*(a|b)p^*(c|b)}\right]$$
$$=\sum_{a,b,c} p^*(a\wedge b\wedge c)\log\frac{p^*(a\wedge b)p^*(a\wedge b\wedge c)p^*(b)p^*(b)}{p^*(a)p^*(b)p^*(b)p^*(a\wedge b)p^*(c\wedge b)}$$
$$=\sum_{a,b,c} p^*(a\wedge b\wedge c)\log\frac{p^*(a\wedge b\wedge c)}{p^*(a)p^*(c\wedge b)}=I(A,\{B,C\}).$$

By enumerating the parents D_i of C_i as $D^1,\ldots,D^k$, we can iterate the above relation to get

$$I(C_i,D_i)=I(C_i,D^1)+I(C_i,D^2|D^1)+$$
$$I(C_i,D^3|\{D^1,D^2\})+\ldots+I(C_i,D^k|\{D^1,\ldots,D^{k-1}\}).$$

Therefore,

$$\sum_{i=1}^N I(C_i,D_i)=\sum_{i=1}^N\sum_j I(C_i,D^j|\{D^1,\ldots,D^{j-1}\}),$$

and the cross entropy distance between the network distribution and the target distribution is minimised just when the sum of the arrow weights is maximised. ■

Note that this result is independent of choice of enumeration of the variables, as can be seen from the proof.

There are various ways one might try to find a constrained network with maximum or close to maximum weight, but perhaps the simplest is a greedy adding-arrows strategy: start off with the discrete graph and at each stage find and weigh the arrows whose addition would ensure that the dag structure and constraints $\mathcal{C}$ remain satisfied, and add one with maximum weight. If more than one best arrow exists we can spawn several new graphs by adding each best arrow to the previous graph, and we can constantly prune the number of graphs by eliminating those which no longer have maximum weight. We stop the algorithm when no more arrows can be added.[45]

Given this algorithm and its justification, we now have answers to our two questions of this section. We seek out a good arrow to add by finding the arrow with

[44] This much is a straightforward generalisation of the proof of [Chow & Liu 1968] that the best tree-based approximation is the maximum weight spanning tree.

[45] See [Williamson 2000b] and [Williamson 2000] for analyses of the performance of this algorithm, which turns out to be remarkably effective for a greedy approach.

maximum conditional mutual information weight. We limit the complexity of the network by imposing constraints on the network.

Thus in stage two of the two-stage methodology we can improve the causal network obtained in stage one by adding arrows — these arrows link causally related variables or more generally probabilistically dependent variables, and a good strategy is to add the weightiest arrow which does not violate constraints on the complexity of the network. The conditional mutual information weighting is a measure of conditional dependence and so in effect the strategy is to add an arrow between two nodes that are most (conditionally) dependent. The resulting graph will not necessarily reflect the true causal relations amongst the variables, and so stage two corresponds more closely to the abstract foundations for Bayesian networks than any causal interpretation.

7 CONCLUSION

While the independence assumption poses significant problems for a straightforward objective or subjective interpretation of Bayesian networks, independence can be though of as a means of determining a rational belief function from an agent's background knowledge. Thus Bayesian networks can be given firm foundations by adopting a two-stage approach, whereby one first adopts a subjective causal interpretation which may then be dropped as the network is refined in order to better approximate a target objective probability function. These foundations appeal to information-theoretic notions and assumptions about causality which are somewhat less contentious than the independence assumption. Stage one is justified by maximum entropy considerations while an adding-arrows strategy for stage two can be justified by minimising cross entropy relative to the objective distribution. This approach is not subject to many of the problems that beset the objective or subjective interpretations considered in §2 and §3: we do not need to worry about individuation of variables, and stage two can be used to compensate for the presence of accidental and extra-causal dependencies and any discrepancies between the subjective network and an objective causal network. The advantage over the abstract approach is that we don't require a database of past case data to determine a network — stage one makes use of causal and probabilistic background knowledge. The two-stage methodology can be viewed as a way of integrating background knowledge (including qualitative causal knowledge) with machine learning techniques (of which the adding-arrows strategy is one example).[46]

Department of Philosophy, King's College, London.

[46]Thanks to David Corfield, Donald Gillies and Jeff Paris for helpful comments, and the UK Arts and Humanities Research Board for funding this research.

BIBLIOGRAPHY

[Andersson et al. 1996] S. Andersson, D. Madigan & M. Perlman: 'An alternative Markov property for chain graphs', Proceedings of the 12th Conference on Uncertainty in Artificial Intelligence, Portland OR: Morgan Kaufmann, pages 40-48.

[Arntzenius 1992] Frank Arntzenius: 'The common cause principle', Philosophy of Science Association 1992 (2), pages 227-237.

[Binder et al. 1997] John Binder, Daphne Koller, Stuart Russell & Keiji Kanazawa: 'Adaptive probabilistic networks with hidden variables', Machine Learning 29, pages 213-244.

[Buntine 1996] Wray Buntine: 'A guide to the literature on learning probabilistic networks from data', IEEE Transactions on Knowledge and Data Engineering 8(2), pages 195-210.

[Butterfield 1992] Jeremy Butterfield: 'Bell's theorem: what it takes', British Journal for the Philosophy of Science 43, pages 41-83.

[Cartwright 1989] Nancy Cartwright: 'Nature's capacities and their measurement', Oxford: Clarendon Press.

[Chow & Liu 1968] C.K. Chow & C.N. Liu: 'Approximating discrete probability distributions with dependence trees', IEEE Transactions on Information Theory IT-14, pages 462-467.

[Fisher 1935] Ronald Fisher: 'The design of experiments', Edinburgh: Oliver & Boyd.

[van Fraassen 1980] Bas C. van Fraassen: 'The scientific image', Clarendon Press, Oxford.

[Frydenberg 1990] M. Frydenberg: 'The chain graph Markov property', Scandanavian Journal of Statistics 17, pages 333-353.

[Glymour 1997] Clark Glymour: 'A review of recent work on the foundations of causal inference', [McKim & Turner 1997], pages 201-248.

[Glymour & Cooper 1999] Clark Glymour & Gregory F. Cooper(eds.): 'Computation, causation, and discovery', Cambridge, Massachusetts: The M.I.T. Press.

[Grünwald 2000] Peter Grünwald: 'Maximum entropy and the glasses you are looking through', Proceedings of the 16th conference of Uncertainty in Artificial Intelligence, Stanford University, Morgan Kaufmann, pages 238-246.

[Hausman 1998] Daniel M. Hausman: 'Causal asymmteries', Cambridge: Cambridge University Press.

[Hausman 1999] Daniel M. Hausman: 'The mathematical theory of causation', review of [McKim & Turner 1997], British Journal for the Philosophy of Science 50, pages 151-162.

[Healey 1991] Richard Healey: 'Review of Paul Horwich's "Asymmetries in time"', The Philosophical Review 100, pages 125-130.

[Heckerman et al. 1999] David Heckerman, Christopher Meek & Gregory Cooper: 'A Bayesian approach to causal discovery', in [Glymour & Cooper 1999], pages 141-165.

[Herskovitz 1991] Edward Herskovitz: 'Computer-based probabilistic-network construction', PhD Thesis, Stanford University.

[Humphreys 1997] Paul Humphreys: 'A critical appraisal of causal discovery algorithms', in [McKim & Turner 1997], pages 249-263.

[Humphreys & Freedman 1996] Paul Humphreys & David Freedman: 'The grand leap', British Journal for the Philosophy of Science 47, pages 113-123.

[Jitnah 1999] Nathalie Jitnah: 'Using mutual information for approximate evaluation of Bayesian networks', PhD Thesis, School of Computer Science and Software Engineering, Monash University.

[Jordan 1998] Michael I. Jordan(ed.): 'Learning in Graphical Models', Cambridge, Massachusetts: The M.I.T. Press 1999.

[Kwoh & Gillies 1996] Chee-Keong Kwoh & Duncan F. Gillies: 'Using hidden nodes in Bayesian networks', Artificial Intelligence 88, pages 1-38.

[Lad 1999] Frank Lad: 'Assessing the foundation for Bayesian networks: a challenge to the principles and the practice', Soft Computing 3(3), pages 174-180.

[Lemmer 1993] John F. Lemmer: 'Causal modeling', in Proceedings of the 9th Conference on Uncertainty in Artificial Intelligence, San Mateo: Morgan Kaufmann, pages 143-151.

[Lemmer 1996] John F. Lemmer: 'The causal Markov condition, fact or artifact?', SIGART 7(3).

[McKim & Turner 1997] Vaughn R. McKim & Stephen Turner: 'Causality in crisis? Statistical methods and the search for causal knowledge in the social sciences', University of Notre Dame Press.

[Mill 1843] John Stuart Mill: 'A system of logic, ratiocinative and inductive: being a connected view of the principles of evidence and the methods of scientific investigation', New York: Harper & Brothers, eighth edition, 1874.

[Neapolitan 1990] Richard E. Neapolitan: 'Probabilistic reasoning in expert systems: theory and algorithms', New York: Wiley.

[Oliver & Smith 1990] R.M. Oliver & J.Q. Smith: 'Influence diagrams, belief nets and decision analysis', Chichester: Wiley.

[Papineau 1992] David Papineau: 'Can we reduce causal direction to probabilities?', Philosophy of Science Association 1992 (2), pages 238-252.

[Paris 1994] Jeff Paris: 'The uncertain reasoner's companion', Cambridge: Cambridge University Press.

[Paris 1999] Jeff Paris: 'Common sense and maximum entropy', Synthese 117, pages 73-93.

[Paris & Vencovská 1997] Jeff Paris & Alena Vencovská: 'In defense of the maximum entropy inference process', International Journal of Automated Reasoning 17, pages 77-103.

[Paris & Vencovská 2001] J.B. Paris & A. Vencovská: 'Common sense and stochastic independence', this volume.

[Pearl 1988] Judea Pearl: 'Probabilistic reasoning in intelligent systems: networks of plausible inference', San Mateo, California: Morgan Kaufmann.

[Pearl 2000] Judea Pearl: 'Causality: models, reasoning, and inference', Cambridge University Press.

[Pearl & Dechter 1996] J. Pearl & R. Dechter: 'Identifying independencies in causal graphs with feedback', Proceedings of the 12th Conference of Uncertainty in Artificial Intelligence, Portland OR: Morgan Kaufmann.

[Pearl et al. 1990] Judea Pearl, Dan Geiger & Thomas Verma: 'The logic of influence diagrams', in [Oliver & Smith 1990], pages 67-87.

[Price 1992] Huw Price: 'The direction of causation: Ramsey's ultimate contingency', Philosophy of Science Association 1992 (2), pages 253-267.

[Reichenbach 1956] Hans Reichenbach: 'The direction of time', Berkeley & Los Angeles, University of California Press, reprinted 1971.

[Richardson 1996] T. Richardson: 'A discovery algorithm for directed cyclic graphs', Proceedings of the 12th Conference of Uncertainty in Artificial Intelligence, Portland OR: Morgan Kaufmann, pages 454-461.

[Robins & Wasserman 1999] James M. Robins & Larry Wasserman: 'On the impossibility of inferring causation from association without background knowledge', in [Glymour & Cooper 1999], pages 305-321.

[Rolnick 1974] William B. Rolnick: 'Causality and physical theories', New York: American Institute of Physics.

[Salmon 1980] Wesley C. Salmon: 'Probabilistic causality', in [Salmon 1998], pages 208-232.

[Salmon 1984] Wesley C. Salmon: 'Scientific explanation and the causal structure of the world', Princeton: Princeton University Press.

[Salmon 1998] Wesley C. Salmon: 'Cauality and explanation', Oxford: Oxford University Press.

[Savitt 1996] Steven F. Savitt: 'The direction of time', British Journal for the Philosophy of Science 47, pages 347-370.

[Scheines 1997] Richard Scheines: 'An introduction to causal inference', in [McKim & Turner 1997], pages 185-199.

[Schlegel 1974] Richard Schlegel: 'Historic views of causality', in [Rolnick 1974], pages 3-21.

[Smith 1990] J.Q. Smith: 'Statistical principles on graphs', in [Oliver & Smith 1990], pages 89-120.

[Sober 1988] Elliott Sober: 'The principle of the common cause', in James H. Fetzer (ed.): 'Probability and causality: essays in honour of Wesley C. Salmon', pages 211-228.

[Spirtes 1995] P. Spirtes: 'Directed cyclic graphical representation of feedback models', Proceedings of the 11th Conference on Uncertainty in Artificial Intelligence, Montreal QU: Morgan Kaufmann, pages 491-498.

[Spirtes et al. 1993] Peter Spirtes, Clark Glymour & and Richard Scheines: 'Causation, Prediction, and Search', Lecture Notes in Statistics 81, Springer-Verlag.

[Spirtes et al. 1997] Peter Spirtes, Clark Glymour & Richard Scheines: 'Reply to Humphreys and Freedman's review of 'Causation, prediction, and search", British Journal for the Philosophy of Science 48, pages 555-568.

[Sucar et al. 1993] L.E. Sucar, D.F. Gillies & D.A. Gillies: 'Objective probabilities in expert systems', Artificial Intelligence 61, pages 187-208.

[Sundaram 1996] Rangarajan K Sundaram: 'A first course in optimisation theory', Cambridge: Cambridge University Press.

[Verma & Pearl 1991] T. Verma & J. Pearl: 'Equivalence and synthesis of causal models', Los Angeles, Cognitive Systems Laboratory, University of California.

[Wermuth et al. 1994] N. Wermuth, D. Cox & J. Pearl: 'Explanations for multivariate structures derived from univariate recursive regressions', Center of Survey Research and Methodology, ZUMA, Mannheim, FRG, revised 1998.

[Williamson 1999] Jon Williamson: 'Does a cause increase the probability of its effects?', philosophy.ai report pai_jw_99_d, http://www.kcl.ac.uk/philosophy.ai.

[Williamson 2000] Jon Williamson: 'A probabilistic approach to diagnosis', Proceedings of the Eleventh International Workshop on Principles of Diagnosis (DX-00), Morelia, Michoacen, Mexico, June 8-11 2000.

[Williamson 2000b] Jon Williamson: 'Approximating discrete probability distributions with Bayesian networks', in Proceedings of the International Conference on Artificial Intelligence in Science and Technology, Hobart Tasmania, 16-20 December 2000, pages 106-114.

PETER M. WILLIAMS

PROBABILISTIC LEARNING MODELS

1 INTRODUCTION

The purpose of this review is to provide a brief outline of some uses of Bayesian methods in artificial intelligence, specifically in the area of neural computation.

Prior to the 1980s, much of knowledge representation in Artificial Intelligence was concerned with rule-based or expert systems. The aim was to write rules and facts, expressing human knowledge in some domain, in a quasi-logical language. Although achieving some successes, this approach came to be seen by the 1980s as cumbersome in adapting to changing circumstances, partly owing to the explosion in the list of rules and exceptions needed to cover novel cases. In the mid 1980s a new paradigm emerged, referred to variously as parallel distributed processing, connectionism or neural computing, largely through the publication of Rumelhart and McClelland [1986]. One of the motivating ideas was that living creatures are programmed by experience, rather than by spelling out every step in a process, and that representations of human knowledge and learning need not to be restricted to areas in which rule-based algorithms can be found. An important tool in implementing this programme was the *neural network* which, in a very rudimentary way, might be seen as sharing some of the characteristics of the brain.

Although having its origins in the biological and cognitive sciences, significant contributions to neural computation were soon made by physicists and statisticians. Furthermore applications were made to practical problems outside the field of the biosciences. The list includes **prediction** (weather and utility demand forecasting, medical diagnosis, derivative and term-structure models in finance) **navigation and control** (aircraft landing, plant monitoring, autonomous robots) **pattern recognition** (speech and signal processing, hand-written character recognition, finger-print identification, mineral exploration) etc. Possibly the reasons for this widespread interest is that neural networks can be used wherever a linear model is used, that they can model non-linear relationships, and that they require no detailed model of the underlying process.

During the ten years or so following the publication of [Rumelhart and McClelland, 1986] much attention was given to improving the algorithms used for fitting neural network models. At the same time, it was appreciated that there were close links with established statistical machine learning and pattern recognition methods. As the subject has developed, it has grown closer to information theory, statistics, image and signal processing, as regards both its engineering and neuroscience concerns. Inevitably its maturing expression in statistical form has led to an application of the Bayesian approach.

Neural computing methods have been applied to both supervised and unsupervised learning. The practical applications mentioned above have been largely in

D. Corfield and J. Williamson (eds.), Foundations of Bayesianism, 117–134.

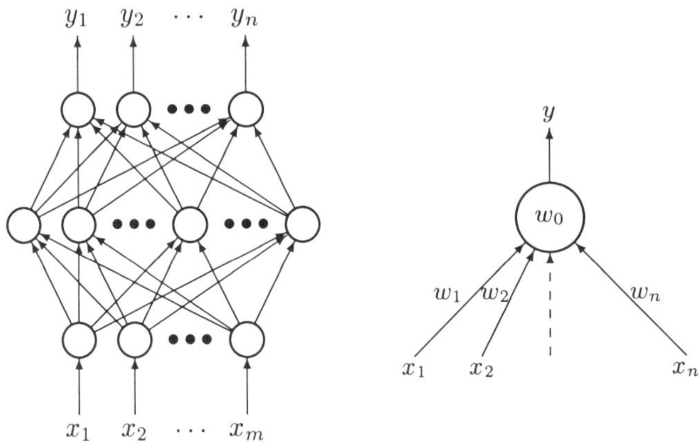

Figure 1. The diagram on the left shows a layered feed-forward network with inputs $x_1, \ldots, x_m$ and outputs $y_1, \ldots, y_n$ separated by a single layer of hidden units. In general there can be any number of hidden layers. The diagram on the right shows an individual processing unit, with input weights $w_1, \ldots, w_n$ and bias w_0.

the area of supervised learning, which includes classification, regression and interpolation. In the past five years, however, much work at the forefront of neural computation has been in the area of unsupervised learning, which traditionally includes clustering and density estimation. Both of these areas will be touched on briefly.

2 NEURAL NETWORKS

The simplest form of neural network provides a way of modelling the relationship between an independent variable $\mathbf{x}$ and a dependent variable $\mathbf{y}$. For example, $\mathbf{x}$ could be financial data up to a certain time and $\mathbf{y}$ could be a future stock index, exchange rate, option price etc. Or $\mathbf{x}$ could represent geophysical features of a mining prospect and $\mathbf{y}$ could represent mineralization at a certain depth. In general $\mathbf{x}$ and $\mathbf{y}$ can be any vectors of continuous or discrete numerical quantities.

Such a network implements an input-output mapping

$$\mathbf{y} = f(\mathbf{x}, \mathbf{w})$$

from $\mathbf{x} = (x_1, \ldots, x_m)$ to $\mathbf{y} = (y_1, \ldots, y_n)$. The mapping depends on the values of the connection strengths or weights $\mathbf{w}$ in the network. The connections between processing elements can be arbitrary but a layered (non-cyclic) architecture, as shown on the left of Figure 1, is commonest.

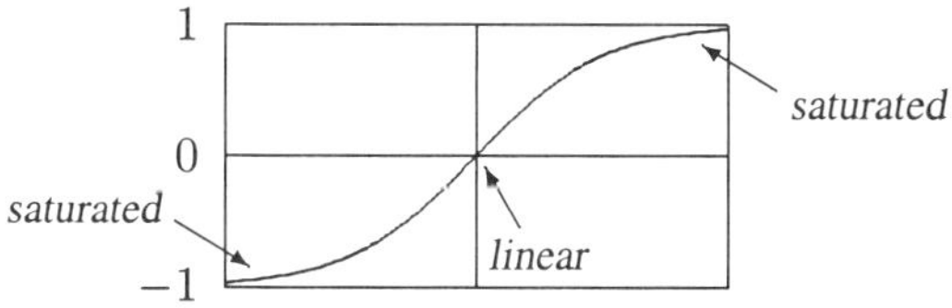

Figure 2. Diagram showing the hyperbolic tangent squashing function. Other functions of similar form may be used. Typically such functions are monotonic, approximately linear in the central range and saturate to asymptotic values at the extremes.

Each non-input unit in the network has input weights $w_1, \ldots, w_n$ and a bias w_0 as shown on the right in Figure 1. For hidden units, namely those that are neither input nor output units,

$$y = \tanh(w_0 + w_1 x_1 + \cdots + w_n x_n)$$

where the transfer function, in this case the hyperbolic tangent, squashes the output into the interval $(-1, 1)$ as shown in Figure 2. For output units, we assume a direct linear relationship:

$$y = w_0 + w_1 x_1 + \cdots + w_n x_n$$

A network of the type shown in Figure 1 can have several internal layers of hidden units connecting input units and output units. A feedforward neural network is therefore a composition of linear and non-linear transformations like a multi-layer sandwich

——— linear ———
.... squash ....
——— linear ———
.... squash ....
——— linear ———

in this case having two hidden layers. Without non-linear squashing, the sandwich would collapse, by composition, to a single linear transformation. With the interposed non-linearities a neural network becomes a universal approximator capable of modelling an arbitrary continuous function [Hornik, 1993; Barron, 1994; Ripley, 1996].

2.1 *Model fitting*

Suppose we have a training set of pairs of observations $(\mathbf{x}_1, y_1), \ldots, (\mathbf{x}_N, y_N)$ where each $\mathbf{x}_i$ is a vector of inputs and each y_i is now a single scalar target output

$(i = 1, \ldots, N)$. Then least squares fitting consists of choosing weights $\mathbf{w}$ to minimise the data misfit

$$(1) \quad E(\mathbf{w}) = \frac{1}{2}\sum_{i=1}^{N}\Big(y_i - f(\mathbf{x}_i, \mathbf{w})\Big)^2$$

where $f(\mathbf{x}, \mathbf{w})$ is the function computed by the network for a given set of weights $\mathbf{w}$. Since the gradient .$(\mathbf{w})$ is easily computed using the so-called backpropagation algorithm, standard optimisation techniques such as conjugate gradient or quasi-Newton methods [Williams, 1991; Bishop, 1995a] can be applied to minimise (1).

To make the link with Bayesian methods, recall first that least squares fitting is equivalent to maximum likelihood estimation, assuming Gaussian noise. To see this, suppose the target variable has a conditional normal distribution

$$p(y \,|\, \mathbf{x}) = \frac{1}{\sqrt{2\pi}\sigma} \exp -\frac{1}{2}\left(\frac{y - \mu(\mathbf{x})}{\sigma}\right)^2$$

where $\mu(\mathbf{x})$ is the conditional mean, for a given input $\mathbf{x}$, and where the variance σ^2 is assumed, for the present, to be constant. Then, assuming independence, the negative log likelihood of the data $(\mathbf{x}_1, y_1), \ldots, (\mathbf{x}_N, y_N)$ is proportional to

$$(2) \quad \frac{1}{2}\sum_{i=1}^{N}\Big(y_i - \mu(\mathbf{x}_i, \mathbf{w})\Big)^2$$

up to an additive constant. Comparison of (1) and (2) shows that least squares fitting is equivalent to maximum likelihood estimation of the weights, if the network output $f(\mathbf{x}, \mathbf{w})$ is understood to compute the conditional mean $\mu(\mathbf{x})$.

More generally, there is no need to assume that σ is constant for all inputs $\mathbf{x}$. If we allow the network to have two outputs, we can allow them to compute the two input-dependent parameters $\mu(\mathbf{x})$ and $\sigma(\mathbf{x})$ of the predictive distribution for y. It is more convenient, however, in order to have an unconstrained parametrisation, to model $\log \sigma(\mathbf{x})$ rather than $\sigma(\mathbf{x})$, so that the network can be visualised schematically as in Figure 3. The negative log likelihood of the data can now be written as

$$(3) \quad L(\mathbf{w}) = \frac{1}{2}\sum_{i=1}^{N}\left\{\log \sigma(\mathbf{x}_i, \mathbf{w})^2 + \left(\frac{y_i - \mu(\mathbf{x}_i, \mathbf{w})}{\sigma(\mathbf{x}_i, \mathbf{w})}\right)^2\right\}$$

and this can be considered as the generalised error function. Maximum likelihood fitting is obtained by minimising $L(\mathbf{w})$ where $\nabla L(\mathbf{w})$ can also be computed by backpropagation.[1]

[1] In the general multivariate case the conditional density $p(\mathbf{y} \,|\, \mathbf{x})$ for the target variable $\mathbf{y} = (y_1, \ldots, y_n)$ is proportional to $|\Sigma|^{-1/2} \exp\left\{-\frac{1}{2}(\mathbf{y} - \boldsymbol{\mu})^{\mathrm{T}}\Sigma^{-1}(\mathbf{y} - \boldsymbol{\mu})\right\}$ where $\boldsymbol{\mu}$ is vector of conditional means and Σ is the conditional covariance matrix. We can then model $\boldsymbol{\mu}(\mathbf{x})$ and $\log \Sigma(\mathbf{x})$ as functions of $\mathbf{x}$ in ways that depend on the outputs of a neural network when $\mathbf{x}$ is given as input [Williams, 1999]. This permits modelling of the full conditional correlations in multivariate data. Applications to heteroskedastic (time-dependent) volatility in financial time series are given in [Williams, 1996].

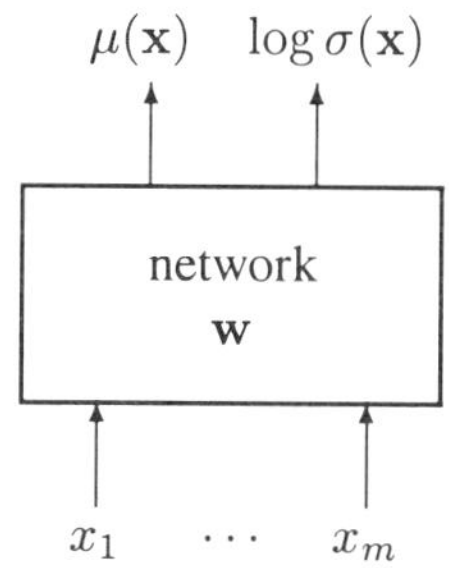

Figure 3. Schematic representation of a neural network with input $\mathbf{x} = (x_1, \ldots, x_m)$ and weights $\mathbf{w}$, whose output is interpreted as computing the mean and log variance of the target variable.

2.2 *Need for regularisation*

Neural networks, of the type considered here, differ from linear statistical models in being universal approximators, capable of fitting arbitrary continuous non-linear functions. This means (i) that there is greater potential for overfitting the data, leading to poor generalisation outside the training sample, and (ii) that the error surface can be of great complexity. Neural network modelling therefore calls for special techniques in particular (i) for some form of stabilisation, or regularisation, and (ii) for some form of integration over multiple local minima. The second of these is discussed in Section 2.6.

Possible solutions to the overfitting problem include

(a) limiting the complexity of the network;

(b) stopping training early before overfitting begins;

(c) adding extra terms to the cost function to penalise complex models.

The first of these (a) amounts to a form of hard structural stabilisation. For example, the number of weights might be limited to a certain proportion of the number of training items, using various rules of thumb for the exact proportion. A deeper treatment of this approach, including analysis of error bounds, forms part of statistical learning theory [Vapnik, 1998].

The second approach (b) observes that, at early stages of training, the network rapidly fits the broad features of the training set. For small initial values of the weights, the neural network is in fact close to being a linear model (see the central linear segment of Figure 2). As training progresses the network uses more resources to fit details of the training set. The aim is to stop training before the model begins to fit the noise. There are several ways of achieving this including monitoring performance on a test set (but see [Cataltepe *et al.*, 1999]).

The third approach (c) is a form of Tikhonov regularisation [Tikhonov and Arsenin, 1977; Bishop, 1995b]. In the case of neural networks, (c) often takes the form of weight decay [Hinton, 1986]. This adds an extra term to the cost function

$$(4) \quad E(\mathbf{w}) = L(\mathbf{w}) + \lambda R(\mathbf{w})$$

where $L(\mathbf{w})$ expresses the data misfit and $R(\mathbf{w})$ is a regularising term that penalises large weights. The aim becomes minimisation of the overall objective function (4) where λ is a regularising parameter that determines a balance between the two terms, the first expressing misfit and the second complexity. There remains the problem of locating this balance by fixing an appropriate value for λ. This is often chosen by some form of cross-validation. But performance on a test set can be noisy. Different test sets may lead to different values of λ. We therefore examine various Bayesian solutions to this problem.

2.3 *Bayesian approach*

Consider the case of Section 2.1 where we have training data D corresponding to observed pairs $(\mathbf{x}_1, y_1), \ldots, (\mathbf{x}_N, y_N)$. Suppose the aim is to choose the most probable network weights $\mathbf{w}$ given the data, in other words to maximise the posterior probability density $p(\mathbf{w}|D)$. Using Bayes' theorem we have

$$(5) \quad p(\mathbf{w}|D) \propto p(D|\mathbf{w})\, p(\mathbf{w})$$

where $p(D|\mathbf{w})$ is the likelihood of the data and $p(\mathbf{w})$ is the prior over weights. Maximising (5) is the same as minimising its negative logarithm

$$(6) \quad E(\mathbf{w}) = L(\mathbf{w}) - \log p(\mathbf{w}) + \text{constant}$$

where the negative log likelihood $L(\mathbf{w}) = -\log p(D|\mathbf{w})$ is given by (3) in the case of regression with Gaussian noise.

Now suppose that $\mathbf{w}$ has a Laplace prior $p(\mathbf{w}|\lambda) \propto \prod_i \exp -\lambda|w_i|$ where $\lambda > 0$ is a scale parameter [Williams, 1995]. Then, ignoring terms not depending on $\mathbf{w}$, (6) becomes

$$(7) \quad E_\lambda(\mathbf{w}) = L(\mathbf{w}) + \lambda\|\mathbf{w}\|_1$$

where $\|\mathbf{w}\|_p = (\sum_i |w_i|^p)^{1/p}$ is the L_p norm ($p \geq 1$) of the weight vector $\mathbf{w}$. Comparison of (7) with (4) shows that the regularising term $\lambda R(\mathbf{w})$ corresponds to the negative logarithm of the prior. The same is true assuming a Gaussian prior for weights $p(\mathbf{w}|\lambda) \propto \prod_i \exp -(\lambda/2)|w_i|^2$ when (7) becomes

$$(8) \quad E_\lambda(\mathbf{w}) = L(\mathbf{w}) + (\lambda/2)\|\mathbf{w}\|_2^2$$

The difficulty remains, however, that λ is still unknown.[2]

[2] In practice it may be assumed that different classes of weights in the network have different typical, if unknown, scales. The classes are chosen to ensure invariance under suitable transformations of input and output variables. For simplicity we deal here with a single class.

2.4 *Eliminating* λ

One approach is to eliminate λ by means of integration [Buntine and Weigend, 1991; Williams, 1995]. Consider λ to be a hyperparameter with prior $p(\lambda)$ so that

$$(9) \quad p(\mathbf{w}) = \int p(\mathbf{w}|\lambda)\, p(\lambda)\, d\lambda$$

which no longer depends on λ and can be substituted directly into (6). Since $\lambda > 0$ is a scale parameter, it is natural to use a non-informative Jeffreys' prior [1961] for which $p(\lambda)$ is proportional to $1/\lambda$. It is then straightforward to integrate (9) and substitute into (6) to give the objective function

$$(10) \quad E(\mathbf{w}) = L(\mathbf{w}) + W \log ||\mathbf{w}||_p$$

where W is the total number of weights and $p = 1$ for the Laplace prior or $p = 2$ for the Gaussian prior. (10) involves no adjustable parameters so that regularisation of network training, conceived as an optimisation problem, is automatic.[3]

This approach to the elimination of λ is independent of the form taken by the likelihood term $L(\mathbf{w})$, which will depend on the appropriate statistical model for the data. Discrete models, for example, correspond to classification. Non-Gaussian continuous models are discussed in [Bishop and Legleye, 1995; Williams, 1998b] for example.

2.5 *The evidence approach to* λ

An alternative approach to determining of λ is to use its most probable value given the data. This is the value of λ which maximises $p(\lambda|D)$. Since $p(\lambda|D)$ is proportional to $p(D|\lambda)p(\lambda)$, this is the same as choosing λ to maximise $p(D|\lambda)$, assuming that $p(\lambda)$ is relatively insensitive to λ. In this approach $p(D|\lambda)$ is called the *evidence* for λ. $p(D|\lambda)$ can be expressed as an integral over weight space

$$(11) \quad p(D|\lambda) = \int p(D|\mathbf{w}, \lambda)\, p(\mathbf{w}|\lambda)\, d\mathbf{w}$$

and, under suitable simplifying assumptions, this integration can be performed analytically. Specifically, assume that the integrand $p(D|\mathbf{w}, \lambda)$, which is proportional to the posterior distribution of weights $p(\mathbf{w}|D, \lambda)$, can be approximated by a Gaussian in the neighbourhood of a maximum $\mathbf{w} = \mathbf{w}_{\mathrm{MP}}$ of the posterior density. In the case of a Gaussian weight prior, it can then be shown [MacKay, 1992b] that, at a maximum of $p(D|\lambda)$, λ must satisfy

$$(12) \quad \lambda\, ||\mathbf{w}_{\mathrm{MP}}||_2^2 = \sum_i \frac{\nu_i}{\lambda + \nu_i}$$

[3] Instead of using the improper Jeffreys prior, we could use a proper conjugate prior. This is the gamma distribution, for either the Laplace or Gaussian weight priors, with shape and scale parameters $\alpha, \beta > 0$ say. The regularising term is then $(W + \alpha)\log(||\mathbf{w}||_1 + \beta)$ for the Laplace weight prior and $(W/2 + \alpha)\log(||\mathbf{w}||_2^2 + 2\beta)$ for the Gaussian prior. Both reduce to $W \log ||\mathbf{w}||_p$ $(p = 1, 2)$ as α, β approach zero.

where ν_i are the eigenvalues of the Hessian of $L(\mathbf{w}) = -\log p(D|\mathbf{w}, \lambda)$ evaluated at $\mathbf{w} = \mathbf{w}_{\text{MP}}$. (12) can be used as a re-estimation formula for λ, using λ^{old} on the right and λ^{new} on the left, in iterative optimisation of (8).[4]

In the case of a Laplace prior, the evidence approach is essentially equivalent to the integration approach of the previous section [Williams, 1995, Appendix]. MacKay [1994] argues that, in the case of a Gaussian prior, the evidence approach provides better results in practice. MacKay [1994] also discusses the range of validity of the various approximations involved in this approach.

2.6 *Integration methods*

The preceding discussion has assumed that the aim is to minimise $E(\mathbf{w})$ in (6). This is the same as finding a maximum, at $\mathbf{w} = \mathbf{w}_{\text{MP}}$ say, of the posterior distribution $p(\mathbf{w}|D)$. Predictions for the target value y, given inputs $\mathbf{x}$, could then be made using the distribution

$$(13)\quad p(y|\mathbf{x}, D) = p(y|\mathbf{x}, \mathbf{w}_{\text{MP}}).$$

This can be unsatisfactory, however, since it amounts to assuming that the posterior distribution for $\mathbf{w}$ can be adequately approximated by a delta function at $\mathbf{w} = \mathbf{w}_{\text{MP}}$. In practice, for a general neural network model, there may be several non-equivalent modes of the posterior distribution, i.e. local minima of $E(\mathbf{w})$ as was noted in Section 2.2, and each may extend over a significant region.

The correct procedure, from a Bayesian point of view, is to make predictions using an integral over weight space

$$(14)\quad p(y|\mathbf{x}, D) = \int p(y|\mathbf{x}, \mathbf{w})\, p(\mathbf{w}|D)\, d\mathbf{w}$$

where the predictive distribution $p(y|\mathbf{x}, \mathbf{w})$, corresponding to a particular value of $\mathbf{w}$, is weighted by the posterior probability of $\mathbf{w}$.[5]

In some cases (14) can be integrated analytically if suitable assumptions are made. For example, suppose the statistical model $p(y|\mathbf{x}, \mathbf{w})$ is Gaussian and that the posterior distribution $p(\mathbf{w}|D)$ can be adequately approximated by a Gaussian centered at $\mathbf{w} = \mathbf{w}_{\text{MP}}$. Then the posterior distribution is again Gaussian with a variance in which the intrinsic process noise is augmented by an amount, corresponding to model uncertainty, which increases with the dispersion of $p(\mathbf{w}|D)$ around $\mathbf{w}_{\text{MP}}$.

---

[4]Note that since the Hessian is evaluated at a maximum of $p(\mathbf{w}|D, \lambda)$, rather than at a maximum of $p(D|\mathbf{w}, \lambda)$, the eigenvalues may not all be positive. Furthermore, since $\mathbf{w}_{\text{MP}}$ depends on λ, the derivation of (12) strictly speaking ignores terms involving $d\nu_i/d\lambda$ [MacKay, 1992a].

[5]To emphasise a danger in (13) note that (10) will have a local minimum at $\mathbf{w} = \mathbf{0}$, even for sufficiently small $\beta > 0$ as defined in footnote 3. This mode of $p(\mathbf{w}|D)$, however, will normally only have very local extent, so that it will contribute little to the integral in (14), except in cases where there is little detectable coupling between any y_i and $\mathbf{x}_i$ in D, when the opinion implied by $\mathbf{w} = \mathbf{0}$ that $p(y|\mathbf{x}, D) \equiv p(y|D)$ would deserve some weight.

An interesting alternative is to approximate $p(\mathbf{w}|D)$ by a more tractable distribution $q(\mathbf{w})$ obtained by minimising the Kullback-Leibler divergence

$$\int q(\mathbf{w}) \log \frac{q(\mathbf{w})}{p(\mathbf{w}|D)}\, d\mathbf{w}.$$

This is known as ensemble learning [Hinton and van Camp, 1993; MacKay, 1995; Barber and Bishop, 1998]. Typically q takes the form of a Gaussian whose parameters may be fitted using Bayesian methods [MacKay, 1995; Barber and Bishop, 1998]. The advantage of this approach is that the approximating Gaussian is fitted to $p(\mathbf{w}|D)$ globally rather than locally at $\mathbf{w} = \mathbf{w}_{\mathrm{MP}}$. More generally, q may be assumed to have free form, or to be a product of tractable distributions of fixed form.

It is normally still implied, however, that there is essentially only one significant mode for the posterior distribution. In some cases it may be necessary to attempt the integral in (14) using numerical methods. In general (14) can be approximated by

$$(15)\quad p(y|\mathbf{x}, D) \approx \frac{1}{M}\sum_{i=1}^{M} p(y|\mathbf{x}, \mathbf{w}_i)$$

provided $\{\mathbf{w}_1, \ldots, \mathbf{w}_M\}$ is a sample of weight vectors which is representative of the posterior distribution $p(\mathbf{w}|D)$. The problem is to generate the $\{\mathbf{w}_i\}$ by searching those regions of the high-dimensional weight space where $p(\mathbf{w}|D)$ is large and extends over a non-negligible region. This problem has been studied by Neal [1992; 1996] who has developed extensions of the Metropolis Monte Carlo method specifically adapted to neural networks. This method involves a large number of successive steps through weight space but, to achieve a given error bound, only a much smaller number of visited locations need be retained for predictive purposes. A less efficient method, though one that is simple and often effective, is to choose the $\{\mathbf{w}_i\}$ as local minima obtained by some standard optimisation method from sufficiently many independently chosen starting points.

Note that (15) expresses the resulting predictive distribution as a finite mixture of distributions. The variation between these distributions expresses the extent of model uncertainty. For example, the variance of the mixture distribution for y will be the mean of the predicted variances plus the variance of the predicted means. Specifically if μ_i and σ_i^2 are the mean and variance according to $\mathbf{w}_i$, the predicted mean is $\langle \mu_i \rangle$ and the predicted variance is $\langle \sigma_i^2 \rangle + \{\langle \mu_i^2 \rangle - \langle \mu_i \rangle^2\}$ where $\langle \mu_i \rangle$ is the average of $\mu_1, \ldots, \mu_M$ etc. The first term $\langle \sigma_i^2 \rangle$ represents modelled noise and the second term $\langle \mu_i^2 \rangle - \langle \mu_i \rangle^2$ represents model uncertainty.

3 KERNEL-BASED METHODS

3.1 Gaussian processes

The previous discussion took as its basis the idea of a prior $p(\mathbf{w})$ over network weights $\mathbf{w}$. Now, for a given input vector $\mathbf{x}$, the prior $p(\mathbf{w})$ over weights will induce a prior $p(y)$ over the output $y = f(\mathbf{x}, \mathbf{w})$ computed by the network, assuming for simplicity that the network has a single output unit. This is because y depends on network weights $\mathbf{w}$ as well as on the input $\mathbf{x}$, so that uncertainty in the weights induces uncertainty in the output, for given $\mathbf{x}$.

More generally, if we consider the outputs $y_1, \ldots, y_N$ computed for different inputs $\mathbf{x}_1, \ldots, \mathbf{x}_N$, the prior distribution over weights $p(\mathbf{w})$ determines a joint prior $p(y_1, \ldots, y_N)$ over the outputs calculated for these inputs. Given the sometimes opaque role of individual weights in a network, it can be argued that it may be more natural to specify such a prior directly. Predictions can then be made using conditionalisation. Writing $\mathbf{y}_N = (y_1, \ldots, y_N)$ for the observed values at $\mathbf{x}_1, \ldots, \mathbf{x}_N$, the predictive conditional distribution for y_{N+1} at a new point $\mathbf{x}_{N+1}$ is given by

$$(16) \quad p(y_{N+1} \,|\, \mathbf{y}_N) = \frac{p(\mathbf{y}_{N+1})}{p(\mathbf{y}_N)}$$

where, by assumption, the numerator and denominator on the right are known.

Neal [1994; 1996] has shown that, for neural networks with independent and identically distributed priors over weights, the prior $p(y_1, \ldots, y_N)$ converges, for any N, to a multivariate Gaussian as the number of hidden units tends to infinity.[6] Such a family of variables $y(\mathbf{x})$ is called a **Gaussian process**.[7] For a Gaussian process, the conditional predictive distribution (16) for y_{N+1} is also Gaussian with mean and variance which can be expressed as follows. If $\mathbf{y}_N$ has covariance matrix Σ_N and if the covariance matrix for $\mathbf{y}_{N+1}$ is written in the form

$$\Sigma_{N+1} = \begin{pmatrix} \Sigma_N & \boldsymbol{\sigma} \\ \boldsymbol{\sigma}^{\mathrm{T}} & \sigma \end{pmatrix}$$

then the predictive distribution for y_{N+1} has mean $\langle y_{N+1} \rangle = \boldsymbol{\sigma}^{\mathrm{T}} \boldsymbol{\Sigma}_N^{-1} \mathbf{y}_N$ and variance $\sigma - \boldsymbol{\sigma}^{\mathrm{T}} \boldsymbol{\Sigma}_N^{-1} \boldsymbol{\sigma}$. Notice that $\langle y_{N+1} \rangle$ takes the form of a linear combination $\langle y_{N+1} \rangle = \alpha_1 y_1 + \cdots + \alpha_N y_N$ of observed values. The weighting coefficients α_i automatically take account of the correlations between values of $y(\mathbf{x})$ at different input locations $\mathbf{x}_1, \ldots, \mathbf{x}_{N+1}$. As might be expected, the predicted variance $\sigma - \boldsymbol{\sigma}^{\mathrm{T}} \boldsymbol{\Sigma}_N^{-1} \boldsymbol{\sigma}$ is always less than the prior variance σ.

---

[6] Explicit forms for the resulting covariance functions are derived in [Williams, 1998a]

[7] Gaussian processes are already used in Wiener-Kolmogorov time series prediction [Wiener, 1949] and in Matheron's approach to geostatistics [Matheron, 1965].

To implement this model it is necessary to model the covariance matrix $\mathbf{\Sigma}_N$. Decisions must also be made about modelling the mean, if trends or drifts are permitted. Writing

$$\mathbf{\Sigma}_N = \left\{K(\mathbf{x}_i, \mathbf{x}_j, \boldsymbol{\lambda})\right\}_{i,j=1}^{N}$$

where $\boldsymbol{\lambda}$ is the vector of parameters of the model, the process is said to be stationary if $K(\mathbf{x}, \mathbf{x}', \boldsymbol{\lambda})$ depends only on the separation $\mathbf{x} - \mathbf{x}'$ for any $\mathbf{x}, \mathbf{x}'$. For example, Gibbs and MacKay [1997] consider a stationary process

$$K(\mathbf{x}, \mathbf{x}', \boldsymbol{\lambda}) = \lambda_1 \exp\left\{-\frac{1}{2}\sum_{k=1}^{m}\left(\frac{x_k - x'_k}{\alpha_k}\right)^2\right\} + \lambda_2 + \lambda_3 \delta(\mathbf{x}, \mathbf{x}')$$

where $\boldsymbol{\lambda} = (\lambda_1, \alpha_1, \ldots, \alpha_m, \lambda_2, \lambda_3)$ is the vector of adjustable parameters and m is the dimension of $\mathbf{x}$. Williams and Rasmussen [1996] include a further linear regression term $\lambda_4\, x^{\mathrm{T}} x'$. The Bayesian predictive distribution now becomes

$$p(y_{N+1}|\mathbf{y}) = \int p(y_{N+1}|\mathbf{y}, \boldsymbol{\lambda})\, p(\boldsymbol{\lambda}|\mathbf{y})\, d\boldsymbol{\lambda}$$

which must be integrated by Monte Carlo methods or by searching for maxima of $p(\boldsymbol{\lambda}|\mathbf{y}) \propto p(\mathbf{y}|\boldsymbol{\lambda})\, p(\boldsymbol{\lambda})$ and using the most probable values of the hyperparameters [Williams and Rasmussen, 1996; Neal, 1998].

Considerable attention has been paid recently to reducing the computational complexity of Gaussian process modelling [Gibbs and MacKay, 1997; Trecate *et al.*, 1999; Williams and Seeger, 2001] and much of this work is also applicable to other kernel based methods.

It should be noted that in geostatistics the "covariogram" for stationary processes, for which $K(\mathbf{x}, \mathbf{x}') = k(\mathbf{x} - \mathbf{x}')$, is frequently estimated empirically from the data. If a parametric model is used, the form is chosen to reflect prior geological or geochemical knowledge [Cressie, 1993]. Here too, Bayesian methods have been applied to the problem of estimating spatial covariance structures [Kitanidis, 1986; Le and Zidek, 1992].

3.2 Support Vector Machines

A recent significant development in classification and pattern recognition has been the introduction of the Support Vector Machine (SVM) [Vapnik, 1998; Cristianini and Shawe-Taylor, 2000]. SVM methods have been applied to regression [Vapnik *et al.*, 1997] and density estimation [Vapnik and Mukherjee, 2000] but we shall concentrate on binary classification.

Suppose a set of training examples $(\mathbf{x}_1, y_1), \ldots, (\mathbf{x}_N, y_N)$ is given where each $\mathbf{x}_i$ is a vector of real-valued inputs and each $y_i \in \{-1, +1\}$ is the corresponding class label ($i = 1, \ldots, N$). Assume initially that the two classes are *linearly separable*, in other words there exists a linear functional f such that $f(\mathbf{x}_i) < 0$

whenever $y_i = -1$ and $f(\mathbf{x}_i) > 0$ whenever $y_i = +1$. The class label of a new item $\mathbf{x}$ can then be predicted by the sign of $f(\mathbf{x})$.

Where such a separating functional f exists, it will not be unique, if only because it is undetermined to within a positive scalar multiple. More importantly, the separating *hyperplane* $\{\mathbf{x} : f(\mathbf{x}) = 0\}$ is generally not unique. The central tenet of the SVM approach is that the optimal hyperplane is one that maximises the minimum distance between the hyperplane and any example in the training data. The optimal hyperplane can then be found by convex optimisation methods.

For data that is not directly linearly separable, an embedding of the input features $\mathbf{x} \mapsto \Phi(\mathbf{x})$ into some inner-product space $\mathcal{H}$ may allow the resulting training examples $\{(\Phi(\mathbf{x}_i), y_i)\}$ to be linearly separated in $\mathcal{H}$. In that case it turns out that detailed information about Φ and its range are not needed. This is because the optimal hyperplane in $\mathcal{H}$, where distance is defined by the inner product, depends on Φ only through inner products $K(\mathbf{x}, \mathbf{x}') = \Phi(\mathbf{x}) \cdot \Phi(\mathbf{x}')$. In fact the classifying functional, corresponding to the optimal separating hyperplane, can be expressed as

$$(17) \quad y(\mathbf{x}) = w_0 + \sum_{i=1}^{N} w_i K(\mathbf{x}, \mathbf{x}_i)$$

where $w_0, w_1, \ldots, w_N$ are the model parameters and K is the *kernel* function. The practical problem therefore becomes one of choosing a suitable kernel K rather than an embedding Φ. Suitable kernels, namely those deriving from some implicit embedding Φ, include $K(\mathbf{x}, \mathbf{y}) = (\mathbf{x} \cdot \mathbf{y} + 1)^p$ and $K(\mathbf{x}, \mathbf{y}) = \exp -\lambda||\mathbf{x} - \mathbf{y}||^2$.

To avoid overfitting, or in case the dataset is still not linearly separable, the extra constraint that $|w_i| < C$ for $i = 1, \ldots, N$ is imposed. C corresponds to a penalty for misclassification and is chosen by the user. The SVM typically leads to a sparse model in the sense that most of the coefficients w_i vanish. The training items $\mathbf{x}_i$ for which the corresponding w_i is non-zero are called *support vectors* and, typically, lie close to the decision boundary.

Support Vector Machines have their origins in the principle of structural risk minimisation [Vapnik, 1979; Williamson *et al.*, 1998]. The aim is to place distribution-independent bounds on expected generalisation error [Bartlett and Shawe-Taylor, 1999; Vapnik and Chapelle, 2000]. The motivation and results are somewhat different from those associated with Bayesian analysis. In the case of classification, for example, the optimal hyperplane depends only on the support vectors, which are extreme values of the dataset, whereas likelihood-based methods depend on all the data. A Bayesian approach always aims to conclude with a probability distribution over unknown quantities, whereas an SVM typically offers point estimates for regression or hard binary decisions for classification. Bayesian methods have nonetheless been applied to the problem of kernel selection in [Seeger, 2000] where support vector classification is interpreted as efficient approximation to Gaussian process classification. By contrast, the problem of model selection based on error bounds derived from cross-validation is discussed in [Chapelle and Vapnik, 2000; Vapnik and Chapelle, 2000]. A Bayesian treatment of a generalised linear model of identical functional form to the SVM is introduced in [Tipping, 2000]. This approach, using ideas of automatic relevance determination [Neal, 1996], provides probabilistic predictions. It also yields a sparse representation, but one using relevance vectors, which are prototypical examples of classes, rather than support vectors which are close to the decision boundary or, in cases of misclassification, on the wrong side of it.

The SVM performs well in practice and is becoming a popular method. This is partly due to the fact that, once parameters of the model are somehow fixed, SVM training finds a global solution, whereas neural network training, when considered as an optimisation process, may find multiple local minima, as discussed previously.

4 UNSUPERVISED LEARNING

So far we have dealt with labelled data. We assumed given a collection of observations $\{(\mathbf{x}_1, y_1), \ldots, (\mathbf{x}_N, y_N)\}$ where each $\mathbf{x}_n$ represents some known features of the nth item, and y_n is a class label, or value, associated with $\mathbf{x}_n$. The presence of the labels makes this a case of supervised learning. The aim is to use the examples to predict the value of y on the basis of $\mathbf{x}$ for a general out-of-sample case. More generally, the problem is to model the conditional probability distribution $p(y|\mathbf{x})$.

Now suppose we are only given the features $\{\mathbf{x}_1, \ldots, \mathbf{x}_N\}$ without corresponding labels. What can be learned about the $\mathbf{x}_n$? This question, which at first sight seems perplexing, can be given a sense if it is interpreted as a search for interesting features of the data. For example, are there clusters in the data? Are there outliers which appear novel or unusual? What is the intrinsic dimensionality of the data? Can the data be visualised, or otherwise analysed, in a lower dimensional space? Do the data exhibit latent structure, which would help to explain how they were generated?

Some aspects of unsupervised learning relate to qualitative ideas of concept formation studied in cognitive science and the philosophy of science. In quantitative terms, however, unsupervised learning often corresponds to some form of probability density estimation; and special interest attaches to the structures that particular density estimators might exhibit. When considered as a statistical problem, this makes possible the application of Bayesian methods. We consider two examples.

4.1 The Generative Topographic Mapping

The first example concerns a case where density estimation is useful for clustering and for providing low-dimensional representations enabling visualisation of high dimensional data.

Suppose the data lives in a possibly high dimensional space D. If the dataset is $\{\mathbf{x}_1, \ldots, \mathbf{x}_N\}$, where each $\mathbf{x}_n$ is a d-dimensional real vector, then $D = \mathbf{R}^d$.

A common form of density estimation uses a mixture of Gaussians. For a finite mixture, the density at $\mathbf{x} \in D$ may be estimated by

$$(18) \quad p(\mathbf{x}) = \sum_{i \in I} p(\mathbf{x}|i)P(i)$$

where I is some finite index set, $P(i)$ is the weight of the ith component and $p(\mathbf{x}|i)$ is a Gaussian density. In the simplest case, each Gaussian will be spherical with the same dispersion, so that each covariance matrix is the same scalar multiple, β^{-1} say, of the identity. Then

$$p(\mathbf{x}|i) = \left(\frac{\beta}{2\pi}\right)^{d/2} \exp\left\{-\frac{\beta}{2}\|\mathbf{x} - \mathbf{m}_i\|^2\right\}$$

where $\mathbf{m}_i$ is the centre of the ith mixture component.

The idea of the Generative Topographic Mapping [Bishop *et al.*, 1998b; Bishop *et al.*, 1998a] is to embed the index set I in a linear topological space L, so that each $i \in I$ becomes the index of an element $\mathbf{u}_i \in L$, and to require that the mapping $\mathbf{u}_i \mapsto \mathbf{m}_i$ is the restriction of a *continuous* map $\mathbf{u} \mapsto \mathbf{m}_\mathbf{u}$ from L to D. L is now referred to as the *latent* space. For computational reasons, the mixture is still considered to be finite so that we can write (18) as

$$(19) \quad p(\mathbf{x}) = \sum_{i \in I} p(\mathbf{x}|\mathbf{u}_i)P(\mathbf{u}_i)$$

where the weights of the mixture are interpreted as a prior distribution over L concentrated on a certain finite subset.

The purpose of the continuity of the mapping $\mathbf{u} \mapsto \mathbf{m}_\mathbf{u}$ is seen when the mapping from L to D is, in a sense, inverted using Bayes' theorem

$$(20) \quad P(\mathbf{u}_i|\mathbf{x}) = \frac{p(\mathbf{x}|\mathbf{u}_i)P(\mathbf{u}_i)}{p(\mathbf{x})}$$

The significance of (20) is that, for any possible data point $\mathbf{x} \in D$, there is a discrete probability distribution $P(\mathbf{u}_i|\mathbf{x})$ over L concentrated on those elements of latent space which correspond to components of the Gaussian mixture. Typically $P(\mathbf{u}_i|\mathbf{x})$ will be large when $\mathbf{x}$ is close to the centre of the mixture component generated by $\mathbf{u}_i$. In that case $\mathbf{u}_i$ is said to have large responsibility for $\mathbf{x}$. If we write $\mathbf{m}_\mathbf{x}$ for the mean in L of the posterior distribution (20), the mapping $\mathbf{x} \mapsto \mathbf{m}_\mathbf{x}$ maps data points in D to elements of the latent space L. Normally the mapping is continuous, so that points that are close in D should map to points that are close in L.

An important application of the GTM is to data-visualisation. Suppose that L has low dimension, specifically $L = \mathbf{R}^2$, and that the elements of L having positive prior probability lie on a regular two-dimensional grid. Then each element $\mathbf{x}_n$ of the dataset $\{\mathbf{x}_1, \ldots, \mathbf{x}_N\}$ will determine, by the mapping $\mathbf{x} \mapsto \mathbf{m}_\mathbf{x}$, an element $\mathbf{u}_n$ in the convex hull of the regular grid in L. Each data point is therefore visualisable by its representative in a rectangular region of the two dimensional latent space L. Furthermore, it might be expected that distinct populations in data space would be represented by distinct clusters in latent space.

This account has said nothing of the way in which the mapping $\psi : \mathbf{u} \mapsto \mathbf{m_u}$ is characterised. Typically ψ is taken to be a generalised linear regression model fitted, together with the noise parameter β, using the EM algorithm [Bishop *et al.*, 1998b]. A more flexible manifold-aligned noise model is described in [Bishop *et al.*, 1998a] together with methods of Bayesian inference for hyperparameters. [Bishop *et al.*, 1998b] also provides a discussion of the relationship between the GTM and Kohonen's earlier well-known self-organising feature map [Kohonen, 1995].

4.2 *Blind source separation*

The generative topographic map can be viewed as an instance of a generative model in which observations are modelled as noisy expressions of the state of underlying latent variables [Roweis and Ghahramani, 1999]. Several techniques for modelling multivariate datasets can be viewed similarly, including principal component analysis (PCA) and independent component analysis (ICA). Whereas conventional PCA resolves a data item linearly into uncorrelated components, ICA attempts to resolve it into fully statistically independent components [Common, 1994].

ICA has been applied to the problem of blind source separation. Suppose there are M statistically independent sources and N sensors. The sources might be auditory, e.g. speakers in a room, and the sensors microphones. Each sensor receives a mixture of the sources. The task is to recover the hidden sources from the observed sensors. This problem arises in many areas, for example medical signal and image processing, speech processing, target tracking etc. Separation is said to be blind since, in the most general form of the problem, nothing is assumed known about the mixing process or the sources, apart from their mutual independence. In this sense ICA can be considered to be an example of unsupervised learning.

In the simple linear case, it is assumed that the sensor outputs $\mathbf{x}$ can be expressed as $\mathbf{x} = \mathbf{A}\mathbf{s}$ where $\mathbf{s}$ are the unknown source signals and $\mathbf{A}$ is an unknown mixing matrix. One approach to the problem is to recover the source signals by a linear transformation $\mathbf{y} = \mathbf{W}\mathbf{x}$ where $\mathbf{W}$ is the separating matrix [Amari *et al.*, 1996]. The intention is that $\mathbf{y}$ should coincide with the original $\mathbf{s}$ up to scalar multiplication and permutation of channels. The idea is to fit $\mathbf{W}$ by minimising the Kullback-Leibler divergence

$$\int p(\mathbf{y}) \log \frac{p(\mathbf{y})}{\prod_{m=1}^{M} p(y_m)} \, d\mathbf{w}$$

between the joint distribution for $\mathbf{y}$ and the product of its marginals over the source channels $m = 1, \ldots, M$. The minimum value of zero is achieved only if $\mathbf{W}$ can be chosen so that the resulting distribution $p(\mathbf{y})$ factorises over channels, in which case the components $y_1, \ldots, y_M$ of $\mathbf{y} = \mathbf{W}\mathbf{x}$ are independent.

Various algorithms have been proposed for solving this problem at varying degrees of generality [Bell, 1995; Amari *et al.*, 1996; Attias and Schreiner, 1998]. A recent Bayesian approach uses the ideas of ensemble learning mentioned previously [Miskin and MacKay, 2001]. The idea is that the separating matrix $\mathbf{W}$ may be ill-defined if the data is noisy. The approach instead is to provide a method for approximating the posterior distribution over all possible sources. Significantly, the method allows Bayesian model selection techniques to determine the number of sources M which is generally unknown in advance.

5 CONCLUSION

Statistical methods are increasingly guiding principled research in neural information processing, in both its engineering and neuroscientific forms. Improved algorithms and computational resources are making possible the analysis of high dimensional datasets using powerful non-linear models. This is appropriate in view of the complexity of the systems with which the field is now dealing. Inevitably there is a danger of using over-complex models which fail to distinguish between signal and noise. Bayesian methods are proving invaluable in providing model selection techniques for matching complexity of the model to information content of the data in both supervised and unsupervised learning.

School of Cognitive and Computing Sciences, University of Sussex, UK.

BIBLIOGRAPHY

[Amari *et al.*, 1996] S. Amari, A. Cichocki, and H. H. Yang. A new learning algorithm for blind signal separation. In D. S. Touretzky, M. C. Mozer, and M. E. Hasselmo, editors, *Advances in Neural Information Processing Systems 8*, pages 757–763. MIT Press, 1996.

[Attias and Schreiner, 1998] H. Attias and C. E. Schreiner. Blind source separation and deconvolution: The dynamic component analysis algorithm. *Neural Computation*, 10(6):1373–1424, 1998.

[Barber and Bishop, 1998] David Barber and Christopher M. Bishop. Ensemble learning for multi-layer networks. In M. I. Jordan, M. J. Kearns, and S. A. Solla, editors, *Advances in Neural Information Processing Systems 10*, pages 395–401. MIT Press, 1998.

[Barron, 1994] A. R. Barron. Approximation and estimation bounds for artificial neural networks. *Machine Learning*, 14:115–133, 1994.

[Bartlett and Shawe-Taylor, 1999] P. Bartlett and J. Shawe-Taylor. Generalization performance of support vector machines and other pattern classifiers. In B. Schölkopf, C. J. C. Burgess, and A. J. Smola, editors, *Advances in Kernel Methods—Support Vector Learning*, pages 43–54. MIT Press, 1999.

[Bell, 1995] A. J. Bell and T. J. Sejnowski. An information-maximization approach to blind separation and blind deconvolution. *Neural Computation*, 7(6):1129–1159, 1995.

[Bishop, 1995a] C. M. Bishop. *Neural Networks for Pattern Recognition*. Oxford University Press, 1995.

[Bishop, 1995b] Chris M. Bishop. Training with noise is equivalent to Tikhonov regularization. *Neural Computation*, 7(1):108–116, 1995.

[Bishop and Legleye, 1995] C. M. Bishop and C. Legleye. Estimating conditional probability densities for periodic variables. In G. Tesauro, D. Touretzky, and T. Leen, editors, *Advances in Neural Information Processing Systems 7*, pages 641–648. MIT Press, 1995.

[Bishop *et al.*, 1998a] Christopher M. Bishop, Markus Svensén, and Christopher K. I. Williams. Developments of the generative topographic mapping. *Neurocomputing*, 21:203–224, 1998.

[Bishop *et al.*, 1998b] Christopher M. Bishop, Markus Svensén, and Christopher K. I. Williams. GTM: The generative topographic mapping. *Neural Computation*, 10(1):215–234, 1998.

[Buntine and Weigend, 1991] Wray L. Buntine and Andreas S. Weigend. Bayesian back-propagation. *Complex Systems*, 5:603–643, 1991.

[Cataltepe *et al.*, 1999] Zehra Cataltepe, Yaser S. Abu-Mostafa, and Malik Magdon-Ismail. No free lunch for early stopping. *Neural Computation*, 11(4):995–1009, 1999.

[Chapelle and Vapnik, 2000] Olivier Chapelle and Vladimir Vapnik. Model selection for support vector machines. In S. A. Solla, T. K. Leen, and K-B Müller, editors, *Advances in Neural Information Processing Systems 12*, pages 230–236. MIT Press, 2000.

[Common, 1994] P. Common. Independent component analysis: A new concept. *Signal Processing*, 36:287–314, 1994.

[Cressie, 1993] Noel A. C. Cressie. *Statistics for Spatial Data*. Wiley, revised edition, 1993.

[Cristianini and Shawe-Taylor, 2000] N. Cristianini and J. Shawe-Taylor. *An Introduction to Support Vector Machines*. Cambridge University Press, 2000.

[Gibbs and MacKay, 1997] Mark Gibbs and David J. C. MacKay. Efficient implementation of Gaussian processes. Technical report, Cavendish Laboratory, Cambridge, 1997.

[Hinton, 1986] G. E. Hinton. Learning distributed representations of concepts. In *Proceedings of the Eighth Annual Conference of the Cognitive Sicence Society (Amherst, 1986)*, pages 1–12. Hillsdale: Erlbaum, 1986.

[Hinton and van Camp, 1993] G. E. Hinton and D. van Camp. Keeping neural networks simple by minimizing the description length of the weights. In *Proceedings of the Sixth Annual Conference on Computational Learning Theory*, pages 5–13, 1993.

[Hornik, 1993] K. Hornik. Some new results on neural network approximation. *Neural Computation*, 6(8):1069–1072, 1993.

[Jeffreys, 1961] H. Jeffreys. *Theory of Probability*. Oxford, third edition, 1961.

[Kitanidis, 1986] Peter K. Kitanidis. Parameter uncertainty in estimation of spatial functions: Bayesian analysis. *Water Resources Research*, 22(4):499–507, 1986.

[Kohonen, 1995] T. Kohonen. *Self-Organizing Maps*. Springer, 1995.

[Le and Zidek, 1992] Nhu D. Le and James V. Zidek. Interpolation with uncertain spatial covariances: A Bayesian alternative to Kriging. *Journal of Multivariate Analysis*, 43:351–374, 1992.

[MacKay, 1992a] David J. C. MacKay. Bayesian interpolation. *Neural Computation*, 4(3):415–447, 1992.

[MacKay, 1992b] David J. C. MacKay. A practical Bayesian framework for backpropagation networks. *Neural Computation*, 4(3):448–472, 1992.

[MacKay, 1994] David J. C. MacKay. In G. Heidbreder, editor, *Maximum Entropy and Bayesian Methods, Santa Barbara 1993*, Dordrecht, 1994. Kluwer.

[MacKay, 1995] D. J. C. MacKay. Developments in probabilistic modelling with neural networks—ensemble learning. In *Neural Networks: Artificial Intelligence and Industrial Applications*, pages 191–198. Springer, 1995.

[Matheron, 1965] G. Matheron. *La Théorie des Variables Regionalisées et ses Applications*. Masson, 1965.

[Miskin and MacKay, 2001] J. W. Miskin and D. J. C. MacKay. Ensemble learning for blind source separation. In S. Roberts and R. Everson, editors, *Independent Component Analysis: Principles and Practice*. Cambridge University Press, 2001.

[Neal, 1992] Radford M. Neal. Bayesian training of backpropagation networks by the hybrid Monte Carlo method. Technical Report CRG-TR-92-1, Department of Computer Science, University of Toronto, April 1992.

[Neal, 1994] Radford M. Neal. Priors for infinite networks. Technical Report CRG-TR-94-1, Department of Computer Science, University of Toronto, 1994.

[Neal, 1996] Radford M. Neal. *Bayesian Learning for Neural Networks*. Lecture Notes in Statistics No. 118. Springer-Verlag, 1996.

[Neal, 1998] R. M. Neal. Regression and classification using Gaussian process priors. In J. M. Bernardo et al, editor, *Bayesian Statistics 6*, pages 475–501. Oxford University Press, 1998.

[Ripley, 1996] B. D. Ripley. *Pattern Recognition and Neural Networks*. Cambridge University Press, 1996.

[Roweis and Ghahramani, 1999] Sam Roweis and Zoubin Ghahramani. A unifying review of linear gaussian models. *Neural Computation*, 11(2):305–345, 1999.

[Rumelhart and McClelland, 1986] D. E. Rumelhart and J. L. McClelland. *Parallel Distributed Processing: Explorations in the Microstructure of Cognition*. MIT Press, 1986.

[Seeger, 2000] Matthias Seeger. Bayesian model selection for Support Vector machines, Gaussian processes and other kernel classifiers. In S. A. Solla, T. K. Leen, and K-B Müller, editors, *Advances in Neural Information Processing Systems 12*, pages 603–609. MIT Press, 2000.

[Tikhonov and Arsenin, 1977] A. N. Tikhonov and V. Y. Arsenin. *Solutions of Ill-Posed Problems*. John Wiley & Sons, 1977.

[Tipping, 2000] Michael E. Tipping. The relevance vector machine. In S. A. Solla, T. K. Leen, and K.-B. Müller, editors, *Advances in Neural Information Processing Systems 12*, pages 652–658. MIT Press, 2000.

[Trecate *et al.*, 1999] Giancarlo Ferrari Trecate, C. K. I. Williams, and M. Opper. Finite-dimensional approximation of Gaussian processes. In M. J. Kearns, S. A. Solla, and D. A. Cohn, editors, *Advances in Neural Information Processing Systems 11*, pages 218–224. MIT Press, 1999.

[Vapnik, 1979] V. Vapnik. *Estimation of Dependences Based on Empirical Data*. Nauka, Moscow, 1979. English translation: Springer Verlag, New York, 1982.

[Vapnik, 1998] Vladimir Vapnik. *Statistical Learning Theory*. John Wiley, 1998.

[Vapnik and Chapelle, 2000] V. Vapnik and O. Chapelle. Bounds on error expectation for support vector machines. *Neural Computation*, 12(9):2013–2036, 2000.

[Vapnik and Mukherjee, 2000] Vladimir N. Vapnik and Sayan Mukherjee. Support vector method for multivariate density estimation. In S. A. Solla, T. K. Leen, and K-B Müller, editors, *Advances in Neural Information Processing Systems 12*, pages 659–664. MIT Press, 2000.

[Vapnik *et al.*, 1997] Vladimir Vapnik, Steven E. Golowich, and Alex Smola. Support vector method for function approximation, regression estimation, and signal processing. In Michael C. Mozer, Michael I. Jordan, and Thomas Petsche, editors, *Advances in Neural Information Processing Systems 9*, pages 281–287. MIT Press, 1997.

[Wiener, 1949] N. Wiener. *Extrapolation, Interpolation, and Smoothing of Times Series*. MIT Press, 1949.

[Williams, 1998a] Christopher K. I. Williams. Computation with infinite neural networks. *Neural Computation*, 10(5):1203–1216, 1998.

[Williams and Rasmussen, 1996] Christopher K. I. Williams and Carl Edward Rasmussen. Gaussian processes for regression. In Michael C. Mozer David S. Touretzky and Michael E. Hasselmo, editors, *Advances in Neural Information Processing Systems 8*, pages 514–520. The MIT Press, 1996.

[Williams and Seeger, 2001] Christopher K. I. Williams and Matthias Seeger. Using the Nyström method to speed up kernel machines. In T. K. Leen, T. G. Diettrich, and V. Tresp, editors, *Advances in Neural Information Processing Systems 13*. The MIT Press, 2001.

[Williams, 1991] P. M. Williams. A Marquardt algorithm for choosing the step-size in backpropagation learning with conjugate gradients. Cognitive Science Research Paper CSRP 229, University of Sussex, February 1991.

[Williams, 1995] P. M. Williams. Bayesian regularization and pruning using a Laplace prior. *Neural Computation*, 7(1):117–143, 1995.

[Williams, 1996] P. M. Williams. Using neural networks to model conditional multivariate densities. *Neural Computation*, 8(4):843–854, 1996.

[Williams, 1998b] P. M. Williams. Modelling seasonality and trends in daily rainfall data. In M. I. Jordan, M. J. Kearns, and S. A. Solla, editors, *Advances in Neural Information Processing Systems 10*, pages 985–991. The MIT Press, 1998.

[Williams, 1999] P. M. Williams. Matrix logarithm parametrizations for neural network covariance models. *Neural Networks*, 12(2):299–308, 1999.

[Williamson *et al.*, 1998] R. C. Williamson, J. Shawe-Taylor, P. L. Bartlett and M. Anthony. Structural risk minimization over data-dependent hierarchies. *IEEE Transactions on Information Theory*, 44(5):1926–1940, 1998.

PART II

LOGIC, MATHEMATICS AND BAYESIANISM

COLIN HOWSON

THE LOGIC OF BAYESIAN PROBABILITY

For the last eighty or so years it has been generally accepted that the theory of Bayesian probability is a theory of partial belief subject to rationality constraints. There is also a virtual consensus that both the measure of belief and the constraints to which it is subject can only be provided via utility theory. It is easy to see why this should be so. The underlying idea, accepted initially by both de Finetti and Ramsey in their seminal papers ([1964] and [1931] respectively, though the paper 1964, first published in 1937, built on earlier work), but going back at least as far as Bayes' Memoir [1763], is that an agent's degree of belief in or uncertainty about a proposition A can be assessed by their rate of substitution of a quantity of value for a conditional benefit [S if A is true, 0 if not]. The natural medium of value is, of course, money, but the obvious difficulties with sensitivity to loss and the consequent diminishing marginal value of money seem to lead, apparently inexorably, to the need to develop this idea within an explicit theory of utility. This was first done only in this century, by Ramsey [1931]; today it is customary to follow Savage [1954] and show that suitable axioms for preference determine a reflexive and transitive ordering 'at least as probable as' and thence, given a further assumption about how finely the state space can be partitioned, a unique probability function.

The results of these various endeavours have all the hallmarks of a vigorously progressing research-programme. For all that, I do not myself think that it is the right way to provide a foundation for epistemic probability. I believe that the current state of utility theory itself is far from satisfactory, but underlying that concern is the feeling that one should not need a general theory of rational preference in order to talk sensibly about estimates of uncertainty and the laws these should obey. These estimates are *intellectual judgments*, and they are constrained by rules of consistency. In support of this view is an elementary mathematical fact seldom highlighted but of much significance. The probability of a proposition A is the expected value of its indicator function, that is the function defined on the space of relevant possibilities which takes the value 1 on those possible states of affairs that make A true, and 0 on the others. In other words, *probability is expected truth-value.* Truth and its logical neighbourhood are surely the right focus, not rationality (whatever that is).

These considerations suggest a view of mathematical uncertainty well-represented if not prominent among the seventeenth and eighteenth century pioneers, which is that the laws of epistemic probability are, in Leibniz's words, 'une nouvelle espèce de logique'[1]. And James Bernoulli, in the *Ars Conjectandi*, talked of measuring the probability of an uncertain proposition A in terms of the number of 'proofs' of A relative to the number of 'proofs' of not-A. The sort of logical

[1] 'I have more than once said that we should have a new kind of logic which would treat degrees of probability' (*New Essays*, bk. IV, ch. XVI).

D. Corfield and J. Williamson (eds.), Foundations of Bayesianism, 137–159.

analysis of probability proposed by Leibniz and Bernoulli was never properly developed, however. In retrospect it is easy to see the factors that hindered it: firstly, rapid technical development, and exploration of the problem-solving power of the mathematical theory, were primary, relegating 'philosophical' investigation to a relatively low-priority task; secondly, a satisfactory theory of *deductive* logic did not arrive until the beginning of the twentieth century; thirdly, probability became a puzzlingly equivocal notion, with two seemingly quite different aspects. These S-D. Poisson [1823] labelled respectively *chance*, signifying a property of events generated by repeatable random devices, measured by long-run frequencies (this is more or less Carnap's probability$_2$), and *probabilité*, signifying warranted degree of certainty relative to an agent's knowledge-state (cf. Carnap's probability$_1$).

The list of factors is not complete. Possibly the most influential, and one that seemed to many to prove decisive, was a principle additional to the standard probability axioms which was regarded as indispensable for actually determining probability-values. The principle is usually known now by the name Keynes (an advocate of it) gave it: the *Principle of Indifference*. Enough has been written about the difficulties and paradoxes attending the use of this principle (see, for example, [Howson and Urbach, 1993, Ch. 4] to make it unnecessary to undertake another long discussion here. But the connection between the principle and the logical programme is intimate, and when stated in the context of modern semantics it is very plausible. Ironically, therefore, it is also in such a setting that the basic problem with the principle is most easily identified and its gravity appreciated. Thus, suppose that a sentence B admits n models distinct up to isomorphism, in r of which A is also true. Then it seems obvious that there is a logical interpretation of a conditional probability $P(A|B)$ evaluated according to the rule: $P(A|B) = r/n$.[2] Such an interpretation is practically explicit in Bolzano [1850, Sec 66 et seq.], and fully explicit a century later in Carnap [1950], for whom this function (denoted by $c^{\dagger}$) explicated formally the idea of a *partial entailment* of A by B, measured by the proportion of B's models which are also models of A (though Carnap abandoned this measure almost immediately because of its inability to deliver a type of induction by enumeration; see [Howson and Urbach, 1993, Ch. 4].

Of course, there is a problem when B does not admit only finitely many models. In such cases it may still nevertheless be possible to exploit a 'natural' metric structure, such as when the possibilities are parametrizable by real numbers in a compact interval (such a structure was famously exploited by Bayes when he derived a conditional, 'posterior', probability distribution for a binomial parameter in his [1763]). However, it was in precisely this type of case that serious problems with the Principle of Indifference first became evident, with the discovery of what were called the 'paradoxes of geometrical probability', where 'geometrical'

[2] Laplace of course enunciated in effect just this rule when he defined the probability of an event to be the number of cases favourable to the event divided by the number of all possible cases, where these cases are 'equally possible'. The proviso has been much commented on, but in the semantic context its meaning is clear enough: that is certainly how Bolzano and the German school understood it later.

was the traditional word referring to the real-number continuum, or some compact subset thereof (the best-known of these 'paradoxes' is Bertrand's chord problem; see [Kac and Ulam, 1968, pp. 37–39]). This underlying problem is that how the elementary possibilities are conceived is not absolute but relative to some conceptual frame — in effect a *language* — and depending on how this is chosen the probability-values will themselves vary.

The mathematical subtleties of continuous possibility spaces rather conceal this point by suggesting that it is only in such spaces that real problems arise (still unfortunately a common point of view), so here is a very elementary example which shows that the problem lies right at the heart of the idea of taking Laplacean ratios to compute probabilities. Consider two simple first-order languages with identity and a one-place predicate symbol Q, and no other relation or function symbols. Language 1 has no individual names (constants), and language 2 has two individual names a, b. In both languages there are identical sentences

A: *Something has the property Q*

B: *There are exactly 2 individuals.*

There are only three models of B in language 1 distinguishable up to isomorphism: one containing no, one containing one and one containing two instances of Q. Two of these satisfy A. In language 2, on the other hand, the fact that the individuals can be distinguished by constants means that there are more than three distinct models of B: allowing for the possibility that the constants might name the same individual there are eight, six of which satisfy A. Using the Laplacean definition of $P(A|B)$ above we therefore get different answers for the value of $P(A|B)$ depending on which language we use. In language 1 the value is 2/3 and in Language 2 it is 3/4 (cf. Maxwell–Boltzmann vs. Bose–Einstein statistics). Relative to each language the models are of course all 'equally possible'.

To sum up: the 'equal possibilities' of the Principle of Indifference are equal relative to some conceptual frame, or language. This can in general be chosen in different ways and, depending on the choice, the ratios of numbers of favourable to possible cases *for the same event or proposition* may vary — where they can be computed at all. Moreover, not only does there seem to be no non-arbitrary way of determining the 'correct' language, or even being able to assign any meaning to the notion of a correct language, but in continuous spaces entirely equivalent frames will exist, related by one-to-one bicontinuous transformations, which will generate different probabilities.

This intractable knot was eventually untied, or rather cut, only in the last century, in the move from objectively to subjectively measured uncertainty, for with that move was jettisoned the Principle of Indifference, as inappropriate in a theory merely of consistent degrees of belief. Unfortunately there was no further systematic development of the idea that the probability axioms are no more than consistency constraints, at any rate within an explicitly logical setting: Ramsey went on to pioneer the development of subjective probability as a subtheory of utility theory (we shall see why shortly), and de Finetti employed the idea of penalties like Dutch Books to generate the probability laws. De Finetti's Dutch Book argument and its extension to scoring rules generally also lead in the wrong direction, of financial prudence in unrealistic circumstances (you always agree to take either side of a bet with any specified stake at your personal betting quotient). So despite its promising starting idea, from a strictly *logical* point of view Ramsey's and de Finetti's work represented a dead end. In what follows I propose to go back to the beginning, and combine the intuitions of Leibniz and J. Bernoulli with the conceptual apparatus of modern logic. We shall then get a 'rational reconstruction' of history closer to their intentions than to the actual course of events.

1 DEGREE OF BELIEF

There is a long tradition of informally expressing one's uncertainty about a proposition A in the odds which one takes to reflect the currently best estimate of the chances for and against A's being true. Mathematically, odds are a short step from probabilities, or at any rate the probability scale of the unit interval. The step is taken by normalising the odds, by the rule p = odds/(1+odds). p is called the *betting quotient* associated with those odds. The p-scale has the advantage as an uncertainty measure that it is both bounded and symmetrical about even-money odds (unlike the odds scale itself where the even-odds point, unity, is close to one end of the scale (0) and infinitely far from the other). Since the seventeeth century betting quotients as the measure of uncertainty have been called probabilities, and for the time being I shall do so myself (I am quite aware that they have not yet been shown to be probabilities in the technical sense: that will come later). Note that the inverse transformation gives odds $= p/(1-p)$.

To determine how such probabilities should be evaluated in specific cases was the function of the Principle of Indifference, a function which, as we have seen, it was unable to discharge consistently. However, abandoning the Principle, as Ramsey saw, seems to leave behind only *beliefs* about chances. This appears to signal a move into mathematical psychology, and particularly into measurement theory to develop techniques for measuring partial belief. Such at any rate was the programme inaugurated and partly carried out by Ramsey in his pioneering study 'Truth and Probability' [1931]. According to Ramsey the empirical data of partial belief are behaviourally-expressed preferences for rewards which depend on the outcomes of uncertain events: for example, in bets. Though Ramsey's idea seems to have a scope far beyond ordinary betting, as he pointed out we can always think in a general context of bets not just against human opponents but also against Nature. But preferences among bets will normally depend not only on the odds but also on the size of the stake: for large stakes there will be a natural disinclination to risk a substantial proportion of one's fortune, while for very small ones the odds will not matter overmuch. The only answer to this problem, Ramsey believed, was to give up the idea that odds, at any rate money odds, could measure uncertainty, and invoke instead a very general theory of rationally contrained preference: in other words, axiomatic utility theory.

But invoking the elaborate apparatus of contemporary utility theory, with its own more or less serious problems, seems like taking a hammer — and a hammer of dubious integrity and strength — to crack a nut. Why not simply express your beliefs by reporting the probabilities you feel justified by your current information, in the traditional way? The answer usually given is that to do so begs questions that only full-blown utility theory can answer. These relate to what these probabilities actually *mean*. They are odds, or normalised odds, so presumably they should indicate some property of bets at those odds. For example, I would be taken to imply that in a bet on A at any other odds than $p/(1-p)$, where p is my probability of A, I think one side of that bet would be positively disadvantaged, given what I know about A and the sorts of things that I believe make A more likely to be true than false — or not as the case may be. Thus my 'personal probability' (that terminology is due to Savage) determines what I believe to the *fair odds* on A: i.e. those odds which I believe give neither side an advantage calculable on the basis of my existing empirical knowledge. This is where the objections start.

The first is that your assessment of which odds, or betting quotients, do or do not confer advantage is a judgment which cannot be divorced from considerations of your own — hypothetical or actual — gain or loss and how you value these;

> for to ask which of two "equal" betters ask has the advantage is to ask which of them has the preferable alternative. [Savage, 1954, p. 63]

Granted that, we seem after all ineluctably faced with the task of developing a theory of uncertainty as part of a more general theory of preference, i.e. utility theory: precisely what it was thought could be avoided. But should we grant it? One hesitates to dismiss summarily a considered claim of someone with the authority of Savage, but nonetheless it simply is not true. Here is a simple counterexample (due in essence to [Hellman, 1997, p. 195]): imagine the bettors to be coprophiliacs and the stakes measures of manure. One's own preferences are *irrelevant* to judging fairness. They have only seemed relevant because gambles are traditionally paid in money and money is a universal medium of value.

Nevertheless, a Savagean objector might continue, to compute advantage you still need to know how the bettors themselves evaluate the payoffs, (a) in isolation and (b) in combination with what are perceived by those parties to be the chances of the payoffs; and both (a) and (b) may vary depending on the individual. For example, one party may be risk-averse and the other risk-prone, and a fair bet between such antagonists will be quite different from a fair bet between two risk-neutral ones. The answer to this is that the concept of advantage here is that of *bias*: on such a criterion a bet is fair simply if the money odds match what are perceived to be the fair odds, whatever the beliefs or values of the bettors. This is easily seen to imply an expected value criterion. For suppose R and Q are the sums of money staked, and that the odds measure of your uncertainty is $p : (1-p)$. The money odds $R : Q$ match your fair odds just in case $pQ = (1-p)R$, i.e.

(1) $pQ - (1-p)R = 0$

tells us that the bet is fair just when what is formally an expected value is equal to 0.

Now things do not look at all promising, however, for (1) seems to lead immediately to the notorious St Petersburg Problem. For those unfamiliar with it, this involves a denumerably infinite set of bets where a fair coin is repeatedly tossed and the bettor pays 1 ducat to receive 2^n ducats if the nth toss is the first to land heads, $n = 1, 2, \ldots$. Given no other information, the coin's assumed equal tendency to land heads or tails will presumably determine the fair odds. In that case the associated probability of getting the first head at the nth toss is 2^{-n}, and the expected value of each bet is clearly 0. Yet everyone intuitively feels that no bettor would be wise in accepting even a large finite number of these bets, let alone all of them — which would of course mean staking an infinite sum of money. The *inequity* of such bets, according to practically all commentators from Daniel Bernoulli onwards, is due to the diminishing marginal utility of money, and in particular the inequality in value between losing and gaining the same sum: the loss outweighs the gain, the more noticeably the larger the sum involved. In the St Petersburg game you are extremely likely to lose most of your 100 ducats if you accept the first 100 of the bets, a considerable sum to lose on a trifle. Your opponent could even less afford to pay out if the 100th bet won. Either way you would be silly to accept the bets even though they are fair by the criterion of money expectation. Nowadays it is taken for granted that the only solution to the problem is to use a utility function which is not only concave (like Daniel Bernoulli's logarithmic function) but also bounded above (unlike the logarithm).

I do not think we should worry too much about the St Petersburg Problem, for it begs a question in its turn, namely that a bet cannot be fair which it would be highly imprudent for one side to accept. But this is exactly what is being questioned, and is I think just false. A contract between a prince and a pauper is not unfair just because one can pay their debt easily and the other cannot. That is to confuse two senses of fairness: as lack of bias (the sense intended here), in which payoffs are balanced against probabilties according to (1), and as lack of differential impact those payoffs will have taking into consideration the wealth of the players. And indeed these quite distinct ideas have become confused in the Bayesian literature, to the extent that probability has become almost uniformly regarded as necessarily a subtheory of a theory of prudent behaviour.

The idea that the expected money-gain principle was vulnerable to the St Petersburg Problem was already challenged over two centuries ago by Condorcet, who pointed out that in repeated bets at odds $2^n - 1 : 1$ on heads landing first on the nth toss against, the average gain converges in probability to the expected value, i.e. 0, while, to quote Todhunter reporting Condorcet, "if any other ratio of stakes be adopted a proportional advantage is given to one of the players" [Todhunter, 1865, pp. 392–393]. Of course, this argument relies on (a) the full apparatus of probability theory (the quick modern proof would use Chebychev's Inequality) and

(b) assuming that the trials are uncorrelated with constant probability, neither of which assumption is appropriate here. But one doesn't need all that anyway: the moments equation (1) itself is a sufficient answer.

To sum up: the agent's probability is the odds, or the betting quotient, they currently believe fair, with the sense of 'fair' that there is no calculable advantage to either side of a bet at those odds. Despite opposing the widely accepted view that subjective probability can be coherently developed only within a theory of utility, this view is, I believe, quite unexceptionable, and certainly not vulnerable to what are usually taken to be decisive objections to it. Not only is it unexceptionable: it will turn out to deliver the probability axioms in a way that is both elegant and fully consonant with the idea that they are nothing less than conditions of consistency, and a complete set of such conditions at that.

2 CONSISTENCY

Now we can move on to the main theme of this paper. Ramsey claimed that the laws of probability are, with the probability function interpreted as degree of belief, laws of consistency and so of the species of logic. Unfortunately, as we saw, Ramsey then proceeded to divert the theory into the alien path of utility, where 'consistent' meant something like 'rational'. But rationality has nothing essentially to do with logic, except at the limits. Can we give the idea of consistent assignments of fair betting quotients an authentically logical meaning? The answer is that we can. We proceed in easy stages. A traditional sense of consistency for assignments of numbers is equation-consistency, or *solvability*. A set of equations is consistent if there is at least one single-valued assignment of values to its variables. The variables evaluated in terms of betting quotients are propositions. Correspondingly, we can say that an assignment of fair betting quotients is consistent just in case it can be solved in a analogous sense, the sense of being extendable to a single-valued assignment to all the propositions in the language determined by them (this is the notion of consistency Paris appeals to in a recent work on the mathematical analysis of uncertainty [Paris, 1994, p. 6]). But what, it might be asked, has the notion of consistency as solvability to do with logical consistency? Everything, it turns out. For *deductive consistency itself is nothing but solvability.* To see why, it will help to look at deductive consistency in a slightly different sort of way, though one still equivalent to the standard account, as a property not directly of sets of sentences *but of truth-value assignments.*

According to the standard (classical) Tarskian truth-definition for a first or higher-order language conjunctions, disjunctions and negations are homomorphically mapped onto a Boolean algebra of two truth-values, $\{T, F\}$, or $\{1, 0\}$ or however these elements are to be signified (here T or 1 signifies 'true' and F or 0 signifies 'false'). Now consider *any* attribution of truth-values to some set Σ of sentences of L, i.e. *any* function from Σ to truth-values. We can say that this assignment is consistent if it is capable of being extended to a function from the entire set of sentences of L to truth-values which satisfies those homomorphism constraints. For propositional languages the language of equation-solvability is sometimes explictly used: formulas can be regarded as representing boolean polynomial equations [Halmos, 1963, p. 8] in the algebra of two truth-values, and sets of them are consistent just in case they have a simultaneous solution.

The theory of 'signed' semantic tableaux or trees is a syntax perfectly adapted to seeing whether such equations are soluble and if so, finding all the solutions to them. ('Signing' a tableau just means appending Ts and Fs to the constituent sentences. The classic treatment is Smullyan [1968, pp. 15–30], a simplified account is in [Howson, 1997b].) Here is a very simple example:

$$\begin{array}{l} A\ \ T \\ A \rightarrow B\ \ T \\ B\ \ F \end{array}$$

The tree rule for $[A \rightarrow B\ \ T]$ is the binary branching

$F\ A \qquad B\ T$

Appending the branches beneath the initial signed sentences results in a *closed tree*, i.e. one on each of whose branches occurs a sentence to which is attached both a T and an F. A soundness and completeness theorem for trees [Howson, 1997b, pp. 107–111] tells us that any such tree closes if and only if the initial assignment of values to the three sentences $A, A \rightarrow B$ and B is inconsistent, i.e. unsolvable over L subject to the constraints of the general truth-definition.

To sum up: in deductive logic (semantic) consistency can be equivalently defined in the equational sense of a truth-value assignment being solvable, i.e. extendable to a valuation over all sentences of L satisfying the general rules governing truth-valuations. By a natural extension of the more familiar concept we can call such an extension a *model* of the initial assignment. Note that this sense of consistency does not pick out a different concept from the more usual one of a property of sets of sentences. Indeed, the two are essentially equivalent, as can be seen by noting that an assignment of truth-values to a set Σ of sentences is consistent in the solvability sense above just in case the set obtained from Σ by negating each sentence in Σ assigned F is consistent in the standard (semantic) sense.

We have become accustomed to understand by consistency deductive consistency and thereby something that exists, so to speak, only in a truth-centred environment. That this is not necessarily the case is now clear, for deductive consistency is seen to be merely an application of a much more general (and older) idea of consistency as solvability, having nothing necessarily to do with truth at all, but merely with assignments of values, not necessarily and indeed not usually truth-values, to variables in such a way that does not result in overdetermination.

What deductive and probabilistic consistency do have in common however is that the variables in question are propositional, and to proceed further we need to specify the language relative to which an assignment of fair betting quotients is solvable (if it is), subject to the appropriate constraints. For the sake of definitieness let us start, as in deductive logic, with a language relative to which the class of propositions will be determined. In fact, we can employ just the same sort of language, a first order language. Let L be one such, without identity. Let Ω = the class of structures $\Im$ interpreting the extralogical vocabulary of L. For any sentence A of L let $\text{Mod}(A) = \{\Im : A \text{ is true in } \Im\}$. Let $F = \{\text{Mod}(A)\text{: } A \text{ a sentence of } L\}$. Following Carnap [1971, pp. 35–37] F is the set of *propositions* of L. Note that F is a Boolean algebra isomorphic to the Lindenbaum sentence algebra of L [Paris, 1994, p. 34]. In fact, it will be better to work in a rather more extensive class of propositions, because F as it stands represents merely the propositions expressible by single sentences of L. But it is well-known that the mathematical theories incorporated in any minimally acceptable theory of physics, for example, are not expressible by single sentences of a first order language: they are not finitely axiomatisable (even the simplest of all mathematical theories, the theory of identity investigated by Euclid over 2000 years ago, is not finitely axiomatisable). The customary closing off of F under denumerably infinite unions (disjunctions) and intersections (conjunctions), generating the Borel field $B(F)$, (more than) allows such theories to be treated on a par with their finitely axiomatisable cousins.

In Ω and $B(F)$ we have two of the three ingredients of what mathematicians call a *probability space*. The third is a probability function defined on $B(F)$. Finding this will be the next task (in what follows I shall use $A, B, C, \ldots$ now to denote members of $B(F)$). The first step on the way is to determine the appropriate constraints on solutions of assignments of fair betting quotients. These will function like the purely *general* rules of truth in classical truth-definitions, as analytic properties of truth. In the context of fair betting quotients the constraints should presumably be analytic of the notion of fairness as applied to bets. At this point it is helpful to transform the payoff table

A	
T	Q
F	$-R$

into the well-known (betting-quotient, stake) 'coordinates' introduced by de Finetti in his seminal paper [1964]. The stake S is $R + Q$ and the betting quotient p^* is of course just R/S, and the table above becomes

A	
T	$S(1 - p^*)$
F	$-p^*S$

Where I_A is the indicator function of A, the bet can now be represented as a random quantity $S(I_A - p^*)$, and the equation (1) now transforms to

$$pS(1 - p^*) - p^*S(1 - p) = 0 \tag{1'}$$

where p is your fair betting quotient. Clearly, the left hand side is equal to 0 just when $p = p^*$, which is merely a different way of stating that a fair bet is one in which your estimate of the fair odds is identical with the money odds. Besides bets like the above there are also so-called *conditional bets*, i.e. bets on a proposition A which require the truth of some proposition B for the bet to go ahead: if B is false the bet on A is annulled. The bet is called a conditional bet on A given B. A betting quotient on A in a conditional bet is called a conditional betting quotient. A conditional bet on A given B with stake S and conditional betting quotient p clearly has the form $I_B S(I_A - p)$. If your uncertainty about A is registered by your personal fair betting quotient on A then your uncertainty, your conditional uncertainty, on A *on the supposition that* B *is true* will plausibly be given by your conditional fair betting quotient on A given B.

Let (F) be the set of formal constraints other than $0 \leq p \leq 1$ which fair betting quotients, including conditional fair betting quotients, should *in general* satisfy. This general content is contained in the claim that a fair bet is unbiased given the agent's own beliefs. These of course are unspecified, varying as they do from individual to individual. We can quickly infer

(a) If p is the fair betting quotient on A, and A is a logical truth, then $p = 1$; if A is a logical falsehood $p = 0$. Thus logical truth, logical falsehood and entailment relations correspond to the extreme values of fair betting quotients. Similarly if B entails A then the conditional betting quotient on A given B should be 1, and 0 if B entails the negation of A.

(b) Fair bets are invariant under change of sign of stake.

The reason for (a) is not difficult to see. If A is a logical truth (i.e. $A = \Omega$) and p is less than 1 then in the bet $S(I_A - p)$ with betting quotient p, I_A is identically 1 and so the bet reduces to the positive scalar quantity $S(1-p)$ received come what may. Hence the bet is not fair since one side has a manifest advantage. Similar reasoning shows that if A is a logical falsehood then p must be 0. Similar reasoning accounts for the conditions relating to entailment. As to (b), (1′) shows that the condition for a bet to be fair is independent both of the magnitude and sign of S.

But there is something else to (F) besides (a) and (b), a natural closure condition which can be stated as follows: *if a set of fair bets determines a bet on a proposition B with betting quotient q then q is the fair betting quotient on B.* What is the justification for this apart from 'naturalness'? It is well-known by professional bookmakers that certain combinations of bets amount to a bet on some other event, inducing a corresponding relationship between the betting quotients. For example, if A and B are mutually inconsistent then simultaneous bets at the same stake are extensionally the same as a bet on $A \vee B$ with that stake, and if p and q are the betting quotients on A and B respectively, we easily see that $S(I_A - p) + S(I_B - q) = S(I_{A \vee B} - r)$ if and only if $r = p + q$. Now add to this the thesis that if each of a set of bets gives zero advantage then the net advantage of anybody accepting all of them should also be zero (though this thesis is not *provable*, it seems so fundamentally constitutive of the ordinary notion of a fair game that we are entitled to adopt it as a desideratum to be satisfied by any formal explication; and, of course, when it is explicated as zero expected value within the fully developed mathematical theory we have the elementary theorem that expectation is a linear functional and hence all expectations, zero or not, add over sums of random variables). Putting all this together we obtain the closure principle above.

To proceed further, note that bets obey the following arithmetical conditions:

(i) $-S(I_A - p) = S(I_{\neg A} - (1-p))$.

(ii) If $A \& B = \bot$ then $S(I_A - p) + S(I_B - q) = S(I_{A \vee B} - (p+q))$.

(iii) If $\{A_i\}$ is a denumerable family of propositions in $B(F)$ and $A_i \& A_j = \bot$ and p_i are corresponding betting quotients and $\sum p_i$ exists then $\sum S(I_{A_i} - p_i) = S(I_{\vee A_i} - \sum p_i)$.

(iv) If $p, q > 0$ then there are nonzero numbers S, T, W such that $S(I_{A\&B} - p) + (-T)(I_B - q) = I_B W(I_A - p/q)$ (T/S must be equal to p/q). The right hand side is clearly a conditional bet on A given B with stake W and betting quotient p/q.

Closure tells us that if the betting quotients on the left hand side are fair then so are those on the right. The way the betting quotients on the left combine to give those on the right is, of course, just the way the probability calculus tells us that probabilities combine over compound propositions and for conditional probabilities.

Now for the central definition. Let Q be an assignment of personal fair betting quotients to a subset X of $B(F)$. By analogy with the deductive case, we shall say that Q is *consistent* if it can be extended to a single-valued function on all the propositions of L satisfying suitable conditions.

The final stage in our investigation is to generate interesting properties of consistency. If we suggestively signify a fair betting quotient on A by $P(A)$ closure tells us

(i′) $P(\neg A) = 1 - P(A)$.

(ii′) If $A \& B = \bot$ then $P(A \vee B) = P(A) + P(B)$.

(iii′) If $\{A_i\}$ is a denumerable family of propositions in $B(F)$ and $A_i \& A_j = \bot$ and $\sum P(A_i)$ exists then $P(\vee A_i) = \sum P(A_i)$.

(iv′) If $P(A \& B)$ and $P(B) > 0$ then $P(A|B) = P(A\&B)/P(B)$.

It might seem slightly anomalous that in (iv′) both $P(B)$ and $P(A\&B)$ should be positive, but it will turn out that only the former condition need be retained. At any rate, from the equations (i′)–(iv′) and (F) it is a short and easy step to proving the following theorem:

THEOREM 1. *An assignment Q of fair betting quotients (including conditional fair betting quotients) to a subset X of $B(F)$ is consistent (has a model) if and only if Q satisfies the constraints of the countably additive probability calculus; i.e. if and only if there is a countably additive probability function on $B(F)$ whose restriction to X is Q.*

Proof. The proof of the theorem is straightforward. Necessity is a fairly obvious inference from the closure property. It is not difficult to see that the condition that $P(A\&B)$ be positive in (iv′) can be jettisoned once we can assume invariance under sign of stake. Also, given finite additivity, the condition in (ii′) that $\sum P(A_i)$ exists is provable (using the Bolzano–Weierstrass Theorem). For sufficiency, all we really have to do is show that closure follows, but this is easy. For suppose P is a probability function on $B(F)$ and that X_i are bets on a finite or countable set of propositions A_i in which the betting quotients are the corresponding probabilities $P(A_i)$. Suppose also that $\sum X_i = S(I_B - q)$ for some proposition B. The expected value relative to P of each X_i is 0, and since expectations are linear functionals the expected value of the sum is also 0. Hence the expected value of $S(I_B - q)$ must be 0 also and so $q = P(B)$. So we have closure. ∎

I pointed out earlier that there is a soundness and completeness theorem for trees (signed or unsigned), establishing an extensional equivalence between a semantic notion of consistency, as a solvable truth-value assignment, and a syntactic notion, as the openness of a tree from the initial assignment. In the theorem above we seem therefore to have an analogous soundness and completeness theorem, establishing an extensional equivalence between a semantic notion of consistency, i.e. having a model, and a syntactic one, deductive consistency with the probability axioms when the probability functor P signifies the fair betting quotients in Q. The deductive closure of the rules of the probability calculus is now seen as the complete theory of generally valid probability-assignments, just as the closure of the logical axioms in a Hilbert-style system is the complete theory of generally valid assignments of truth. My proposal is to take the theorem above as a soundness and completeness theorem for a logic of uncertainty, of the sort Leibniz seems to have had in mind when he called probability 'a new kind of logic'. To complete the discussion we must consider briefly what qualifies a discipline for the title 'logic'.

3 LOGIC (WHAT IS IT?)

Wilfrid Hodges's well-known text on elementary deductive logic tells us that 'Logic is about consistency' [Hodges, 1974]. This raises a question. Should 'consistency' just mean 'deductive consistency', or might there be other species of consistency closely kindred the deductive variety entitling their ambient theories to the status of logic or logics? It may well be the case that logic is about consistency without foreclosing the possibility of there being logics other than deductive. To answer these questions we first need an answer to the question 'What is logic?'.

My own belief is that there is no fact of the matter about what entitles a theory of reasoning to logical status, and one has to proceed as one does in extending common law to new cases, by appeal to precedent and common sense. Here again, of course, one must be selective, but with modern deductive logic in mind I propose — hesitantly — the following criteria for a discipline to be a logic:

(a) Its field is statements and relations between them.

(b) It adjudicates a mode of non-domain-specific reasoning in its field.

(c) Ideally it should incorporate a semantic notion of consistency extensionally equivalent to a syntactic one: it should have a *soundness and completeness theorem*. First order logic famously has a soundness and completeness theorem; so of course do many modal systems. Second order logic does not, but one could argue that that is the exception proving the rule, for it is largely for this reason that second order logic is generally regarded as not being a logic.

(a) and (b) are certainly satisfied by Bayesian, i.e. evidential, probability theory: any factual statement whatever can be in the domain of such a probability function. An interesting fact implicit in the discussion above is that the the statements involved here are not tied to any language: a Borel field does not of course have to be generated by the sets of model-structures of a first, or indeed any, order language. It can be completely language-free, and still nevertheless be regarded as a set of propositions, propositions in the most general sense of a class of subsets of a possibility-space, closed under the finite and countable Boolean operations. This creates a great deal of freedom, e.g. to assign probabilities to the measurable subsets of Euclidean n-space, a freedom not available in any of the classical logical languages..

As to (c), the theorem above establishes an extensional equivalence between a semantic notion, having a model, and a syntactic one (the probability calculus is a purely syntactic theory, of a function assigning values to elements of an algebra), and as such, I have claimed, is in effect a soundness and completeness theorem for Bayesian logic. The question is whether fulfilling (a)–(c) is sufficient to warrant the title 'logic'. It is of course quite impossible in principle to prove this, just as it is impossible in principle to prove the Church Turing Thesis that the partial recursive functions exhaust the class of computable (partial) functions. In the latter case 'computable function', and in the former 'logic', have no precise mathematical definition. In addition, new theories of reasoning are increasingly marketed under the general title 'logic' as the information technology revolution gets under way (whereas even twenty years ago most logic texts were produced in university philosophy departments, now probably most are produced in computer science departments). Under any reasonable definition of 'theory of general reasoning' the rules of evidential probability would qualify as such, and hence as logic in this broad, liberal construal. But what has been shown is that a much tighter criterion, having an authentic semantics provably equivalent extensionally to an equally authentic syntax, applies to evidential probability in much the same sort of way that it applies to first order logic. Of course, it is open to anyone to deny that a completeness result is essential to a genuine logic; this is, of course, just what advocates of second order logic say (assuming that only full models are counted). Be that as it may, there seems little doubt that the completeness, i.e. axiomatisability, of first order logic has been a major factor in its widespread acceptance not only as logic but pretty well as *the* (classical) logic.

I have made consistency the focus of my discussion. It might well be objected that central to logic is the idea of *consequence*, or deduction. In a recent collection [Gabbay, 1994] dedicated to the discussion of what is to count as logic we find that this is a view common to almost all the authors. For example, 'logic is the science of deduction' [Hacking, 1994, p. 5]; 'a logical system is a pair $(\mid\!\sim, S_{\mid\sim})$ where $S_{\mid\sim}$ is a proof theory for $\mid\!\sim$ [$\mid\!\sim$ is a consequence relation]'. [Gabbay, 1994, p. 181]; 'Logic is concerned with what follows from what i.e. with logical consequence' [Aczel, 1994, p. 262]; and so on. I think that the reflections above show that this view, though widespread, is nevertheless incorrect. It arose because traditionally logic has been about conditions necessary and sufficient for the preservation of just one of the values in a two-valued system, the truth-value 'true'. In this sense, and indeed quite naturally, logic has traditionally been *deterministic*. It is true that there have been proposals for various sorts of many-valued logics, discrete and continuous, but even there the tendency has been to retain as far as possible something like a traditional concept of consequence. Even Adams's explicitly probabilistic system does this [Adams, 1998]. I believe that it is misguided because it is in effect a denial of the freedom such a mutli-valued system affords to get away from what is, I believe, nothing more than an artifact of two-valued systems. Of course, even in the account proposed here there is a consequence relation, but it is only the trivial one of *deductive* consequence from the probability axioms, telling us that if such and such are the probability-values of a specified set of propositions, then so and so is the probability of some other proposition. Williamson, discussing the account I have given above, points out that a relation of probabilistic consequence emerges naturally by analogy with the usual deductive notion of semantic consequence. A sentence A is a semantic consequence of a set Σ of sentences iff every model of Σ is a model of A. This transforms to: an assignment $r(A)$ is a consequence of an assignment $q(B_1), ..., q(B_n)$ iff every probability function extending q also extends r, i.e. iff every model, in the sense I have given above, of q is a model of r [Williamson, 2001]. But this does not generate any notion of consequence between A and $B_1, \ldots, B_n$ themselves. As Williamson notes, it generates a notion of probabilistic consequence, but only in the deductive sense above: $r(A)$ is a consequence of $q(B_1), \ldots, q(B_n)$ iff $P(A) = q(A)$ follows deductively from the probability axioms together with the 'assumption formulas' $P(B_1) = q(B_1), \ldots, P(B_n) = q(B_n)$.

Contemporary discussions of the relation between probability and formal deductive logic take a quite different approach to the one I regard as implicit in the theorem above. Some, e.g. [Gaifman, 1964; Scott and Krauss, 1970], take the logical aspect of probability to be exhausted by defining a probability function on the sentences of a suitable formal language, either a standard first order language or an infinitary one (as with Scott and Krauss), and showing how standard measure-theoretic arguments have to be correspondingly modified, in particular the extension theorem that states that a countably additive probability function on a field of sets has a unique countably additive extension on the Borel closure [Kolmogorov, 1956] p. 17). Gaifman provides an analogue of this for finitely additive probability functions defined on the class of sentences of a first order language with equality, showing that if a condition that has consequently come to be known as the Gaifman condition is satisfied then there is a unique extension from the quantifier-free sentences to all the sentences of the language (the Gaifman condition states that the supremum of the probabilities of all the disjunctions of instances of a formula is equal to the probability of its existential quantification; in terms of the Lindenbaum algebra the Gaifman condition is that probabilities commute with suprema).

Others, like Fagin and Halpern [1988], and in a different way Heifetz and Mongin [forthcoming], incorporate probability into the *syntax* of a formal language. 'Pulling down' probability into the object language is of course very much in the spirit of modal logic, and indeed Heifetz and Mongin introduce what they call a modal operator, 'the probability is at least α that ...', for all rational α in [0,1], which they interpret according to a generalisation of Kripke semantics incorporating a probability distribution over a class of possible worlds. What distinguishes my own account most of all from the modal one(s) and the others that I have mentioned is that they take the probability axioms as pretty much given: in Gaifman and Scott and Krauss probability is just a weakening of the usual two-valued semantic valuation, while in Heifetz and Mongin the axioms are, as in Savage, derivative from a set of rationality constraints over preferences. I believe I have shown in the foregoing that the probability axioms for epistemic probability are naturally, if not inescapably, interpreted as being of the same general species of consistency constraint as the rules of deductive logic itself, and to that extent *intrinsically* logical in nature.

We now move to consider some collateral benefits accruing from construing the (epistemic) probability calculus as a species of logical axioms. Where a title, like 'logic', is already uncontroversially bestowed elsewhere, discharging some genuine explanatory function should be a condition of its extension to a putative new case. As we shall see below, there are in addition to (a)–(c) above other interesting points of contact, or near-contact, with classical deductive logic, and a logical understanding of the rules of probability will, I hope, be seen to bring with it a considerable explanatory bonus.

The following topics in Bayesian probability are all regarded as problematic to some extent or other: countable additivity; strict coherence versus coherence;

rationality; completeness; conditionalisation; sharp versus fuzzy or interval-valued probability; inductive inference; penalties for infringing the rules. The logical view at the least usefully illuminates these and at best solves them. I shall deal with them in turn.

3.1 *Countable Additivity*

There has been a good deal of controversy concerning the status of the principle of countable additivity within the theory of epistemic probability. Most workers in the field, including, famously, de Finetti himself, reject it, while a much smaller number accept it. I do not think that it is necessary to go into the details of the protagonists' arguments. The theorem above shows that it must be adopted within any adequate view of the rules of probability as consistency constraints. The fact is that if we wish to assign probabilities as widely as possible then consistency in their assignment over compound propositions can then be guaranteed only by adding countable additivity to the axioms.

3.2 *Rationality*

I started out by remarking that the recent history of subjective probability has tended to neglect the logical aspect identified by Leibniz, favouring instead a rationality interpretation of the constraints as prudential criteria of one type or another. The trouble with adopting this line is that it is very difficult to *demonstrate* in any uncontroversial and non-question-begging way that violation of any of the constraints is positively irrational. Take the requirement of transitivity for preferences, for example: it is not evident that certain types of intransitive preference are necessarily irrational, especially when it is considered that the comparisons are always pairwise (for a counterexample see Hughes [1981]). The logical view, on the other hand, need not in principle be troubled by links with rationality of only doubtful strength, since logic is not *about* rational belief or action as such. Thus, deductive logic is about the conditions which sets of sentences must satisfy to be capable of being simultaneously true (deductive consistency), and the conditions in which the simultaneous truth of some set of sentences necessitates the truth of some given sentence (deductive consequence): in other words, it specifies the conditions regulating what might be called consistent truth-value assignments. This objectivism is nicely paralleled in the interpretation of the probability axioms as the conditions regulating the assignment of consistent fair betting quotients.

3.3 *Completeness*

Under the aspect of logic the probability axioms are as they stand complete. Hence any extension of them — as in principles for determining 'objective' prior probabilities — goes beyond pure logic. This should come as something of a relief: the principles canvassed at one time or another for determining 'objective' priors have been the Principle of Indifference, symmetry principles including principles of invariance under various groups of transformations, simplicity, maximum entropy and many others. All these ideas have turned out to be more or less problematic: at one extreme inconsistent, at the other, empty. It is nice not to have to recommend any.

3.4 *Coherence versus Strict Coherence*

A hitherto puzzling qustion posed first by Shimony [1955] and then repeated by Carnap [1971, pp. 111–114] is easily answered if we accept that the probability axioms are laws of consistency. Consider a set of bets on n propositions $A_1, ..., A_n$ with corresponding betting quotients p_i. The classic Dutch Book argument shows that a necessary and sufficient condition for there being, for every set of stakes S_i, a distribution of truth-values over the A_i such that for that distribution there is a non-negative gain to the bettor (or loss: reverse the signs of the stakes), is obedience to the probability axioms. However, if we substitute 'positive' for 'non-negative' we also get an interesting result: the necessary and sufficient condition now becomes that the probability function is in addition *strictly positive*, i.e. it takes the value 0 only on logical falsehoods. Which of these two Dutch Book arguments should we take to be the truly normative one: that we should always have the possibility of a positive gain, or that we should always have the possibility of a non-negative gain? It might seem that the second is the more worthwhile objective: what is the point of going to a lot of trouble computing and checking probability values just to break even? On the other hand, strictly positive probability functions are very restrictive. There can be no continuous distributions, for example, so a whole swathe of standard statistics seems to go out of the window. There does not seem to be a determinately correct or incorrect answer to the question of what to do, which is why it is a relief to learn that the problem is purely an artifact of the classic Dutch Book argument. Give up the idea that the probability laws are justified in those terms and the problem vanishes. Indeed, we now have a decisive objection to specifying *any* conditions additional to the probability axioms: the laws of probability as they stand are complete.

3.5 *Unsharp Probabilities*

We seldom if ever have personal probabilities, defined by Bayes' procedure of evaluating uncertain options, which can be expressed by an exact real number. My value for the probability that it will rain some time today is rather vague, and the value 0.7, say, is no more than a very rough approximation. In the standard Bayesian model the probability function takes real-number values. But if we are trying to use the model to understand agents' actual cognitive decisions it would seem useful if not mandatory to assume that they have more or less diffuse probabilities — because they mostly if not invariably do in the real world.

If I am correct then probabilistic and deductive models of reasoning are intimately related, suggesting that considerations which prove illuminating in one can be profitably transferred, *mutatis mutandis*, to the other. So ask: what corresponds in deductive models to consistent probability-values? Answer: truth-values. Well, deductive models, or at any rate the standard ones, equally fail to be realistic through incorporating 'sharp' truth-values, or what comes to the same thing, predicates having sharp 'yes'/'no' boundaries. Thus, it is assumed in standard deductive logic that for each predicate Q and individual a in the domain, a definitely has Q or it definitely does not. An equivalent way of stating the assumption is in terms of *characteristic functions*: the characteristic function of Q is a function f_Q on the domain of individuals such that for each a, $f_Q(a) = 1$ (i.e. a has Q) or $f_Q a) = 0$ (a does not have Q). No intermediate values are permitted. And this apparatus is used to model reasoning in natural languages which by nature are highly unsharp, except in the very special circumstances when technical vocabularies are employed. There are actually good functional reasons why natural predicates are not sharp: their flexibility in extending beyond know cases is an indispensable feature of the adaptive success of natural languages. Not surprisingly the modelling of these languages by artificially sharp ones results in 'paradoxes' of the Sorites type (whose classic exemplar is the Paradox of the Heap).

Such unpalatable results have prompted the investigation of more accurate methods of modelling informal deductive reasoning by means of 'vague' predicates, and in particular the use of so-called 'fuzzy' ones, where $\{0,1\}$-valued characteristic functions are replaced by continuous functions, with appropriate rules for their use. The analogue for blunting 'sharp' probability values is to replace them with unsharp, interval-valued ones, and the theory of these is well-understood (see [Walley, 1991]). But 'sharp' probability models find their justification in investigations of how evidence, in the form of reports of observations, should affect estimates of the credibility of hypotheses. It is quite difficult to answer this and related questions, e.g. how sensitive y is to x where the data are particularly numerous or varied or both, without using a theory that can say things like 'Suppose the prior value of the probability is x', and then use the machinery of the point-valued probability calculus (in particular Bayes' Theorem) to calculate that the posterior value is y. So we need a fairly strong theory which will tell us things like this; and in the standard mathematical theory of probability we have a very rich theory indeed. At the same time, the model is not too distant from reality; it is quite possible to regard it as a not-unreasonable approximation in many applications, for example where the results obtained are robust across a considerable range of variation in the probability parameters. Many of the limiting theorems in particular have this property.

Similar sorts of considerations apply to the usual formal models of deductive reasoning. There are non-sharp models, but it is partly the sharpness itself of the more familiar structures that explains why they still dominate logical investigations: nearly all the deep results of modern logic, like the Completeness Theorem for the various formulations of first order logic, and the limitative theorems of

Church, Gödel, Tarski etc., are derived within 'sharp' models. Much more could be written on this subject, but space and time are limited and enough has, I hope, now been said to convey why sharp models are not the unacceptable departures from a messier reality that at first sight they might seem to be.

3.6 Conditionalisation

If the rules of the probability calculus are a complete set of consistency constraints, what is the status of conditionalisation, which is not one of them, though it is standardly regarded as a 'core' Bayesian principle? Recall that conditionalisation is the rule that

$$\frac{P(B|A) = r \quad P'(A) = 1}{P'(B) = r}$$

where '$P'(A) = 1$' signifies an exogenous 'learning' of A; P is your probability function up to that point, and P' after. There is a well-known Dutch Book argument for this rule due to David Lewis (reported in [Teller, 1973]). I have given detailed reasons elsewhere [Howson and Urbach, 1993, Ch. 6] why I believe any Dutch Book argument in the 'dynamic' context to be radically unsound and I shall not repeat them all here. What I will do is show how consideration of a corresponding, obviously unsound, deductive analogue enables us to translate back and see why the probabilistic dynamic rule should in principle be unsound too. Consider first a possibly slightly unfamiliar — though sound — version of modus ponens, where v is a truth-valuation and $r = 1$ (true) or 0 (false):

$$\text{(2)} \quad \frac{v(A \to B) = r \qquad v(A) = 1}{v(B) = r}.$$

But now suppose v and w are distinct and consider

$$\frac{v(A \to B) = r \qquad w(A) = 1}{w(B) = r}.$$

This 'dynamic' version of (2), where v and w represent earlier and later valuations, is clearly invalid. Indeed, suppose A says that B is false, i.e. $A = \neg B$, and you now (v) accept $\neg B \to B$ (i.e. $v(\neg B \to B) = 1$): it means accepting B), but later (w) accept $\neg B$ (i.e. $w(\neg B) = 1$). If you try to 'conditionalise' and infer B (i.e. $w(B) = 1$) you will obviously be inconsistent.

Here is a probabilistic analogue of that counterexample. Let A say not that B is false but that B is going to be less than 100% probable: A = "$P'(B) < 1$", where P' is your probability function at some specified future time t (e.g. you are imagining that you will be visited by Descartes's demon). Further, suppose $P(B) = 1$, and suppose $P(A) > 0$. If you are a consistent probablistic reasoner $P(B|A) = 1$. But suppose at t it is true that $P'(B) < 1$, and you realise this by introspection (i.e. $P'(A) = 1$). If you try to conditionalise on A and infer

$P'(B) = P(B|A)$ you will be inconsistent. In the deductive 'dynamic' modus ponens, only if $v(A \to B) = w(A \to B)$ can you validly pass from $w(A) = 1$ and $v(A \to B) = r$ to $w(B) = r$ ($r = 0$ or 1): i.e.

$$\text{(3)} \quad \frac{w(A \to B) = v(A \to B) \qquad w(A) = 1}{w(B) = v(A \to B)}$$

which is just a substitution instance of (2). This suggests the analogous rule

$$\text{(4)} \quad \frac{P'(B|A) = P(B|A) \qquad P'(B) = 1}{P'(B) = P(B|A)}$$

Indeed this rule is valid, but you don't need a new rule to tell you so: *it already follows from the probability calculus*! In other words, conditionalisation is only conditionally valid, and the conditions for validity are already supplied by the probability calculus. Similar considerations apply to 'probability kinematics'.

The rather striking similarities between (3) and (4) above suggest a further point of contact between Bayesian probability and deductive logic. Do not these rules imply that P(B|A) is the probability of a type of sentential conditional? But, as is well-known, this question seemed to have been answered firmly in the negative by David Lewis with his so-called 'triviality theorem' [Lewis, 1973]. In spite of the apparent finality of the answer there have been attempts to bypass it via various types of non-Boolean propositional logics incorporating a conditional B⇒A satisfying the so-called Adams Principle $P(B|A) = P(A \Rightarrow B)$ (from [Adams, 1975]). Long before Lewis's result de Finetti had suggested a non-Boolean, three-valued logic of conditionals [de Finetti, 1964, Section 1] satisfying that condition, with values 'true', 'false'and 'void', where a conditional is 'void' if the antecedent is false (cf. called-off bets). This is not the place for a discussion of these attempts and I refer the reader to Milne [1997] for a relatively sympathetic account of this and other theories of non-Boolean conditionals, and to Howson [1997a] for a somewhat less sympathetic discussion.

3.7 *Inductive Inference*

This is a vast topic and I can only sketch here the account I have given at length elsewhere [Howson, 2000, Ch. 8]. The logical view of the principles of subjective probability exhibits an extension of deductive logic which still manages to remain non-ampliative. It therefore also respects Hume's argument that there is no sound inductive argument from experiential data that does not incorporate an inductive premise, and it also tells us what the inductive premise will look like: *it will be a probability assignment that is not deducible from the probability axioms*. Far from vitiating Bayesian methodology I believe this Humean view, which arises naturally from placing epistemic probability in an explictly logical context, strengthens it against many of the objections commonly brought against it (see Howson *loc. cit.*).

3.8 Sanctions

A logical view of the probability axioms is all very well, but what sanction attaches to infringing them on this view? Indeed, doesn't the analogy with deductive logic break down at precisely this point? There is after all an obvious sanction to breaking the rules of deductive consistency: what you say cannot be true if it is inconsistent. Perhaps surprisingly, however, there seems to be as much — or as little — sanction in both cases. In the probabilistic one there certainly are sanctions: not just the (usually remote) theoretical possibility of being Dutch-Booked were you to bet indiscriminately at inconsistent betting quotients (note that there is no presumption in the earlier discussion that you will and certainly not that you ought to bet at your fair betting quotients, or indeed that you ought to do anything at all), but those arising in general from accepting fallacious arguments with probabilities: we have only to look at the notorious Harvard Medical School Test to see what these might be (see [Howson, 2000, pp. 52–54]). Morover, probabilistic inconsistency is as self-stultifying as deductive: as we saw in the previous chapter, inconsistency means that you differ *from yourself* in the uncertainty value you attach to propositions, just as deductive inconsistency means that you differ from yourself in the truth-values you attach to them.

As to the positive sanctions penalising deductive inconsistency, these on closer inspection turn out to be less forceful than they initially seemed. Here is Albert rehearsing them:

> Logical [deductive] consistency serves a purpose. Beliefs cannot possibly be true if they are inconsistent. Thus, if one wants truth, logical consistency is necessary. An analogous argument in favor of Bayesianism would have to point out some advantage of coherence unavailable to those relying on non-probabilistic beliefs and deductive logic alone. Such an argument is missing. [Albert, 2001, p. 366]

The two principal assertions here are both incorrect. Firstly, logical consistency is *not* necessary for truth. False statements are well known to have true consequences, lots of them, and inconsistent statements the most of all since every statement follows from a contradiction. Current science may well be be inconsistent (many distinguished scientists think it is), but it has nevertheless provided a rich bounty of truths. The benefits of maintaining deductive consistency are more complex and less direct than is often supposed, but the principal penalty is the same in both the deductive and the probabilistic case: inconsistency amounts to evaluating the same proposition in incompatible ways. It is self-stultifying. More generally, the practical penalties attaching to deductive inconsistency are not obviously greater than those attaching to probabilistic inconsistency. What practical consequences flow from either will vary with one's situation.

4 CONCLUSION

In the foregoing pages I have tried to carry through Leibniz's programme for understanding the rules of evidential probability as a species of logic as authentic as those of deductive logic. In both cases these rules are conditions of solvability of value-assignments, with similar completeness properties. I have also argued that this leads to a quite different, and much more coherent and fruitful, view of Bayesian probability than the usual one which is as a theory of prudential rationality. One thing that Lakatos's well-known theory of scientific research programme emphasises is that a programme may be almost written off yet eventually come back to win the field (see Lakatos [1970]). It is early perhaps to predict that this will happen with the logical programme, but I hope I have shown that its resources are more than adequate to the task.

ACKNOWLEDGEMENT

I would like to thank Oxford University Press for allowing me to reproduce passages from *Hume's Problem: Induction and the Justification of Belief*, C. Howson, 2000.

Department of Philosophy, Logic and Scientific Method, London School of Economics, London, UK.

BIBLIOGRAPHY

[Aczel, 1994] P. Aczel. Schematic consequence. In [Gabbay, 1994, pp. 261–273].

[Adams, 1975] E. W. Adams. *The Logic of Conditionals*, Reidel, Dordrecht, 1975.

[Adams, 1998] E. W. Adams. *A Primer of Probability Logic*, CSLI, Stanford, 1998.

[Albert, 2001] M. Albert. Bayesian learning and expectations formation: anything goes. This volume, pp. 347–368.

[Anscombe and Aumann, 1963] F. J. Anscombe and R. J. Aumann. A definition of subjective probability, *Annals of Mathematical Statistics*, **34**, 199–205, 1963.

[Bayes, 1763] T. Bayes. An essay towards solving a problem in the doctrine of chances, *Philosophical Transactions of the Royal Society of London*, 1763.

[Bernoulli, 1715] J. Bernoulli. *Ars Conjectandi*, Basel, 1715.

[Bolzano, 1850] B. Bolzano. *Theory of Science*, 1850

[Carnap, 1971] R. Carnap. A basic system of inductive logic. In *Studies in Inductive Logic and Probability*, Volume I, R. Carnap and R. C. Jeffrey, eds. pp. 33–167. University of California Press, 1971.

[Carnap, 1950] R. Carnap. *The Logical Foundations of Probability*, Chicago: University of Chicago Press, 1950.

[Couturat, 1901] L. Couturat. *La Logique de Leibniz*, Paris, 1901.

[de Finetti, 1964] B. de Finetti. Foresight; its Logical Laws, its Subjective Sources', *Studies in Subjective Probability*, H. Kyburg and H. Smokler, eds. pp. 93–159. Wiley, 1964. (de Finetti's paper was published originally in 1937 in French.)

[Fagin and Halpern, 1988] R. Fagin and J. Y. Halpern. Reasoning about Knowledge and Probability: Preliminary Report. In *Proceedings of the Second Conference on Theoretical Aspects of Reasoning about Knowledge*, M. Y. Vardi, ed. pp. 277–293. Morgan Kaufmann, 1988.

[Gabbay, 1994] D. M. Gabbay, ed. *What is a Logical System*?, Oxford: Oxford University Press, 1994

[Gabbay, 1994a] D. M. Gabbay. What is a logical system? In [Gabbay, 1994, pp. 179–217].
[Gaifman, 1964] H. Gaifman. Concerning measures in first order calculi. *Israel Journal of Mathematics*, **2**, 1–18, 1964.
[Hacking, 1994] I. Hacking. What is logic? In [Gabbay, 1994, pp. 1–35].
[Halmos, 1963] P. Halmos. *Lectures on Boolean Algebras*, Van Nostrand, Princeton,, 1963.
[Heifetz and Mongin, forthcoming] A. Heifetz and P. Mongin. The Modal Logic of Probability, forthcoming.
[Hellman, 1997] G. Hellman. Bayes and beyond. *Philosophy of Science*, **64**, 190–205, 1997.
[Hodges, 1974] W. Hodges. *Logic*, Harmondsworth: Penguin Books, 1974.
[Howson, 1997a] C. Howson. Logic and probability', *British Journal for the Philosophy of Science*, **48**, 517–531, 1997.
[Howson, 1997b] C. Howson. *Logic With Trees*, London: Routledge, 1997..
[Howson, 2000] C. Howson. *Hume's Problem: Induction and the Justification of Belief*, Oxford: Oxford University Press, 2000.
[Howson and Urbach, 1993] C. Howson and P. Urbach. *Scientific Reasoning: the Bayesian Approach*, 2nd edition, Chicago: Open Court, 1993.
[Hughes, 1980] R. I. G. Hughes. Rationality and intransitive preferences, *Analysis*, **40**, 132–134, 1980.
[Kac and Ulam, 1968] M. Kac and S. Ulam. *Mathematics and Logic*, New York: Dover, 1968.
[Kolmogorov, 1956] A. N. Kolmogorov. *Foundations of the Theory of Probability*, New York: Chelsea, 1956.
[Lakatos, 1970] I. Lakatos. Falsification and the methodology of scientific research programmes. In *Criticism and the Growth of Knowledge*, I. Lakatos and A. Musgrave, eds. pp. 91–197. Cambridge: Cambridge University Press, 1970.
[Lewis, 1973] D. Lewis. Probabilities of conditionals and conditional probabilities. *Philosophical Review*, vol. LXXXV, 297–315, 1973.
[Milne, 1997] P. M. Milne. Bruno de Finetti and the logic of conditional events. *British Journal for the Philosophy of Science*, **48**, 195–233, 1997.
[Paris, 1994] J. Paris. *The Uncertain Reasoner's Companion. A Mathematical Perspective*, Cambridge: Cambridge University Press, 1994.
[Poisson, 1823] S.-D. Poisson. *Recherches sur la probabilité des jugements en matière civile et en matière criminelle*, Paris, 1823.
[Ramsey, 1931] F. P. Ramsey. Truth and probability. In *The Foundations of Mathematics*, R.B. Braithwaite, ed. London: Kegan Paul, 1931.
[Savage, 1954] L. J. Savage. *The Foundations of Statistics*, New York: Wiley, 1954.
[Scott and Krauss, 1970] D. Scott and P. Krauss. Assigning probabilities to logical formulas. In *Aspects of Inductive Logic*, J. Hintikka and P. Suppes, eds. pp. 219–264, 1970.
[Shimony, 1955] A. Shimony. Coherence and the axioms of confirmation. *Journal of Symbolic Logic*, **20**, 1–28, 1955.
[Smullyan, 1968] R. Smullyan. *First Order Logic*, New York: Dover, 1968.
[Teller, 1973] P. Teller. Conditionalisation and Observation, *Synthese*, **26**, 218–58, 1973.
[Todhunter, 1865] I. Todhunter. *A History of the Mathematical Theory of Probability*, Cambridge and London, 1865.
[Walley, 1991] P. Walley. *Statistical Reasoning with Imprecise Probabilities*, London: Chapman and Hall, 1991.
[Williamson, 2001] J. Williamson. Probability logic. In 'Handbook of the Logic of Inference and Argument: The Turn Toward the Practical', D. Gabbay, R. Johnson, H.J. Ohlbach & J. Woods, eds. Elsevier, 2001.

MARIA CARLA GALAVOTTI

SUBJECTIVISM, OBJECTIVISM AND OBJECTIVITY IN BRUNO DE FINETTI'S BAYESIANISM

The paper will focus on Bruno de Finetti's position, which combines Bayesianism with a strictly subjective interpretation of probability. For de Finetti, probability is always subjective and expresses the degree of belief of the evaluating subject. His perspective does not accommodate a notion of "objective chance" in the way other subjectivists, including Frank Ramsey, do. To de Finetti's eyes, objectivism, namely the idea that probability depends entirely on some aspects of reality, is a distortion, and the same holds for the idea that there exists an absolute notion of objectivity, to be grounded on objective facts. For him there is no problem of objectivity beyond that of the evaluation of probabilities in a Bayesian framework. This is a complex procedure, which includes subjective elements as well as the consideration of objective elements like observed frequencies.

1 DE FINETTI'S SUBJECTIVISM

Bruno de Finetti used to call his perspective "subjective Bayesianism" [de Finetti, 1969, p. 3], to stress that in his conception Bayes' scheme is assigned a central role, and that it goes hand in hand with a subjective view of probability. Inspired by what we would today call a radically "anti-realist" philosophy, de Finetti finds in the Bayesian approach a way of combining empiricism and pragmatism. The resulting position is not only incompatible with any perspective based on an objective notion of probability, neither can it be assimilated to other subjective views of probability. While being opposed both to frequentism and logicism, taken as "objective" views of probability, de Finetti's perspective strays from Ramsey's subjectivism in important respects.

De Finetti entrusted his philosophy of probability, called "probabilism", to the paper "Probabilismo",[1] which he regarded as his philosophical manifesto. Its starting point is a refusal of the notion of truth, and the related notions of determinism and "immutable and necessary" laws. In their place, de Finetti reaffirms a conception of science seen as a human activity, a product of thought, having as its main tool probability. "...no science - says de Finetti - will permit us to say: this fact will come about, it will be thus and so because it follows from a certain law, and that law is an absolute truth. Still less will it lead us to conclude skeptically: the absolute truth does not exist, and so this fact might or might not come about, it may go like this or in a totally different way, I know nothing about it. What we can say is this: *I foresee* that such a fact will come about, and that it will happen

[1] The paper was written in 1929 and published in 1931: see [de Finetti, 1931b].

D. Corfield and J. Williamson (eds.), Foundations of Bayesianism, 161–174.

in such and such a way, because past experience and its scientific elaboration by human thought make this forecast seem reasonable to me" [de Finetti, 1931b, p. 170, English edition]. Probability is precisely what makes a forecast possible. And since a forecast is always referred to a subject, being the product of his experience and convictions, "the logical instrument that we need is the subjective theory of probability". In other words, probabilism represents for de Finetti an escape from the antithesis between absolutism and skepticism, and at its core one finds the subjective notion of probability.

Following the subjectivist approach, probability "means degree of belief (as actually held by someone, on the ground of his whole knowledge, experience, information) regarding the truth of a *sentence,* or *event* E (a fully specified 'single' event or sentence, whose truth or falsity is, for whatever reason, unknown to the person)" [de Finetti, 1968, p. 45]. According to de Finetti, one can show not only that this notion of probability is the only non contradictory one, but also that it covers all uses of probability in science and everyday life. This program is realized in two steps: firstly, an operational definition of probability is worked out, secondly, it is argued that the notion of objective probability is reducible to that of subjective probability.

The operational definition moves along well known lines: probability is defined in terms of betting quotients, namely the degree of probability assigned by an individual to a certain event is identified with the betting quotient at which he *would be* ready to bet a certain sum on its occurrence. The fundamental and unique criterion one must obey to avoid sure losses is that of coherence. The individual in question should be thought of as one in a condition to bet whatever sum against any gambler whatsoever, free to choose the betting conditions, like someone holding the bank at a gambling-casino. Probability can be defined as the fair betting quotient he would attach to his bets. Coherence is a sufficient condition for the fairness of a betting system, and a behaviour conforming to coherence satisfies the principles of probability calculus, which can be derived from the notion of coherence defined in the specified way. This result was certainly grasped by Ramsey, but is fully worked out only by de Finetti in "Sul significato soggettivo della probabilità" [de Finetti, 1931a].

Here de Finetti, in addition to the quantitative introduces a qualitative definition of subjective probability, based on the relation of "at least as probable as". He then argues that it is not essential to adopt a quantitative notion of probability expressed by a real number; the latter is the most common way of talking about probability, and also the simplest one, but is in no way the only one. This illustrates the role that de Finetti assigns to betting quotients within his theory: they offer an apt device for measuring probability and defining it operationally, but they do not represent an essential component of the notion of probability, which is a primitive notion, expressing "the psychological sensation of an individual" [de Finetti, 1931a, p. 302].

This point has been overlooked by the literature. The idea that probability can be defined in various ways is a central feature of de Finetti's perspective, where the scheme of bets represents only a convenient device for talking about probability in a way that makes it understandable to the "man in the street". Also, in his *Theory of Probability* de Finetti points out that the scheme of bets is just a useful tool, leading to "simple and useful insights" [de Finetti, 1970a, p. 180, English edition]. In addition to the scheme of bets, he adopts another way of measuring probability by means of scoring rules based on penalties, which is shown to be equivalent to the first one. Something more will be said on this method in the following pages. It is worth noting that the autonomous value assigned by the author to the notion of probability marks a difference between his position and that of the other major supporters of subjectivism, namely F.P. Ramsey and L.J. Savage.[2] Unlike these authors, de Finetti does not see probability as strictly connected with utility and claims that probability and utility have "different 'cogent values': an indisputable value in the case of probability, a rather uncertain value in the case of ...utility" [de Finetti, 1955, p. 7].

The second part of de Finetti's program amounts to the reduction of objective to subjective probability. This is done by means of the so-called "representation theorem", which was obtained by de Finetti already in 1928, though its best known formulation is contained in "La prévision: ses lois logiques, ses sources subjectives" [de Finetti, 1937]. This result is crucial, because it gives applicability to subjective probability by bridging degrees of belief and observed frequencies. The fundamental notion here is that of "exchangeability", which can be defined as follows: events belonging to a sequence are *exchangeable* if the probability of h successes in n events is the same, for whatever permutation of the n events, and for every n and $h \leq n$. The representation theorem says that the probability of exchangeable events can be represented as follows: imagine the events were probabilistically independent, with a common probability of occurrence p. Then the probability of a sequence with h occurrences in n would be $p^h(1-p)^{n-h}$. But if the events are only exchangeable, the sequence has a probability $\omega_h^{(n)}$, representable according to de Finetti's representation theorem as a *mixture* over the $p^h(1-p)^{n-h}$ with varying values of p:

$$\omega_h^{(n)} = \int p^h(1-p)^{n-h} f(p)dp.$$

Here $f(p)$ is a uniquely defined density for the variable p, or in other words, it gives the weights $f(p)$ for the various values $p^h(1-p)^{n-h}$ in the above mixture.

In order to understand de Finetti's position, it is useful to start by considering how an objectivist would proceed when assessing the probability of an unknown event. An objectivist would assume an objective success probability p. But its value would in general remain unknown. One could give weights to the possible values of p, and determine the weighted average. The same applies to the probability of a sequence with $h$successes in n independent repetitions. Note that because of independence it does not matter where the successes appear. De Finetti focuses on the latter, calling *exchangeable* those sequences where the places of

[2]See [Ramsey, 1926] and [Savage, 1954].

successes don't make a difference in probability. These need not be independent sequences. An objectivist who wanted to explain subjective probability, would say that the weighted averages are precisely the subjective probabilities. But de Finetti proceeds in the opposite direction, with his representation theorem. It says in his interpretation: *starting* from the subjective judgment of exchangeability, one can show that there is only one way of giving weights to the possible values of the unknown objective probabilities. According to this interpretation, objective probabilities become useless and subjective probability can do the whole job.

In the course of a comment on the notion of exchangeability, de Finetti reaffirms that the latter represents the correct way of expressing the idea that is usually conveyed by the phrase "independent events with constant but unknown probability". If we take an urn of unknown composition, says de Finetti, the above phrase means that, relative to each of all possible compositions of the urn, the events can be seen as independent with constant probability. Then he points out that "...what is unknown is the composition of the urn, not the probability: the latter is always known and depends on the subjective opinion about the composition, which opinion is modified as new drawings are made, and observed frequencies are taken into account" [de Finetti, 1995, p. 214]. It should not pass unnoticed that for de Finetti subjective probability, being the expression of the feelings of the subjects evaluating it, is always definite and known: "Probability as *degree of belief* is surely known by anyone" [de Finetti, 1973, p. 356].

An example, taken from the article "Logical Foundations and Measurement of Subjective Probability" illustrates in what sense "the concept of unknown probability... must be seen as fictitious and misleading" [de Finetti, 1970b, p. 144]. The example compares the processes of Bayes-Laplace and Pólya:

> "It is well-known that the processes of Bayes-Laplace and Pólya are identical as probabilistic models although very different in the way they are produced. Bayes-Laplace model is a Bernoulli process: in the drawing of balls (with replacement) from an urn containing white and black balls in an unknown proportion, the probability distribution of this proportion is uniform over the interval (0, 1). A Pólya process (contagious probabilities) consists in drawing balls from an urn containing, in the beginning, two balls, one white and one black, and where after each draw, not only is the ball drawn replaced, but also another one of the same color is added. After $N = W + B$ drawings (W = number of white, B = number of black) there are $N + 2$ balls $(W + 1)$ white and $(B + 1)$ black; the probability of the next trial is $(W + 1)/(N + 2)$. But, surprisingly enough, this is the same that happens in the Bayes-Laplace model: that is the famous Laplace succession rule. What is the lesson? In the Bayes-Laplace version it is correct to call 'unknown probability' the 'unknown proportion' (which has a real existence). The wording would be: 'the probability of each trial conditional to the knowledge of the unknown proportion

> and given the fact that my subjective opinion agrees with the standard assumption that the drawings are stochastically independent and that all the balls have equal probability'. In the Pólya version it is formally possible to think of a fictitious urn of Bayes-Laplace type existing in some supposed world of Platonic ideas ... But that, outside Platonism, is obviously a pointless fiction. In conclusion, the recourse to concepts like 'objective unknown probability' in a problem is neither justified nor useful for intrinsic reasons. It may correspond to something realistic under particular factual features, not of a probabilistic model, but of a specific device"(ibid.).

From a philosophical point of view, the reduction of objective to subjective probability is to be seen in a pragmatic perspective. It is performed in the same pragmatic spirit that inspires the operational definition of subjective probability in terms of coherent betting quotients, and complements the latter. If such a reduction is based on consideration of the role played by objective probability in statistical reasoning, it is again the role played by subjective probability in life and science that gives an operational basis for its definition. "Probability - says de Finetti - is actually already *defined* implicitly by the role played, with respect to the decisional criterion of an individual, by the fact that he evaluates it in a certain way" [de Finetti, 1963, p. 66].

The representation theorem does not serve only the purpose of reducing objective to subjective probability; it also shows how subjective probability can be applied to statistical inference. In this connection the representation theorem plays a vital role within subjectivism, a role whose importance can hardly be overrated. According to de Finetti, statistical inference can be entirely performed by exchangeability in combination with Bayes' rule. If the notion of probability as degree of belief is grounded in an operational definition, probabilistic inference — taken in a subjective sense — is grounded in Bayes' theorem. Therefore, de Finetti's probabilism is intrinsically Bayesian; one could say that for him Bayesianism represents the crossroads where pragmatism and empiricism meet subjectivism. He thinks that one needs to be Bayesian in order to be a subjectivist, but on the other hand subjectivism is a choice to be made if one embraces a pragmatist and empiricist philosophy.

As reflected by the article "Initial Probabilities: A Prerequisite for any Valid Induction" the shift from prior (or initial) to posterior (or final) probabilities, is considered by de Finetti the cornerstone of statistical inference.[3] In this connection he takes a "radical approach" by which "all the assumptions of an inference ought to be interpreted as an overall assignment of initial probabilities" [de Finetti, 1969, p. 9]. Though this shift is given a subjective interpretation, in the sense that going from prior to posterior assessments involves a shift from one subjective probability to another, it also involves consideration of objective factors.

[3]On the problem of the choice of initial probabilities de Finetti wrote a joint paper with Savage: see [de Finetti and Savage, 1962].

Before we face this issue, it is worth noting that for de Finetti updating one's mind in view of new evidence does not mean changing opinion: "If we reason according to Bayes' theorem we do *not* change opinion. We keep the same opinion and we update it to the new situation. If yesterday I said 'Today is Wednesday', today I say 'It is Thursday'. Yet I have not changed my mind, for the day following Wednesday is indeed Thursday" [de Finetti, 1995, p. 100]. If the idea of *correcting* previous opinions is completely alien to this perspective, so is the notion of self-correcting procedure, which occupies a central place within the perspective of other authors, such as Hans Reichenbach.[4] De Finetti's attitude is grounded in the conviction that there are no "correct" and "rational" probability assignments: "The subjective theory... — he says — does not contend that the opinions about probability are uniquely determined and justifiable. Probability does not correspond to a self-proclaimed 'rational' belief, but to the effective personal belief of anyone" [de Finetti, 1951, p. 218]. Incidentally, we might notice that his attitude in this connection marks a sharp difference from the logicism of Rudolf Carnap and Harold Jeffreys,[5] who believe that there are "correct" probability evaluations. In this sense, logicism attributes to probability theory a normative aspect which is absent from subjectivism.

2 OBJECTIVISM AND OBJECTIVITY

De Finetti's subjective Bayesianism is intransigent, even dogmatic. Not only is subjective Bayesianism the sole acceptable way of addressing probabilistic inference and the whole of statistical methodology, but it makes any form of "objectivism" look silly. In de Finetti's words:

> "The whole of subjective statistics is based on this simple theorem of probability calculus [Bayes' theorem]. Consequently, subjective statistics has a very simple and general foundation. Moreover, being grounded only on the basic axioms of probability, subjective statistics does not depend on those definitions of probability that would narrow its range of application (like, for instance, the definitions based on the idea of equally probable events). Nor — once one endorses this view — is there any need to resort to empirical formulae, in order to characterize inductive reasoning. Objectivist statisticians, on the contrary, make extensive use of empirical formulae. The need to do so stems only from their refusal to admit the use of initial probability $P(E)$. They reject the use of initial probability because they reject the idea of a probability that depends on the state of information. However, by doing so they distort everything: not only do they make probability an objective entity...they even make it a *theological* entity: they claim

[4]See [Reichenbach, 1949].

[5]See [Carnap, 1950] and [Jeffreys, 1931; Jeffreys, 1939].

that 'true' probability exists, outside us, independently of a person's judgment" [de Finetti, 1995, p. 99].

This passage highlights a main feature of de Finetti's position, namely his refusal of objective probability, which is deemed not only useless, but even meaningless, like all metaphysical notions. Throughout his life, de Finetti held that "probability does not exist". This claim, which appears in capital letters in the Preface to the English edition of his *Theory of Probability*, is the *leit-motiv* of his production. "Objective probability never exists" he says in "Il significato soggettivo della probabilità" [de Finetti, 1931a], and almost fifty years later he opens the article "Probabilità" in the "Einaudi Encyclopedia" with the words: "is it true that probability 'exists'? What could it be? I would say no, it does not exist" [de Finetti, 1980, p. 1146]. Such aversion to "objective" probability is inspired by the desire to keep probability free from metaphysical "contaminations".[6]

De Finetti's refusal to attach an "objective" meaning to probability ends with a denial of the notions of "chance" and "physical probability". No doubt, the lack of consideration for the notions of "chance" and "physical probability" represents a limitation of de Finetti's perspective.[7] Spurred by his anti-realism, de Finetti never paid much attention to the use made of probability in science, in the conviction that science is just a continuation of everyday life and subjective probability is all that is needed. Only the volume *Filosofia della probabilità*, containing the text of a course given by de Finetti in 1979, includes a few remarks to the effect that probability distributions belonging to statistical mechanics can be taken as more solid grounds for subjective opinions [de Finetti, 1995, p. 117]. These remarks suggest that late in his life de Finetti might have entertained the idea that when probability assignments are strictly related to scientific theories, they acquire a special meaning.

The road to a more flexible form of subjectivism, which can accommodate these concepts, has been paved by the other "father" of modern subjectivism, Frank Ramsey. He defines "chance" and "probability in physics" in terms of "systems of beliefs" making reference to theories accepted by the scientific community. Ramsey thought that the probabilities we encounter in physics are derived from physical theories. Their objective character descends from the objectivity ascribed to theories that are commonly accepted as true. Within Ramsey's perspective, this idea is combined with a pragmatic approach to theories and truth that would have been quite congenial to de Finetti, had he been acquainted with it.[8] In fact, his remarks contained in *Filosofia della probabilità* lean in the same direction. But de Finetti did not grasp the insights of Ramsey's philosophy, though he knew about his subjective definition of probability, to which the French probabilist Maurice

[6]See [Galavotti, 1989] for an exposition of the anti-metaphysical and anti-realist basis of de Finetti's subjectivism.

[7]This is argued in [Galavotti, 1995-96] and [Galavotti, 1997].

[8]For a comparison between the philosophy of probability of Ramsey and de Finetti see [Galavotti, 1991]. For Ramsey's notion of chance see [Galavotti, 1995; Galavotti, 1999].

Fréchet called his attention around 1937. In the Cambridge of the Twenties another Bayesian often praised by de Finetti, Harold Jeffreys, put forward the idea that one can make sense of physical probability in an epistemic framework, holding a position akin to that of Ramsey.[9] To be sure, Jeffreys was a logicist more than a subjectivist. More recently, however, the idea that subjectivism should be flexible enough to accommodate for a notion of physical probability has been heralded by statisticians as well as philosophers, as testified, for instance, by the work of I.J. Good and R.C. Jeffrey.[10]

Having refused the notion of "objective" probability and denied that there are "correct" probability assignments, the radical subjectivist de Finetti still faces the problem of objectivity of probability evaluations. Let us examine his position on this issue. His point of departure is the conviction that the process through which probability judgments are obtained is more complex than is supposed by the other interpretations of probability, which define probability on the basis of a unique criterion. While subjectivists distinguish between the definition and the evaluation of probability, and do not mix them up, upholders of the other interpretations confuse them: they look for a unique criterion - be it frequency, or symmetry - and use it as grounds for both the definition and the evaluation of probability. In so doing, they embrace a "rigid" attitude towards probability, an attitude which consists "in defining (in whatever way, according to whatever conception) the probability of an event, and in univocally determining a function" [de Finetti, 1933, p. 740]. On the contrary, subjectivists adopt an "elastic" approach, which "consists in demonstrating that all functions f have all the necessary and sufficient properties to represent probability evaluations (also in this case, defined according to whatever conception, in whatever way) which are not intrinsically contradictory, leaving to a second (extra-mathematical) stage the discussion and analysis of reasons and criteria for the choice of a particular among all possible ones" (ibid., p. 741). In other words, for the subjectivist all coherent functions are admissible; far from being committed to a single rule or method, the choice of one particular function is seen as the result of a complex and largely context-dependent procedure, which necessarily involves subjective elements.

The explicit recognition of the role played by subjective elements within the complex process of the formation of probability judgments is for de Finetti a prerequisite for the appraisal of objective elements: "Subjective elements — he says — will noways (*sic*) destroy the objective elements nor put them aside, but bring forth the implications that originate only after the conjunction of both objective and subjective elements at our disposal" [de Finetti, 1973, p. 366]. To be sure, Bayesian subjectivism requires that objective elements also be taken into account, but such objective elements are not seen as the only basis for judgment. "Subjectivism — de Finetti says — is one's *degree of belief* in an outcome, based on an evaluation making the best use of all the information available to him and his

[9] See [Jeffreys, 1955].

[10] See [Good, 1965] and [Jeffrey, 1997].

own skill...Subjectivists...believe that every evaluation of probability is based on available information, including objective data" [de Finetti, 1974b, p. 16]. In conclusion, "Every probability evaluation essentially depends on two components: (1) the objective component, consisting of the evidence of known data and facts; and (2) the subjective component, consisting of the opinion concerning unknown facts based on known evidence" [de Finetti, 1974a, p. 7].

De Finetti warns that the objective component of probability judgments, namely factual evidence, is in many ways context-dependent: evidence must be collected carefully and skillfully, its exploitation depends on the judgment on what elements are relevant to the problem under consideration, and can be useful to the evaluation of related probabilities. In addition, the collection and exploitation of evidence depends on economic considerations varying in practical cases. So, one can say that the collection and exploitation of factual evidence involves subjective elements of various sorts. Equally subjective is the decision on how to let objective elements influence belief. Typically, one relies on information regarding frequencies. For de Finetti frequencies, like symmetry considerations, are useful and important ingredients of probability evaluations, *provided that* they are not used uncritically as automatic rules and simply equated with probability. Those who do so, namely frequentists, are simply committed to "superstition":

> "There is no worse conceptual distortion than that owing to which, starting from the premise that any sequence can occur, one defines probability in terms of a property (that of exhibiting a certain frequency) pertaining only to a portion of all sequences... when we define probability in terms of frequency, we define it thoughtlessly. The only objective thing is the set of all possible sequences, but it does not say anything concerning their probability. The probability of sequences can only be the feeling we had before and which characterized our expectation. Here we have a perversion of language, logic and common sense. Such a logical mistake is unacceptable, because the set of all possible sequences (which is logically determined) cannot be confused with probability (which, on the contrary, is subjective)" [de Finetti, 1995, pp. 140–141].

Keeping in mind the distinction between the definition and the evaluation of probability, one can make good use of frequencies within probability evaluations. It is precisely in this connection that exchangeability enters the stage, giving the reason "why expected future frequencies should be guessed according to past observed frequencies", and thereby creating a strong connection between "subjectivistic and objectivistic interpretations" [de Finetti, 1970b, p. 143]. As already stressed, de Finetti assigns to exchangeability a subjective interpretation, according to which exchangeability represents a

> "directly observable property of the probability evaluation. It means that for every set of n of the events concerned, the probability that

> *all* events occur is the same; it depends only on n... Under such a clear subjective condition (and a few side restrictions to avoid special cases, such as that of repeated trials with a known probability that remains unchanged), one is perfectly free to improve the evaluation of probabilities for any future events according to the frequency of the observed ones. This improvement generally entails modifying the initial evaluation... so as to approach gradually the obtained frequency of the events observed up to that time" [de Finetti, 1974a, p. 12].

Therefore exchangeability allows probability judgments to be improved in view of observed frequencies in an empiricist fashion fully in tune with the subjectivist approach.

Since the present contribution is meant to be historically oriented, it is not out of place to make a brief digression on the origin of exchangeability. It is de Finetti's merit to have combined the subjective notion of probability in terms of coherent beliefs with that of exchangeability. In so doing, he was able to guarantee the applicability of subjective probability to practical situations, including those encountered within experimental science. Exchangeability is the missing aspect in Ramsey's perspective, which could have made him see the link between degrees of belief and observed frequencies. It can be conjectured that by the time of his death Ramsey came very close to recognizing such a link. Evidence that he was intrigued by the relationship between frequency and degree of belief is offered by his note "Miscellaneous Notes on Probability",[11] where he ponders over the idea that "degree γ of belief means acting appropriately to a frequency γ", of which he says that "it is [the]...one which makes calculus of frequencies applicable to degrees of belief" [Ramsey, 1991, p. 275]. The justification of this claim lies precisely with exchangeability, a property that Ramsey knew through his teacher and colleague in Cambridge William Ernest Johnson, a logicist who is seen as a forerunner of Carnap's inductive logic, which assigns a privileged role to the same probabilistic property he calls "symmetry".[12] Ramsey himself made use of this property, named by him "equiprobability of all permutations", in a short note called "Rule of Succession" in [Ramsey, 1991], which contains a derivation of Laplace's Rule of Succession from the property in question. However, Ramsey was unable to connect it with degree of belief in the way de Finetti did.

Going back to the evaluation of probabilities, when a considerable amount of information about frequencies is available, it influences probability assignments through the assumption of exchangeability. Often, however, this kind of information is scant, and in this case the problem of how to obtain good probability evaluations is open. The problem is addressed by de Finetti in a number of works, especially starting from the Sixties. The approach adopted is based on penalty methods, like the so called "Brier's rule", named after the meteorologist Brier who applied it to weather forecasts. De Finetti did extensive work on scoring rules,

[11] The note was written in 1928 and appears in [Ramsey, 1991].

[12] On this point see [Zabell, 1988].

partly in cooperation with Savage. He even tested the goodness of this method through an experiment among his students, who were asked to make forecasts on the results of soccer matches in the Italian championship. A simple description of such methods, referred to the case of three possible results (as with soccer matches) is the following:

> "everybody participating in...[an experiment on probabilistic forecasts] is asked to indicate the probabilities of (for instance) the three possible results — victory, or draw, or defeat — of the home-team in a specified football match; say, e.g. 50% - 30% - 20% . A scoring rule indicates how much the participant is to be penalised depending on the 'distance' between the assessed probabilities and the actual result. The simplest and most practical scoring rule is Brier's rule; if someone indicates as his own opinion $P(E) = p$, the score (i.e. the penalisation) is the square of the distance between forecast and result: $p^2 (= (p-0)^2)$ if the result is 0 (E does not happen), and $(1-p)^2$ if the result is 1 (E does happen). The fact that Brier's rule is a *proper* one is proved since for a person indicating as his probability assessment a $\bar{p}$ different from his own effective opinion p, expected penalisation is increased by $(p - \bar{p})^2$. Analogously, when the possible results are three (as in football) indicating them with the vertices of an equilateral triangle, and the forecast with the centre of gravity of weights indicating the probabilities of each vertex, a proper scoring rule is $(P - P_0)^2$ (square of the distance between probability assessment and effectual result" [de Finetti, 1981, p. 55].

Scoring rules of this kind are based on the idea that the device in question should oblige those who make probability evaluations to be as accurate as they can and, if they have to compete with others, to be honest. In fact any deviation from the true p, on the part of the person who evaluates probability, enables the opponent to engage him in a disadvantageous bet. Such rules play a twofold role within de Finetti's approach. In the first place, they offer a suitable tool for an operational definition of probability. As recollected, in his late works de Finetti adopted such a device to define subjective probability. In addition, these rules offer a method for improving probability evaluations made both by a single person and by several people, because they can be employed as methods for exercising "self-control", as well as a "comparative control" over probability evaluations [de Finetti, 1980, p. 1151].

De Finetti assigns these methods, which are quite widespread among Bayesian statisticians, a straightforward interpretation in tune with subjectivism:

> "The objectivists, who reject the notion of personal probability because of the lack of verifiable consequences of any evaluation of it, are faced with the question of admitting the value of such a 'measure of success' as an element sufficient to soften their fore-judgments.

> The subjectivists, who maintain that a probability evaluation, being a measure of someone's beliefs, is not susceptible of being proved or disproved by the facts, are faced with the problem of accepting some significance of the same 'measure of success' as a measure of the 'goodness of the evaluation'" [de Finetti, 1962a, p. 360].

The following remark further clarifies de Finetti's position: "though maintaining the subjectivist idea that no fact can prove or disprove belief, I find no difficulty in admitting that any form of comparison between probability evaluations (of myself, of other people) and actual events may be an element influencing my further judgment, of the same status as any other kind of information" (ibid.). De Finetti's work on scoring rules is in tune with a widespread attitude among Bayesians, an attitude that has given rise to a vast literature on "well-calibrated" estimation methods.

If these methods provide us with useful devices for improving probability evaluations, a whole series of elements seem to be relevant in this connection. De Finetti mentions an array of such elements, including:

1. "degree of competence or care in forecasts concerning different subject matters, epochs or regions;
2. optimistic or pessimistic attitudes...
3. degree of influence of the most recent facts...
4. degree of deviation from statistical standards, according to specific knowledge of each item...
5. stability or flexibility (evolutionary or oscillating) of opinions without a change in the available information, by thinking about or by the influence of another's opinions...
6. conscious or unconscious adaptation of the opinion to standard patterns of statistical theory and practice...." ([de Finetti, 1970b, pp. 141–142].

To sum up, the evaluation of probability is seen as a most complex procedure, resulting from the concurrence of all sorts of factors. Starting from the recognition of the fact that probability is subjective, and that there is no unique, "rational" way of assessing probability, one can make room for a whole array of elements that can influence probability evaluations, suggesting various ways of ameliorating them. De Finetti's remarks in this connection may sound very general, but his warnings against pre-confectioned recipes for the evaluation of probabilities should be taken seriously.

In a paper dealing with issues related to economic theory, de Finetti discusses the "dangerous mirage" of "identifying objectivity and objectivism" [de Finetti, 1962b, p. 344] and exhorts "to fight against the ambush of pseudo-objectivity which is concealed under the false shield of 'objectivism', boasting of it as if it were a chrism of 'scientificity'" (ibid., p. 360). Since objectivism is nothing but a conceptual distortion, an absolute idea of objectivity grounded on it can only be a chimera. A more viable notion of objectivity lies with a "deep analysis of problems", aimed at avoiding hasty judgments, superficial intuitions and careless conclusions, to form evaluations which are the best one can attain in the light of the available information. Such a deep analysis of problems will include consideration of objective elements, in the awareness that, taken by themselves, these are neither necessary not sufficient to guarantee objectivity in an absolute sense. This is because absolute objectivity does not exist: "only a honest reflection, careful of facts and other people's ideas can lead to the maximum possible objectivity" (ibid., p. 367).

Department of Philosophy, University of Bologna, Italy.

BIBLIOGRAPHY

[Carnap, 1950] R. Carnap. *Logical Foundations of Probability*, Chicago: Chicago University Press, 1950. Second edition with modifications 1962.

[de Finetti, 1931a] B. de Finetti. Sul significato soggettivo della probabilità", *Fundamenta mathematicae* **17**, pp. 298–329, 1931.

[de Finetti, 1931b] B. de Finetti. Probabilismo. *Logos*, pp.163-219, 1931, reprinted in B. de Finetti, *La logica dell'incerto,* Milano, Il Saggiatore, 1989, pp. 3–70. English translation in *Erkenntnis*, **31**, pp. 169–223, 1989.

[de Finetti, 1933] B. de Finetti. Sul concetto di probabilità. *Rivista italiana di statistica, economia e finanza* **5**, pp. 723–747, 1933.

[de Finetti, 1937] B. de Finetti. La prévision: ses lois logiques, ses sources subjectives. *Annales de l'Institut Henri Poincaré* **7**, pp.1–68, 1937. English translation in H. E. Kyburg and H. E. Smokler (eds.), *Studies in Subjective Probability*, New York-London, Wiley, pp. 95–158, 1964.

[de Finetti, 1951] B. de Finetti. Recent Suggestions for the Reconciliation of Theories of Probability. In *Proceedings of the Second Berkeley Symposium on Mathematical Statistics and Probability*, Berkeley, University of California Press, pp. 217–225, 1951.

[de Finetti, 1955] B. de Finetti. La probabilità e il comportamento di fronte all'incertezza, *Assicurazioni* **1-2**, pp. 3–15, 1955.

[de Finetti, 1962a] B. de Finetti. Does it Make Sense to Speak of 'Good Probability Appraisers'? In I.J. Good *et al.* (eds.), *The Scientist Speculates. An Anthology of Partly-Baked Ideas*, New York, Basic Books, pp. 357–364, 1962.

[de Finetti, 1962b] B. de Finetti. Obiettività e oggettività: critica a un miraggio. *La Rivista Trimestrale*, **1**, pp. 343–367, 1962.

[de Finetti, 1963] B. de Finetti. La decisione nell'incertezza. *Scientia* **98**, pp. 61–68, 1963.

[de Finetti, 1968] B. de Finetti. Probability: the Subjectivistic Approach. In R. Klibansky (ed.), *La philosophie, contemporaine*, Firenze, La Nuova Italia, pp. 45–53, 1968.

[de Finetti, 1969] B. de Finetti. Initial Probabilities: a Prerequisite for any Valid Induction. *Synthèse* **20**, pp. 2–16, 1969.

[de Finetti, 1970a] B. de Finetti. *Teoria delle probabilità*. Einaudi, Torino, 1970. English edition: *Theory of Probability*, New York, Wiley, 1975.

[de Finetti, 1970b] B. de Finetti. Logical Foundations and Measurement of Subjective Probability. *Acta Psychologica*, **34**, pp. 129–145, 1970.

[de Finetti, 1973] B. de Finetti. Bayesianism: Its Unifying Role for Both the Foundations and the Applications of Statistics. *Bulletin of the International Statistical Institute, Proceedings of the 39th Session*, pp. 349–368, 1973.

[de Finetti, 1974a] B. de Finetti. The Value of Studying Subjective Evaluations of Probability. In C.-A.S. Staël von Holstein (ed.), *The Concept of Probability in Psychological Experiments*, Dordrecht-Boston, Reidel, pp. 1–14, 1974.

[de Finetti, 1974b] B. de Finetti. The True Subjective Probability Problem. In C.-A.S. Staël von Holstein (ed.), *The Concept of Probability in Psychological Experiments*, Dordrecht-Boston, Reidel, pp.15–23, 1974.

[de Finetti, 1980] B. de Finetti. Probabilità. In *Enciclopedia Einaudi*, Torino, Einaudi, **X**, pp. 1146–1187, 1980.

[de Finetti, 1981] B. de Finetti. The Role of 'Dutch Books' and of 'Proper Scoring Rules'. *British Journal for the Philosophy of Science* **32**, pp. 55–56, 1981.

[de Finetti, 1995] B. de Finetti. *Filosofia della probabilità*. Milano, Il Saggiatore, 1995.

[de Finetti and Savage, 1962] B. de Finetti and L.J. Savage. Sul modo di scegliere le probabilità iniziali. *Biblioteca del Metron, serie C: Note e commenti*, Roma, Istituto di Statistica dell'Università, pp. 82–154, 1962.

[Galavotti, 1989] M.C. Galavotti. Anti-realism in the Philosophy of Probability: Bruno de Finetti's Subjectivism. *Erkenntnis*, **31**, pp. 239–261, 1989.

[Galavotti, 1991] M.C. Galavotti. The Notion of Subjective Probability in the Work of Ramsey and de Finetti. *Theoria*, **57**, pp. 239–259, 1991.

[Galavotti, 1995] M.C. Galavotti. F.P. Ramsey and the Notion of 'Chance'. In J. Hintikka and K. Puhl (eds.), *The British Tradition in 20th Century Philosophy. Proceedings of the 17th International Wittgenstein Symposium*, Wien, Hölder-Pichler-Tempsky, pp. 330–340, 1995.

[Galavotti, 1995-96] M.C. Galavotti. Operationism, Probability and Quantum Mechanics. *Foundations of Science*, **1**, pp. 99–118, 1995-96.

[Galavotti, 1997] M.C. Galavotti. Probabilism and Beyond. *Erkenntnis*, **45**, pp. 253–265, 1997.

[Galavotti, 1999] M.C. Galavotti. Some Remarks on Objective Chance (F.P. Ramsey, K.R. Popper and N.R. Campbell). In M.L. Dalla Chiara *et al.* (eds.), *Language, Quantum, Music*, Dordrecht-Boston, Kluwer, pp. 73–82, 1999.

[Good, 1965] I.J. Good. *The Estimation of Probabilities*. Cambridge, Mass., The M.I.T. Press, 1965.

[Jeffrey, 1997] R.C. Jeffrey. Unknown Probabilities. In D. Costantini and M.C. Galavotti (eds.), *Probability, Dynamics and Causality*, Dordrecht-Boston, Kluwer, pp. 327–335, 1997.

[Jeffreys, 1931] H. Jeffreys. *Scientific Inference*. Cambridge, Cambridge University Press, 1931. Third edition with modifications 1973.

[Jeffreys, 1939] H. Jeffreys. *Theory of Probability*. Oxford, Clarendon Press, 1939. 2nd edition 1948. Third edition with modifications 1961.

[Jeffreys, 1955] H. Jeffreys. The Present Position in Probability Theory. *British Journal for the Philosophy of Science*, **5**, pp. 275–289, 1955. Also in Jeffreys, H. and Swirles, B. (eds.), *Collected Papers of Sir Harold Jeffreys on Geophysics and Other Sciences*, London-Paris-New York, Gordon and Breach Science Publishers, **VI**, pp. 421–435, 1971-1977.

[Ramsey, 1926] F.P. Ramsey. Truth and Probability. In Ramsey 1931, pp. 156–198, 1926.

[Ramsey, 1931] F.P. Ramsey. *The Foundations of Mathematics and Other Logical Essays*. R.B. Braithwaite, ed. London, Routledge and Kegan Paul, 1931.

[Ramsey, 1991] F.P. Ramsey. *Notes on Philosophy, Probability and Mathematics*. Ed. by M.C. Galavotti, Naples, Bibliopolis, 1991.

[Reichenbach, 1949] H. Reichenbach. *The Theory of Probability*. Berkeley-Los Angeles, University of California Press, 1949. Second edition 1971. English translation of with modifications of *Wahrscheinlichkeitslehre,* Leyden, Sijthoff, 1935.

[Savage, 1954] L.J. Savage. *The Foundations of Statistics*. New York, John Wiley and Sons, 1954.

[Zabell, 1988] S.L. Zabell. Symmetry and its Discontents. In B. Skyrms and W.L. harper (eds.), *Causation, Chance, and Credence*, Dordrecht-Boston, Kluwer, **I**, pp. 155–190, 1988.

DAVID CORFIELD

BAYESIANISM IN MATHEMATICS

INTRODUCTION

I shall begin by giving an overview of the research programme named in the title of this paper. The term 'research programme' suggests perhaps a concerted effort by a group of researchers, so I should admit straight away that since I have started looking investigating the idea that plausible mathematical reasoning is illuminated by Bayesian ideas, I have not encountered in the literature anyone else who has thought to develop the views of the programme's founder, the Hungarian mathematician, George Pólya. I should further admit that Pólya never termed himself a Bayesian as such. Motivation for the programme may, therefore, be felt sorely necessary. Let us begin, then, with three reasons as to why one might want to explore the possibility of a Bayesian reconstruction of plausible mathematical reasoning:

(a) To acquire insight into a discipline one needs to understand how its practitioners reason plausibly. Understanding how mathematicians choose which problems to work on, how they formulate conjectures and the strategies they adopt to tackle them requires considerations of plausibility. Since Bayesianism is widely considered to offer a model of plausible reasoning, it provides a natural starting point. Furthermore, Pólya has already done much of the spadework with his informal, qualitative type of Bayesianism.

(b) The computer has only recently begun to make a serious impact on the way some branches of mathematics are conducted. A precise modelling of plausibility considerations might be expected to help in automated theorem proving and automated conjecture formation, by providing heuristics to guide the search and so prevent combinatorial explosion. Elsewhere, computers are used to provide enormous quantities of data. This raises the question of what sort of confirmation is provided by a vast number of verifications of a universal statement in an infinite domain. It also suggests that statistical treatments of data will become more important, and since the Bayesian approach to statistics is becoming increasingly popular, we might expect a Bayesian treatment of mathematical data, especially in view of its construal of probability in terms of states of knowledge, rather than random variables.

(c) The plausibility of scientific theories often depends on the plausibility of mathematical results. This has always been the case, but now we live in an era where for some physical theories the *only* testable predictions are mathematical ones. If we are to understand how physicists decide on the plausibility of their theories, this must involve paying due consideration to the effect of verifying mathematical predictions.

D. Corfield and J. Williamson (eds.), Foundations of Bayesianism, 175–201.

Now, if one decides to treat plausible and inductive reasoning in the sciences in Bayesian terms, it seems clear that one would want to do the same for mathematics. After all, it would appear a little extravagant to devise a second calculus. In any case, Bayesianism is usually presented by its proponents as capable of treating all forms of uncertain reasoning. This leads us to conclude that Bayesianism in science requires Bayesianism in mathematics. Once this is accepted, one must respond in two ways according to the discoveries one makes while examining Bayesianism in mathematics:

I Bayesianism cannot be made to work for mathematics, therefore Bayesianism cannot give a complete picture of scientific inference.

II Some forms of Bayesianism can be made to work for mathematics, therefore one of these must be adopted by Bayesian philosophers to give a more complete picture of scientific inference.

The arguments presented in this paper indicate that the antecedent of I is false and the antecedent of II true, opening the prospect of an expanded, but modified, Bayesianism.

In this paper there is only space to treat a part of the motivation given above. The first two sections question which varieties of the many forms of Bayesianism are able to accommodate mathematical reasoning. Many Bayesians hold it as a tenet that logically equivalent sentences should be believed with equal confidence and any evidence should have an equal impact on their degrees of belief. However, such an assumption plays havoc with any attempt to throw light on mathematical reasoning. In section 1 I argue that if a Bayesian modelling of plausible reasoning in mathematics is to work, then the assumption of logical omniscience must be dropped.

In Pólya's version, we have only the right to specify the direction of change in the credence we give to a statement on acquiring new information, not the magnitude. However, Edwin Jaynes demonstrated that one of the central grounds for this decision on the part of Pólya to avoid quantitative considerations was wrong. In section 2 I consider whether there is anything amiss with a quantitative form of Bayesianism in mathematics.

One criticism often made of Bayesian philosophy of science is that it does not help very much in anything beyond toy problems. While it can resolve simple issues, such as accounting for how observing a white tennis shoe provides no confirmation for the law 'all ravens are black', it provides no insight into real cases of theory appraisal and confirmation. Everything rests on the assignment of priors, but how an expert could be considered to go about this is enormously complicated. Recognising what is correct in this criticism, I think there is still useful work to be done. In section 3 I shall be looking in particular at: reasoning by analogy; choice of proof strategy (for automated theorem proving); and, large scale induction (particularly enumerative induction).

1 PROBABILITY THEORY AS LOGIC

In his *Mathematics and Plausible Reasoning* (Pólya [1954a; 1954b]), Pólya considers mathematics to be the perfect domain in which to devise a theory of plausible reasoning. After all, where else could you find such unequivocal instances of facts satisfying general laws? As a noted mathematician actively engaged in research, he delightfully conveys inferential patterns by means of examples of his own use of plausible reasoning to generate likely conjectures and workable strategies for their proof. Now, such plausible reasoning in mathematics is, of course, necessary only because mathematics does not emerge as it appears on the pages of a journal article or textbook, that is, in its semi-rigorous deductive plumage. Indeed, it is due to the failure of what we might call "logical omniscience", the capacity to know immediately the logical consequences of a set of hypotheses, that mathematicians are forced to resort to what might be called a guided process of trial and error, not so dissimilar to that employed in the natural sciences.

In the second of the two volumes mentioned above, Pólya works his account of plausible reasoning into a probabilistic mould. While he did not name himself as such, we can thus reasonably view Pólya as a member of the Bayesian camp and, indeed, as a pioneer who influenced some later prominent Bayesians. Certainly, Edwin Jaynes learned from his work, and it is clear that Judea Pearl has read him closely. So here we have something of a paradox: plausible mathematical reasoning, the subject of Pólya's analysis, was an important source of ideas for some of the leading figures of Bayesianism, and yet it is necessitated by the fact that people involved in this most rigorous branch of knowledge are not able to uphold one of the widely held tenets of Bayesianism, namely, that logically equivalent statements should receive identical degrees of belief, or alternatively, that tautologies should be believed with degree of belief set at 1.

Logical omniscience comes as part of a package which views Bayesianism as an extension of deductive logic, as for example in Howson (this volume). In another of its manifestations, we hear from Jaynes the motto 'probability theory as logic'. For him: "Aristotelian deductive logic is the limiting form of our rules for plausible reasoning, as the robot becomes more and more certain of its conclusions" [Jaynes, forthcoming, Ch 2, p. 11].[1] Here we are to imagine a robot who reasons perfectly in Bayesian terms, handicapped only by the imperfections of its data and the incompleteness of the set of hypotheses it is considering.

We have then a tension when it comes to mathematical reasoning: if Bayesianism is to be seen as an extension of deductive logic, in the sense that the premises are now not required to be known with certainty, then one should consider the two inferential calculi to be similar in as many respects as possible. Since deductive logic is held as a regulating ideal, as, for example, when we say:

(1) If A is true and A entails B, then B is true,

[1]References to Jaynes are from his unfinished book — *Probability Theory: The Logic of Science* - available at http://bayes.wustl.edu . This is soon to appear in print.

should we not have

(2) If $Pr(A) = p$ and A entails B, then $Pr(B) \geq p$?

However, making this assumption raises a few problems. For one thing it implies that any consequence of a given axiomatised mathematical theory should be believed at least as strongly as that theory. Then, assuming Wiles is correct, to mimic the ideal rational agent you must set Pr(Fermat's Last Theorem) no lower than Pr(ZFC set theory), indeed no lower than your degree of belief in whichever system you feel confident can cope with arithmetic. There is of course the question as to how one might want to interpret Pr(ZFC set theory), but for statements whose logical complexity is the same as that of Fermat's Last Theorem, all one needs is the consistency of ZFC for truth to entail proof. And if you were to pitch this at 0.5, say, then this would provide a minimum for all provable truths of arithmetic, along with those of just about any other branch of mathematics, of this logical complexity.

If a mathematician suddenly became endowed with such omniscience, it would not be the end of mathematics, there is far more to mathematics than truth and provability, but one may safely predict that she would be much in demand. The logicistic conceptions of mathematics are accurate enough that the discipline would become unrecognisable. Without the sixty years leading up to Wiles' work, we would have known that (ZFC is consistent & Fermat's Last Theorem) is logically equivalent to (ZFC is consistent). And when finding a proof of a result we knew by omniscience to be correct, we could check up on the validity of lemmas rather than risk wasting time on false ones. How different a picture we gain from Pólya's representation of mathematics as a fallibly practised discipline and as the perfect place to investigate inductive and plausible reasoning.

So logical omniscience is an assumption that we cannot hold on to if we wish to investigate plausible reasoning in mathematics, which if Pólya was correct is perhaps what the Bayesian should be doing. But what prevents us from dropping this assumption? Two of the most common justifications for Bayesianism are Cox's theorem and the Dutch Book argument. Cox's theorem merely assumes that logical equivalence implies equality of probabilities. On the other hand, Dutch Book style arguments or those based on the preference for some linearly valued commodity attempt to justify it by claiming that if an agent offers different betting quotients on what are in fact logically equivalent sentences, then stakes can be set so that they will necessarily lose. But then isn't it surprising that there are many instances in the past where mathematicians have bet? Indeed, in view of the definitive way mathematical statements, even universal ones, may be settled, they would seem to make at least as good propositions to wager on as statements from the natural sciences.

Surely it is reasonable to prefer a bet on the trillionth decimal digit of π being between 0 and 8, than one at the same odds on its being 9. If, however, 9 is the correct digit, then it follows as a "mere" calculation from one of the series expansions for π. That is, "$\pi = 4(1 - 1/3 + 1/5 - 1/7 + \ldots)$" and "$\pi = 4(1 -$

$1/3 + 1/5 - 1/7 + \ldots$) & the trillionth decimal place of π is 9" would be logically equivalent and so to be believed with the same confidence, and so the second bet should be preferred. But, mathematicians spend their working lives making decisions on the basis of the level of their confidence in the truth of mathematical propositions. We would not want to brand them as irrational for devoting time to an attempted proof of their hunch that a certain statement follows from a set of assumptions merely because the hunch turns out to be wrong.

There is a suggestion in the writings of several Bayesians that (2) only holds when we *come to know* about the logical relationship between two propositions.

> Given two propositions A, B it may happen that one is true if and only if the other is true; we then say that they have the same *truth value*. This may be only a simple tautology (i.e., A and B are verbal statements which obviously say the same thing), or it may be that only after immense mathematical labors is it proved that A is the necessary and sufficient condition for B. From the standpoint of logic it does not matter; *once it is established*, by any means, that A and B have the same truth value, then they are logically equivalent propositions, in the sense that any evidence concerning the truth of one pertains equally well to the truth of the other, and they have the same implications for any further reasoning.
>
> Evidently, then, it must be the most primitive axiom of plausible reasoning that two propositions with the same truth-value are equally plausible. [Jaynes, forthcoming, Ch. 1, p. 6] (second emphasis mine)

In this less rigid framework we might say that if A is known by the agent to entail B, then she should ensure that she has $Pr(B) \geq Pr(A)$. In other words, we are generalising from an interpretation of deductive logic no stronger than:

(3) 'If I judge A to be true and I judge A to entail B, then I should judge B to be true.'

Opposed to the 'probability as logic' position are the subjectivists, whose number include followers of de Finetti. Here the accent is on uncertainty:

> The only relevant thing is uncertainty — the extent of our knowledge and ignorance. The actual fact of whether or not the events considered are in some sense *determined*, or known by other people, and so on, is of no consequence. [de Finetti, 1974, p. xi]

Since probability is seen as a measure of an individual's uncertainty, it is no wonder that de Finetti permits non-extreme degrees of belief about mathematical facts, even those which are decidable. Indeed, this probabilistic treatment seems to extend to even very accessible truths:

> Even in the field of tautology (i.e. of what is true or false by mere definition, independently of any contingent circumstances) we always

> find ourselves in a state of uncertainty. In fact, even a single verification of a tautological truth (for instance, of what is the seventh, or billionth, decimal place of π, or of what are the necessary or sufficient conditions for a given assertion) can turn out to be, at a given moment, to a greater or lesser extent accessible or affected with error, or to be just a doubtful memory. [de Finetti, 1974, p. 24]

Presumably then for de Finetti one may be rational and yet have a degree of belief in '91 is prime' less than 1. Perhaps you are unsure so you set it to 0.6. If so, when I ask you for $Pr(7 \times 13 = 91)$ you had better give me an answer no greater than 0.4. But then can't I force you to realise that you have an inconsistent betting quotient by making you see that 7×13 really is the same as 91, or is it just a case where I should allow you to alter your betting quotient after this lesson? More radically still, should one be expected to know that the correctness of this product contradicts the claim that 91 is prime?

In his article, *Slightly More Realistic Personal Probability*, Ian Hacking [1967] sets out a hierarchy of strengths of Bayesianism. These strengths he correlates with ways of saying whether a statement can be possibly true. At the weaker end we find a position he terms 'realistic personalism', where non-zero probabilities will be attributed by a subject to any statement not known by them to be false, knowledge being taken in a very strict sense: "a man can know how to use *modus ponens*, can know the rule is valid, can know p, and can know $p \supset q$, and yet not know q, simply because he has not thought of putting them together" [Hacking, 1967, p. 319]. At the stronger end we find logical omniscience and divine knowledge. Now clearly the coherence provided by realistic personalism is not enough to equip you for a life as a gambler. For instance, it is advisable not to advertise the odds at which you would accept either side of a wager on a decidable mathematical proposition on a mathematics electronic bulletin board. But Dutch Book arguments do not work to prove your irrationality on the grounds that someone may know more than you. If they do know more than you, you will lose whether the subject of your bet is mathematics, physics or the date of the next general election.

However, there is a point here: surely you can be criticised for betting on a proposition whose truth value you know you could discover with a modicum of effort, perhaps by the tap of a single button on the computer in front of you. As Hacking points out [Hacking, 1967, pp. 323–4], besides coherence one needs a principle which calls on you to maximize expected subjective utility. Information acquired for free can only help increase this, and so inexpensive reasoning or information gathering is generally a good thing. But this principle is not needed solely by a personalism weaker than that based on logical omniscience. Where the presupposition of logical omniscience forces you to reason, and indeed reason unreasonably much, it does not require you even to look down to note the colour of your socks before you bet on it. Only some principle of expected utility does this. But then surely you should allow this principle to be the rationale for your logical reasoning as well, rather than relying on the very unreasonable idealisation of logical omniscience which offers little more by way of advice than to be as perfect a mathematician as you can be.

Even admitting that we should not assume logical omniscience when we consider mathematics, it might be thought that this assumption is not too unrealistic for other walks of life. After all, doesn't the uncertainty which necessitates plausible reasoning in ordinary life and the natural sciences arise for other reasons – uncertainty in data, inaccessibility of object of study, incompleteness of background knowledge? You might think that it would count as the least of your worries that your logical and mathematical powers are not quite perfect. Hence, an assumption in standard Bayesian treatments of scientific inference that logically equivalent sentences should be accorded the same degree of belief. However, in many situations in science the uncertainty of mathematical knowledge plays an important part, as I have explained in a companion paper, not least in the area of mathematical predictions, a phenomenon as yet largely ignored by philosophers, where physicists gain confidence that they are on the right track when purely mathematical conjectures arising from their work turn out to be correct. Plausibility of scientific statements depends on uncertain mathematical knowledge.

To give briefly an indication of this, we hear of the mathematical physicist, Edward Witten, that he

> ...derived a formula for Donaldson invariants on Kähler manifolds using a twisted version of supersymmetric Yang-Mills theory in four dimensions. His argument depends on the existence of a mass gap, cluster decomposition, spontaneous symmetry breaking, asymptotic freedom, and gluino condensation. While none of this is rigorous by mathematical standards, the final formula is correct in all cases which can be checked against rigorous mathematical computations. [Freed and Uhlenbeck, 1995, p. 2]

Such confirmation increases your confidence in the power of your physical modelling. The more surprising the verified mathematical conjecture the greater the boost to your confidence.

It is interesting to wonder why nobody (at least to my knowledge) has taken Pólya up on his Bayesianism in mathematics. What is the underlying intuition behind the avoidance of a Bayesian treatment of plausible and inductive reasoning in mathematics? We can begin to understand what is at stake when we read Mary Hesse's claim that "...since mathematical theorems, unlike scientific laws, are matters of *proof*, it is not likely that our degree of belief in Goldbach's conjecture is happily explicated by probability functions." [Hesse, 1974, p. 191]. There are two responses to this. First, while it is true that the nature of mathematics is characterised like no other discipline by its possession of deductive proof as a means of attaining the highest confidence in the trustworthiness of its results, proofs are never perfectly secure. Second, and more importantly, what gets overlooked here is the prevalence in mathematics of factors other than proof for changing degrees of belief.

The lack of attention plausible mathematical reasoning has received reflects the refusal of most anglophone philosophers of mathematics to consider the way mathematical research is conducted and assessed. On the basis of this refusal, it is very easy then to persist in thinking of mathematics merely as a body of established truths. As classical deductive logic may be captured from a probability calculus which allows propositions to have probabilities either 0 or 1, the belief that mathematics is some kind of elaboration of logic and that the mathematical statements to be considered philosophically are those known to be right or wrong go hand in hand. We could say in fact that mathematics has suffered philosophically from its success at accumulating knowledge since this has deflected philosophers' attention from mathematics as it is being developed. But one has only to glance at one of the many survey articles in which mathematicians discuss the state of play in their field, to realise the vastness of what they know to be unknown but are very eager to know, and about which they may be thought to have degrees of belief equal neither to 0 nor to 1.[2]

We shall see in section 3 how mathematical evidence comes in very different shapes and sizes. But even remaining with 'proved' or well-established statements, although there would appear to be little scope for plausible reasoning, there are a number of ways that less than certain degrees of belief can be attributed to these results. David Hume described this lack of certainty well:

> There is no Algebraist nor mathematician so expert in his science, as to place entire confidence in his proof immediately on his discovery of it, or regard it as any thing, but a mere probability. Every time he runs over his proofs, his confidence encreases; but still more by the approbation of his friends; and is rais'd to its utmost perfection by the universal assent and applauses of the learned world. Now 'tis evident, that this gradual encrease of assurance is nothing but the addition of new probabilities, and is deriv'd from the constant union of causes and effects, according to past experience and observation. [Hume, 1739, pp. 180–1]

Perfect credibility may be difficult to achieve for proofs taking one of a number of non-standard forms, from humanly-generated unsurveyable proofs to computer-assisted proofs to probabilistic proofs. These latter include tests for the primality of a natural number, n. Due to the fact that around half of the numbers less than n are easily computed "witnesses" to its being composite, if such is the case, a small sample can quickly show beyond any set level of doubt whether n is prime.

While a certain amount of suspicion surrounds the latter type of 'proof', from the Bayesian perspective, one can claim that all evidence shares the property that it

[2] I mean to exclude here the immense tracts of totally uninteresting statements expressible in the language of ZFC in which one will never care to have a degree of belief. An idea of the plans in place for the expansion of (interesting) mathematics can be gleaned from the following claim: 'It is clear... that the set-based mathematics we know and love is just the tip of an immense iceberg of n-categorical, and ultimately ω-categorical, mathematics. The prospect of exploring this huge body of new mathematics is both exhilarating and daunting.' [Baez and Dolan, 1999, p. 32].

produces changes in some degrees of belief. The absence of any qualitative difference in the epistemic import of different types of proof has recently been noted by Don Fallis [1997], who considers many possible ways of distinguishing epistemically between deductive proofs and probabilistic proofs and finds none of them adequate. He draws the conclusion, therefore, that there is no such difference. Fallis centres his discussion around 'proofs' which involve clever ways of getting strands of DNA to react to model searches for paths through graphs, putting beyond reasonable doubt the existence or non-existence of such paths. Despite there being here a reliance on biochemical knowledge, Fallis still sees no qualitative difference as regards the justificatory power of this type of proof. Confidence in mathematical statements is being determined by natural scientific theory. This appears less surprising when you consider how complicated, yet well-modelled, configurations of silicon can be used to generate evidence for mathematical propositions.

Fallis's point may be expressed in Bayesian terms as follows.[3] The acceptability of a mathematical statement is dependent solely on your rational degree of belief in that statement conditionalised on all the relevant evidence. Whatever level you set yourself (0.99 or 0.99999) the type of evidence which has led you there is irrelevant. A ten thousand page proof may provide as much support as a probabilistic proof or the non-appearance of a counter-example. To contemplate the reliability of a result in a particular field we should think of someone from outside the field asking a specialist for their advice. If the trustworthy expert says she is very certain that the result may be relied upon, does it matter to the enquirer how the specialist's confidence arises? This depiction could be taken as part of a larger Bayesian picture. The very strong evidence we glorify with the name 'proof' is just as much a piece of evidence as is a verification of a single consequence. Bayesianism treats in a uniform manner not just the very strong evidence that Fallis considers, but all varieties of partial evidence. Let us now see what we are to make of this partial evidence.

2 QUANTITATIVE BAYESIANISM

Pólya understood plausible inference to be quite different from deductive logic. In his eyes [Pólya, 1954b, pp. 112–116], deductive logic is:

(a) Impersonal — independent of the reasoner;

(b) Universal – independent of the subject matter;

(c) Self-sufficient – nothing beyond the premises is needed;

(d) Definitive – the premises may be discarded at the end of the argument.

[3] He points out (private communication), however, that he is not necessarily committed to a Bayesian analysis of his position which assumes that one's rational degree of belief is all that really matters in mathematical justification.

On the other hand, plausible inference is characterised by the following properties:

(a) The direction of change in credibility is impersonal, but the strength may be personal;

(b) It can be applied universally, but domain knowledge becomes important for the strength of change, so there are practical limitations;

(c) New information may have a bearing on a plausible inference, causing one to revise it;

(d) The work of plausible inference is never finished as one cannot predict what new relevant information may arise.

One of the principal differences seems to be that in the deductive case nobody requires of you that you maximise the set of deductive consequences of what you hold to be certain. If you are asked whether you know the truth status of a statement, you search about for a proof or disproof of it from what you already know. If you find nothing, you just admit your ignorance, and no-one can accuse you of anything worse than stupidity if you have overlooked such a proof or disproof. We do not go around blaming ourselves for not having known before Wiles that Fermat's Last Theorem is provable, even though the resources were in some sense available. Deductive logic is there to safeguard you from taking a false step, not from omitting to take a correct step. On the other hand, we may use plausible inference to argue about the plausibility of *any* statement based on what we know at present.[4] The question is how to think about the way we go about arriving at degrees of belief on the basis of what we already know.

It is clear that the strength of a mathematician's belief in the correctness of a result has an impact on their practice: Andrew Wiles would hardly have devoted seven years to Fermat's Last Theorem had he not had great faith in its veracity. No doubt we could give a complicated Bayesian reconstruction of his decision to do so in terms of the utility of success, the expected utility of lemmas derived in a failed attempt, and so on. For a more simple example, we may give a Bayesian reconstruction of the following decision of the French Academy:

> The impossibility of squaring the circle was shown in 1885, but before that date all geometers considered this impossibility as so "probable" that the Académie des Sciences rejected without examination the, alas!, too numerous memoirs on this subject that a few unhappy madmen sent in every year. Was the Académie wrong? Evidently not,

[4]Jaynes [forthcoming, Ch. 10, p. 21] has a similar view on the difference between deductive logic and probability theory as logic: "Nothing in our past experience could have prepared us for this; it is a situation without parallel in any other field. In other applications of mathematics, if we fail to use all of the relevant data of a problem, the result will be that we are unable to get any answer at all. But probability theory cannot have any such built-in safety device, because in principle, the theory must be able to operate no matter what our incomplete information might be".

> and it knew perfectly well that by acting in this manner it did not run the least risk of stifling a discovery of moment. The Académie could not have proved that it was right, but it knew well that its instincts did not deceive it. If you had asked the Academicians, they would have answered: "We have compared the probability that an unknown scientist should have found out what has been vainly sought for so long, with the probability that there is one madman the more on earth, and the latter has appeared to us the greater. [Poincaré, 1905, pp. 191–2]

These alternatives, being mad and being right, were hardly exhaustive. Leaving aside the person's sanity we can contrast the probability that their proof is correct with the probability that it is incorrect.

$$\Pr(\text{proof correct}|\ \text{author unknown}, I) =$$

$$\Pr(\text{proof correct}|\ \text{author unknown, true}, I)\cdot\Pr(\text{true}|\text{author unknown}, I) +$$
$$\Pr(\text{proof correct}|\ \text{author unknown, false}, I)\cdot\Pr(\text{false}|\text{author unknown}, I)$$

$$= \Pr(\text{proof correct}|\ \text{author unknown, true}, I)\cdot\Pr(\text{true}|\ \mathrm{I})$$

where I is the background knowledge.

Substituting reasonable estimates of the Académie's degrees of belief will lead to a very small value for this last expression because its two factors are small. On the other hand, a submitted proof of the possibility of squaring the circle by a known mathematician, or a submitted proof of its impossibility by an unknown author would presumably have been dealt with more tolerantly.

Notice that this reconstruction would not seem to require one to go beyond vague talk of very high or very low probabilities. By contrast, when it comes to offering a betting ratio for the trillionth decimal digit of π being 9, it would seem to be eminently reasonable to propose precisely 1/10, and yet neither the coherence of realistic personalism nor any requirement to maximize expected subjective utility imposes this value upon you. What appears to determine this value is some form of the principle of indifference based on our background knowledge. With a simple grasp of the idea of a decimal expansion we simply have no reason to believe any single digit more likely than any other. Those who know a little more may have heard that there is no statistical evidence to date for any lack of uniformity in the known portion of the expansion, probably rendering them much less likely to be swayed in their betting ratio by a spate of 9s occurring shortly before the trillionth place. So, unless some dramatic piece of theoretical evidence is found, it seems that most mathematicians would stick with the same betting ratio until the point when they hear that computers have calculated the trillionth place.[5]

The issue to be treated in this section is whether we require a quantitative, or even algorithmic, form of Bayesianism to allow us to explicate plausible mathematical reasoning, or whether, like Pólya, we can make do with a qualitative form

[5] As of 1999 they had reached the 206 billionth.

of it. First, it will be helpful for us to contrast Pólya's position with that of Jaynes. For Jaynes, Pólya was an inspiration. Indeed, he

> ... was the original source of many of the ideas underlying the present work. We show how Pólya's principles may be made quantitative, with resulting useful applications. [Jaynes, forthcoming, Ch. 1, p. 3]

As is well known, Jaynes was at the objectivist end of the Bayesian spectrum. In other words, his aim was to establish principles (maximum entropy, transformation groups, etc.) applicable in as many situations as possible, in which a reasonable being could rationally decide on their prior probabilities. Pólya, on the other hand, reckoned that one would have to stay with a qualitative treatment (e.g., if A is analogous to B and B becomes more likely, then A becomes somewhat more likely), in that the direction of changes to confidence might be determined but not their strength. But Jaynes claimed that this decision was based on a faulty calculation made by Pólya when he was considering the support provided to Newton's theory of gravitation by its prediction of the existence and location of a new planet, now called Neptune. The incorrect calculation occurred when Pólya was discussing the boost to confidence in Newtonian gravitation brought about by the observation of a previously unknown planet precisely where calculations predicted it to be, based on observed deviations in Uranus's orbit.

Pólya takes Bayes theorem in the form,

$$\text{Pr(Newt. Grav.|Neptune)} = \frac{\text{Pr(Newt. Grav.)}\cdot\text{Pr(Neptune|Newt. Grav.)}}{\text{Pr(Neptune)},}$$

where Pr(Neptune) corresponds to a scientist's degree of belief that the proposed planet lie in the predicted direction. For the purposes of the calculation, he estimates Pr(Neptune) in two ways. First, he calculates the probability of a point lying within one degree of solid angle of the predicted direction, and arrives at a figure of $0.00007615 \approx 1/13100$. Second, on the grounds that the new planet might have been expected to lie on the ecliptic, he uses the probability of a point on a circle lying within one degree of the specified position, yielding a value for Pr(Neptune) of 1/180. He then argues that Pr(Newtonian Gravitation) must be less than Pr(Neptune), otherwise Bayes' theorem will lead to a posterior probability greater then 1, but that it is unreasonable to imagine a scientist's degree of belief being less than even the larger figure of 1/180, since Newtonian Gravitation was already well-confirmed by that point. He concludes, "We may be tempted to regard this as a refutation of the proposed inequality." [Pólya, 1954b, p. 132] and suggests we return to a safer qualitative treatment.

However, as Jaynes points out, Pólya's calculations were in fact of the prior to posterior odds ratio of two theories: on the one hand, Newtonian gravitation, and on the other, a theory which predicts merely that there be another planet, firstly anywhere and secondly on the ecliptic. Indeed, from the confirmation, Newtonian gravitation is receiving a boost of 13100 or 180 *relative* to the theory that there is one more planet somewhere. Pólya had forgotten that if Pr(Newtonian Gravitation) is already high then so too would Pr(Neptune) be.

We are told by Jaynes that Pólya realised his mistake and went on to participate vigorously in the former's lectures at Stanford University in the 1950s. However, Pólya had given several further arguments against quantitative plausible reasoning, so even if Jaynes could block this particular argument, one would need to confront the others. Reading through them, however, one notes that Pólya is making fairly standard points: the incomparability of evidence and conjectures, problems with the principle of indifference, etc.

Could it be that your background predisposes you to adopt a certain type of Bayesianism? The physicist relies on symmetry considerations pertaining to the mechanisms producing the data, the philosopher of science on vaguer considerations of theory evaluation, while the economist must integrate a mass of data with her qualitative, quasi-causal understanding of the economy. Are disputes among Bayesians like the blind men feeling different parts of an elephant?

Bayesianism applied to reasoning in the natural sciences appears to fall into two rather distinct categories:

(i) analysis of data from, say, nuclear magnetic resonance experiments or astrophysical observations;

(ii) plausible reasoning of scientists by philosophers of science (e.g., [Franklin, 1986]).

We may wonder how strong the relation is between them. Rosenkrantz [1977] attempted a unified treatment, and he indicates by his subtitle 'Towards a Bayesian Philosophy of Science' that a treatment of history and philosophy of science issues alongside statistical issues should be 'mutually enriching' [Rosenkrantz, 1977, p. xi].

Jaynes himself was less sure about how far one could take the historical reconstructions of scientific inference down a Bayesian route. After his discussion of Pólya's attempt to quantify Neptune discovery he claims:

> But the example also shows clearly that in practice the situation faced by the scientist is so complicated that there is little hope of applying Bayes' theorem to give quantitative results about the relative status of theories. Also there is no need to do this, because the real difficulty of the scientist is not in the reasoning process itself; his common sense is quite adequate for that. The real difficulty is in learning how to formulate new alternatives which fit better the facts. Usually, when one succeeds in doing this, the evidence for the new theory soon becomes so overwhelming that nobody needs probability theory to tell him what conclusions to draw. [Jaynes, forthcoming, Ch. 5, p. 17]

This note occurs in a chapter entitled 'Queer uses of probability', by which he intends that at present we have no rational means for ascribing priors. So, despite his professed debt to *Mathematics and Plausible Reasoning*, we find two poles of Bayesianism represented by Jaynes and Pólya. For Jaynes, any rational agent possessing the same information will assign identical probability functions. For Pólya, two experts with the same training may accord different changes to their degrees of belief on discovery of the same fact. One imagines a machine making plausible inferences, the other emphasises the human aspect.

Jaynes:

> ... instead of asking, "How can we build a mathematical model of human common sense?" let us ask, "How could we build a machine which would carry out useful plausible reasoning, following clearly defined principles expressing an idealized common sense? [Jaynes, forthcoming, Ch. 1, p. 5]

Pólya:

> A person has a background, a machine has not. Indeed, you can build a machine to draw demonstrative conclusions for you, but I think you can never build a machine that will draw plausible inferences. [Pólya, 1954b, p. 116]

Perhaps it is the lack of exactitude which steers Jaynes away from modelling scientific reasoning. After a lifetime investigating how symmetry considerations allow the derivation of the principles of statistical mechanics, it must be difficult to adapt to thinking about plausibility in complex situations.

But if a physicist might be excused, what of a philosopher? John Earman, while discussing how a physicist's degrees of belief in cosmological propositions were affected by the appearance of General Relativity on the scene, tells us

> But the problem we are now facing is quite unlike those allegedly solved by classical principles of indifference or modern variants thereof, such as E. T. Jaynes's maximum entropy principle, where it assumed that we know nothing or very little about the possibilities in question. In typical cases the scientific community will possess a vast store of relevant experimental and theoretical information. Using that information to inform the redistribution of probabilities over the competing theories on the occasion of the introduction of the new theory or theories is a process that is, in the strict sense of the term, *a*rational: it cannot be accomplished by some neat formal rules, or, to use Kuhn's term, by an algorithm. On the other hand, the process is far from *ir*rational, since it is 'informed by reasons. But the reasons, as Kuhn has emphasized, come in the form of persuasions rather than proof. In Bayesian terms, the reasons are marshalled in the guise of plausibility arguments. The deployment of plausibility arguments is an art form for which there currently exists no taxonomy. And in view of the limitless variety of such arguments, it is unlikely that anything

> more than a superficial taxonomy can be developed. [Earman, 1992, p. 197]

This seems a rather pessimistic analysis for a professed Bayesian. Does the 'limitless variety' of these arguments mean that we should not expect to find patterns among them? Despite the talk of their deployment being an 'art form', Earman does allow himself to talk about the objective quality of these plausibility arguments. Indeed, he claims that:

> Part of what it means to be an "expert" in a field is to possess the ability to recognize when such persuasions are good and when they are not. [Earman, 1992, p. 140]

Interestingly, it is Pólya the "expert" in mathematics who believes that it is possible to extract the patterns of good plausibility arguments from his field.

So, out of the three, Jaynes, Pólya and Earman, representatives of three different types of Bayesianism, it is Pólya who believes one can say something quite concrete about plausible reasoning. All realise that plausible reasoning is a very complex process. Neither Jaynes nor Earman cannot see a way forward with plausible scientific reasoning. This leaves Pólya who gets involved with real cases of (his own) mathematical reasoning, which he goes on to relate to juridical reasoning and reasoning about one's neighbour's behaviour. Is he right to claim that mathematics provides a better launch pad to tackle everyday reasoning than does science?

If we want a fourth Bayesian to complete the square, we might look to the computer scientist Judea Pearl. Like Pólya, Pearl believes we can formulate the principles of everyday common sense reasoning, and like Jaynes he thinks Bayesian inference can be conducted algorithmically. To be able to do the latter requires a way of encoding prior information efficiently to allow Bayesian inference to occur. For Pearl [Pearl, 2000], (*this volume*) humans store their background information efficiently in the form of causal knowledge. The representation of this causal knowledge in a reasonably sparse Bayesian network is the means by which a machine can be made to carry out plausible reasoning and so extend our powers of uncertain reasoning.

In his earlier work Pearl [1988] expressed his appreciation of Pólya's ideas, and yet found fault with his restriction to the elucidation of *patterns* of plausible reasoning rather than a *logic*. He considers Pólya's loose characterisation of these patterns not to have distinguished between evidence factoring through consequences and evidence factoring through causes. For instance, Pólya asserts that when B is known to be a consequence of A, the discovery that B holds makes it more likely that A holds. This, however, is a well known fallacy of causal reasoning. I see that the sprinkler on my lawn is running and that the grass is wet, but this does make it more probable to me that it has rained recently even though wet grass is a consequence of it having done so. But one need not remain with causal stories to reveal this fallacy. A consequence of a natural number being divisible by four is that it is even. I find that a number I seek is either 2 or 6. Although I have learnt that it is even, this discovery reduces the probability of its being divisible by 4 to zero. Essentially, what Pólya overlooked was the web-like nature of our beliefs, only departing from patterns involving two propositions when he considered the possibility of two facts having a common ground. In Bayesian networks, converging arrows are equally important but must be treated differently.

It remains to be seen whether the techniques of Bayesian networks may illuminate scientific inference. Now we shall turn our attention to examine what Bayesianism has to say about certain aspects of mathematical reasoning.

3 WHAT MIGHT BE ACHIEVED BY BAYESIANISM IN MATHEMATICS

Varieties of mathematical evidence may be very subtle, lending support to Earman and Jaynes' scepticism. Pólya [1954b, p. 111] himself had the intuition that two mathematicians with apparently similar expertise in a field might have different degrees of belief in the truth of a result and treat evidence for that result differently. Even though each found a consequence of the result equally plausible, the establishment of this consequence could have an unequal effect on their ratings of the likelihood of the first result being correct. The complex blending of the various kinds of evidence experienced through a mathematician's career would explain the differences in these reactions, some of which might be attributable to aspirations on the part of each of them either to prove or disprove the theorem. But Pólya goes further to suggest that such differences of judgement are based on "still more obscure, scarcely formulated, inarticulate grounds" [Pólya, 1954b, p. 111].

Such appeals to the inexpressible, or at least to the imprecisely expressed, are not at all uncommon. For example, the mathematician Sir Michael Atiyah asserts that

> ... it is hard to communicate understanding because that is something you get by living with a problem for a long time. You study it, perhaps for years. You get the feel of it and it is in your bones. [Atiyah, 1984, p. 305]

Such comments may have been devised by mathematicians to give an air of mystery to their practice. A sceptic could point out that doctors have done likewise in the past by alleging that diagnosis requires some profound intuitive faculty of divination, an attractive image shattered by the successful construction of expert systems which have shown physicians to be replaceable in some situations, by machines using propositionally encoded evidence. However, the success of artificial intelligence in some areas of medical diagnosis may be contrasted with the extreme difficulty in getting computers to do anything that might be termed creative in mathematics.[6] The essential point does not concern whether or not

[6]A possible exception is the recent successful automated solution of the Robbins problem (see http://www.mcs.anl.gov/~mccune/), drawn to my attention by Don Fallis.

mathematicians in fact rely on non-propositional knowledge, so much as whether there might be something about this type of knowledge which is indispensable to doing mathematics.

Certainly, evidence for the correctness of a statement may be very subtle. It may even arise through an experience of failure. In [Corfield, 1997] I pointed out the inaccuracy on Lakatos's part of his notion of lemma-incorporation, the idea that faulty proofs are generally repaired by some simple insertion of a lemma. As I explained there, while proving the so-called 'duality theorem' Poincaré had come to realise that an assumption he was making about the way differential manifolds intersect was invalid in general. However, he still believed that the general strategy could be made to work of constructing for a given set of manifolds of equal dimension a manifold of complementary dimension which intersected each of the members of the set exactly once. He just needed to have the intersections occur in a more controlled fashion. One can only guess how this experience impacted on his degree of belief in the duality theorem. It is quite probable that even though the initial proof was found to be wrong, the experience of near success with a variant of a strategy gave him hope that another variant would work. It must also happen, however, that mathematicians are discouraged by such setbacks.

Evidence can also involve the non-discovery of something as Sherlock Holmes well knew when he built his case on the observation of a dog that did not bark. The classic example of the unsurveyable human-generated kind of proof at the present time is the proof of the classification of finite simple groups into 5 infinite families and 26 sporadic outsiders. How does one's degree of belief in this result depend on such potentially flawed lengthy evidence? Fallis [1997] has Gorenstein, the driving force behind the collective proof, confessing that confidence is boosted less by the proof itself than by the fact that no other such groups have been found. Similarly, remarks are often to be heard concerning the consistency of ZFC that one would have expected to have encountered a contradiction by now.

We should also remember that evidence for mathematical propositions comes from sources which have only recently become available. The use of computers to fill in the gaps of human proofs has become acceptable, but computers are used in many other ways in mathematics. For example, they provide evidence for conjectures via calculations made on samples, and they produce visual evidence in dynamical systems theory, as in the drawing of attractors or power spectra. Reliance on computer evidence raises some novel issues. Oscar Lanford is attributed with pointing out that

> ... in order to justify a computer calculation as part of a proof..., you must not only prove that the program is correct (and how often is that done?) but you must understand how the computer rounds numbers, and how the operating system functions, including how the time-sharing system works. [Hirsch, 1994, p. 188]

Moreover, if more than one piece of computer evidence is being considered, how do we judge how similar they are for conditionalising purposes? This would require one to know the mathematics behind any similarities between the algorithms utilised.

It is clear then that any account of mathematical inference will require a very expressive language to represent all the various forms of evidence which impact on belief in mathematical propositions. The Bayesian wishing to treat only propositions couched in the language of the object level might hope to be able to resort to Jeffrey conditionalisation, but this comes at the price of glossing over interesting features of learning. Concerning scientific inference, Earman [1992, pp. 196–8] asserts that many experiences will cause the scientist to undergo *non-Bayesian* shifts in their degrees of belief, i.e., ones unaccountable for by any form of algorithmic conditionalisation. These shifts, the resetting of initial probabilities, are very common, he claims, arising from the expansion of the theoretic framework or from the experience of events such as "[n]ew observations, even of familiar scenes; conversations with friends; idle speculations; dreams..." [Earman, 1992, p. 198]. One might despair of making any headway, but taking Pólya as a guide we may be able to achieve something. While recognising that making sense of plausible reasoning in mathematics will not be easy, I believe that three key areas of promise for this kind of Bayesianism in mathematics are analogy, strategy and enumerative induction.

3.1 Analogy

Before turning to a probabilistic analysis of plausible reasoning in the second volume of *Plausible Reasoning*, Pólya had devoted the first volume [1954a], as its subtitle suggests, to the themes of analogy and induction. Analogies vary as to their precision. When vague they contribute to what he called the *general atmosphere* surrounding a mathematical conjecture, which he contrasts to pertinent *clear facts*. While verifications of particular consequences are straightforwardly relevant facts, the pertinence of analogical constructions may be hard to discern precisely. Nevertheless, mathematicians, succh as Gian-Carlo Rota, take them to be vitally important:

> The enrapturing discoveries of our field systematically conceal, like footprints erased in the sand, the analogical train of thought that is the authentic life of mathematics. [Kac *et al.*, 1986, p. ix]

Let us illustrate this with an example. At the present time the vast majority of mathematicians have a high degree of belief in the Riemann Hypothesis. Recall that the Riemann zeta function is defined as the analytic continuation of $\zeta(s) = \Sigma n^{-s}$ summed over the natural numbers, and that the hypothesis claims that if s is a zero of $\zeta(s)$, then either $s = -2, -4, -6$, ..., or the real part of s equals 1/2. Many roots have been calculated (including the first 1.5 billion zeros in the upper complex plane along with other blocks of zeros), all confirming the theory, but despite this "overwhelming numerical evidence, no mathematical proof is in sight." [Cartier, 1992, p. 15]. As Bayesians have explained, there are limits to the value of showing that your theory passes tests which are conceived to be very similar. If, for example, a further 100 million zeros of the zeta function are found to have their real part equal to 1/2, then little change will occur in mathematicians' degrees of belief, although a little more credibility would be gained if this were true of 100 million zeros around the 10^{20} th, which is precisely what has happened.

In this example the clear facts making up the numerical evidence can lend only limited credence by themselves. After all, there are 'natural' properties of the natural numbers which are known to hold for exceedingly long initial sequences. What counts in addition beyond evidential facts, however numerous, is the credibility of stronger results, general consequences and analogies. Indeed, if an analogy is deemed strong enough, results holding for one side of it are thought to provide considerable support for their parallels. Concerning the Riemann conjecture (RC), we are told that:

> There is impressive numerical evidence in its favour but certainly the best reason to believe that it is true comes from the analogy of number fields with function fields of curves over finite fields where the analogue of RC has first been proved by A. Weil. [Deninger, 1994, p. 493]

This analogy[7] was postulated early in this century as a useful way of providing a halfway house across an older analogy, developed by Dedekind and Weber, from algebraic number fields to function fields over the complex numbers. However, the translation techniques between the three domains have still not been perfected. The more geometric side of the analogy Deninger mentions was able to absorb cohomological techniques, allowing Weil to prove the Riemann hypothesis analogue in 1940. An extraordinary amount of effort has since been expended trying to apply cohomology to number theory (Weil, Grothendieck, Deligne, etc.) with the establishment of the standard Riemann hypothesis as one of its aims.

How should we judge how analogous two propositions, A and B, are to each other? For Pólya [1954b, p. 27] it correlates to the strength of your "hope" for a common ground from which they both would naturally follow. Increase in confidence in A will then feed up to the common ground, H, and back down to B.[8] If analogy is to be treated anywhere, I believe mathematics will provide a good location, since there are plenty of excellent examples to be found there. In Pólya's principal example, Euler noticed that the function sin x/x resembles a polynomial in several respects: it has no poles; it has the right number of zeros, which do not accumulate; it behaves symmetrically at $\pm\infty$. On the other hand, unlike a polynomial, sin x/x remains bounded. Even with this disanalogy, it seemed plausible that polynomials and sin x/x would share other properties. One notable feature of complex polynomials is that any one of them may be expressed as a product

[7] See also [Katz and Sarnak, 1999], in particular the table on page 12.

[8] Notice here the flavour of a Bayesian network: H pointing to both A and B.

of factors of the form (1 − x/root), taken over all of its roots. Might this property also apply to sin x/x? Well, the roots of this function are $\pm\pi, \pm 2\pi, \pm 3\pi, \ldots$, suggesting that we should have

$$\frac{\sin x}{x} = \left(1 - \frac{x^2}{\pi^2}\right)\left(1 - \frac{x^2}{4\pi^2}\right)\left(1 - \frac{x^2}{9\pi^2}\right)\cdots,$$

On the other hand, expanding the function as a Taylor series, we have

$$\sin x/x = 1 - x^2/6 + x^4/120 - \ldots$$

Equating coefficients of x^2 suggests then that

$$1 + 1/4 + 1/9 + 1/16 + \ldots = \pi^2/6.$$

Even after checking this sum to several decimal places Euler was not absolutely confident in the result, but he had in fact solved a famous problem by analogical reasoning.

It might be that what is happening here is something similar to what Pearl [2000] has termed the "transfer of robust mechanisms to new situations". We have a mechanism that links factorisation of a function to its zeros. We find it applies for complex polynomials and wonder whether it may be extended. Features of polynomials that may be required in the new setting are that they have the right number of zeros, they remain bounded on compact sets, they behave similarly at $\pm\infty$. Might the mechanism be expected to work for a non-polynomial function possessing these features, such as sin x/x? What if you force the variable measuring the number of roots to be infinite? We may find it hard to estimate quantitatively the similarity between a function like sin x/x and a complex polynomial, but it is clear that tanx/x or exp x are less similar, the former having poles, the latter having no zeros and asymmetric behaviour at $\pm\infty$, and indeed the mechanism does fail for them.

In this case, once the realisation that an analogy was possible, it didn't cost much to work through the particular example. Euler hardly needed to weigh up the degree of similarity since calculations of the sum quickly convinced him. However, to develop a general theory of the expansion of complex functions did require greater faith in the analogy. This paid off when further exploration into this mechanism allowed mathematicians to form a very general result concerning entire complex functions, providing the "common ground" for the analogues.

3.2 *Strategy*

Moving on now to strategy, the title of Deninger's paper—*Evidence for a Cohomological Approach to Analytic Number Theory*—is also relevant to us. His aim in this paper is to increase our degree of belief that a particular means of thinking about a field will lead to new results in that field. This is a question of strategy.

At a finer level one talks of tactics. Researchers from the AI community working on automated theorem proving, have borrowed these terms. One tactic devised by Larry Wos [Wos and Pieper, 1999] involves thinking in terms of how probable it is that the computer can reach the target theorem from a particular formula generated from the hypotheses during the running of the programme. This tactic takes the form of a weighting in the search algorithm in favour of formulas which have a syntactical form matching the target.

Elsewhere, researchers in Edinburgh are interested in the idea of the choice of tactics [Bundy, 1999]. There is an idea of likening mathematics to a game of bridge where the mathematician, like the declarer, has some information and a range of strategies to achieve their goal (finesse, draw trumps, squeeze). Of course, there is a difference. In bridge, you are in the dynamic situation where you cannot try out every strategy, as the cards get played. This forces you to pay very close attention to which tactics have the best chance of working. In mathematics, on the other hand, with a computer it does not cost you much to try things out, although one does risk combinatorial explosion. At present, probabilities are being used for their computer bridge player, they are not yet being used for their automated theorem prover. While the computer has a small repertoire of syntactical tactics (rippling, resonance, heat, etc.) there is less need for an assessment of the chance of each working, but presumably the number of proof techniques will grow.

These automated proof strategies are at present syntactically based. Naturally, semantic considerations play the dominant role for human mathematician. Pólya was active in this area too. To give a brief flavour of his ideas, if when planning to solve a problem, any of the following should increase your confidence in your plan [Pólya, 1954b, pp. 152–153]:

- Your plan takes all relevant information into account.
- Your plan provides for a connection between what is known and what is unknown.
- Your plan resembles some which have been successful in problems of this kind.
- Your plan is similar to one that succeeded in solving an analogous problem.
- Your plan succeeded in solving a particular case of the problem.
- Your plan succeeded in solving a part of the problem.

3.3 *Enumerative induction*

Besides the incorrect Bayesian calculation of the confirmation provided by the observation of Neptune, Pólya does resort to a quantitative sketch in another place [Pólya, 1954b, pp. 96–7]. Here he outlines how one might think through the boost to the credibility of Euler's formula for a polyhedron (vertices − edges +

faces = 2) known to hold for some simple cases, when it is found to be true of the icosahedron. (12 − 30 + 20 = 2). His approach is to reduce the problem to the chances of finding three numbers in the range 1 to 30 with the property that the second is equal to the sum of the other two, i.e., (V − 1) + (F − 1) = E. The proportion of these triples is around 1 in 60, providing, he argues, a boost of approximately 60 to the prior probability of Euler's conjecture. Here again we see the same problem that Jaynes located in the Neptune calculation. The ratio of the likelihood of the Euler conjecture compared to that of its negation is 60.

In any case Pólya's construction can only be viewed as sketchy. It is not hard to see that the number of edges will always be at least as great as one and a half times the number of faces or the number of vertices. (For the latter, for example, note that each edge has two ends, but at least three of these ends coincide at a vertex). Thus one should have realised that there are further constraints on the possible triples and hence that the likelihood ratio due to the evidence for the Euler formula should have been in comparison to better informed rival conjecture, and so not so large. But the interesting point is that Pólya goes on to say that:

> If the verifications continue without interruption, there comes a moment, sooner or later when we feel obliged to reject the explanation by chance. [Pólya, 1954b, p. 97]

The question then arises as to whether one is justified in saying such a thing on the basis of a finite number of verifications of a law covering an infinite number of cases. This will hinge on the issue of the prior probability of such a law.

Now, consider Laplace's rule of succession. If you imagine yourself drawing with replacement from a bag of some unknown mixture of white and black balls, and you have seen m white balls, but no black balls, the standard use of the principle of indifference suggests that the probability that the next n will be positive is

$$(m + 1)/(m + n + 1).$$

As $n \to \infty$, this probability tends to zero. In other words, if verifying a mathematical conjecture could be modelled in this fashion, no amount of verification could help you raise your degree of belief above zero.

This accords with the way Rosenkrantz [1977] views the situation. He considers the particular case of the twin prime conjecture: that there are an infinite number of pairs of primes with difference 2. He mentions that beyond the verification of many cases, there are arguments in analytic number theory which suggest that you can form an estimate for the number of twin primes less than n and show that it diverges. He then continues:

> Now if Popper's point is that no examination of 'positive cases' could ever raise the probability of such a conjecture to a finite positive value, I cannot but agree. Instances alone cannot sway us! But if his claim is that *evidence of any kind* (short of proof) can raise the probability of a

> general law to a finite positive value, I emphatically disagree. On the cited evidence for the twin prime conjecture, for example, it would seem to me quite rational to accept a bet on the truth of the conjecture at odds of, say 100:1, that is to stake say \$ 100 against a return of \$ 10 000 should the conjecture prove true. [Rosenkrantz, 1977, p. 132]

So for Rosenkrantz, with no background knowledge, the principle of indifference forces a universal to have zero, or perhaps an infinitesimal (something also considered by Pólya), prior probability. However, other considerations may determine a positive probability.

> Subject-specific arguments usually underlie probability assessments in mathematics. [Rosenkrantz, 1977, p. 90]

In support of this view, returning to the Euler conjecture, we should note that there was background knowledge. For polygons, it is a trivial fact that there is a linear relation between the number of vertices and the number of edges, namely, V = E. Hence, a simple linear relation might be expected one dimension higher.

Is it always this kind of background knowledge which gives the prior probability of a conjecture a 'leg-up'? Do we ever have a situation with no background knowledge, i.e., where a *general atmosphere* is lacking? Consider the case of John Conway's 'Monstrous Moonshine', the conjectured link between the j-function and the monster simple group. The j-function arose in the nineteenth century from the study of the parameterisation of elliptic curves. It has a Fourier expansion in $q = \exp(2\pi i\tau)$:

$$j(\tau) = 1/q + 744 + 196884q + 21493760q^2 + 864299970q^3 + ...$$

One day while leafing through a book containing this expansion, a mathematician named John Mackay observed that there was something familiar about the third coefficient of this series. He recalled that 196,883 was the dimension of the smallest non-trivial irreducible representation of what was to become known as the monster group, later confirmed to be the largest of the 26 sporadic finite simple groups. Better still, adding on the 1 dimension of the trivial representation of the monster group results in equality.

In view of the very different origins of these entities, the j-function from nineteenth century work on elliptic curves and the monster group from contemporary work in finite group theory, if one had asked a mathematician how likely she thought it that there be some substantial conceptual connection between them or common ground explaining them both, the answer would presumably have been "vanishingly small". In Bayesian terms, Pr(connection/ numerical observation) is considerably greater than Pr(connection), but the latter is so low that even this unlikely coincidence does not bolster it sufficiently to make it credible. Naturally, McKay was told that he was 'talking nonsense'. He then went on, however, to observe that the second nontrivial representation has dimension 21296876. A quick calculation revealed that the fourth coefficient of the j-function could be expressed as: 21493760 = 21296876 + 196883 + 1. In fact every further coefficient of the j-function turns out to be a simple sum of the dimensions of the monster's representations. At this point the question of whether there is some connection has been all but answered—it has become a near certainty. Conway challenged the mathematics community to resolve this puzzle.

Fourier expansion in $q = \exp(2\pi i\tau)$:

$$j(\tau) = 1/q + 744 + 196884q + 21493760q^2 + 864299970q^3 + \ldots$$

196884	$196883 + 1$
21493760	$21296876 + 196883 + 1$
864299970	$842609326 + 21296876 + 196883 + 196883 + 1 + 1$
...	

The answer eventually arrived through a construction by Richard Borcherds, a student of Conway, which earned him a Fields' Medal. Borcherds managed to spin a thread from the j-function to the 24-dimensional Leech lattice, and from there to a 26-dimensional space-time inhabited by a string theory whose vertex algebra has the monster as its symmetry group.

So why does the monster group–j-function connection become so likely by the time you have seen three or four of the sums, even with a minuscule prior, when other inductions are less certain after billions of verifications? Would we find consensus on how the number of instances affects one's confidence? Surely most people would agree that it was a little reckless on Fermat's part to conjecture publicly that $2^{2^n} + 1$ is prime after verifying only 5 cases (and perhaps a check on divisibility by low primes for the sixth).

n	0	1	2	3	4	5
$2^{2^n} + 1$	3	5	17	257	65537	$4294967297 = 641 \times 6700417$

Is it possible to use Bayes' theorem, even merely suggestively?

Let us return to the case of the Riemann hypothesis (RH). If we have a prior degree of belief for RH, how can 1.5 billion verifications affect it? Surely they must, but then is there some asymptotic limit? One might want to factor the RH as follows

$$\Pr(\text{RH} \mid \text{Data}) = \Pr(\text{RH}|p = 1, \text{Data}) \cdot \Pr(p = 1|\text{Data}),$$

where p denotes the limiting proportion, if this exists, of the zeros that lie on the line, taking the zeros in the order of increasing size of modulus.

For the second factor we might then have started with a prior distribution over p according to the weighted sum of the exhaustive set of hypotheses about p: non-convergent p; p in $[0, 1)$; $p = 1$. Then if one can imagine some element of independence between the zeros, e.g., the fact that the nth zero lies on the line provides no information on the $(n + 1)$th, then the confirmation provided by the 1.5 billion zeros should push the posterior of $p = 1$ to take up nearly all the probability accorded to convergent p. This kind of assumption of independence has been used by mathematicians to make conjectures about the distribution of primes, so may be appropriate here. We might also consider that 1.5 billion positive instances provides an indication that p is convergent. Again, however, this consideration would depend on experience in similar situations.

For the first factor, out of all the functions you have met for which their zeros have held for a large initial section and the proportion of cases is 1, you are wondering what proportion are universally true. It is clear, then, that again much would depend on prior experience. For example, something that would be kept in mind is that the function $\pi(x)$, defined as the number of primes less than x, is known to be less than a certain function, denoted $\mathrm{li}(x)$, up to 10^{12}, and that there is good evidence that this is so up to 10^{30}. But it is known not to hold somewhere before 10^{400}. Indeed, there appears to be a change close to 1.4×10^{316}.

Returning finally to 'Monstrous Moonshine', perhaps we should look harder for a reliance on background knowledge. First, it is worth remembering that the dimensions of the monster group's representations and the coefficients of the j-function were not 'made up'. They come from 'natural' mathematical considerations. Imagine in the monstrous moonshine case if the two sides were not 'interesting' entities or that you knew for a fact that these numbers were randomly generated, wouldn't you take more convincing? Similar considerations are discussed by Paris [Paris *et al.*, 2000], who wish to justify some 'natural' prior distribution of probability functions over n variables.

> ...what in practice I might claim to know, or at least feel justified in believing, is that the data I shall receive will come from some real world 'experiment', some natural probability function; it will not simply have been made up. And in this case, according to my modeling, I do have a prior distribution for such functions. [Paris *et al.*, 2000, p. 313]

Evidence for the fact that background knowledge is coming into play in this case is provided by the fact that on presentation of the example to an audience of non-mathematicians they found the numerical coincidences not at all convincing. Despite the fact that a mathematician has no knowledge of a reason for a connection between these two mathematical entities, some slight considerations must play a role. Indeed, what seemed to disappoint the non-mathematicians was the need to include multiples of the dimensions of the irreducible representations. A mathematician, on the other hand, is well aware that in general a group representation is a sum of copies of irreducible ones. For example, the right regular representation, where the group acts on a vector space with basis its own elements, is such a sum where the number of copies of each irreducible representation is equal to its dimension. Behind the addition of dimensions are sums of vector spaces. Second, a mathematician would know that the j-function arises as a basic function, invariant under the action of the modular group. This offers the possibility that group theory might shed some light on the connection.

4 CONCLUSION

We have covered a considerable stretch of ground here. Clearly much work remains to be done on Pólya's research programme, but I think we can allow ourselves a little more optimism than Earman. I have isolated the following areas as potentially useful to study in a Bayesian light: (1) Analogy; (2) Strategy choice; and, (3) The use of large computations to increase plausibility of conjectures. Elsewhere I shall consider two additional areas: (4) Mathematical predictions in physics; and, (5) The use of stochastic ideas in mathematics (random graphs, random matrices, etc.). It is important to note that we need not necessarily arrive at some quantitative, algorithmic Bayesian procedure to have made progress. If Bayesianism in mathematics suggests interesting questions in the philosophy of mathematics, then I think we can say that it has served its purpose.

Department of Philosophy, King's College London.

BIBLIOGRAPHY

[Atiyah, 1984] M. Atiyah. An Interview with Michael Atiyah. *Mathematical Intelligencer*, **6**(1), 1984. Reprinted in *Collected Works, vol. 1: Early Papers, General Papers*, pp. 297–307, Oxford: Oxford University Press, 1988.

[Baez and Dolan, 1999] J. Baez and J. Dolan. Categorification. In *Higher Order Category Theory*, E. Tetzler and M. Kapranov, eds. pp. 1–36. American Mathematical Society, Providence, RI, 1999.

[Bundy, 1999] A. Bundy. Proof planning methods as schemas. *Journal of Symbolic Computation* **11**, 1–25, 1999.

[Cartier, 1992] P. Cartier. An introduction to compact Riemann surfaces. In *From Number Theory to Physics*,M. Waldschmidt, P. Moussa, J.-M. Luck and C. Itzykson, eds. Springer-Verlag, Berlin, 1992.

[Corfield, 1997] D. Corfield. Assaying Lakatos's philosophy of mathematics. *Studies in History and Philosophy of Science*, **28**, 99–121, 1997.

[Deninger, 1994] C. Deninger. Evidence for a cohomological approach to analytic number theory. In *First European Congress of Mathematics, Vol. 1,*, A. Joseph *et al.* eds. pp. 491–510. Birkhaüser, Basel, 1994.

[Earman, 1992] J. Earman. *Bayes or Bust?: A Critical Examination of Bayesian Confirmation Theory*, MIT Press, Cambridge, MA, 1992.

[Fallis, 1997] D. Fallis. The epistemic status of probabilistic proof. *Journal of Philosophy* **94**, 165–186, 1997.

[de Finetti, 1974] B. de Finetti. *Theory of Probability: A Critical Introductory Treatment.* Translated by A. Machi and A. Smith. Wiley, London, 1974.

[Franklin, 1986] A. Franklin. *The Neglect of Experiment*, Cambridge University Press, Cambridge, 1986.

[Freed and Uhlenbeck, 1995] D. Freed and K. Uhlenbeck, eds. *Geometry and Quantum Field Theory*, American Mathematical Society, 1995.

[Hacking, 1967] I. Hacking. Slightly more realistic personal probability, *Philosophy of Science* **34**, 311–325, 1967.

[Hesse, 1974] M. Hesse. *The Structure of Scientific Inference*. MacMillan, London, 1974.

[Hirsch, 1994] M. Hirsch. Responses to "Theoretical Mathematics", by A. Jaffe and F. Quinn', *Bulletin of the American Mathematical Society* **30**, 187–191, 1994.

[Hume, 1739] D. Hume. *A Treatise of Human Nature*. Clarendon Press, Oxford, 1739.

[Jaynes, forthcoming] E. Jaynes. *Probability Theory: The Logic of Science*, Cambridge University Press, forthcoming.

[Kac *et al.*, 1986] M. Kac, G.-C. Rota and J. Schwartz. *Discrete Thoughts: Essays in Mathematics, Science, and Philosophy*. Birkhaüser, Boston, 1986.

[Katz and Sarnak, 1999] N. Katz and P. Sarnak. Zeroes of zeta functions and symmetry. *Bulletin of the American Mathematical Society*, 36(1): 1–26, 1999.

[Paris *et al.*, 2000] J. Paris, P. Watton and G. Wilmers. On the structure of probability functions in the natural world. *International Journal of Uncertainty, Fuzziness and Knowledge-Based Systems*, 2000.

[Pearl, 1988] J. Pearl. *Probabilistic Reasoning in Intelligent Systems*. San Mateo, CA: Morgan Kaufman, 1988.

[Pearl, 2000] J. Pearl. *Causality: Models, Reasoning and Inference*, Cambridge University Press, 2000.

[Poincaré, 1905] H. Poincaré. *Science and Hypothesis*. Dover Publications, New York, 1905.

[Pólya, 1954a] G. Pólya. *Mathematics and Plausible Reasoning: Induction and analogy in mathematics, vol. 1*, Princeton University Press, Princeton, 1954.

[Pólya, 1954b] G. Pólya. *Mathematics and Plausible Reasoning: Patterns of plausible inference, vol. 2*, Princeton University Press, Princeton, 1954.

[Rosenkrantz, 1977] R. Rosenkrantz. *Inference, Method and Decision. Towards a Bayesian Philosophy of Science*. Reidel, Boston, 1977.

[Wos and Pieper, 1999] L. Wos and G. W. Pieper. *A fascinating country in the world of computing: Your guide to automated reasoning*. Singapore: World Scientific, 1999.

J. B. PARIS & A. VENCOVSKÁ

COMMON SENSE AND STOCHASTIC INDEPENDENCE

1 INTRODUCTION

In this paper we shall extend the results in [Paris and Vencovská, 1990] and [Paris, 1999] on common sense belief formation from (finite) knowledge bases of *linear* probabilistic constraints to include also the case of polynomial *non-linear* constraints and in particular constraints expressing stochastic independence. Indeed our results will be seen to extend to entirely general sets of constraints provided their solution sets are closed.

To start with however we shall recall the context, assumptions, definitions and conclusion of these earlier papers. Briefly, we assumed that the degrees of belief that an agent assigns to sentences of a particular propositional language satisfy the standard Kolmogorov axioms for probability (i.e. 'belief equals probability') and considered the situation where such beliefs were to be assigned *solely* on the basis of a finite set of linear constraints on these beliefs/probabilities (the so called *Watts Assumption*, that the knowledge base is *all the agent's knowledge*, see [Paris, 1994]). The question we considered in [Paris and Vencovská, 1990] and [Paris, 1999] was *how* such an agent should assign these beliefs if s/he is to act in accordance with 'common sense'. In other words we were interested in what common sense 'processes' (if any) might exist for assigning beliefs on the basis of such linear knowledge bases.

We gave an answer to this question by formulating a number of common sense principles of belief formation which, as we argued, any such process (formally called an *inference process*) should satisfy, and then went on to show that there was precisely one inference process, the *Maximum Entropy Inference Process*, which satisfied all these principles. Hence, if one accepts, not unreasonably, that within this context obedience to these principles is a necessary condition for common sense, then the adoption of this inference process is identifiable with acting according to common sense and hence, we would argue, with acting 'intelligently'.

Formally, using the notation of [Paris, 1999], let L stand for the countably infinite set of propositional variables $p_1, p_2, ..., p_n, ...,\ n \in \mathbb{N}$, let SL denote the sentences of L built up using the connectives $\wedge, \vee, \neg$ and let $SL(p_{i_1}, ..., p_{i_n})$ denote the set of sentences of the finite sublanguage of L with (distinct) propositional variables $p_{i_1}, ..., p_{i_n}$. We will use θ, φ, ψ etc. to denote sentences.

We say that w is a *probability function* on SL (and similarly for $SL(p_{i_1}, ..., p_{i_n})$) if $w : SL \to [0, 1]$ and for all $\theta, \varphi \in SL$,

$$(P1)\quad \text{if } \models \theta \text{ then } w(\theta) = 1,$$
$$(P2)\quad \text{if } \models \neg(\theta \wedge \varphi) \text{ then } w(\theta \vee \varphi) = w(\theta) + w(\varphi).$$

D. Corfield and J. Williamson (eds.), Foundations of Bayesianism, 203–240.

As shown, for example, in [Paris, 1994], simple consequences of ($P1$-2) are that for $\theta, \varphi \in SL$ ($SL(p_{i_1}, ..., p_{i_n})$),

(i) If $\models \theta$ then $w(\theta) = 1$ and $w(\neg\theta) = 0$.

(ii) If $\theta \models \varphi$ then $w(\theta) \leq w(\varphi)$, and if $\theta \equiv \varphi$ then $w(\theta) = w(\varphi)$.

(iii) $w(\theta \vee \varphi) = w(\theta) + w(\varphi) - w(\theta \wedge \varphi)$.

In what follows $w, w_0, etc.$ will be used for probability functions. Notice that if $\psi \in SL(p_{i_1}, ..., p_{i_n})$ then, by the Disjunctive Normal Form Theorem,

$$\psi \equiv \bigvee_{j=1}^{r} \alpha_{k_j},$$

for some distinct *atoms* α_{k_j} of $SL(p_{i_1}, ..., p_{i_n})$, that is sentences of the form

$$\pm p_{i_1} \wedge \pm p_{i_2} \wedge ... \wedge \pm p_{i_n}$$

where $\pm p$ stands for p or $\neg p$. So by ($P1$-2) and (ii), for w a probability function on $SL(p_{i_1}, ..., p_{i_n})$,

$$(1) \quad w(\psi) = \sum_{j=1}^{r} w(\alpha_{k_j}).$$

By (i) then,

$$\sum_{j=1}^{2^n} w(\alpha_j) = 1,$$

and we see that w is determined by the 2^n−ary vector $\langle w(\alpha_1), ..., w(\alpha_{2^n})\rangle \in \mathbb{D}_n$ where

$$\mathbb{D}_n = \{ \langle x_1, ..., x_{2^n}\rangle \mid x_1, ..., x_{2^n} \geq 0, \ \sum_{j=1}^{2^n} x_j = 1 \},$$

and conversely every $\vec{x}$ in this set determines a unique probability function w (with $w(\alpha_j) = x_j$). In what follows we shall, where convenient, identify probability functions and points in $\mathbb{D}_n$, the necessary ordering of the atoms being left implicit.

For w a probability function on SL (and similarly $SL(p_{i_1}, ..., p_{i_n})$) and $\varphi \in SL$, $w(\varphi) \neq 0$, the conditional probability function $w(-|\varphi) : SL \longrightarrow [0, 1]$ is defined by

$$w(\theta|\varphi) = \frac{w(\theta \wedge \varphi)}{w(\varphi)}.$$

To avoid any problems in conditionals with possibly zero denominators we shall adopt the convention in what follows that expressions of the form

$$\sum_{i=1}^{r} a_i w(\theta_i|\varphi) = b,$$

stand for

$$\sum_{i=1}^{r} a_i w(\theta_i \wedge \varphi) = bw(\varphi).$$

For the purposes of this paper it will be convenient to formulate the notion of a linear knowledge base, i.e. a knowledge base of linear constraints, as in [Paris and Vencovská, 1990] rather than adopt the special case simplifications of [Paris, 1999]. Thus we define a *Linear Knowledge Base* on $SL(p_{i_1}, ..., p_{i_n})$ to be a finite set of constraints,

$$\left\{ \sum_{i=1}^{r} a_{ij} w(\theta_i) = b_j \mid j = 1, ..., m \right\},$$

where the $\theta_i \in SL(p_{i_1}, ..., p_{i_n})$ and the a_{ij}, b_j are real, which is consistent, i.e. satisfied by some probability function w_0 on $SL(p_{i_1}, ..., p_{i_n})$ (or, equivalently, any larger language).

Let $LCL(p_{i_1}, ..., p_{i_n})$ denote the set of such knowledge bases. Notice that if $\{p_{i_1}, ..., p_{i_n}\} \subseteq \{p_{j_1}, ..., p_{j_s}\}$ then $LCL(p_{i_1}, ..., p_{i_n}) \subseteq LCL(p_{j_1}, ..., p_{j_s})$. Let $LCL = \bigcup_{i_1,...,i_n \in \mathbb{N}} LCL(p_{i_1}, ..., p_{i_n})$.

Define an *inference process* to be a function N such that for any finite non-empty subset $\{p_{i_1}, ..., p_{i_n}\}$ of L and $K \in LCL(p_{i_1}, ..., p_{i_n})$, $N(\{p_{i_1}, ..., p_{i_n}\}, K)$ is a probability function on $SL(p_{i_1}, ..., p_{i_n})$ which satisfies K.

In what follows we shall be particularly interested in the *Maximum Entropy Inference Process*, ME, which is defined as follows: Given a language with propositional variables $\{p_{i_1}, ..., p_{i_n}\}$, $K \in LCL(p_{i_1}, ..., p_{i_n})$, let $\alpha_1, ..., \alpha_{2^n}$ run through the atoms of $SL(p_{i_1}, ..., p_{i_n})$. Then $ME(\{p_{i_1}, ..., p_{i_n}\}, K)$ is that probability function w on $SL(p_{i_1}, ..., p_{i_n})$ satisfying K for which the entropy,

$$E(w(\alpha_1), w(\alpha_2), ..., w(\alpha_{2^n})) = -\sum_{i=1}^{2^n} w(\alpha_i) \log w(\alpha_i),$$

is maximal.[1]

Following [Paris and Vencovská, 1990] the 'common sense' principles of uncertain reasoning (for linear knowledge bases) referred to above may now be stated formally as desiderata on an inference process N defined on LCL. With an eye on future extensions the formulation of the principles given here will at times differ slightly from that given in [Paris and Vencovská, 1990] and [Paris, 1999].

[1]In what follows we shall often use *concavity* of E, i.e. the property that for $\vec{a}, \vec{b} \in \mathbb{D}_n$, $E(k_1\vec{a} + k_2\vec{b}) \geq k_1 E(\vec{a}) + k_2 E(\vec{b})$ whenever $k_1, k_2 \geq 0$ and $k_1 + k_2 = 1$, the inequality becoming strict whenever both k_1 and k_2 are (moreover) non-zero.

Irrelevant Information Principle
Let $K_1 \in LCL(p_{i_1},...,p_{i_n})$, $K_2 \in LCL(p_{j_1},...,p_{j_m})$ with $\{i_1,...,i_n\} \cap \{j_1,...,j_m\} = \emptyset$. Then for $\theta \in SL(p_{i_1},...,p_{i_n})$,

$$N(\{p_{i_1},...,p_{i_n}\},K_1)(\theta) = N(\{p_{i_1},...,p_{i_n},p_{j_1},...,p_{j_m}\},K_1 \cup K_2)(\theta).$$

The principle of Irrelevant Information as presented here provides use with a very valuable simplification. Namely, by taking K_2 to be empty we see that for N satisfying this principle $N(\{p_{i_1},...,p_{i_n},p_{j_1},...,p_{j_m}\},K_1)(\theta)$ does not depend on the particular overlying language $\{p_{i_1},...,p_{i_n},p_{j_1},...,p_{j_m}\}$ chosen (a property known as *Language Invariance* in earlier papers [Paris, 1994], [Paris and Vencovská, 1989], [Paris and Vencovská, 1990], [Paris and Vencovská, 1997]). Since we are interested in inference processes satisfying this principle, we shall henceforth therefore omit explicit mention of the argument $\{p_{i_1},...,p_{i_n}\}$ of N whenever this does not cause confusion.

Equivalence Principle
If $K_1, K_2 \in LCL(p_{i_1},...,p_{i_n})$ are equivalent in the sense that a probability function w satisfies K_1 just if it satisfies K_2, then $N(K_1) = N(K_2)$.

In other words N may be thought of as a choice function on the sets $V(K) \subseteq \mathbb{D}_n$ of probability functions satisfying K.

Renaming Principle
Let $\alpha_1,...,\alpha_{2^n}$ and $\beta_1,...,\beta_{2^n}$ be the atoms of $SL(p_{i_1},p_{i_2},...,p_{i_n})$ and $SL(p_{j_1},p_{j_2},...,p_{j_n})$ respectively and let $K_1 \in LCL(p_{i_1},p_{i_2},...,p_{i_n})$, $K_2 \in LCL(p_{j_1},p_{j_2},...,p_{j_n})$ be respectively

$$\left\{ \sum_{i=1}^{2^n} a_{ij} w(\alpha_i) = b_j \mid j = 1,2,...,r \right\},$$

$$\left\{ \sum_{i=1}^{2^n} a_{ij} w(\beta_i) = b_j \mid j = 1,2,...,r \right\}.$$

Then $N(K_1)(\alpha_i) = N(K_2)(\beta_i)$ for $i = 1,2,...,2^n$.

Relativization Principle
Suppose that $\varphi \in SL(p_{i_1},...,p_{i_n})$ and $K_1, K_2 \in LCL(p_{i_1},...,p_{i_n})$ are respectively the sets of constraints[2]

$$\left\{ \sum_{i=1}^{r} a_{ij} w(\theta_i \wedge \varphi) = b_j | j = 1,...,m \right\} \quad \cup \quad \{w(\varphi) = c\},$$

[2] According to what looks clearest we shall sometimes specify sets of constraints, as here, as actual sets whilst on other occasions simply listing their elements.

$$\left\{\sum_{i=1}^{k} d_{ij} w(\psi_i \wedge \neg\varphi) = e_j | j = 1, ..., s\right\} \quad \cup \quad \{w(\neg\varphi) = 1 - c\}.$$

Then for $\theta \in SL(p_{i_1}, ..., p_{i_n})$, $N(K_1)(\theta \wedge \varphi) = N(K_1 \cup K_2)(\theta \wedge \varphi)$.

Obstinacy Principle
If $K_1, K_2 \in LCL(p_{i_1}, ..., p_{i_n})$ and $N(K_1)$ satisfies K_2, then $N(K_1) = N(K_1 \cup K_2)$.

Independence Principle
Let $K \in LCL$ be the set of constraints

$$\{w(p_1 \wedge p_2) = a, \ w(p_1 \wedge p_3) = b, \ w(p_1) = c\}$$

where $c > 0$. Then $N(K)(p_1 \wedge p_2 \wedge p_3) = \frac{ab}{c}$.

Continuity Principle
If $K, K_m \in LCL$ for $m \in \mathbb{N}$ and $\lim_{m\to\infty} \partial(V(K_m), V(K)) = 0$, where ∂ is the (Blaschke) distance between non-empty subsets X, Y of $\mathbb{D}_n$ defined by

$$\begin{aligned}\partial(X, Y) = \inf\{\delta > 0 | &\forall x \in X \exists y \in Y \ |x - y| \leq \delta \text{ and} \\ &\forall y \in Y \exists x \in X \ |x - y| \leq \delta\},\end{aligned}$$

then $\lim_{m\to\infty} N(K_m) = N(K)$.

Informal justifications for these principles are given in [Paris and Vencovská, 1990] and [Paris, 1999]. All these principles hold for ME. Indeed, by combining, and sharpening slightly, the results in [Paris, 1999], [Paris and Vencovská, 1990], [Paris and Vencovská, 1997], [Paris and Vencovská, 1996] we have the following result characterizing ME (for three alternate characterizations with a similar 'axiomatic' flavour see the work of Shore and Johnson, [Shore and Johnson, 1980], Csiszár, [Csiszár, 1989], and Kern-Isberner [Kern-Isberner, 1998]).

THEOREM 1. *Let N be an inference process which satisfies the principles of Irrelevant Information, Equivalence, Renaming, Relativization, Obstinacy, Independence and Continuity on LCL. Then N agrees with ME on LCL.*

Put another way this theorem says that in the case where an agent's knowledge base is in LCL and $\theta \in SL$ then there is one and only one belief value (probability) that the agent can assign to θ which is consistent with the agent acting (as an inference *process*) common sensically.

Interesting as this result might be it is certainly not beyond criticism (see for example [Paris and Vencovská, 1997]). In particular some might feel that it applies to an excessively limited class of knowledge bases. It is true, as detailed further in [Paris, 1999] for example, that when asserting 'the working knowledge they use' experts usually express it in the form of linear constraints, indeed in the even simpler forms $w(\theta) = b$ or $w(\theta|\phi) = b$. However one might object that in practice an expert is surely aware of other forms of constraints, in particular constraints expressing stochastic independences, and that our conclusion would have more relevance if it could be generalized to this case. In the sections that follow we shall look at what happens if we widen our knowledge bases to include also such *non-linear* constraints.

2 STOCHASTIC INDEPENDENCE

Our first observation on this goal of widening our knowledge bases to include also constraints expressing independence is that if we do it naively then 'common sense' proves false! The following example originates with Paul Courtney, see [Courtney, 1992] or [Paris, 1994].

> A certain jam factory receives jars and lids from two manufactures, C and D say, each individually supplying the same number of jars as lids. At the factory all the lids and all the jars from the two manufacturers go to two *independent* input points. Unfortunately due to some oversight years ago the jars and lids from manufacturer C are marginally smaller than those from manufacturer D so that if, at the final capping, a lid from one manufacturer is put on a jar from the other manufacturer it leaks, visibly. Wise old Professor Marmalade is well aware of all this from his years of experience moon-lighting as a cleaner on the factory floor, and he knows that the probability that any one pot of jam will leak is 3/8. So what probability should he give to the next bottle coming down the production line being of the larger variety? Clearly from what he knows there is complete symmetry here between 'larger' and 'smaller', so in accord with the spirit of the principles (precisely the Renaming Principle) he should give the same probability to it being larger as he would give to it being smaller. So he has to give the answer 1/2. Unfortunately this answer is not actually consistent with his knowledge base!

In fact the constraints as given are satisfied by just two probability functions, which give the probabilities 1/4 and 3/4 respectively to 'next bottle coming down the line is large' and there is no solution which gives this event probability 1/2. Nor, in view of the large/smaller symmetry involved, is it reasonable to hope that appealing to some further common sense principle could break the deadlock. The fact is that there are only two solutions (even without invoking common sense) and there is no rational way to pick between them. If wise old Professor Marmalade had done the calculations and was required to give a figure it is hard to see that he could improve upon tossing a fair coin to pick between the two options. In summary then, on the basis of this one contrived example it seems that if we allow in also non-linear constraints then 'common sense' may only be capable of cutting down the possibilities to within a number of different, but equally acceptable or reasonable, choices. This being the case it seems we should therefore relax our requirement that inference processes are single valued, allowing them instead to take as value a non-empty set of probability functions and consider again appropriate common sense principles.

In the sections that now follow the plan is to tread precisely this path with the intention of deriving an analogous result to Theorem 1.

3 EXTENDED INFERENCE PROCESSES AND COMMON SENSE PRINCIPLES

We define an *Extended Knowledge Base* on $SL(p_{i_1}, ..., p_{i_n})$ to be a finite set of constraints,

$$\{ f_j(w(\theta_1), ..., w(\theta_m)) = 0 \mid j = 1, ..., r \},$$

where the $\theta_i \in SL(p_{i_1}, ..., p_{i_n})$ and the f_j are polynomials with real coefficients, which is consistent, i.e. satisfied by some probability function w_0 on $SL(p_{i_1}, ..., p_{i_n})$ (or, equivalently, any larger language). We shall use $ECL(p_{i_1}, ..., p_{i_n})$ to denote the set of such knowledge bases. For $K \in ECL(p_{i_1}, ..., p_{i_n})$, let $V^{\{p_{i_1},...,p_{i_n}\}}(K)$ denote the set of probability functions on $SL(p_{i_1}, ..., p_{i_n})$ satisfying K. When it is clear from the context we shall omit the specification of $\{p_{i_1}, ..., p_{i_n}\}$ and write simply $V(K)$.

Let $ECL = \bigcup_{i_1,...,i_n \in \mathbb{N}} ECL(p_{i_1}, ..., p_{i_n})$.

Now define an *extended inference process* to be a function N such that for any finite non-empty subset $\{p_{i_1}, ..., p_{i_n}\}$ of L and $K \in ECL(p_{i_1}, ..., p_{i_n})$, $N(\{p_{i_1}, ..., p_{i_n}\}, K)$ is a non-empty set of probability functions on $SL(p_{i_1}, ..., p_{i_n})$ each of which satisfies K.

As in the linear case we shall formulate desirable, common sense, properties of extended inference processes as (extended) principles. In this it is useful to think of N as the property of some agent and $N(\{p_{i_1}, ..., p_{i_n}\}, K)$ as an agent's choice of most reasonable, or acceptable, or *preferred* solutions to the constraints K. The intention of each of the principles is that they place restrictions on this choice of most preferred solutions which the agent should obey if s/he is to be considered to be acting 'in accord with common sense'.

Prior to introducing these principles, the following terminology will be useful. For M a set of probability functions on $SL(p_{k_1}, ..., p_{k_r})$, $\{p_{i_1}, ..., p_{i_n}\} \subseteq \{p_{k_1}, ..., p_{k_r}\}$ and $\varphi \in SL(p_{k_1}, ..., p_{k_r})$ let $M \upharpoonright \{p_{i_1}, ..., p_{i_n}\}$ and $M \upharpoonright \varphi$ denote the set of probability functions from M restricted to $SL(p_{i_1}, ..., p_{i_n})$ and the set of probability functions from M restricted to $\{\theta \wedge \varphi \mid \theta \in SL(p_{k_1}, ..., p_{k_r})\}$ respectively.

Irrelevant Information Principle
Let $K_1 \in ECL(p_{i_1}, ..., p_{i_n})$, $K_2 \in ECL(p_{j_1}, ..., p_{j_m})$ with $\{i_1, ..., i_n\} \cap \{j_1, ..., j_m\} = \emptyset$. Then

$$N(\{p_{i_1},...,p_{i_n}\},K_1) = N(\{p_{i_1},...,p_{i_n},p_{j_1},...,p_{j_m}\},K_1 \cup K_2) \upharpoonright \{p_{i_1},...,p_{i_n}\}.$$

The justification for this principle is (as in the LCL case given in [Paris, 1999]) that since the knowledge provided by K_2 refers to an entirely disjoint language it should be irrelevant as far as the restriction of one's (i.e. a common sense agent's[3]) chosen preferred probability functions on $\{p_{i_1},...,p_{i_n}\}$ are concerned.

As in the linear case it follows that an extended inference N process satisfying this principle is *language invariant*, i.e. $N(\{p_{i_1},...,p_{i_n},p_{j_1},...,p_{j_m}\},K_1) \upharpoonright \{p_{i_1},...,p_{i_n}\}$ does not depend on the particular overlying language $\{p_{i_1},...,p_{i_n}, p_{j_1},...,p_{j_m}\}$ chosen. Consequently we omit the first argument of N whenever this does not cause confusion.

Equivalence Principle
If $K_1, K_2 \in ECL(p_{i_1},...,p_{i_n})$ are equivalent in the sense that a probability function w satisfies K_1 just if it satisfies K_2, then $N(K_1) = N(K_2)$.

The justification for this principle is that it is common sense that one's choice of preferred solutions should depend only on the choices available and not on how they are presented (packaged!).

Notice that for N satisfying Equivalence and $K \in ECL(p_{i_1},p_{i_2},...,p_{i_n})$ N is essentially a choice function on the subsets $V(K)$ of $\mathbb{D}_n$. Consequently when $X \subseteq \mathbb{D}_n$ and we know there is a $K \in ECL$ such that $V(K) = X$, we may sometimes write $N(X)$ in place of $N(K)$. Related to this point we shall sometimes use w to denote an element of $N(K)$ (when we want to think of $N(K)$ as a set of probability functions) and sometimes as $\vec{w}$ (when we want to think of $N(K)$ as a subset of $\mathbb{D}_n$).

Henceforth we shall assume that N satisfies Irrelevant Information and Equivalence without further mention.

Renaming Principle
Let $\alpha_1,...,\alpha_{2^n}$ and $\beta_1,...,\beta_{2^n}$ be the atoms of $SL(p_{i_1},p_{i_2},...,p_{i_n})$ and $SL(p_{j_1}, p_{j_2},...,p_{j_n})$ respectively and let $K_1 \in ECL(p_{i_1},p_{i_2},...,p_{i_n})$, $K_2 \in ECL(p_{j_1}, p_{j_2},...,p_{j_n})$ be respectively

$$\{ f_j(w(\alpha_1),...,w(\alpha_{2^n})) = 0 \mid j = 1,2,...,r \},$$

$$\{ f_j(w(\beta_1),...,w(\beta_{2^n})) = 0 \mid j = 1,2,...,r \}.$$

Then $w \in N(K_1) \Leftrightarrow w\sigma \in N(K_2)$, where σ is the bijection from the atoms of $SL(p_{j_1},...,p_{j_n})$ to the atoms of $SL(p_{i_1},...,p_{i_n})$ given by $\sigma(\beta_k) = \alpha_k$ for $k = 1,...,2^n$.

[3]In what follows 'one' should be thought of as just a personalized common sense agent.

The justification for this principle is that σ is no more than a renaming of 'possible worlds' and one's choice of preferred solutions to K should not be affected by this immaterial change.

In applying the Renaming Principle (and certain later principles) it is useful to notice that by (1) any set of constraints in ECL is equivalent (in the sense of having the same solutions) to a set of constraints only involving $w(\alpha)$ for α atoms from some finite sublanguage of L and, by our standing assumption of Equivalence, N agrees on these two sets of constraints.

For applications of Renaming it is also useful to introduce a little notation. For $K \in ECL(p_{i_1}, p_{i_2}, ..., p_{i_n})$,

$$K = \{ \ f_j(w(\alpha_1), ..., w(\alpha_{2^n})) = 0 \mid j = 1, 2, ..., r \ \}$$

and σ a permutation of $\{1, ..., 2^n\}$ let

$$\sigma K = \{ \ f_j(w(\alpha_{\sigma(1)}), ..., w(\alpha_{\sigma(2^n)})) = 0 \mid j = 1, 2, ..., r \ \}.$$

Similarly for $\vec{u} = \langle u_1, ..., u_{2^n} \rangle \in \mathbb{D}_n$ let $\sigma\vec{u} = \langle u_{\sigma(1)}, ..., u_{\sigma(2^n)} \rangle$. With this notation in place Renaming gives that

$$\sigma\vec{u} \in N(K) \Leftrightarrow \vec{u} \in N(\sigma K).$$

It turns out that in the extension from LCL to ECL Renaming has actually lost a lot of its former power, a point we return to in the concluding section of this paper[4].

Relativization Principle

Suppose that $\varphi \in SL(p_{i_1}, ..., p_{i_n})$ and $K_1, K_2 \in ECL(p_{i_1}, ..., p_{i_n})$ are respectively the sets of constraints[5]

$$\{f_j(w(\theta_1 \wedge \varphi), ..., w(\theta_r \wedge \varphi)) = 0 \mid j = 1, ..., m\} \cup \{w(\varphi) = c\},$$

$$\{g_j(w(\psi_1 \wedge \neg\varphi), ..., w(\psi_t \wedge \neg\varphi)) = 0 \mid j = 1, ..., s\} \cup \{w(\neg\varphi) = 1 - c\}.$$

Then $N(K_1) \restriction \varphi = N(K_1 \cup K_2) \restriction \varphi$.

Notice that (for $c > 0$) we can equivalently express K_1 as

$$\{f_j(cw(\theta_1|\varphi), ..., cw(\theta_r|\varphi)) = 0 \mid j = 1, ..., m\} \cup \{w(\varphi) = c\},$$

and similarly for K_2. This leads to the justification for this principle, namely that conditioned on φ both K_1 and $K_1 \cup K_2$ contain the same knowledge so, conditioned on φ, the choices of preferred solutions should also agree.

[4]However as Lemma 5 will shortly show enough remains that we can still derive the classical 'principle of indifference'.

[5]In expressions like these we may, for the sake of elegance, write $w(\varphi) = c$ rather than the formally correct $w(\varphi) - c = 0$ etc..

Obstinacy Principle
If $K_1, K_2 \in ECL(p_{i_1}, ..., p_{i_n})$ and $N(K_1) \cap V(K_2) \neq \emptyset$ then $N(K_1 \cup K_2) = N(K_1) \cap V(K_2)$.

To justify this principle suppose on the contrary that it failed. First consider the case that there was a preferred solution w of K_1 which satisfied K_2 but was not a preferred solution of $K_1 \cup K_2$. Let w' be a preferred solution of $K_1 \cup K_2$. Then one would have the situation where w was at least as preferred as w' as a solution of K_1 and both satisfied K_2, but when K_2 was explicitly stated this preference was lost! The other case is similar. Suppose there was a preferred solution w' of $K_1 \cup K_2$ which was not a preferred solution of K_1. Let w be a preferred solution of K_1 which satisfied K_2. Then similarly as before w would be strictly preferred to w' as a solution of K_1 and both satisfied K_2, but when K_2 was explicitly stated this strict preference was lost!

Put another way, learning something one already believed should not cause one to change one's beliefs!

In what follows the following immediate consequences of Obstinacy (and of our standing assumption of Equivalence) will be useful. Firstly, if $w \in N(K_1)$ satisfies K_2 then $w \in N(K_1 \cup K_2)$. Secondly, if $V(K_1) \supseteq V(K_2)$ and $N(K_1) \cap V(K_2) \neq \emptyset$ then $N(K_2) = N(K_1) \cap V(K_2)$.

Independence Principle
Let $K \in ECL$ be the set of constraints

$$\{ w(p_1 \wedge p_2) = a, \; w(p_1 \wedge p_3) = b, w(p_1) = c \}$$

where $c > 0$. Then for $w \in N(K)$, $w(p_1 \wedge p_2 \wedge p_3) = \frac{ab}{c}$.

Notice that this set of constraints K is equivalent to

$$\{ cw(p_2|p_1) = a, \; cw(p_3|p_1) = b, \; w(p_1) = c \}.$$

The justification for this principle is that the knowledge provides no interrelationship between the conditional beliefs in p_2, p_3 given p_1 (nor any other relationships) and in consequence p_2 and p_3 should be treated as independent (i.e. stochastically independent) given p_1.

In [Paris, 1999] a somewhat weaker, and possibly more easily acceptable, form of this principle was given for the linear case. We would surmise that at the cost of further complicating the proof a similar weakening would also suffice here.

Continuity Principle
If $K, K_m \in ECL$, $\vec{u}_m \in N(K_m)$ for $m \in \mathbb{N}$, $\lim_{m\to\infty} \partial(V(K_m), V(K)) = 0$ and $\lim_{m\to\infty} \vec{u}_m = \vec{u}$ then $\vec{u} \in N(K)$.

The justification for this is that the property of being a preferred solution should not die in the limit. Notice that this is not the same as saying that if $\vec{u} \in V(K)$ is a preferred solution of K and K' is 'close to' K then K' should have a preferred solution close to $\vec{u}$. Indeed it will follow easily from the main theorem of this paper that we cannot possibly hope to have both in general for non-linear constraints. [It does hold in the linear case.]

We end this section by proving some basic properties which hold for any extended inference process N satisfying the principles of Irrelevant Information, Equivalence, Renaming, Relativization, Obstinacy and Independence. At this point the first-time reader may wish to jump to the next section and refer back to these lemmas (and the notion of an N-solution given prior to Lemma 4) as and when required.

For the rest of this section assume that N satisfies these principles.

LEMMA 2. *Assume that $K \in ECL(p_{i_1}, ..., p_{i_n})$, $\vec{\rho} = \langle \rho_1, ..., \rho_{2^n} \rangle \in N(K)$ and $j, k, l, m \in \{1, ..., 2^n\}$ (j, k, l, m distinct) are such that*

$$V(K \cup \{w(\alpha_i) = \rho_i \mid i \notin \{j,k,l,m\}\}) = V(K_1 \cup \{w(\alpha_i) = \rho_i \mid i \notin \{j,k,l,m\}\}),$$

where K_1 is the set of constraints

$$\{w(\alpha_j)+w(\alpha_k)+w(\alpha_l)+w(\alpha_m) = c,\ w(\alpha_j)+w(\alpha_k) = a,\ w(\alpha_j)+w(\alpha_l) = b\}$$

and $c = \rho_j + \rho_k + \rho_l + \rho_m$, $a = \rho_j + \rho_k$, $b = \rho_j + \rho_l$. Then $\rho_j \rho_m = \rho_k \rho_l$.

Proof. Without loss of generality we may assume j, k, l, m are $1, 2, 3, 4$ respectively and the $i_1, i_2, ..., i_n$ are $1, 2, ..., n$. By a remark following the introduction of the Obstinacy Principle for extended inference processes, $\vec{\rho} \in N(K \cup \{w(\alpha_i) = \rho_i \mid 4 < i \leq 2^n\})$ so $\vec{\rho} \in N(K_1 \cup \{w(\alpha_i) = \rho_i \mid 4 < i \leq 2^n\})$. Now consider K_2 consisting of constraints (in the extended language $L(p_1, p_2, ..., p_{2^n})$),

$$w(p_1 \wedge p_2) = a,\ w(p_1 \wedge p_3) = b,\ w(p_1) = c,$$

$$w(\neg p_j \wedge p_1) = 0,\ w(\neg p_j \wedge \neg p_1 \wedge p_2 \wedge p_3) = \rho_j, \quad 4 \leq j \leq 2^n,$$

$$w(\neg p_1 \wedge \neg p_i \wedge \neg p_j) = 0, \quad 4 \leq i < j \leq 2^n.$$

By Irrelevant Information there is a probability function in

$$N(K_1 \cup \{w(\alpha_i) = \rho_i \mid 4 < i \leq 2^n\} \cup \{w(\neg p_{n+1}) = 0, ..., w(\neg p_{2^n}) = 0\})$$

whose values on the α_j are ρ_j, $1 \leq j \leq 2^n$. But there is a clear renaming of the atoms which sends this set of constraints to K_2 and hence using the Renaming Principle we see that

$$\begin{aligned}
\rho_1 &= w(p_1 \wedge p_2 \wedge p_3 \wedge \bigwedge_{4 \leq j \leq 2^n} p_j), \\
\rho_2 &= w(p_1 \wedge p_2 \wedge \neg p_3 \wedge \bigwedge_{4 \leq j \leq 2^n} p_j), \\
\rho_3 &= w(p_1 \wedge \neg p_2 \wedge p_3 \wedge \bigwedge_{4 \leq j \leq 2^n} p_j),
\end{aligned}$$

$$\rho_4 = w(p_1 \wedge \neg p_2 \wedge \neg p_3 \wedge \bigwedge_{4 \leq j \leq 2^n} p_j),$$

$$\rho_r = w(\neg p_1 \wedge p_2 \wedge p_3 \wedge \neg p_r \wedge \bigwedge_{4 \leq j \leq 2^n, j \neq r} p_j), \quad \text{for } 4 < r \leq 2^n,$$

for some $w \in N(K_2)$. Now using Relativization (with $\varphi = p_1$) and Equivalence, we can see that the same is true if we replace K_2 by K_3, where K_3 is obtained from K_2 by replacing $w(\neg p_j \wedge \neg p_1 \wedge p_2 \wedge p_3) = \rho_j$ by $w(\neg p_j \wedge \neg p_1) = 0$ for $4 \leq j \leq 2^n$ and dropping $w(\neg p_1 \wedge \neg p_i \wedge \neg p_j) = 0$ for $4 \leq i < j \leq 2^n$, i.e. taking

$$K_3 = \{w(p_1 \wedge p_2) = a,\ w(p_1 \wedge p_3) = b,\ w(p_1) = c,\ w(\neg p_j) = 0 \mid 4 \leq j \leq 2^n\}.$$

Using Irrelevant Information (to ignore the $w(\neg p_j) = 0$) and then Independence gives (provided $\rho_1 + \rho_2 + \rho_3 + \rho_4 \neq 0$)

$$\rho_1 = \frac{(\rho_1 + \rho_2)(\rho_1 + \rho_3)}{\rho_1 + \rho_2 + \rho_3 + \rho_4},$$

which simplifies to $\rho_1\rho_4 = \rho_2\rho_3$ (either way), as required. ∎

LEMMA 3. *Let* $K_1 \in ECL(p_{i_1}, ..., p_{i_n})$, $K_2 \in ECL(p_{j_1}, ..., p_{j_m})$ *with* $\{i_1, ..., i_n\} \cap \{j_1, ..., j_m\} = \emptyset$. *Let* $\alpha_1, ..., \alpha_{2^n}$ *and* $\beta_1, ..., \beta_{2^m}$ *be the atoms of* $SL(p_{i_1}, ..., p_{i_n})$ *and* $SL(p_{j_1}, ..., p_{j_m})$ *respectively, so* $\alpha_i \wedge \beta_j$ $(i = 1, ..., 2^n,\ j = 1, ..., 2^m)$ *are the atoms of* $SL(p_{i_1}, ..., p_{i_n}, p_{j_1}, ..., p_{j_m})$. *Then* $\vec{u} = \langle u_{11}, ..., u_{12^m}, ..., u_{2^n 2^m}\rangle$ *is in* $N(K_1 \cup K_2)$ *just if there are* $\vec{\rho} = \langle \rho_1, ..., \rho_{2^n}\rangle \in N(K_1)$ *and* $\vec{\tau} = \langle \tau_1, ..., \tau_{2^m}\rangle \in N(K_2)$ *such that*

$$u_{ij} = \rho_i \tau_j, \quad i = 1, ..., 2^n,\ \ j = 1, ..., 2^m,$$

where ρ_i, τ_j *and* u_{ij} *pertain to* α_i, β_j *and* $\alpha_i \wedge \beta_j$ *respectively.*

Proof. Assume $\vec{u} \in N(K_1 \cup K_2)$. Then by Irrelevant Information,

$$\left\langle \sum_{j=1}^{2^m} u_{1j}, ..., \sum_{j=1}^{2^m} u_{2^n j} \right\rangle$$

is equal to some $\vec{\rho} \in N(K_1)$ and

$$\left\langle \sum_{i=1}^{2^n} u_{i1}, ..., \sum_{i=1}^{2^n} u_{i2^m} \right\rangle$$

is equal to some $\vec{\tau} \in N(K_2)$. By a remark following the introduction of the Obstinacy Principle for extended inference processes, $\vec{u}$ is in $N(K)$ where K is obtained from $K_1 \cup K_2$ by adding the constraints

$$\sum_{j=1}^{2^m} w(\alpha_i \wedge \beta_j) = \rho_i, \quad i = 1, ..., 2^n,$$

$$\sum_{i=1}^{2^n} w(\alpha_i \wedge \beta_j) = \tau_j, \quad j = 1, ..., 2^m,$$

and the constraints

$$w(\alpha_i \wedge \beta_j) = u_{ij} \quad \langle i, j \rangle \notin \{\langle p, r \rangle, \langle q, r \rangle, \langle p, s \rangle, \langle q, s \rangle\},$$

where $p, q \in \{1, ..., 2^n\}$, $r, s \in \{1, ..., 2^m\}$, $p \neq q$, $r \neq s$ are fixed, but arbitrary. By Lemma 2 it follows that $u_{pr}u_{qs} = u_{qr}u_{ps}$, which together with $\sum_{j=1}^{2^m} u_{ij} = \rho_i$ and $\sum_{i=1}^{2^n} u_{ij} = \tau_j$ yields that $u_{ij} = \rho_i \tau_j$, as required.

Conversely, let $\vec{\rho} \in N(K_1)$ and $\vec{\tau} \in N(K_2)$. By Obstinacy and Irrelevant Information,

$$N(K_1 \cup K_2 \cup \{\sum_{j=1}^{2^m} w(\alpha_i \wedge \beta_j) = \rho_i \mid i = 1, ..., 2^n\})$$

is equal to

$$N(K_1 \cup K_2) \cap \{w \mid \sum_{j=1}^{2^m} w(\alpha_i \wedge \beta_j) = \rho_i, \quad i = 1, ..., 2^n\}.$$

By Equivalence, the former is

$$N(\{w(\alpha_i) = \rho_i \mid i = 1, ..., 2^n\} \cup K_2),$$

which by Irrelevant Information must contain a w satisfying

$$(2) \quad \sum_{i=1}^{2^n} w(\alpha_i \wedge \beta_j) = \tau_j, \quad j = 1, ..., 2^m.$$

Consequently, $N(K_1 \cup K_2)$ contains w satisfying both (2) and

$$\sum_{j=1}^{2^m} w(\alpha_i \wedge \beta_j) = \rho_i, \quad i = 1, ..., 2^n,$$

and just as above this w can be shown to satisfy $w(\alpha_i \wedge \beta_j) = \rho_i \tau_j$. ■

Paralleling the development in [Paris and Vencovská, 1990] we now introduce some matrix notation which will considerably simplify later proofs, in particular that of the first key lemma, Lemma 7. [In the proof of the main Theorem 18 there will be three key lemmas, Lemmas 7, 15, 16, each of which can be seen as successively stronger special cases of the theorem.]

First note that any $K \in LCL$ corresponds to some system of linear equations

$$B\langle w(\alpha_1), ..., w(\alpha_{2^n})\rangle^T = \langle b_1, ..., b_m \rangle^T$$

where B is a $m \times 2^n$-matrix with real coefficients such that $B\vec{w}^T = \vec{b}^T$ forces $\sum_{i=1}^{2^n} w_i = 1$ and there is some $\vec{w} \geq 0$ satisfying these equations.

Conversely suppose that B is an $m \times r$ matrix such that the system of equations $B\vec{w}^T = \vec{b}^T$ forces $\sum_{i=1}^{r} w_i = 1$ for all solutions $\vec{w}$ with $\vec{w} \geq 0$ and at least one such solution exists. Let $\alpha_{j_1}, ..., \alpha_{j_r}$ be some (distinct) atoms of some $SL(p_{i_1}, ...p_{i_n})$. Then the set K of constraints

$$B\langle w(\alpha_{j_1}), ..., w(\alpha_{j_r})\rangle^T = \vec{b}^T$$

is in LCL and the set of solutions $\langle w(\alpha_{j_1}), ..., w(\alpha_{j_r})\rangle$, for $w \in N(K)$, of $B\vec{w}^T = \vec{b}^T$ is, by Irrelevant Information, Equivalence and Renaming, independent of the particular atoms $\alpha_{j_1}, ..., \alpha_{j_r}$ and overlying language $L(p_{i_1}, ...p_{i_n})$ chosen. We call these solutions (unambiguously) the *N-solutions* of $B\vec{w}^T = \vec{b}^T$. Notice also that if in addition $w_k = 0$ for all solutions $\vec{w}$ of $B\vec{w}^T = \vec{b}^T$ with $\vec{w} \geq 0$ then all N-solutions will have kth coordinate 0 and the N-solutions with this coordinate removed will be precisely the N-solutions of the system of equations $B'\vec{w'}^T = \vec{b}^T$ where B' is the $m \times (r-1)$ matrix formed by removing the kth column from B and $\vec{w'}$ is $\vec{w}$ with the kth coordinate omitted. We shall use these facts repeatedly and without further mention in what follows.

The next lemma shows that N satisfies the extension to this context of Shore and Johnson's System Independence property, see [Shore and Johnson, 1980].

LEMMA 4. *Suppose that $\vec{w} = \langle w_1, ..., w_k\rangle$ and $\vec{z} = \langle z_1, ..., z_r\rangle$ and that the system of linear equations*

$$(3) \quad \begin{pmatrix} B & 0 \\ 0 & C \end{pmatrix} (\vec{\mathrm{w}}, \vec{\mathrm{z}})^{\mathrm{T}} = (\vec{\mathrm{b}}, \vec{\mathrm{d}})^{\mathrm{T}}$$

implies that $\sum_{i=1}^{k} w_i = a$ and $\sum_{i=1}^{r} z_i = 1 - a$ for some $1 > a > 0$. Suppose further that (3) has a solution with $\vec{w} \geq 0$, $\vec{z} \geq 0$. Then $\vec{\lambda}$ is an N-solution of

$$(4) \quad \mathrm{B}\vec{\mathrm{w}}^{\mathrm{T}} = \mathrm{a}^{-1}\vec{\mathrm{b}}^{\mathrm{T}}$$

if and only if there is some $\vec{\delta}$ such that $a\vec{\lambda}, \vec{\delta}$ is an N-solution of (3).

Proof. Let n be such that $2^n \geq k+r$ and let $\alpha_1, ..., \alpha_{2^n}$ be atoms of $SL(p_1, ..., p_n)$. Then (after discarding zeros) (4) corresponds to K_1 defined by

$$B\langle w(\alpha_1), ..., w(\alpha_k)\rangle^T = a^{-1}\vec{b}^T,$$

$$w(\alpha_i) = 0 \text{ for } i = k+1, ..., 2^n,$$

and (3) corresponds to K defined by

$$B\langle w(\alpha_1 \wedge p_{n+1}), ..., w(\alpha_k \wedge p_{n+1})\rangle^T = \vec{b}^T,$$

$$C\langle w(\alpha_{k+1} \wedge \neg p_{n+1}), ..., w(\alpha_{k+r} \wedge \neg p_{n+1})\rangle^T = \vec{d}^T,$$

$$w(\alpha_i \wedge \neg p_{n+1}) = 0 \text{ for } i = 1, ..., k,$$

$$w(\alpha_i \wedge p_{n+1}) = 0 \text{ for } i = k+1, ..., k+r,$$

$$w(\alpha_i \wedge p_{n+1}) = w(\alpha_i \wedge \neg p_{n+1}) = 0 \text{ for } i = k+r+1, ..., 2^n.$$

Note that $B\langle w(\alpha_1 \wedge p_{n+1}), ..., w(\alpha_k \wedge p_{n+1})\rangle^T = \vec{b}^T$ implies $\sum_{i=1}^k w(\alpha_i \wedge p_{n+1}) = w(p_{n+1}) = a$. Let $\psi = \bigvee_{i=1}^k \alpha_i$. We need to prove that

(5) $\vec{\lambda} = \langle \lambda_1, ..., \lambda_k \rangle \in N(K_1) \restriction \psi \Leftrightarrow a\vec{\lambda} \in N(K) \restriction \psi \wedge p_{n+1}$

(where the atoms of $SL(p_1, ..., p_n)$ and $SL(p_1, ..., p_n, p_{n+1})$ are assumed to be ordered as $\alpha_1, ..., \alpha_{2^n}$ and $\alpha_1 \wedge p_{n+1}, ..., \alpha_{2^n} \wedge p_{n+1}, \alpha_1 \wedge \neg p_{n+1}, ..., \alpha_{2^n} \wedge \neg p_{n+1}$ respectively). In other words, (5) amounts to saying that there is some $\vec{\tau} = \langle \tau_1, ..., \tau_{2^n} \rangle \in N(K_1)$ such that $\tau_i = \lambda_i$ for $i = 1, ..., k$ if and only if there is some $\vec{\rho} = \langle \rho_1, ..., \rho_{2^{n+1}} \rangle \in N(K)$ such that $\rho_i = a\lambda_i$ for $i = 1, ..., k$.
Let $K_2 = K_1 \cup \{w(p_{n+1}) = a\}$. By Lemma 3,

$$\vec{\lambda} \in N(K_1) \restriction \psi \Leftrightarrow a\vec{\lambda} \in N(K_2) \restriction \psi \wedge p_{n+1}.$$

Consequently, any $\vec{\rho} \in N(K_2)$ satisfies

$$B\langle \frac{\rho_1}{a}, ..., \frac{\rho_k}{a} \rangle^T = a^{-1}\vec{b}^T,$$

i.e. $B\langle \rho_1, ..., \rho_k \rangle^T = \vec{b}^T$. It follows, by Obstinacy, that $N(K_2) = N(K_3)$, where K_3 consists of K_2 and

$$B\langle w(\alpha_1 \wedge p_{n+1}), ..., w(\alpha_k \wedge p_{n+1})\rangle^T = \vec{b}^T.$$

Now K_3 is equivalent to

$$B\langle w(\alpha_1 \wedge p_{n+1}), ..., w(\alpha_k \wedge p_{n+1})\rangle^T = \vec{b}^T,$$

$$B\langle w(\alpha_1 \wedge \neg p_{n+1}), ..., w(\alpha_k \wedge \neg p_{n+1})\rangle^T = a^{-1}(1-a)\vec{b}^T,$$

$$w(\alpha_i \wedge p_{n+1}) = w(\alpha_i \wedge \neg p_{n+1}) = 0 \text{ for } i = k+1, ..., 2^n,$$

$$w(p_{n+1}) = w(\psi \wedge p_{n+1}) = a.$$

By Equivalence and Relativization, $N(K_3) \restriction \psi \wedge p_{n+1} = N(K) \restriction \psi \wedge p_{n+1}$. The result follows. ■

Our next result shows that N satisfies Laplace's 'principle of indifference'.

LEMMA 5. *Assume that $K \in ECL(p_{i_1}, ..., p_{i_n})$ is of the form*

$$\left\{ f_j \left(\sum_{\alpha \in A_1} w(\alpha), ..., \sum_{\alpha \in A_r} w(\alpha) \right) = 0 \mid j = 1, ..., m \right\},$$

where the $A_1, ..., A_r$ form a partition of the atoms of $SL(p_{i_1}, ..., p_{i_n})$. Then for any $w \in N(K)$, $w(\alpha) = w(\alpha')$ whenever α and α' are in the same A_j.

Proof. We first prove a special case of this lemma, namely that

$$N(\{p_1\}, \emptyset) = \{\langle 1/2, 1/2 \rangle\}.$$

Suppose $\langle a, 1-a \rangle \in N(\{p_1\}, \emptyset)$. Then by Lemma 3 there is $w \in N(\{p_1, p_2\}, \emptyset)$ such that $w(p_1) = w(p_2) = a$ and $w(p_1 \wedge p_2) = a^2$, $w(p_1 \wedge \neg p_2) = a(1-a) = w(\neg p_1 \wedge p_2)$, $w(\neg p_1 \wedge \neg p_2) = (1-a)^2$. By Renaming then there is also $w_1 \in N(\{p_1, p_2\}, \emptyset)$ with $w_1(p_1 \wedge p_2) = a^2$, $w_1(p_1 \wedge \neg p_2) = (1-a)^2$, $w_1(\neg p_1 \wedge p_2) = w_1(\neg p_1 \wedge \neg p_2) = a(1-a)$. Hence $w_1(p_1) = a^2 + (1-a)^2$, $w_1(p_2) = a^2 + a(1-a) = a$. Again by Lemma 3 we must have $w_1(p_1 \wedge p_2) = a(a^2 + (1-a)^2)$. But we already know that $w_1(p_1 \wedge p_2) = a^2$ so this forces a to be one of $0, 1/2, 1$.

To see that $a \neq 1$ (and hence by Renaming that $a \neq 0$) suppose on the contrary $a = 1$. Then by Lemma 3 again there is $w_2 \in N(\{p_1, p_2\}, \{w(p_2) = 1/3\})$ with $w_2(p_1 \wedge p_2) = 1/3$, $w_2(p_1 \wedge \neg p_2) = 2/3$, $w_2(\neg p_1 \wedge p_2) = w_2(\neg p_1 \wedge \neg p_2) = 0$. By Renaming there is a $w_3 \in N(\{p_1, p_2\}, \{w(p_2) = 1/3\})$ with $w_3(p_1 \wedge p_2) = 1/3$, $w_3(\neg p_1 \wedge \neg p_2) = 2/3$, $w_3(p_1 \wedge \neg p_2) = w_3(\neg p_1 \wedge p_2) = 0$. But this means that by Irrelevant Information

$$\langle w_3(p_1), w_3(\neg p_1) \rangle = \langle 1/3, 2/3 \rangle \in N(\{p_1\}, \emptyset),$$

which we have already shown is impossible. Thus $N(\{p_1\}, \emptyset) = \{\langle 1/2, 1/2 \rangle\}$.

To now prove the general case as stated in the lemma let $\vec{\tau} \in N(K)$ and without loss of generality let α_1, α_2 be distinct elements of A_1 where $\alpha_1, ..., \alpha_{2^n}$ run through the atoms of $SL(p_{i_1}, ..., p_{i_n})$. Clearly it is enough to show that $\tau_1 = \tau_2$. Let $K_1 \in LCL(p_{i_1}, ..., p_{i_n})$ be the set of *linear* constraints

$$\{w(\alpha_1) + w(\alpha_2) = \tau_1 + \tau_2\} \cup \{w(\alpha_j) = \tau_j \mid 3 \leq j \leq 2^n\}.$$

By Obstinacy $\vec{\tau} \in N(K_1)$ and by Lemma 4 there is a $\langle \lambda_1, \lambda_2 \rangle \in N(\{p_1\}, \emptyset)$ such that

$$(\tau_1 + \tau_2)\langle \lambda_1, \lambda_2 \rangle = \langle \tau_1, \tau_2 \rangle.$$

But as we have already shown the only possibility here is $\lambda_1 = 1/2 = \lambda_2$, giving $\tau_1 = \tau_2$, as required. ■

In the next section we shall show that the principles of Irrelevant Information, Equivalence, Renaming, Relativization, Obstinacy, Independence and Continuity collected in this section are consistent in the sense that there is at least one extended inference process satisfying them.

THE EXTENDED MAXIMUM ENTROPY PROCESS (EME)

The Extended Maximum Entropy Process (EME) is defined as follows: if $K \in ECL(p_{i_1}, ..., p_{i_n})$ and $\alpha_1, ..., \alpha_{2^n}$ run through the atoms of $SL(p_{i_1}, ..., p_{i_n})$, then $EME(\{p_{i_1}, ..., p_{i_n}\}, K)$ is the set of probability functions w on $SL(p_{i_1}, ..., p_{i_n})$ satisfying K for which the entropy $E(w(\alpha_1), ..., w(\alpha_{2^n}))$ is maximal. Note that the $V^{\{p_{i_1}, ..., p_{i_n}\}}(K)$ are non-empty and compact, so the $EME(\{p_{i_1}, ..., p_{i_n}\}, K)$ are well defined and non-empty. However, unless $V^{\{p_{i_1}, ..., p_{i_n}\}}(K)$ is convex, $EME(\{p_{i_1}, ..., p_{i_n}\}, K)$ may contain more than one point.

THEOREM 6. *EME satisfies the Irrelevant Information, Equivalence, Renaming, Relativization, Obstinacy, Independence and Continuity Principles.*

Proof. *Irrelevant Information:* This is proved like the corresponding property for ME, see [Paris, 1994]. Using the notation of the statement of the principle but with EME in place of N assume that $\vec{\rho} = \langle \rho_1, ..., \rho_{2^n} \rangle \in EME(K_1)$, $\vec{\tau} = \langle \tau_1, ..., \tau_{2^m} \rangle \in EME(K_2)$ and $\vec{\gamma} = \langle \gamma_{11}, ..., \gamma_{12^m}, ..., \gamma_{2^n 2^m} \rangle \in EME(K_1 \cup K_2)$ where the atoms are ordered in the obvious way so that the atom for γ_{ij} corresponds to the conjunction of the atoms for ρ_i and τ_j. Notice that $\langle \rho_1\tau_1, \rho_1\tau_2, ..., \rho_1\tau_{2^m}, \rho_2\tau_1, ..., \rho_{2^n}\tau_{2^m} \rangle \in V(K_1 \cup K_2)$, $\langle \gamma_{1.}, ..., \gamma_{2^n.} \rangle \in V(K_1)$ and $\langle \gamma_{.1}, ..., \gamma_{.2^m} \rangle \in V(K_2)$, where $\gamma_{i.} = \sum_j \gamma_{ij}$ and $\gamma_{.j} = \sum_i \gamma_{ij}$. It suffices to prove that

$$E(\vec{\rho}) = E(\gamma_{1.}, ..., \gamma_{2^n.}),$$

$$E(\rho_1\tau_1, ..., \rho_1\tau_{2^m}, ..., \rho_{2^n}\tau_{2^m}) = E(\vec{\gamma}).$$

But this follows from the assumed properties of $\vec{\rho}, \vec{\tau}, \vec{\gamma}$ since they give

$$\begin{aligned} -\sum_i \sum_j \rho_i\tau_j \log(\rho_i\tau_j) &= -\sum_i \rho_i \log \rho_i - \sum_j \tau_j \log \tau_j \\ &\geq -\sum_i \gamma_{i.} \log \gamma_{i.} - \sum_j \gamma_{.j} \log \gamma_{.j} \\ &= -\sum_i \sum_j \gamma_{ij} \log \gamma_{ij} \\ &\quad + \sum_{i,j} \gamma_{i.}\gamma_{.j} \left[\frac{\gamma_{ij}}{\gamma_{i.}\gamma_{.j}} \log \left(\frac{\gamma_{ij}}{\gamma_{i.}\gamma_{.j}} \right) \right] \\ &\geq -\sum_i \sum_j \rho_i\tau_j \log(\rho_i\tau_j), \end{aligned}$$

using the fact that by convexity of $x \log x$ on $[0, 1]$

$$\sum_{i,j} \gamma_{i.}\gamma_{.j} \left[\tfrac{\gamma_{ij}}{\gamma_{i.}\gamma_{.j}} \log \left(\tfrac{\gamma_{ij}}{\gamma_{i.}\gamma_{.j}} \right) \right] \geq$$
$$\left(\sum_{ij} \gamma_{i.}\gamma_{.j} \tfrac{\gamma_{ij}}{\gamma_{i.}\gamma_{.j}} \right) \log \left(\sum_{ij} \gamma_{i.}\gamma_{.j} \tfrac{\gamma_{ij}}{\gamma_{i.}\gamma_{.j}} \right) = \log(1) = 0.$$

Equivalence, *Renaming* and *Obstinacy* are immediate.
Relativization: Without loss of generality, assume that $\varphi = \bigvee_{i=1}^{p} \alpha_i$. For all $\langle \rho_1, ..., \rho_p \rangle$, $\langle \rho'_1, ..., \rho'_p \rangle \in V(K_1) \upharpoonright \varphi$ and all $\langle \tau_{p+1}, ..., \tau_{2^n} \rangle$,

$$\langle \rho_1, ..., \rho_p, \tau_{p+1}, ..., \tau_{2^n} \rangle \in V(K_1) \Leftrightarrow \langle \rho'_1, ..., \rho'_p, \tau_{p+1}, ..., \tau_{2^n} \rangle \in V(K_1)$$

and similarly for all $\langle \rho_1, ..., \rho_p \rangle$, $\langle \rho'_1, ..., \rho'_p \rangle \in V(K_1 \cup K_2) \upharpoonright \varphi$ and all $\langle \tau_{p+1}, ..., \tau_{2^n} \rangle$,

$$\langle \rho_1, ..., \rho_p, \tau_{p+1}, ..., \tau_{2^n} \rangle \in V(K_1 \cup K_2) \Leftrightarrow$$
$$\langle \rho'_1, ..., \rho'_p, \tau_{p+1}, ..., \tau_{2^n} \rangle \in V(K_1 \cup K_2).$$

It follows that maximizing entropy in $V(K_1)$ or in $V(K_1 \cup K_2)$ amounts (in either case) to independently maximizing both $\sum_{i=1}^{p} w(\alpha_i) \log w(\alpha_i)$ and $\sum_{i=p+1}^{2^n} w(\alpha_i) \log w(\alpha_i)$ so since $V(K_1) \upharpoonright \varphi = V(K_1 \cup K_2) \upharpoonright \varphi$,

$$EME(K_1) \upharpoonright \varphi = EME(K_1 \cup K_2) \upharpoonright \varphi,$$

as required.
Independence: Here we have the same K as in the linear case and working out the (unique) solution maximizing the entropy gives the answer.
Continuity: Let $K_m, K, \vec{u}^m, \vec{u}$ be as in the formulation of the Continuity Principle and assume that $\vec{u} \notin EME(K)$. Then there is some $\vec{v} \in V(K)$ such that $E(\vec{v}) > E(\vec{u})$ and consequently by the continuity of E, for large m there must be $\vec{v}^m \in V(K_m)$ such that $E(\vec{v}^m) > E(\vec{u}^m)$, contradiction. ■

4 THE EXTENDED INEVITABILITY THEOREM

Henceforth let N be an extended inference process satisfying the (extended) principles of Irrelevant Information, Equivalence, Relativization, Obstinacy, Independence and Continuity. Our plan now is to show the main theorem of this paper (Theorem 18) that this determines N uniquely as EME, a result analogous to Theorem 1 but now for this extended class of knowledge bases.

The proof of Theorem 18 will take us some time and proceeds via a string of lemmas. Our first lemma is a special case of Theorem 18.

LEMMA 7. *If l is a rational line then* $N(l) = \{ME(l)\}$.

By a *line* here we understand a set of constraints l in LCL of the form

$$(w(\alpha_i) - a_i)(b_j - a_j) = (w(\alpha_j) - a_j)(b_i - a_i), \quad 1 \leq i, j \leq 2^n,$$

where $\vec{a}, \vec{b} \in \mathbb{D}_n$, $\vec{a} \neq \vec{b}$, and, as usual, $\alpha_1, \alpha_2, ..., \alpha_{2^n}$ run through the atoms. We say that this line is *rational* if $(b_j - a_j)/(b_i - a_i)$ are rational whenever $b_i \neq a_i$. Hopefully without generating any further confusion we shall also on occasion use

l to denote the straight line (segment), i.e set of points in $\mathbb{D}_n$ satisfying l. Note that this line passes through the points $\vec{a}, \vec{b}$.

The proof of Lemma 7 is rather technical and requires a detailed analysis of the proof of Theorem 1 given in [Paris and Vencovská, 1990]. The reader may be forgiven for omitting it on the first reading and skipping directly to the preamble prior to Lemma 10. A second alternative would be to simply accept the following *Uniqueness Principle*, from which Lemma 7 follows by a somewhat shorter proof.

Uniqueness Principle

If l is a rational line then $N(l)$ is a single point.

After all Theorem 1 shows that restricted to linear knowledge bases ME is the unique common sense inference process so that in this sense the (unique) point of maximum entropy on a line (in $\mathbb{D}_n$) can be justified as the preferred solution to a set of linear constraints specifying that line. Our reason for preferring not to introduce this principle as such is that, entirely reasonable as it is, it is actually derivable from our other principles and even at the cost of some technical difficulties we would wish to minimize our assumptions.

Proof of Lemma 7 We need to introduce some notation. For $\vec{\delta} = \langle \delta_1, ..., \delta_m \rangle$ with $\delta_k \neq 0$ let

$$D(\vec{\delta}, k) = \begin{pmatrix} -\delta_k & & & \delta_1 & & & \\ & \ddots & & \vdots & & 0 & \\ & & -\delta_k & \delta_{k-1} & & & \\ & & & \delta_{k+1} & -\delta_k & & \\ & 0 & & \vdots & & \ddots & \\ & & & \delta_m & & & -\delta_k \end{pmatrix}$$

That is, $D(\vec{\delta}, k)$ is a $(m-1) \times m$-matrix with entries ϵ_{ij} where

$$\begin{array}{ll} \epsilon_{ii} = -\delta_k & \text{for } i \leq k-1, \\ \epsilon_{ii+1} = -\delta_k & \text{for } i \geq k, \\ \epsilon_{ik} = \delta_i & \text{for } i \leq k-1, \\ \epsilon_{ik} = \delta_{i+1} & \text{for } i \geq k, \\ \epsilon_{ij} = 0 & \text{otherwise.} \end{array}$$

The main step in our proof of Lemma 7 is provided by the next lemma.

LEMMA 8. *Let $\vec{\delta} = \langle \delta_1, ..., \delta_m \rangle$ be an integer vector satisfying $\sum_{j=1}^m \delta_j = 0$ and $\delta_k \neq 0$ and let $D(\vec{\delta}, k)$ be as above. Let $\vec{d}$ be a vector such that the system $D(\vec{\delta}, k)\vec{w}^T = \vec{d}^T$, $\vec{w} \geq 0$ has a solution and all solutions of $D(\vec{\delta}, k)\vec{w}^T = \vec{d}^T$ satisfy $\sum_{j=1}^m w_j = 1$. Then $D(\vec{\delta}, k)\vec{w}^T = \vec{d}^T$ has a unique N-solution and this agrees with the ME-solution.*

Proof. To start with, assume that $\delta_i \neq 0$ for $i = 1, ..., m$. The proof is by induction on $p(\vec{\delta}) = \sum_{i=1}^m |\delta_i| - m$.

Case $p(\vec{\delta}) = 0$. To deal with this case, we introduce for each $k \in \mathbb{N}$ the $(2k-1) \times 2k$ matrix E_k which has entries d_{ij} where $d_{1j} = 1$ for $1 \leq j \leq 2k$, $d_{ii} = d_{i(i+1)} = 1$ for $2 \leq i \leq 2k-1$ and $d_{ij} = 0$ otherwise. We have $E_1 = (1, 1)$,

$$E_2 = \begin{pmatrix} 1 & 1 & 1 & 1 \\ 0 & 1 & 1 & 0 \\ 0 & 0 & 1 & 1 \end{pmatrix},$$

$$E_3 = \begin{pmatrix} 1 & 1 & 1 & 1 & 1 & 1 \\ 0 & 1 & 1 & 0 & 0 & 0 \\ 0 & 0 & 1 & 1 & 0 & 0 \\ 0 & 0 & 0 & 1 & 1 & 0 \\ 0 & 0 & 0 & 0 & 1 & 1 \end{pmatrix},$$

and so on. These matrices are useful since our system $D(\vec{\delta}, k)\vec{w}^T = \vec{d}^T$ is equivalent (up to renaming, see below) to $E_{\frac{m}{2}}\vec{w}^T = \vec{e}^T$ for an appropriate $\vec{e}$. To see that this indeed is the case, note that $D(\vec{\delta}, k)\vec{w}^T = \vec{d}^T$ is equivalent to $D(\vec{\delta}, 1)\vec{w}^T = \vec{d'}^T$ for an appropriate $\vec{d'}$. Also since $\sum_{i=1}^m \delta_i = 0$ and $p(\vec{\delta}) = 0$, m must be even and half of the δ_i's must be 1 and the other half -1. Assume that $\delta_{2i} = 1$ and $\delta_{2i-1} = -1$ for $i = 1, ..., \frac{m}{2}$ (this assumption can be justified using Renaming, as in [Paris and Vencovská, 1990], p.210). Adding all the equations in $D(\vec{\delta}, 1)\vec{w}^T = \vec{d'}^T$ together produces $\sum_{i=1}^m w_i = 1$ and adding the $i+1$st row to the ith row for $i = 1, ..., m-2$ yields the remaining $m-2$ rows of $E_{\frac{m}{2}}\vec{w}^T = \vec{e}^T$ (where $\vec{e} = \langle 1, e_2, ..., e_{m-1} \rangle$).

Let $\vec{\tau}$, $\vec{\rho}$ be, respectively, an N-solution and *the* ME-solution of the system K_k given by $E_k\vec{w}^T = \vec{e}^T$. If some $\rho_i = 0$ then by the 'Open-mindedness' of ME (see [Paris, 1994, p. 95]) $w_i = 0$ for every solution $\vec{w}$ of $E_k\vec{w}^T = \vec{e}^T$ satisfying $\vec{w} \geq 0$ so $\tau_i = 0$. But then $\vec{\rho}, \vec{\tau}$ with their ith coordinates removed are both solutions of $E_k^-\vec{w}^T = \vec{e}^T$ where E_k^- is the *regular* matrix produced by deleting the ith column from E_k so perforce also $\rho_j = \tau_j$ for $1 \leq j \leq 2k, j \neq i$.

So now assume that $\vec{\rho} > 0$, noting that, in consequence, we must have $\vec{e} > 0$. For $k = 1$ and $k = 2$ the required result follows directly from the Lemma 5 and Independence Principle respectively. For $k \geq 3$ the proof will be far less straightforward and will lean heavily on the following result from [Paris and Vencovská, 1990][6].

LEMMA 9. *Let* $B\vec{w}^T = b^T$ *be a system of linear equations which has a solution* $\vec{w} \geq 0$ *and is such that for any solution* $\vec{w}$, $\sum_j w_j = 1$. *Then the solution* $\vec{\nu}$ *of* $B\vec{w}^T = b^T$ *which maximizes the entropy (the* ME*-solution) is the only solution which satisfies*

(i) $\prod_{\gamma_j \geq 0} \nu_j^{\gamma_j} = \prod_{\gamma_j \leq 0} \nu_j^{-\gamma_j}$ *for all* $\vec{\gamma}$ *such that* $B\vec{\gamma}^T = \vec{0}^T$ *(equivalently, for all* $\vec{\gamma}$ *from some set which spans the kernel of* B*),*

[6] Lemma 10, p.197

(ii) If $B\vec{w}^T = \vec{b}^T$, $\vec{w} \geq 0$ *does not imply* $w_j = 0$ *then* $\nu_j > 0$ *(i.e., in the terminology from [Paris and Vencovská, 1990],* $\vec{\nu}$ *is a positive solution of* $B\vec{w}^T = \vec{b}^T$*).*

First notice that by part (i) of this lemma,

$$(6) \quad \rho_1\rho_3 \ldots \rho_{2k-1} = \rho_2\rho_4 \ldots \rho_{2k},$$

a fact that we shall shortly need to employ.

Let C_k be the $(2k-1) \times (4k-4)$ matrix with entries (E_k, F) where F has entries f_{ij} with $f_{1j} = 1$ for $1 \leq j \leq 2k-4$, $f_{i(i-2)} = 1$ for $3 \leq i \leq 2k-2$ and $f_{ij} = 0$ otherwise. For example, for $k = 4$ we have

$$C_4 = \begin{pmatrix} 1 & 1 & 1 & 1 & 1 & 1 & 1 & 1 & 1 & 1 & 1 & 1 \\ 0 & 1 & 1 & 0 & 0 & 0 & 0 & 0 & 0 & 0 & 0 & 0 \\ 0 & 0 & 1 & 1 & 0 & 0 & 0 & 0 & 1 & 0 & 0 & 0 \\ 0 & 0 & 0 & 1 & 1 & 0 & 0 & 0 & 0 & 1 & 0 & 0 \\ 0 & 0 & 0 & 0 & 1 & 1 & 0 & 0 & 0 & 0 & 1 & 0 \\ 0 & 0 & 0 & 0 & 0 & 1 & 1 & 0 & 0 & 0 & 0 & 1 \\ 0 & 0 & 0 & 0 & 0 & 0 & 1 & 1 & 0 & 0 & 0 & 0 \end{pmatrix}.$$

Assume that $\vec{c} = \langle 1, c_2, \ldots, c_{2k-1} \rangle$ is some vector such that $C_k\vec{w}^T = \vec{c}^T$ has solutions with $\vec{w} \geq 0$[7]. By Lemma 2 the following identities hold for any N-solution of $C_k\vec{w}^T = \vec{c}^T$.

$$\begin{array}{lrcl} (7) & w_2 w_{2k+1} & = & w_1 w_3, \\ (8) & w_{2k+i} w_{2k+i+1} & = & w_1 w_{i+3} \text{ for } 1 \leq i \leq 2k-5, \\ (9) & w_{4k-4} w_{2k} & = & w_1 w_{2k-1}, \end{array}$$

and also

$$\begin{array}{lrcl} (10) & w_2 w_4 & = & w_3 w_{2k+2}, \\ (11) & w_{2k+i} w_{i+4} & = & w_{i+3} w_{2k+i+2} \text{ for } 1 \leq i \leq 2k-6, \\ (12) & w_{4k-5} w_{2k-1} & = & w_{2k-2} w_{2k}. \end{array}$$

Note that the statement that some $\vec{w}$ satisfies (7)-(9) can be reformulated as

$$\prod_{\gamma_j \geq 0} w_j^{\gamma_j} = \prod_{\gamma_j \leq 0} w_j^{-\gamma_j} \quad \text{for } \vec{\gamma} \in \Lambda,$$

where Λ contains the vectors

[7] We use $\vec{w}$ as a variable for vectors both of length $2k$ and of length $4k-4$, relying on the context to prevent confusion.

γ_1	γ_2	γ_3	γ_4	γ_5	$\dots$	γ_{2k-2}	γ_{2k-1}	γ_{2k}	γ_{2k+1}	γ_{2k+2}	γ_{2k+3}	$\dots$	γ_{4k-5}	γ_{4k-4}
1	−1	1	0	0	…	0	0	0	−1	0	0	…	0	0
1	0	0	1	0	…	0	0	0	−1	−1	0	…	0	0
1	0	0	0	1	…	0	0	0	0	−1	−1	…	0	0
⋮					⋱							⋱		
1	0	0	0	0	…	1	0	0	0	0	0	…	−1	−1
1	0	0	0	0	…	0	1	−1	0	0	0	…	0	−1

There are $2k-3$ vectors in Λ, they are all linearly independent and they all satisfy $C_k\vec{\gamma}^T = \vec{0}^T$, so they span the kernel of C_k. By Lemma 9, $C_k\vec{w}^T = \vec{c}^T$ and (7)-(9) determine a unique *positive* solution.

Let $q_1, \dots, q_{2k-4}$ be values calculated from (7) and (8) under the assumption that $\vec{w} = \langle \rho_1, \dots, \rho_{2k}, q_1, \dots, q_{2k-4}\rangle$ satisfies them. We obtain

$$q_{2i} = \frac{\rho_2\rho_4\cdots\rho_{2i+2}}{\rho_3\rho_5\cdots\rho_{2i+1}}, \quad q_{2i-1} = \frac{\rho_1\rho_3\cdots\rho_{2i+1}}{\rho_2\rho_4\cdots\rho_{2i}} \quad \text{for } 1 \le i \le k-2.$$

Then (9) is also satisfied since $\vec{\rho}$ satisfies (6). Let $H = 1 + \sum_{i=1}^{2k-4} q_i$ and let

$$\vec{c} = \left\langle 1, \frac{e_2}{H}, \frac{e_3 + q_1}{H}, \dots, \frac{e_{2k-2} + q_{2k-4}}{H}, \frac{e_{2k-1}}{H} \right\rangle.$$

Note that all entries in $\vec{c}$ are strictly positive, we will refer to them as $\langle 1, c_2, \dots, c_{2k-1}\rangle$. Then $\vec{\mu} = \frac{1}{H}\langle \rho_1, \dots, \rho_{2k}, q_1, \dots, q_{2k-4}\rangle$ is a (*positive*)[8] solution of $C_k\vec{w}^T = \vec{c}^T$ which satisfies (7)-(9). Below, we show that any N-solution of $C_k\vec{w}^T = \vec{c}^T$ has all entries strictly positive (and thus all N-solutions are *positive* in the sense of [Paris and Vencovská, 1990]). It follows that $C_k\vec{w}^T = \vec{c}^T$ has the unique N-solution $\vec{\mu}$. By Obstinacy the system

$$C_k\vec{w}^T = \vec{c}^T, \quad w_{2k+i} = \frac{q_i}{H} \text{ for } 1 \le i \le 2k-4,$$

also has a unique N-solution $\vec{\mu}$. Hence, by Equivalence,

$$E_k\vec{w}^T = \frac{1}{H}\vec{e}^T, \quad w_{2k+i} = \frac{q_i}{H} \text{ for } 1 \le i \le 2k-4,$$

also has a unique N-solution $\vec{\mu}$. By Lemma 4, K_k has a unique N-solution $\vec{\rho}$, which proves what we want to show.

Showing that all the N-solutions of $C_k\vec{w}^T = \vec{c}^T$ are positive requires some effort. Any N-solution $\vec{\tau}$ of this system satisfies (7)-(12). Assume $\tau_i = 0$ for some i. We shall derive a contradiction.

First consider the case in which $\tau_1 \neq 0$, so it must be some other τ_i which is zero. In (i)–(v) below we will eliminate all possibilities.

[8] All coordinates of $\vec{\mu}$ are strictly positive so the $\vec{\mu}$ certainly is positive in the sense of [Paris and Vencovská, 1990], see Lemma 9.

(i) If $\tau_2 = 0$ then $\tau_3 = 0$ by (7), but that is impossible since $\tau_2 + \tau_3 = c_2 \neq 0$.

(ii) If $\tau_3 = 0$ then either $\tau_2 = 0$ or $\tau_{2k+1} = 0$ by (7). The former is impossible as above and the latter implies $\tau_4 = 0$ by (8) so this is also impossible since $\tau_3 + \tau_4 + \tau_{2k+1} = c_3 \neq 0$.

(iii) If for some $1 \leq i \leq 2k - 5$, $\tau_{i+3} = 0$ then either $\tau_{2k+i} = 0$ or $\tau_{2k+i+1} = 0$ by (8). The former implies that $\tau_{i+2} = 0$ by (8) again, or (7) when $i = 1$, in which case $\tau_{i+2} + \tau_{i+3} + \tau_{2k+i}$ would be 0. But that must equal $c_{i+2} \neq 0$, contradiction. On the other hand the latter implies that $\tau_{i+4} = 0$ by (8), or by (9) when $i = 2k-5$, so this is also impossible since $\tau_{i+3} + \tau_{i+4} + \tau_{2k+i+1} = c_{i+3} \neq 0$.

(iv) If $\tau_{2k-1} = 0$ or $\tau_{2k} = 0$ then a contradiction follows just as above for τ_2 and τ_3, using (9) and (8).

(v) If $\tau_j > 0$ for $1 \leq j \leq 2k$ then the remaining τ_{2k+i} for $1 \leq i \leq 2k - 4$ must also be non-zero by (8).

Now assume $\tau_1 = 0$. We shall derive a contradiction by taking a sum of certain lines of $C_k\vec{\tau}^T = \vec{c}^T$ and then argue that $\vec{\tau}$ cannot possibly satisfy the resulting equation. To start with take the sum of all the i + 2nd lines $1 \leq i \leq 2k - 4$ of $C_k\vec{\tau}^T = \vec{c}^T$ for which $\tau_{2k+i} \neq 0$ along with line 2 if $\tau_2 \neq 0$ and line $2k - 1$ if $\tau_{2k} \neq 0$. On the left hand side we obtain a sum which involves all the non-zero τ_i's *at most* once, since by (7)-(9) no two subsequent lines are included. To involve all the non-zero τ_i's *exactly* once, we now add in also various other lines according to the following prescription (a)–(d):

(a) If $\tau_{2k+i} \neq 0$, $\tau_{2k+i+1} = \tau_{2k+i+2} = \ldots = \tau_{2k+i+m-1} = 0$ and $\tau_{2k+i+m} \neq 0$ $(i \geq 1, 2k - 4 - i \geq m > 2)$ add in also lines $i + 4, i + 6, \ldots, i + m$ if m is even and lines $i + 4, \ldots i + m - 1$ if m is odd. This will ensure that all the non-zero τ_is (which had been missed before) from $\tau_{i+4}, \tau_{i+5}, \ldots, \tau_{i+m+1}$ are added exactly once. (If m is even then this is obvious, and if m is odd then it would appear that τ_{i+m+1} may have been missed, but by (11), since $\tau_{2k+i+m} \neq 0$ and $\tau_{2k+i+m-2} = 0$ we have $\tau_{i+m+1} = 0$.)

(b) If $\tau_{2k+1} = \tau_{2k+2} = \ldots = \tau_{2k+m-1} = 0$ and $\tau_{2k+m} \neq 0$, for $2 < m \leq 2k - 4$, add in also the lines $2, 4, \ldots, m$ if m is even and lines $2, 4, \ldots, m - 1$ if m is odd with justification as above in the case of odd m using the fact that $\tau_{m+1} = 0$.

(c) If $\tau_{4k-3-m} \neq 0$ and $\tau_{4k-m-2} = \tau_{4k-m-1} = \ldots = \tau_{4k-4} = 0$ for $2 \leq m < 2k - 3$ add in lines $2k - m + 1, 2k - m + 3, \ldots, 2k - 1$ if m is even and lines $2k - m + 2, 2k - m + 4, \ldots, 2k - 1$ if m is odd (using the fact that $\tau_{2k-m+1} = 0$, which follows from (11) since $\tau_{4k-3-m} \neq 0$ and $\tau_{4k-m-1} = 0$).

(d) If all $\tau_{2k+1},...,\tau_{4k-4}$ are zero, then by (10), $\tau_2 = 0$ or $\tau_4 = 0$. In the former case, add lines $3, 5, ...2k-1$ and in the latter case add lines $2, 5, 7, ...2k-1$.

We now have an equation which has the sum of all the nonzero τ_i's which must be 1 on the left hand side and $\sum_{i:\ line\ i\ was\ chosen} \frac{1}{H}(e_i + q_{i-2})$ on the right hand side (where we set $q_0 = q_{2k-3} = 0$). Since for each $2 \le i \le 2k-1$, we have $\rho_i + \rho_{i+1} = e_i$, and no two subsequent lines were chosen, $\sum_{i:\ line\ i\ was\ chosen} e_i \le 1$. Since all the q_i, $1 \le i \le 2k-4$, are strictly positive and at least one of the lines $3,...,2k-2$ was left out, $\sum_{i:\ line\ i\ was\ chosen} q_i < \sum_{i=1}^{2k-4} q_i$ so considering the definition of H we can see that the righthand side of our equation is strictly less than one, which gives the desired contradiction.

This completes the proof of Lemma 8 in the base case when $p(\vec{\delta}) = 0$. We now turn to the case when $p(\vec{\delta}) > 0$.

Case $p(\vec{\delta}) > 0$. This part of the proof uses the same key ideas as the corresponding argument in [Paris and Vencovská, 1990], pp.211-215 but now adapted to apply to *extended* inference processes.

Assume that Lemma 8 holds for $\vec{\delta}$ with $\delta_i \neq 0$ for all i and $p(\vec{\delta}) < N$ for some $N \geq 1$. Let $\vec{\delta} = \langle \delta_1, ..., \delta_m \rangle$ be such that $p(\vec{\delta}) = N$, $\delta_i \neq 0$ for all i and $D(\vec{\delta}, k)\vec{w}^T = \vec{d}^T$ satisfies the hypotheses of Lemma 8. Without loss of generality we can assume that $k = m$ and $\delta_r \neq \pm 1$ for some $r \neq m$. Consider the set of constraints

$$D(\vec{\delta}, m)\vec{w}^T = \vec{d}^T, \qquad \sum_{j=1}^{|\delta_r|} z_j = 1,$$

corresponding to K given by $C\vec{x}^T = \vec{d}^T$, where

$$C = \begin{pmatrix} \underbrace{-\delta_m, ..., -\delta_m}_{|\delta_r|\ times} & & 0 & & \delta_1, ..., \delta_1 \\ & \underbrace{-\delta_m, ..., -\delta_m}_{|\delta_r|\ times} & & & \vdots\ \vdots \\ & & \ddots & & \\ 0 & & & \underbrace{-\delta_m, ..., -\delta_m}_{|\delta_r|\ times} & \underbrace{\delta_{m-1}, ..., \delta_{m-1}}_{|\delta_r|\ times} \end{pmatrix}$$

and $\vec{x} = \langle x_{11}, ..., x_{1|\delta_r|}, x_{21}, ..., x_{2|\delta_r|}, ..., x_{m|\delta_r|} \rangle$. By Lemmas 3 and 5 the N-solutions of K are precisely

$$\left\langle \underbrace{\frac{\tau_1}{|\delta_r|}, ..., \frac{\tau_1}{|\delta_r|}}_{|\delta_r|\ times}, ..., \underbrace{\frac{\tau_m}{|\delta_r|}, ..., \frac{\tau_m}{|\delta_r|}}_{|\delta_r|\ times} \right\rangle$$

where $\vec{\tau}$ are the N-solutions of $D(\vec{\delta},m)\vec{w}^T = \vec{d}^T$. By Obstinacy, $K \cup \{x_{rj} = x_{r1} \mid 1 \leq j \leq |\delta_r|\}$ has the same N-solutions so by Obstinacy again $G\vec{x}^T = \vec{g}^T$ has the same N-solutions, where G is obtained from C by replacing the rth row

$$\underbrace{0,...,0,}_{(r-1)|\delta_r|\ times} \quad \underbrace{-\delta_m,...,-\delta_m,}_{|\delta_r|\ times} \quad \underbrace{0,...,0,}_{(m-r-1)|\delta_r|\ times} \quad \underbrace{\delta_r,...,\delta_r}_{|\delta_r|\ times}$$

by $|\delta_r|$ new rows

$$\begin{array}{ccccccccccccc}
0 & \ldots & 0 & -\delta_m & 0 & \ldots & 0 & 0 & \ldots & 0 & \delta_r|\delta_r|^{-1} & \ldots & \delta_r|\delta_r|^{-1} \\
0 & \ldots & 0 & 0 & -\delta_m & \ldots & 0 & 0 & \ldots & 0 & \delta_r|\delta_r|^{-1} & \ldots & \delta_r|\delta_r|^{-1} \\
\vdots & & \vdots & \vdots & \vdots & & \vdots & \vdots & & \vdots & \vdots & & \vdots \\
0 & \ldots & 0 & 0 & \ldots & 0 & -\delta_m & 0 & \ldots & 0 & \delta_r|\delta_r|^{-1} & \ldots & \delta_r|\delta_r|^{-1}
\end{array}$$

and

$$\vec{g} = \left\langle d_1, ..., d_{r-1}, \underbrace{d_r|\delta_r|^{-1}, ..., d_r|\delta_r|^{-1}}_{|\delta_r|\ times}, d_{r+1}, ...d_m \right\rangle.$$

Now consider an N-solution $\vec{\tau}$ of $D(\vec{\delta},m)\vec{w}^T = \vec{d}^T$. Let Λ be the set

$$\{\langle i,1\rangle \mid 1 \leq i \leq m\} \cup \{\langle r,j\rangle \mid 1 \leq j \leq |\delta_r|\}.$$

By the above analysis and using Obstinacy,

$$\vec{\eta} = \left\langle \underbrace{\frac{\tau_1}{|\delta_r|}, ..., \frac{\tau_1}{|\delta_r|}}_{|\delta_r\ |times}, ..., \underbrace{\frac{\tau_m}{|\delta_r|}, ..., \frac{\tau_m}{|\delta_r|}}_{|\delta_r\ |times} \right\rangle$$

remains an N-solution of

$$G\vec{x}^T = \vec{g}^T, \quad x_j = \eta_j \ \text{for}\ j \notin \Lambda.$$

By Equivalence, $\vec{\eta}$ is also an N-solution of

$$S \begin{pmatrix} x_{11} \\ \vdots \\ x_{(r-1)1} \\ x_{r1} \\ \vdots \\ x_{r|\delta_r|} \\ x_{(r+1)1} \\ \vdots \\ x_{m1} \end{pmatrix} = \vec{s}^T, \quad x_j = \eta_j \ \text{for}\ j \notin \Lambda$$

where $\vec{s}$ is

$$\langle d_1|\delta_r|^{-1}, ..., d_{r-1}|\delta_r|^{-1}, \underbrace{t, t, ..., t}_{|\delta_r|\ times}, d_{r+1}|\delta_r|^{-1}, ..., d_m|\delta_r|^{-1}\rangle,$$

with $t = (d_r - \delta_r \tau_m(1 - |\delta_r|^{-1}))|\delta_r|^{-1}$, and

$$S = \begin{pmatrix} -\delta_m & & & & & & & & \delta_1 \\ & \ddots & & & & 0 & & & \vdots \\ & & -\delta_m & & & & & & \delta_{r-1} \\ & & & -\delta_m & & & & & \frac{\delta_r}{|\delta_r|} \\ & & & & \ddots & & & & \vdots \\ & & & & & -\delta_m & & & \frac{\delta_r}{|\delta_r|} \\ & & & & & & -\delta_m & & \delta_{r+1} \\ & & 0 & & & & & \ddots & \vdots \\ & & & & & & & -\delta_m & \delta_{m-1} \end{pmatrix}.$$

Let $H = \sum_{s \in \Lambda} \eta_s$. By Lemma 4,

$$\frac{1}{H}\langle \eta_{11}, ..., \eta_{(r-1)1}, \eta_{r1}, ..., \eta_{r|\delta_r|}, \eta_{(r+1)1}, ..., \eta_{m1}\rangle,$$

i.e.

$$\frac{1}{H}\left\langle \frac{\tau_1}{|\delta_r|}, ..., \frac{\tau_{r-1}}{|\delta_r|}, \frac{\tau_r}{|\delta_r|}, ..., \frac{\tau_r}{|\delta_r|}, \frac{\tau_{r+1}}{|\delta_r|}, ..., \frac{\tau_m}{|\delta_r|}\right\rangle,$$

is an N-solution of $S\vec{w}^T = \frac{1}{H}\vec{s}^T$. Now S is $D(\vec{\psi}, m + |\delta_r| - 1)$, where

$$\vec{\psi} = \langle \delta_1, ..., \delta_{r-1}, \frac{\delta_r}{|\delta_r|}, ..., \frac{\delta_r}{|\delta_r|}, \delta_{r+1}, ..., \delta_m\rangle,$$

and $p(\vec{\psi}) = p(\vec{\delta}) - (|\delta_r| - 1) < p(\vec{\delta})$, so by the inductive assumption,

$$\frac{1}{H}\left\langle \frac{\tau_1}{|\delta_r|}, ..., \frac{\tau_{r-1}}{|\delta_r|}, \frac{\tau_r}{|\delta_r|}, ..., \frac{\tau_r}{|\delta_r|}, \frac{\tau_{r+1}}{|\delta_r|}, ..., \frac{\tau_m}{|\delta_r|}\right\rangle$$

is also the (unique) ME-solution of $S\vec{w}^T = \frac{1}{H}\vec{s}^T$.

By Lemma 9, if $\delta_r > 0$ then

$$(13)\quad \prod_{\substack{\delta_i \geq 0 \\ i \neq r}} \left(\frac{\tau_i}{H|\delta_r|}\right)^{\delta_i} \prod_{j=1}^{|\delta_r|} \frac{\tau_r}{H|\delta_r|} = \prod_{\substack{\delta_i \leq 0 \\ i \neq r}} \left(\frac{\tau_i}{H|\delta_r|}\right)^{-\delta_i},$$

and analogously if $\delta_r < 0$. Hence if some $\tau_i = 0$ then there must also be a $\tau_j = 0$ with δ_i, δ_j having different signs. But then referring back to the original system $D(\vec{\delta}, m)\vec{w}^T = \vec{d}^T$ for which $\vec{\tau}$ is an N-solution we see that the equation

$$-\delta_i w_j + \delta_j w_i = 0$$

is derivable from this system. As remarked earlier this forces $D(\vec{\delta}, m)\vec{w}^T = \vec{d}^T$ to have a unique solution, so necessarily the N-solution is unique and agrees with the ME-solution.

On the other hand if no τ_i is zero then since (13) forces

$$\prod_{\delta_i \geq 0} \tau_i^{\delta_i} = \prod_{\delta_i \leq 0} \tau_i^{-\delta_i},$$

by Lemma 9 $\vec{\tau}$ must again be the ME-solution to $D(\vec{\delta}, m)\vec{w}^T = \vec{d}^T$.

This completes the induction step.

It remains to remove the assumption that $\delta_i \neq 0$ for $i = 1, ..., m$. Assume that $D(\vec{\delta}, k)\vec{w}^T = \vec{d}^T$ satisfies the assumptions of Lemma 8. Let $\delta_{i_1}, ..., \delta_{i_r}$ be the nonzero coordinates of $\vec{\delta}$, where $i_1 < ... < i_r$ and $k = i_q$. Let $\vec{\gamma} = \langle \delta_{i_1}, ..., \delta_{i_r} \rangle$. Our system is equivalent to the system of equations

$$D(\vec{\gamma}, q) \begin{pmatrix} w_{i_1} \\ \vdots \\ w_{i_r} \end{pmatrix} = \begin{pmatrix} d_{i_1} \\ \vdots \\ d_{i_{q-1}} \\ d_{i_{q+1}-1} \\ \vdots \\ d_{i_r-1} \end{pmatrix},$$

$$(14) \quad -\delta_k w_i = \begin{cases} d_i & \text{for } i \leq k-1,\ i \notin \{i_1, ..., i_r\}, \\ d_{i-1} & \text{for } i \geq k+1,\ i \notin \{i_1, ..., i_r\}. \end{cases}$$

Thus the system implies $\sum_{j=1}^r w_{i_j} = a$ for some a. By Lemma 4, the N-solutions of

$$(15) \quad D(\vec{\gamma}, q) \begin{pmatrix} w_{i_1} \\ \vdots \\ w_{i_r} \end{pmatrix} = \frac{1}{a} \begin{pmatrix} d_{i_1} \\ \vdots \\ d_{i_{q-1}} \\ d_{i_{q+1}-1} \\ \vdots \\ d_{i_r-1} \end{pmatrix}$$

are precisely the vectors $\langle \tau_{i_1}/a, ..., \tau_{i_r}/a \rangle$ for N-solutions $\langle \tau_1, ..., \tau_m \rangle$ of $D(\vec{\delta}, k)\vec{w}^T = \vec{d}^T$. Since (15) satisfies the assumptions of Lemma 8 and the coordinates of $\vec{\gamma}$ are nonzero such a vector $\langle \tau_{i_1}/a, ..., \tau_{i_r}/a \rangle$ must, by the case already proved, equal the ME-solution of (15). But clearly, by re-running the above derivation, this is $\langle \rho_{i_1}/a, ..., \rho_{i_r}/a \rangle$ where $\vec{\rho}$ is the ME-solution of $D(\vec{\delta}, k)\vec{w}^T = \vec{d}^T$ so $\tau_{i_j} = \rho_{i_j}$ for $j = 1, ..., r$. But this gives that $\vec{\tau} = \vec{\rho}$ since the remaining coordinates are uniquely determined by (14) and finally completes the proof of Lemma 8. ■

Proof of Lemma 7 continued Using Lemma 8 it is straightforward to prove Lemma 7. First notice that the assumption that $\vec{\delta}$ is an integer vector can be replaced by the assumption that it is a rational vector since for a rational $\vec{\delta}$ there is some (non-zero) integer number a such that $a\vec{\delta}$ is an integer vector (by Equivalence, $D(a\vec{\delta}, k)\vec{w}^T = a\vec{d}^T$ has the same N-solutions as $D(\vec{\delta}, k)\vec{w}^T = \vec{d}^T$). Noting that a rational line with $a_k \neq b_k$ is given by the set of constraints

$$\{-w_j + \frac{b_j - a_j}{b_k - a_k} w_k = a_k \frac{b_j - a_j}{b_k - a_k} - a_j \mid j \neq k\}$$

we can see that it corresponds to $D(\vec{\delta}, k)\vec{w}^T = \vec{d}^T$ where $\delta_k = 1$, $\delta_j = \frac{b_j - a_j}{b_k - a_k}$ for $j \neq k$ (with the appropriate $\vec{d}$). Since $\sum_{j=1}^{2^n} a_j = \sum_{j=1}^{2^n} b_j = 1$, the above $\vec{\delta}$ satisfies $\sum_{j=1}^{2^n} \delta_j = 0$. Hence Lemma 8 applies and the ME-solution of $D(a\vec{\delta}, k)\vec{w}^T = a\vec{d}^T$ must be the unique N-solution of $D(a\vec{\delta}, k)\vec{w}^T = a\vec{d}^T$, equivalently, $N(l) = \{ME(l)\}$, as required. ■

The aim now is to build on Lemma 7 to give the second key Lemma 15.

As far as our proof is concerned an important consequence of allowing polynomial constraints is that we can describe any *finite* set of points. Precisely, given $\{\vec{a}^{(1)}, \vec{a}^{(2)}, ..., \vec{a}^{(m)}\} \subset \mathbb{D}_n$ let $K_{\vec{a}^{(1)}, \vec{a}^{(2)}, ..., \vec{a}^{(m)}}$ be the set consisting of the single constraint

$$\prod_{k=1}^{m} \left(\sum_{i=1}^{2^n} (w(\alpha_i) - a_i^{(k)})^2 \right) = 0.$$

Then

$$V(K_{\vec{a}^{(1)}, \vec{a}^{(2)}, ..., \vec{a}^{(m)}}) = \{\vec{a}^{(1)}, \vec{a}^{(2)}, ..., \vec{a}^{(m)}\}.$$

Apart from singletons it is not possible to specify finite sets using just linear constraints. The fact that we now have this ability enables us to express a notion of relative preference ($\prec$) between individual pairs of points. Precisely, define the binary relations $\preceq$, $\prec$ and $\cong$ on $\mathbb{D}_n$ by

$$\begin{aligned} \vec{a} \preceq \vec{b} \quad &\Longleftrightarrow \quad \vec{b} \in N(K_{\vec{a}, \vec{b}}), \\ \vec{a} \prec \vec{b} \quad &\Longleftrightarrow \quad \text{not}(\vec{b} \preceq \vec{a}), \\ \vec{a} \cong \vec{b} \quad &\Longleftrightarrow \quad \vec{a} \preceq \vec{b} \;\&\; \vec{b} \preceq \vec{a}. \end{aligned}$$

Thus $\vec{a} \prec \vec{b}$ just if $\vec{a} \notin N(K_{\vec{a}, \vec{b}})$ and $\vec{a} \cong \vec{b}$ just if $\vec{a}, \vec{b} \in N(K_{\vec{a}, \vec{b}})$.

LEMMA 10. *The relations $\preceq$, $\prec$ are transitive.*

Proof. Suppose that $\vec{a} \preceq \vec{b}$ and $\vec{b} \preceq \vec{c}$, i.e. $\vec{b} \in N(K_{\vec{a},\vec{b}})$ and $\vec{c} \in N(K_{\vec{b},\vec{c}})$. Suppose $\vec{c} \notin N(K_{\vec{a},\vec{b},\vec{c}})$. If $\vec{b} \in N(K_{\vec{a},\vec{b},\vec{c}})$ then

$$\vec{b} \in N(K_{\vec{a},\vec{b},\vec{c}}) \cap \{\vec{b},\vec{c}\} \neq \emptyset,$$

so by Obstinacy

$$\vec{c} \in N(K_{\vec{b},\vec{c}}) = N(K_{\vec{a},\vec{b},\vec{c}}) \cap \{\vec{b},\vec{c}\},$$

contradiction. So now $\vec{b} \notin N(K_{\vec{a},\vec{b},\vec{c}})$ and the only option left is that

$$N(K_{\vec{a},\vec{b},\vec{c}}) = \{\vec{a}\}.$$

But in that case

$$\vec{a} \in N(K_{\vec{a},\vec{b},\vec{c}}) \cap \{\vec{a},\vec{b}\} \neq \emptyset,$$

so by Obstinacy

$$\vec{b} \in N(K_{\vec{a},\vec{b}}) = N(K_{\vec{a},\vec{b},\vec{c}}) \cap \{\vec{a},\vec{b}\},$$

contradiction again. Hence the original assumption, that $\vec{c} \notin N(K_{\vec{a},\vec{b},\vec{c}})$, must be false. But then, by Obstinacy, $\vec{c} \in N(K_{\vec{a},\vec{b},\vec{c}}) \cap \{\vec{a},\vec{c}\} \neq \emptyset$ so $\vec{c} \in N(K_{\vec{a},\vec{c}}) = N(K_{\vec{a},\vec{b},\vec{c}}) \cap \{\vec{a},\vec{c}\}$ as required.

The second part now follows directly. ∎

LEMMA 11. *Suppose $K \in ECL$, $\vec{a},\vec{b} \in V(K)$, $\vec{a} \in N(K)$. Then*

$$\vec{b} \in N(K) \iff \vec{a} \preceq \vec{b}.$$

In other words all the vectors in $N(K)$ are congruent ($\cong$) to each other and these are the maximal elements in $V(K)$ with respect to $\preceq$.

Proof. Since $\vec{a} \in \{\vec{a},\vec{b}\} \cap N(K)$, by Obstinacy and the fact that $\vec{a},\vec{b} \in V(K)$,

$$\{\vec{a},\vec{b}\} \cap N(K) = N(K \cup K_{\vec{a},\vec{b}}) = N(K_{\vec{a},\vec{b}}).$$

Hence $\vec{b} \in N(K)$ just if $\vec{b} \in N(K_{\vec{a},\vec{b}})$, equivalently $\vec{a} \preceq \vec{b}$. ∎

LEMMA 12. *Let $l \in ECL$ be a line. Then $ME(l) \in N(l)$*

Proof. Let l_m be rational lines converging (in the metric ∂) to l as $m \to \infty$. By Lemma 7 $N(l_m) = \{ME(l_m)\}$ and the result follows by Continuity since $ME(l) = \lim_{m\to\infty} ME(l_m)$. ∎

Notice that this lemma does not discount the possibility that $N(l)$ may also contain other points. It is essentially the problem of showing that no such further points can exist that occupies much of our effort in what follows.

LEMMA 13. *Suppose $\vec{a},\vec{b} \in \mathbb{D}_n$ and the function E is increasing on the line segment $[\vec{a},\vec{b}]$ from $\vec{a}$ to $\vec{b}$. Then $\vec{a} \preceq \vec{b}$.*

Proof. We may assume that if $a_i = 0$ then $b_i = 0$ (since otherwise prove the result instead for points $\vec{a'}$ on $[\vec{a}, \vec{b}]$ arbitrarily close to, but not equal to, $\vec{a}$, noting that such points do satisfy this property, and then appeal to Continuity to conclude the result for $\vec{a}$). Let l be the line through the points $\vec{a}$ and $\vec{b}$ and let $\vec{e}$ be the point on l at which E is maximal. By the concavity of the function E and the fact that $E(\vec{a}) < E(\vec{b})$ the point $\vec{e}$ must be on the opposite side of $\vec{b}$ to $\vec{a}$. Let $\vec{s}$ be in $\mathbb{D}_n$ not on l and such that if $a_i = 0$ then $s_i = 0$. Let P be the plane through $\vec{a}, \vec{b}, \vec{s}$. If l is normal to $\nabla E(\vec{b})$ then $\vec{b} = \vec{e}$, in which case we immediately have the result by Lemma 11 since $\vec{b} \in N(l)$ by Lemma 12. Otherwise, let l_1 be the line in the plane P through $\vec{b}$ normal to the vector $\nabla E(\vec{b})$. Notice that since E increases along the line segment $[\vec{a}, \vec{b}]$,$\vec{a}$ lies either on l_1 or on the other side of l_1 to $\vec{e}$. In the first case the result again follows by Lemma 12.

Otherwise, for $\vec{x} \in P$, $\vec{x} \neq \vec{a}$ let $l(\vec{x})$ be the line through $\vec{a}$ and $\vec{x}$. Then $ME(l(\vec{x}))$ is on the other side of l_1 to $\vec{a}$ for $\vec{x} = \vec{b}$ (in this case $ME(l(\vec{x})) = \vec{e}$) and, by the Open-mindedness of ME, on the same side of l_1 as $\vec{a}$ when $\vec{x}$ is an intersection point of l_1 with the boundary of $\mathbb{D}_n$. Hence there is a point $\vec{c}$ on l_1 for which $\vec{c} = ME(l(\vec{c}))$. By Lemmas 12 and 11 applied to lines l, l_1 respectively we obtain that $\vec{a} \preceq \vec{c}$ and $\vec{c} \preceq \vec{b}$. The result follows by the transitivity of $\preceq$. ∎

LEMMA 14. *Suppose $l \in ECL$ is a line and $\vec{a} \in N(l)$. Then $[\vec{a}, ME(l)] \subseteq N(l)$.*

Proof. If $\vec{a} = ME(l)$ then the result is obvious so suppose $\vec{a} \neq ME(l)$. Let $\vec{b}$ be on the line segment $[\vec{a}, ME(l)]$. Then E is increasing on $[\vec{a}, \vec{b}]$ so by Lemma 13, $\vec{a} \preceq \vec{b}$. Therefore, since $\vec{a} \in N(l)$, and $\vec{b} \in V(l)$, by Lemma 11, $\vec{b} \in N(l)$, as required. ∎

A consequence of this is that we can now strengthen Lemma 13 to give the second key lemma referred to earlier.

LEMMA 15. *Suppose $\vec{a}, \vec{b}$ are distinct points in $\mathbb{D}_n$ and the function E is increasing on the line segment $[\vec{a}, \vec{b}]$. Then $\vec{a} \prec \vec{b}$.*

Proof. Let $\vec{c} = (\vec{a} + \vec{b})/2$. By continuity and concavity of E there is a neighbourhood $B = \{\vec{x} \mid |\vec{x} - \vec{c}| < \epsilon\} \cap \mathbb{D}_n$ such that for $\vec{d} \in B$ E is increasing on both $[\vec{a}, \vec{d}]$ and $[\vec{d}, \vec{b}]$. Hence by Lemma 13, $\vec{a} \preceq \vec{d} \preceq \vec{b}$. Now suppose on the contrary that $\vec{b} \preceq \vec{a}$. Then by transitivity $\vec{a} \cong \vec{d}$ so, since $\vec{d}$ was an arbitrary point in B, $\vec{x} \cong \vec{y}$ for any $\vec{x}, \vec{y} \in B$. A continuity argument now shows that there is a rational line l with $ME(l) \in B$. Precisely, let $\vec{e}$ be a rational point in B and let l' be a line through $\vec{e}$ normal to $\nabla E(\vec{e})$, so $ME(l') = \vec{e}$. Let l_m, $m \in \mathbb{N}$, be rational lines through $\vec{e}$ such that $\lim_{m \to \infty} \partial(l', l_m) = 0$. Then by continuity of the inference process ME, $\lim_{m \to \infty} ME(l_m) = ME(l') = \vec{e}$, so for sufficiently large m, $ME(l_m) \in B$ and we can take this line to be l. But now, by Lemmas 7, 11, for points $\vec{x}$ on l and close to $ME(l)$ (and so in B), $\vec{x} \prec ME(l)$, contradiction. ∎

We are now set up to prove the third key lemma in this paper.

LEMMA 16. *Suppose $\vec{a}, \vec{b} \in \mathbb{D}_n$ and $E(\vec{a}) < E(\vec{b})$. Then $N(K_{\vec{a},\vec{b}}) = \{\vec{b}\}$, equivalently $\vec{a} \prec \vec{b}$.*

Proof. Let $\vec{e_i}$ be the vector in $\mathbb{D}_n$ with i'th coordinate 1 and all other coordinates 0. Notice that $\vec{b}$ cannot be such a corner point since the global minima of E occur at these points.

The first thing we need to show is that there is some $\vec{e_i}$ such that on the line segment $[\vec{e_i}, \vec{b}]$ from $\vec{e_i}$ to $\vec{b}$ the function E is (strictly) increasing. For suppose not. Then for each $i = 1, 2, ..., 2^n$ there would (by the concavity of the function E) be a number t_i, $0 < t_i < 1$ such that

$$E((1-t_i)\vec{e_i} + t_i\vec{b}) = E(\vec{b}).$$

In this case however,

$$\begin{aligned}\vec{b} &= \sum_{i=1}^{2^n} b_i\vec{e_i} = \sum_{i=1}^{2^n} \frac{b_i}{1-t_i}\left\{((1-t_i)\vec{e_i} + t_i\vec{b}) - t_i\vec{b}\right\} \\ &= \sum_{i=1}^{2^n} \frac{b_i}{1-t_i}((1-t_i)\vec{e_i} + t_i\vec{b}) - \sum_{j=1}^{2^n} \frac{b_j t_j}{1-t_j}\vec{b},\end{aligned}$$

so

$$\vec{b} = \sum_{i=1}^{2^n} \left\{ \frac{b_i}{1-t_i}\left(1 + \sum_j \frac{t_j b_j}{1-t_j}\right)^{-1} ((1-t_i)\vec{e_i} + t_i\vec{b}) \right\}.$$

But since

$$\sum_i \frac{b_i}{1-t_i}\left(1 + \sum_j \frac{t_j b_j}{1-t_j}\right)^{-1} =$$

$$\left(\sum_i b_i + \sum_j \frac{t_j b_j}{1-t_j}\right) \cdot \left(1 + \sum_j \frac{t_j b_j}{1-t_j}\right)^{-1} = 1,$$

the righthand side of this expression is a (strict) convex combination of vectors with entropy equal to the entropy of $\vec{b}$ so by the concavity of E its entropy must exceed that of $\vec{b}$, contradiction.

Having established this let us assume for the moment that $\vec{b}$ and $\vec{a}$ are in the same corner $\vec{e_k}$, in other words that this property holds for both $\vec{a}$ and $\vec{b}$. If E is increasing on the line segment $[\vec{a}, \vec{b}]$ then we know the result by Lemma 15. So suppose not. In this case the points $\vec{e_k}, \vec{a}, \vec{b}$ must all be distinct. [In what follows everything will take place in the plane determined by these three points, indeed in the convex set determined by the intersection of this plane with $\mathbb{D}_n$.] Also there must be a unique point $\vec{c}$ on this line segment strictly between $\vec{a}$ and $\vec{b}$ such that $E(\vec{c}) = E(\vec{b})$. By concavity the function E must be increasing on $[\vec{a}, \vec{c}]$. Also by concavity for any point $\vec{u}$ on $[\vec{c}, \vec{b}]$, $E(\vec{u}) \geq E(\vec{b})$ and for any $E(\vec{e}_k) \leq y \leq E(\vec{u})$ there is a unique point $\vec{z}$ on the line segment $[\vec{e}_k, \vec{u}]$ at which $E(\vec{z}) = y$ and E is increasing on $[\vec{e}_k, \vec{z}]$. Pick a point $\vec{b}'$ on $[\vec{c}, \vec{b}]$ close to (but not equal to) $\vec{b}$ and let $\vec{d}$ be the point on $[\vec{e}_k, \vec{b}']$ with $E(\vec{d}) = E(\vec{b})$. Notice that E increasing on $[\vec{e}_k, \vec{d}]$ and that $\vec{d}$ is also close to $\vec{b}$ (otherwise consider a limit point of such $\vec{d}$ as $\vec{b}'$ tends to $\vec{b}$).

Now we need to consider two possibilities. First assume that E is increasing on $[\vec{e}_k, \vec{c}]$. In that case, since E is increasing on $[\vec{a}, \vec{c}]$ we can pick an interior point $\vec{c'}$ on $[\vec{e}_k, \vec{c}]$ such that E is increasing on $[\vec{a}, \vec{c'}]$ and such that $E(\vec{c'}) < E(\vec{c})$. For $t \in [0, 1]$ let $\vec{g}(t)$ be that (unique) point on the line segment $[\vec{e}_k, (1-t)\vec{c} + t\vec{d}]$ such that

$$(16)\quad E(\vec{g}(t)) = (1-t)E(\vec{c'}) + tE(\vec{c}).$$

So $\vec{g}(0) = \vec{c'}$ and $\vec{g}(1) = \vec{d}$. Now, E is $\mathcal{C}^\infty$ on any set obtained from the triangle with vertices $\vec{e}_k, \vec{c}, \vec{d}$ by removing a small neighborhood of $\vec{e}_k$. To see this, note that if $x_i = 0$ for some $\vec{x} = \langle x_1, ..., x_{2^n} \rangle$ from such a set then $y_i = 0$ for all $\vec{y} = \langle y_1, ..., y_{2^n} \rangle$ from this set, since any point in the triangle (except $\vec{e}_k$) has to be of the form $k_1\vec{e}_k + k_2\vec{a} + k_3\vec{b}$ where the k_1, k_2, k_3 are non-negative, sum to 1 and k_2 and k_3 are non-zero. So for $x_i = 0$ we have to have $a_i = b_i = 0$ and thus also (considering that the entropy increases all the way along the line from $\vec{e}_k$ to $\vec{a}$) $i \neq k$. It follows by the Implicit Function Theorem that $\vec{g}(t)$ is $\mathcal{C}^\infty$ (see [Goursat, 1904])[9]. Also,

$$\frac{dE(\vec{g}(t))}{dt} = E(\vec{c}) - E(\vec{c'}) > 0.$$

By the uniform continuity of $\vec{g}(t)$ and its derivative on $[0, 1]$ there are numbers $0 = r_1 < r_2 < ... < r_s = 1$ such that for $j = 1, 2, ..., s-1$ and $u \in [0, 1]$, when $t = (1-u)r_j + ur_{j+1}$ both $\vec{g}(t)$ and $\frac{d\vec{g}}{dt}(t)$ are close to $(1-u)\vec{g}(r_j) + u\vec{g}(r_{j+1})$ and $\frac{\vec{g}(r_{j+1}) - \vec{g}(r_j)}{r_{j+1} - r_j}$ respectively. Hence, since

$$(17)\quad \frac{dE(\vec{g}(t))}{dt} = -\sum(1 + \log(g_i(t)))\frac{dg_i}{dt}(t) = E(\vec{c}) - E(\vec{c'}) > 0,$$

where the sum is over the i for which x_i is non-zero when $\vec{x}$ is in the triangle $\vec{e}_k, \vec{c}, \vec{d}$ (excepting $\vec{e}_k$), and the derivative with respect to u of $E((1-u)\vec{g}(r_j) + u\vec{g}(r_{j+1}))$ is

$$-\sum(1 + \log((1-u)g_i(r_j) + ug_i(r_{j+1})))(g_i(r_{j+1}) - g_i(r_j)),$$

[9] Consider $G(t, s) = E(s\vec{e}_k + (1-s)((1-t)\vec{c} + t\vec{d}))$ on $[0, 1] \times [\epsilon, 1]$ where ϵ is positive and small. G is $\mathcal{C}^\infty$. By the above remarks, for each $t \in [0, 1]$ there is a unique $s \in [\epsilon, 1]$, $s = h(t)$, such that $G(t, s) = (1-t)E(\vec{c'}) + tE(\vec{c})$. By the Implicit Function Theorem, h is $\mathcal{C}^\infty$. We have $\vec{g}(t) = h(t)\vec{e}_k + (1 - h(t))((1-t)\vec{c} + t\vec{d})$.

it follows that E is also increasing on the line segment $[\vec{g}(r_j), \vec{g}(r_{j+1})]$.
If E is not increasing on $[\vec{e}_k, \vec{c}]$ then we need to pick $\vec{c'}$ to be on the tangent from $\vec{e}_k$ to the contour line $\{\vec{x} : E(\vec{x}) = E(\vec{c})\}$ close to their meeting point $\vec{m}$, on the same side of $\vec{m}$ as $\vec{e}_k$. Then, again, E is increasing on $[\vec{a}, \vec{c'}]$ and $E(\vec{c'}) < E(\vec{c})$, and we can argue similarly as above, letting $\vec{g}(t)$ for $t \in [0, 1]$ be that (unique) point on the line segment $[\vec{e}_k, (1-t)\vec{m} + t\vec{d}]$ such that (16) holds (note that $\vec{m}$ takes on the role of $\vec{c}$). Omitting details, we can again obtain $0 = r_1 < r_2 < ... < r_s = 1$ with $g(0) = g(r_1) = \vec{c'}$ and $g(1) = g(r_s) = \vec{d}$ such that E is increasing on the line segment $[\vec{g}(r_j), \vec{g}(r_{j+1})]$ for each $j = 1, ..., s-1$.

In summary, what we now have is that E is increasing on each of the line segments

$$[\vec{a}, \vec{c'}], [\vec{c'}, \vec{g}(r_2)], ..., [\vec{g}(r_{s-2}), \vec{g}(r_{s-1})], [\vec{g}(r_{s-1}), \vec{d}]$$

so by Lemmas 15 and 10 $\vec{a} \prec \vec{c'} \prec \vec{d}$ and $\vec{d} \in N(K_{\vec{a},\vec{c'},\vec{d}})$. Since $\vec{d}$ can be arbitrarily close to $\vec{b}$ (keeping $\vec{c'}$ fixed), by Continuity $\vec{b} \in N(K_{\vec{a},\vec{c'},\vec{b}})$, so $\vec{c'} \preceq \vec{b}$. Hence, since $\vec{a} \prec \vec{c'}$, by Lemma 10, $\vec{a} \prec \vec{b}$.

We now have the required result in the case where $\vec{a}, \vec{b}$ are in the same 'corner'. To prove the result in general suppose that $\vec{a}$ is in the corner determined by $\vec{e}_k$ and $\vec{b}$ is in the corner determined by $\vec{e}_j$. Then if τ is the permutation which simply transposes k and j, for each σ, $\sigma\vec{a}$ and $\sigma\tau\vec{b}$ will be in the same corner. Let K' be a set of constraints such that

$$V(K') = \{ \sigma\vec{a}, \sigma\vec{b} \mid \sigma \text{ a permutation of } 1, ..., 2^n \}.$$

If $\sigma\vec{a} \in N(K')$ for some permutation σ then by Obstinacy, $\sigma\vec{a} \in N(K_{\sigma\vec{a},\sigma\tau\vec{b}})$ which contradicts the special case proved above. Hence we must have $\sigma\vec{b} \in N(K')$ for some σ. But then, by Renaming, $\vec{b} \in N(\sigma K') = N(K')$. Finally, by Obstinacy,

$$N(K_{\vec{a},\vec{b}}) = N(K' \cup K_{\vec{a},\vec{b}}) = N(K') \cap \{\vec{a}, \vec{b}\} = \{\vec{b}\},$$

as required. ■

COROLLARY 17. *For* $\vec{a}, \vec{b} \in \mathbb{D}_n$,

$$\vec{a} \preceq \vec{b} \iff E(\vec{a}) \leq E(\vec{b}).$$

Proof. The result from left to right follows by Lemma 16. In the other direction it is enough to show that if $E(\vec{a}) = E(\vec{b})$ then $\vec{a} \preceq \vec{b}$. So suppose $E(\vec{a}) = E(\vec{b})$ and $\vec{a} \neq \vec{b}$ (consequently $\vec{b}$ is not the overall maximum entropy point). Let $\vec{b}_m$ be points such that $\lim_{m\to\infty} \vec{b}_m = \vec{b}$ and each $E(\vec{b}_m) > E(\vec{b})$. Then by Lemma 16 $\vec{b}_m \in N(K_{\vec{a},\vec{b}_m})$ so by Continuity $\vec{b} \in N(K_{\vec{a},\vec{b}})$, equivalently $\vec{a} \preceq \vec{b}$, as required. ■

As an immediate corollary we finally have the main theorem of this paper.

THEOREM 18. *EME is the unique extended inference process satisfying the (extended) principles of Irrelevant Information, Equivalence, Renaming, Relativization, Obstinacy, Independence and Continuity.*

Proof. Let $K \in ECL$ and $\vec{b} \in V(K)$, $\vec{a} \in N(K)$. By Lemma 11 and Corollary 17,

$$\vec{b} \in N(K) \iff \vec{a} \preceq \vec{b} \iff E(\vec{a}) \leq E(\vec{b}).$$

From this it follows that $N(K)$ must be the set of points in $V(K)$ of (global) maximum entropy. ■

5 CONCLUSION AND DISCUSSION

To sum up the main result of this paper we started off by assuming that the belief values agents assigned to sentences were subjective probabilities (in the sense that they satisfy (P1-2)) and that these agents assigned such values solely on the basis of knowledge bases consisting of finite sets of polynomial constraints on the their beliefs, followed if necessary by a final random choice between equally preferred or acceptable alternatives. Indeed, more globally, we identified an agent with a *process*, called an extended inference process, which given a knowledge base K produced, on the basis of K, a set of equally acceptable (to the agent) probability functions satisfying K, the ultimate working choice of one of these being left to the agent[10]. Within this framework we then showed that if an agent (as an extended inference process) was required to abide by the 'common sense' principles of Irrelevant Information, Equivalence, Renaming, Obstinacy, Relativization, Independence and Continuity then on such a knowledge base K the agent's set of equally acceptable alternatives *must* be $EME(K)$. In consequence if the set of possible solutions to K was convex (as is the case for the linear knowledge bases considered in [Paris and Vencovská, 1990] and [Paris and Vencovská, 1997]) then $EME(K)$ is a singleton (the unique global maximum entropy solution) and the agent has no choice.

The first point to note about the proof of this result is that we hardly used the fact that we were allowed polynomial constraints. All we really needed was that we had at least all finite sets of linear constraints (in particular to derive Lemma 7), that the constraints yielded closed (in Euclidean Space) sets $V(K)$ and that these subsets of $\mathbb{D}_n$ included amongst their number all finite sets of probability functions (on the same $SL(p_{i_1}, p_{i_2}, ..., p_{i_n})$). In other words we very largely got the result for polynomial constraints *for free*. Indeed, provided these minimal requirements

[10]This relationship between an extended inference process and the knowledge base is directly analogous to the relationship between an expert system shell and the specific data inputted. And just as with any particular 'real' agent, in practice this shell will never meet more than a finite number of actual data sets. Nevertheless it is in principle structured as a *process* for dealing with *any* of the possible data sets.

are satisfied we can add in any other sorts of constraints that we wish, and still draw the same conclusion.

The idea of allowing very exotic sets of constraints as knowledge bases here may appear from the outside rather unnatural. After all, one might add, how could such a knowledge base ever have arisen? However, be that as it may, we would remark that the choice of allowed knowledge bases is an external matter. Under the standing Watts Assumption (see [Paris, 1994]), that the knowledge base is *all the agent's knowledge*, it is of no concern to the agent why a certain knowledge base should be considered nor where it came from.

This point can be illustrated with an example which is sometimes cited as a criticism of the maximum entropy inference process. Suppose we had allowed also constraints of the form

$$\sum_{j=1}^{r} a_j w(\theta_j) \leq b$$

and consider for L the language with just one propositional variable p the knowledge base

$$K = \{ w(p) \leq 1/2 \}.$$

ME (and EME) here give the single solution $w(p) = 1/2$ which has been criticized as 'unreasonable' on the grounds that the chosen solution is right at one end of the range of possibilities. Surely, the argument goes, the solution $w(p) = 1/4$, nestling as it does comfortably in the middle of the range, is more reasonable. We would maintain however that this answer only appears reasonable (if indeed it ever does) because of some possibly unconscious assumptions about *where* this constraint could possibly have come from in the first place. And perhaps the most obvious such assumption is that we have obtained empirical evidence that some natural, objective, probability lies in the interval $[0, 1/2]$. In that case the answer $1/4$ might initially look appealing[11] for this probability, or more precisely as an estimate of this probability. (Notice that Renaming cannot be employed here, we cannot rename 0 as $1/2$!!) But of course there are other, less immediate perhaps, explanations that one could think up for the origin of the knowledge base $w(p) \leq 1/2$ which would entirely justify $w(p) = 1/2$ or $w(p) = 0$ or indeed any other value in between. Fortunately we don't have to struggle with the thorny problem of which of these explanations is the most likely, *our agent* isn't aware of any of them!

At this point the reader might be inclined to wonder whether the results in this paper have any relevance to the real world. After all in *practice* we have so much 'knowledge' about anything that the Watts Assumption can rarely, if ever, be applicable[12] as far as we are concerned[13]. This is a point well taken. However two

[11]But see [Paris *et al.*, 2000] for an alternate view.

[12]Not to mention the inherent computational difficulties even with this assumption, see [Maung and Paris, 1990].

[13]Given this it is intriguing how we ever come to recognize, and (it seems) largely agree upon, what constitutes common sense at all!

comments are worth making. Firstly, for expert systems, as examples of agents who we would wish to act common-sensically, the assumption is much more pertinent. Secondly, the criticism that the assumption does not in practice apply to higher agents such as ourselves seems based in part on the practical difficulty that we cannot ever hope to specify all our knowledge in a particular domain rather than that our knowledge is somehow unbounded. If we assume that our knowledge is in the form of finitely many constraints (though possibly far more than we could ever elucidate) on our beliefs then the results in this paper would, at least theoretically, apply.

Furthermore even if we do not subscribe to the view that all our 'knowledge' is of this type (see [Paris and Vencovská, 1990] and [Paris and Vencovská, 1993] for an alternative view) the form of Theorem 18, that rationality or common sense requirements may essentially constrain the available patterns of reasoning to a single tightly knit family separated only by a final random choice, is interesting in that it holds out the prospect of similar theorems even in the case of more general and realistic knowledge bases.

A second issue concerning Theorem 18 is our interpretation of what comprises 'common sense'. Whilst, in [Paris, 1999] particularly, we have given some explanations as to why *we* find certain principles common-sensical, as matters currently rest it seems ultimately up to the individual reader to draw his/her own conclusions. In these deliberations however we would urge the reader to consider the rationality of *flaunting* any of these principles (within this framework).

Our case in support of these principles has not been greatly helped by some of the innovations which have occurred in the space between [Paris and Vencovská, 1990] and this current paper. Two notions which earlier seemed entirely reasonable have become incompatible with the other principles.

The first is Renaming as in [Paris, 1999], which as we have already seen in Courtney's Paradox leads to a contradiction in this extended setting. The rethink which that paradox precipitated led to the notion of an extended inference process. Nevertheless one might feel that one aspect of renaming should still prevail, namely that highly symmetric solutions to K should be preferred to less symmetric solutions. Unfortunately with the obvious interpretation of 'symmetric' this proto-principle is inconsistent with the others listed in Theorem 18. To see this let $\vec{a}_1, \vec{a}_2, \vec{a}_3 \in \mathbb{D}_2$ be, respectively,

$$\langle \tfrac{1}{6}, \tfrac{1}{6}, \tfrac{1}{3}, \tfrac{1}{3} \rangle, \ \langle \tfrac{12}{66}, \tfrac{13}{66}, \tfrac{21}{66}, \tfrac{20}{66} \rangle, \ \langle \tfrac{13}{66}, \tfrac{12}{66}, \tfrac{20}{66}, \tfrac{21}{66} \rangle.$$

Then $E(\vec{a}_1) < E(\vec{a}_2) = E(\vec{a}_3)$ so by Lemma 11 and Corollary 17 $\vec{a}_1 \notin EME(K_{\vec{a}_1,\vec{a}_2,\vec{a}_3})$ despite the fact that $\vec{a}_1$ is clearly the most 'symmetric' solution of $K_{\vec{a}_1,\vec{a}_2,\vec{a}_3}$. [Nevertheless not all is lost, as remarked earlier, by Lemma 5 we do still retain the quintessential symmetry principle, Laplace's 'principle of indifference'.]

The second example of an incompatible, yet at the same time ostensibly common-sensical, principle is *Open-mindedness*, the requirement that if it is consistent with

K that the belief in θ is not zero (i.e. $K \cup \{w(\theta) > 0\}$ is consistent) then $w(\theta) > 0$ for $w \in N(K)$. In other words *don't entirely dismiss belief in θ unless you have to*. On the face of it this seems entirely reasonable and it was included as one of the principles in [Paris and Vencovská, 1990], though not in [Paris, 1999] since by then (for linear constraints) it was known to be derivable from the other principles. Unfortunately the Open-mindedness principle is inconsistent with the principles used in the main theorem. To see this it is enough to note that we can find $\vec{a}, \vec{b} \in \mathbb{D}_n$ with a_1 equal to zero, all the b_i strictly greater than zero yet $E(\vec{a}) > E(\vec{b})$. Hence, by Theorem 18, $EME(K_{\vec{a},\vec{b}}) = \{\vec{a}\}$, despite the fact that $a_1 = 0$.

At the very least the inconsistency of these last two principles with the others listed in Theorem 18 provide a clear warning that the title of 'common sense' should not be conferred too lightly [14]. Of course none of the principles of Theorem 18 could ever suffer a similar fate since, being satisfied by EME, we know that as a body they are mutually consistent. Nevertheless this poses the intriguing question whether or not there might be other sets of arguably common sense principles uniquely characterizing other extended inference processes.

Apart from the questions already raised in this discussion Theorem 18 opens up a number of other avenues for exploration. For once we have a characterization of what it means to reason common-sensically, or equivalently, as we would argue, *intelligently*, then we can consider the implications of this for processes such as updating and induction (see for example [Paris and Vencovská, 1992], [Paris and Vencovská, 1998], [Paris and Wafy, 2001], [Paris *et al.*, 2001], which followed from [Paris and Vencovská, 1990]). It would also be interesting to find more natural sets of principles and/or constraint sets K to replace those in Theorem 18 analogous to the simplifications of [Paris and Vencovská, 1990] presented in [Paris and Vencovská, 1997] and [Paris, 1999]. These however remain questions for future research.

Department of Mathematics, University of Manchester, UK.

BIBLIOGRAPHY

[Courtney, 1992] P. Courtney. *Models of Belief*, Doctorial Thesis, Manchester University, 1992.

[Csiszár, 1989] I. Csiszár. Why least squares and maximum entropy? An axiomatic approach to inverse problems, *Mathematics Institute of the Hungarian Academy of Sciences*, Preprint No.19/1989.

[Goursat, 1904] E. Goursat. *A Course in Mathematical Analysis*, Ginn and Company, 1904.

[Kern-Isberner, 1998] G. Kern-Isberner. Characterising the principle of minimum cross-entropy within a conditional-logical framework, *Artificial Intelligence*, **98**:169–208, 1998.

[Maung and Paris, 1990] I. Maung and J. B. Paris. A note on the infeasibility of some inference processes, *International Journal of Intelligent Systems*, **5**:595–604, 1990.

[Paris, 1994] J. B. Paris. *The Uncertain Reasoner's Companion - A Mathematical Perspective*, Cambridge University Press, Cambridge, UK, 1994.

[Paris, 1999] J. B. Paris. Common sense and maximum entropy, *Synthese*, **117**:119–132, 1999.

---

[14] See also concluding section of [Paris, 1999] where we consider the problem of formalizing Van Fraassen's 'Symmetry Slogan' from [van Fraassen, 1989].

[Paris and Vencovská, 1989] J. B. Paris and A. Vencovská. Maximum entropy and inductive inference. In *Maximum Entropy and Bayesian Methods*, Ed. J.Skilling, Kluwer Academic Publishers, pp. 397–403, 1989.

[Paris and Vencovská, 1990] J. B. Paris and A. Vencovská. A note on the inevitability of maximum entropy, *International Journal of Approximate Reasoning*, **4**:183–224, 1990.

[Paris and Vencovská, 1990] J. B. Paris and A. Vencovská. Modelling belief, *Proceedings of the Evolving Knowledge Conference*, British Society for the Philosophy of Science, Reading University Pitman Press, pp. 133–154, 1990.

[Paris and Vencovská, 1992] J. B. Paris and A. Vencovská. A method of updating justifying minimum cross entropy, *International Journal of Approximate Reasoning*, **7**:1–18, 1992.

[Paris and Vencovská, 1993] J. B. Paris and A. Vencovská. A model of belief, *Artificial Intelligence*, **64**:197–241, 1993.

[Paris and Vencovská, 1996] J. B. Paris and A. Vencovská. Some observations on the maximum entropy inference process, *Manchester Centre for Pure Mathematics Technical Report L1-96*, Department of Mathematics, Manchester University, UK, 1996.

[Paris and Vencovská, 1997] J. B. Paris and A. Vencovská. In defence of the maximum entropy inference process, *International Journal of Approximate Reasoning*, **17**:77–103, 1997.

[Paris and Vencovská, 1998] J. B. Paris and A. Vencovská. Proof systems for probabilistic uncertain reasoning, *Journal of Symbolic Logic*, **63**:1007–1039, 1998.

[Paris *et al.*, 2001] J. B. Paris, A. Vencovská and M. Wafy. Some limit theorems for ME, MD and CM^{∞}'. *Manchester Centre for Pure Mathematics Technical Report*, Department of Mathematics, Manchester University, UK, preprint number 2001/9, ISSN 1472-9210.

[Paris and Wafy, 2001] J. B. Paris and M. Wafy. On the emergence of reasons in inductive logic. *Logic Journal of the IGPL*, **9**, 207–216, 2001.

[Paris *et al.*, 2000] J. B. Paris, P. N. Watton and G. M. Wilmers. On the structure of probability functions in the natural world. *International Journal of Uncertainty, Fuzziness and Knowledge-Based Systems*, **8**:311–329, 2000.

[Shore and Johnson, 1980] J. E. Shore and R. W. Johnson. Axiomatic derivation of the principle of maximum entropy and the principle of minimum cross-entropy. *IEEE Transactions on Information Theory*, **IT-26**:26–37, 1980.

[van Fraassen, 1989] B. Van Fraasssen. *Laws and Symmetry*, Clarendon Press, Oxford, UK, 1989.

JAMES CUSSENS

INTEGRATING PROBABILISTIC AND LOGICAL REASONING

1 INTRODUCTION

1.1 Probability and Artificial Intelligence

The ability to perform reasoning with uncertainty is a prerequisite for intelligent behaviour. Consequently there has been considerable Artificial Intelligence (AI) research into representing and reasoning with uncertainty. Given that there have been several centuries of successful applications of probability to uncertain reasoning, it would seem a natural tool for uncertainty in AI. However, in 1969 McCarthy and Hayes produced an influential paper [McCarthy and Hayes, 1969], which proclaimed that probabilities were "epistemologically inadequate" and much early AI work on uncertainty accepted this argument.

More recently though, probabilistic approaches to AI uncertainty handling have been used extensively, indeed they are now dominant, as recent proceedings of the *Uncertainty in Artificial Intelligence* conference demonstrate. The success of Bayesian networks as a computationally feasible way of combining structural and quantitative aspects of probability distributions has been instrumental in this probabilistic resurgence.

An important part of the work on probability and AI investigates the logical formalisation of probabilistic reasoning. The goal is to use a logical analysis to answer (or at least clarify) some of the following questions:

1. What are the semantics behind probabilistic reasoning?

2. How should probabilities be updated in the light of new evidence (which may itself be uncertain)?

3. If 90% of Swedes are Protestants, does it follow that the probability that a particular Swede of unknown religion, say, Henrik Boström, has probability 0.9 of being Protestant?

4. Is the probability calculus alone sufficient for uncertain reasoning? If not, what connections are there with other uncertainty formalisms, e.g. belief functions, fuzzy logic and nonmonotonic logic?

D. Corfield and J. Williamson (eds.), Foundations of Bayesianism, 241–260.

1.2 Halpern's logic of probability

Halpern's *An Analysis of First-Order Logics of Probability* [Halpern, 1990] is an important contribution to the first question, where Halpern presents two approaches to giving semantics to (formalised) probability statements. The statement

The probability that a randomly chosen bird flies is greater than 0.9

is formalised as

(1) $w_x(Flies(x)|Bird(x)) > 0.9$

The semantics of such a formula are defined in terms of a *type-1 probability structure* which is a standard first-order model with an associated probability function μ on its domain. Formula 1 is satisfied by any type-1 probability structure where the probability is greater than 0.9 that a domain element, constrained to be in the extension of $Bird$ and sampled according to μ, is in the extension of $Flies$.

In contrast, the statement

The probability that Tweety (a particular bird) flies is greater than 0.9.

is formalised as

(2) $w(Flies(Tweety)) > 0.9$

For such formulae, semantics are defined in terms of a probability distribution over first-order models ("possible worlds") rather than on some particular first-order model. Formula 2 is satisfied by any such distribution that gives probability mass greater than 0.9 to the set of those models in which $Flies(Tweety)$ is true.

These two semantics loosely correspond to the objective and subjective interpretations of probability found in the philosophical literature. Using this logical formalisation, Halpern and colleagues, in a series of papers, have clarified many of the issues involved in probabilistic reasoning. Most importantly, by connecting the two semantics above we can address the "Protestant Swedes" issue in a rigorous manner [Bacchus *et al.*, 1996]. The clarity of such work is often in contrast to much philosophical work on probability which can be "dry, obscure, contrary to all ordinary ideas, and on top of that prolix"[1] See, for example, the Popper-Miller argument concerning the connections between deductive and probabilistic relations [Popper and Miller, 1987]. The Popper-Miller position is criticised in [Cussens, 1996].

1.3 Probabilistic inference is deduction

In addition to such foundational work, considerable attention has been devoted to the practical issues of incorporating probabilistic and logical inference into some

[1]This is Kant's description of his own *Critique of Pure Reason*.

overarching framework. The alleged inability of logic to represent uncertain information is often cited as a motivation for the augmentation of a logical calculus (e.g. first-order predicate calculus) with additional probabilistic capabilities.

First-order predicate calculus is concerned with valid logical inference: when is it certain that the truth of one statement ensures the truth of another? This focus on certainty may seem at odds with probabilistic reasoning. However, standard probabilistic inference, by which is meant the inference of probabilities from probabilities, is also exclusively concerned with deductive, certain reasoning. It is not an "inductive logic"—whatever that might be. For example given that $P(b|a) = 0.3$ and $P(a) = 0.8$, it follows with completely certainty that $P(a \wedge b) = 0.24$.

A logical formalisation of probabilistic inference can therefore be carried out in the same way as the formalisation of other branches of mathematics, such as arithmetic: relevant axioms are added to those of first-order predicate calculus. Unfortunately, as Gödel's celebrated incompleteness result shows, not even arithmetic has a *complete* axiomatisation. Naturally any axiomatisation for probability (such as supplied by Halpern [1990]) can not hope to be complete. Indeed, Abadi and Halpern also show that it "takes very little to make reasoning about probability highly undecidable" [Abadi and Halpern, 1994]. Despite these rather discouraging results, a number of researchers have described systems which incorporate logic and probability in a restricted setting, and it to this work that the remainder of the paper is devoted. In Section 2, *Stochastic Logic Programs* are proposed as one method of integrating logical and probabilistic reasoning. A number of alternative approaches are then considered in Section 3; and in Section 4, there is a summary and possible future work in this area is considered.

2 LOGLINEAR MODELS FOR FIRST-ORDER REASONING

Here a conservative extension to the logic programming framework is proposed by defining probabilities directly on proofs and hence indirectly on goals and atomic formulae. (This work has been presented previously in Cussens [1999a; 1999b; 2000a; 2000b; 2001a; 2001b].) This conservatism allows us to tie probabilistic and logical concepts very closely. Table 1 lists the linkages which the proposed approach establishes.

2.1 *Probabilistic Constraint Logic Programming*

In [Riezler, 1997], Riezler develops Abney's work [Abney, 1997] defining a loglinear model on the *proofs* of formulae. This requires defining features on these proofs (denoted f_i) and defining the model parameters (denoted λ_i). The probability of any proof ω is then $P(\omega) = Z^{-1} \exp(\sum_i \lambda_i f_i(\omega))$ where Z is a normalising constant. This approach applies very naturally to natural language processing (NLP). In NLP, a proof that a sentence belongs to a language amounts to a parse

Logic	Probability
logical variable	random variable
instantiation	instantiation
relations	joint distributions
queries	queries
ground definitions	probability tables
disjunctive definitions	mixture models
defining relations in terms of other relations	defining distributions in terms of other distributions

Table 1. Linking logic and probability

of that sentence, and the loglinear model can be used to find the most likely parse of any particular sentence.

Riezler extends the improved iterative scaling algorithm of Della Pietra, Della Pietra and Lafferty [Della Pietra *et al.*, 1997] to induce features and parameters for a loglinear model from incomplete data. Incomplete data here consists of just atomic formulae, rather than the proofs of those atomic formulae. In an NLP context this means having a corpus of sentences rather than sentences annotated with their correct parses. The former is a considerably cheaper resource than the latter, so this sort of incomplete data handling is of considerable practical importance.

2.2 *Stochastic Logic Programs*

Riezler's framework allows arbitrary features of proofs, and recent experiments have used features "indicating the number of argument-nodes or adjunct-nodes in the tree, and features indicating complexity, parallelism or branching-behaviour" (Stefan Riezler, personal communication).

Here a special case of Riezler's framework is examined, where *the clauses (rules) used in a proof are the features used to define the probability of that proof.* Eisele [1994] examined this approach from an NLP perspective. Muggleton [1996] introduced *stochastic logic programs (SLPs)*, approaching the issue from a general logic programming perspective, with a view to applications in Inductive Logic Programming. Definition 1 defines SLPs and Figure 1 shows $\mathcal{S}_0$ a simple example SLP.

DEFINITION 1. A *stochastic logic program (SLP)* $\mathcal{S}$ is a definite logic program where some of the clauses are parameterised with non-negative numbers. A *pure SLP* is an SLP where all clauses have parameters, as opposed to an *impure* SLP where not all clauses have parameters. A *normalised SLP* is one where parameters for clauses whose heads share the same predicate symbol sum to one. If this is not the case, then we have an *unnormalised SLP*.

```
0.4:s(X) :- p(X), p(X).      0.3:p(a).     0.2:q(a).
0.6:s(X) :- q(X).            0.7:p(b).     0.8:q(b).
```

Figure 1. $\mathcal{S}_0$: A simple pure, normalised SLP.

Definition 1 generalises that found in [Muggleton, 1996], where Muggleton requires SLPs to be range-restricted and with labels for the same predicate summing exactly to one. Also, Muggleton does not use SLPs to define a loglinear model as is done here.

In [Eisele, 1994] and [Muggleton, 1996], Stochastic Context-Free Grammars (SCFGs) were 'lifted' to feature grammars (FGs) and logic programs (LPs) respectively. SCFGs are context-free grammars where each production is labelled, such that the labels for a particular non-terminal sum to one. The probability of a parse is then simply the product of the labels of all production rules used in that parse. Sentence probabilities are given by the sum of all parses of a sentence. The distributions so defined are special cases of loglinear models where the grammar rules define the features f_i and their labels are the parameters $l_i = \exp(\lambda_i)$. The normalising constant Z is guaranteed to be one. This is because the labels for each non-terminal sum to one and because the context-freeness ensures that *derivations never fail*—when generating a sentence from an SCFG a production rule can always be applied to a nonterminal. Because of this a number of techniques (such as the inside-outside algorithm for parameter estimation [Lari and Young, 1990]) can be applied to SCFGs, but cannot be applied directly to stochastic versions of FGs or LPs. (See Abney [1997] for a demonstration of this.)

In the rest of this section an account of (i) the distributions defined by an SLP and (ii) the relations between SLPs and some other well-known probabilistic models are given. For more details on SLPs including accounts of sampling, inference, parameter estimation and semantic analysis see the following papers: [Cussens, 1999a; Cussens, 1999b; Cussens, 2000a; Cussens, 2000b; Cussens, 2001a; Cussens, 2001b; Muggleton, 1996; Muggleton, 2000]. The presentation of SLPs assumes some familiarity with logic programming concepts.

Defining distributions over derivations

An SLP $\mathcal{S}$ with parameters λ together with a goal G defines up to three different related distributions: $\psi_{(\lambda,\mathcal{S},G)}$, $f_{(\lambda,\mathcal{S},G)}$ and $p_{(\lambda,\mathcal{S},G)}$, defined over derivations, refutations and atomic formulae, respectively. Definition 2 defines the distribution over derivations and is an adaptation of a definition found in [Riezler, 1998].

DEFINITION 2. Let G be a goal, and $\mathcal{S}$ be a pure normalised SLP. $\mathcal{S}$ defines a probability distribution $\psi_{(\lambda,\mathcal{S},G)}(x)$ on the set $D(G)$ (the set of derivations starting with G using $\mathcal{S}$) s.t. for all $x \in D(G)$:

$$\psi_{(\lambda,\mathcal{S},G)}(x) = e^{\lambda \cdot \nu(x)} = \prod_{i=1}^{n} l_i^{\nu_i(x)}$$

$\lambda = (\lambda_1, \ldots, \lambda_n) \in \Re^n$ is a vector of log-parameters where λ_i is the log of l_i, the parameter attached to the ith parameterised clause,

$\nu = (\nu_1, \ldots, \nu_n)$ is a vector of clause counts s.t. for each $\nu_i : D(G) \to N \cup \{\infty\}$, $\nu_i(x)$ is the number of times the ith parameterised clause is used as an input clause in derivation x,

$\lambda \cdot \nu(x)$ is a weighted count s.t. $\lambda \cdot \nu(x) = \sum_{i=1}^{n} \lambda_i \nu_i(x)$,

Usually it will be clear which SLP and goal is being used to define the distribution, so we will often abbreviate $\psi_{(\lambda,\mathcal{S},G)}(x)$ to $\psi_\lambda(x)$.

Defining distributions over refutations

In application of SLPs, the main focus of interest is not $\psi_{(\lambda,\mathcal{S},G)}(x)$ but the conditional distribution $\psi_{(\lambda,\mathcal{S},G)}(x|x \in R(G))$, the distribution over derivations given that each is a refutation. We will denote this distribution by $f_{(\lambda,\mathcal{S},G)}$:

$$f_{(\lambda,\mathcal{S},G)}(x) \stackrel{\text{def}}{=} \psi_{(\lambda,\mathcal{S},G)}(x|x \in R(G))$$

Let

$$Z_{(\lambda,\mathcal{S},G)} \stackrel{\text{def}}{=} \sum_{x \in R(G)} \psi_{(\lambda,\mathcal{S},G)}(x) = \psi_{(\lambda,\mathcal{S},G)}(R(G))$$

then

$$f_{(\lambda,\mathcal{S},G)}(x) = \begin{cases} Z^{-1}_{(\lambda,\mathcal{S},G)} \psi_{(\lambda,\mathcal{S},G)}(x) & \text{if } x \in R(G) \\ 0 & \text{if } x \in D(G) \setminus R(G) \end{cases}$$

$f_{(\lambda,\mathcal{S},G)}(x)$ is a *log-linear model* over refutations, defined for goals where $Z_{(\lambda,\mathcal{S},G)} > 0$. To see this, consider Definition 3, where $f_{(\lambda,\mathcal{S},G)}$ is defined without reference to $\psi_{(\lambda,\mathcal{S},G)}$. The only slight alteration to a standard log-linear model is that it is extended so that derivations which are not refutations have probability zero, rather than being undefined.

DEFINITION 3. Let G be a goal such that $Z_{(\lambda,\mathcal{S},G)} > 0$, and $\mathcal{S}$ be a pure normalised SLP. $\mathcal{S}$ defines a log-linear probability distribution $f_{(\lambda,\mathcal{S},G)}(r)$ on the set $R(G)$ (the set of refutations of G using $\mathcal{S}$) s.t. for all $r \in R(G)$:

$$f_{(\lambda,\mathcal{S},G)}(r) = Z^{-1}_{(\lambda,\mathcal{S},G)} e^{\lambda \cdot \nu(r)} = Z^{-1}_{(\lambda,\mathcal{S},G)} \prod_{i=1}^{n} l_i^{\nu_i(r)}$$

$Z_{(\lambda,\mathcal{S},G)} = \sum_{r \in R(G)} e^{\lambda \cdot \nu(r)}$ is a normalizing constant,

λ, ν and $\lambda \cdot \nu(r)$ are defined as in Definition 2.

We extend the normal log-linear definition so that $f_{(\lambda,\mathcal{S},G)}(x) = 0$, if $x \in D(G) \setminus R(G)$. $f_{(\lambda,\mathcal{S},G)}(x)$ is thus a distribution over the whole of $D(G)$.

The distributions ψ_λ and f_λ are more easily understood by referring to the SLD-tree which underlies them. By way of example, Figure 2 shows an annotated SLD-tree for refutations of the goal $\leftarrow s(X)$ using $\mathcal{S}_0$. There are 6 derivations, of which 4 are successful and 2 are failures. The branches of the tree are labelled with (i) the unification effected by choosing clauses and (ii) the parameters attached to these clauses. Since $\mathcal{S}_0$ is pure and normalised ψ_λ is a probability distribution over derivations, and the tree shows how the probability mass of one is divided up as we move down the tree. To find $\psi_\lambda(x)$ for any derivation x we multiply the parameters on the branches corresponding to that derivation. Both failure derivations have probability 0.084, so $Z_{(\lambda,\mathcal{S}_0,\leftarrow s(X))} = 1 - 2 \times 0.084 = 0.832$. So, for example, if the leftmost refutation is r_1, then $f_{(\lambda,\mathcal{S}_0,\leftarrow s(X))}(r_1) = (0.4 \times 0.3 \times 0.3)/0.832 \approx 0.043$ (The tree assumes that the variable in the two $s/2$ clauses is renamed to X' and X''.)

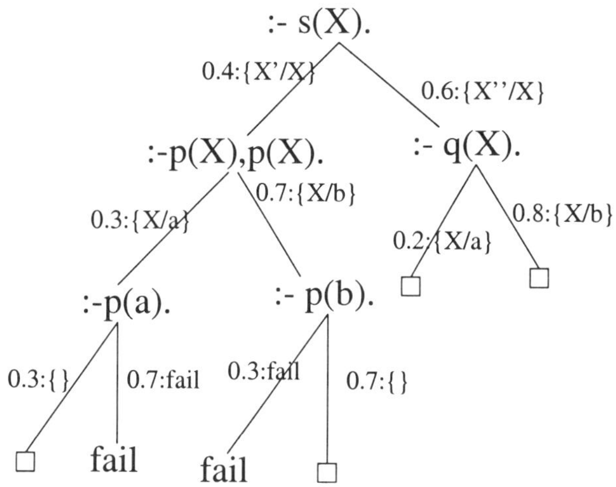

Figure 2. Annotated SLD-tree for $\mathcal{S}_0$.

Defining distributions over atoms

f_λ defines a distribution over atoms via marginalisation. First define the *yield* of a refutation and the *proofs* of an atom.

DEFINITION 4. The *yield* $Y(r)$ of a refutation r of unit goal $G = \leftarrow A$ is $A\theta$ where θ is the computed answer for G using r. The set of *proofs* for an atom y_k is the set $X(y_k) = \{r | Y(r) = y_k\}$. Note that $X(Y(r))$ is the set of all refutations that yield the same atom as r.

We only define yields with respect to unit goals. This is just a convenience, since given a non-unit goal $\leftarrow G_1, \ldots, G_M$, we can always add the clause $A' \leftarrow G_1, \ldots, G_M$, where A' contains all the variables of $G_1, \ldots, G_M$, and then consider yields of $\leftarrow A'$. Note that from a logical perspective a refutation of $\leftarrow A$ with computed answer θ amounts to a proof of $A\theta$, so this choice of terminology is natural.

We now define a distribution $p_{(\lambda,\mathcal{S},G)}$ over atoms in terms of their proofs.

$$(3) \quad p_{(\lambda,\mathcal{S},G)}(y_k) \stackrel{\text{def}}{=} \sum_{r \in X(y_k)} f_{(\lambda,\mathcal{S},G)}(r) = Z^{-1}_{(\lambda,\mathcal{S},G)} \sum_{r \in X(y_k)} e^{\lambda \cdot \nu(r)}$$

If G has t variables, then $p_{(\lambda,\mathcal{S},G)}(y_k)$ defines a t-dimensional distribution over variable bindings for these t variables. Note that we allow non-ground bindings unlike in [Muggleton, 1996; Cussens, 1999b; Muggleton, 2000]. We will see in Section 2.3 how we can use these t-dimensional distributions to encode probabilistic models using other formalisms into SLPs. Returning to the example SLP $\mathcal{S}_0$ we find that it defines a distribution over the sample space $\{s(a), s(b)\}$, where

$$p_{(\lambda,\mathcal{S}_0,\leftarrow s(X))}(s(a)) = (0.4 \times 0.3 \times 0.3 + 0.6 \times 0.2)/0.832 = 0.1875$$

and

$$p_{(\lambda,\mathcal{S}_0,\leftarrow s(X))}(s(b)) = (0.4 \times 0.7 \times 0.7 + 0.6 \times 0.8)/0.832 = 0.8125$$

2.3 *Relations to some existing probabilistic models*

In this section we encode three familiar probabilistic models into SLPs. Considerably more complex SLPs encoding, for example, distributions over a hypothesis space of logic programs are used in [Cussens, 2000b].

Figure 3 shows the *Asia* Bayesian network and an SLP representation of it, where $p_{(\mathcal{S},\lambda,\leftarrow asia(A,T,E,S,L,B,X,D))}$ gives the joint distribution represented by the Bayesian net. The equation

$$\begin{aligned} & P(A,T,E,S,L,B,X,D) \\ = \; & P(A)P(S)P(T|A)P(L|S)P(B|S)P(E|T,L)P(D|E,B)P(X|E) \end{aligned}$$

is directly encoded using an *impure, unnormalised* SLP, with each of the 8 conditional probability tables defined by a single predicate. Since E is a function of T and L, we only need 4 unparameterised clauses to encode $P(E|T,L)$ as opposed to the 8 that would be required if $P(E|T,L)$ were encoded as the other conditional probability distributions are. It is clear that any Bayesian net with discrete variables can be represented by an SLP in this manner.

The translation from Bayesian net to SLP is problematic in that the directionality of Bayesian nets is obscured. In contrast, the mapping between Markov nets and SLPs is transparent. Figure 4 shows a Markov net derived from the *Asia*

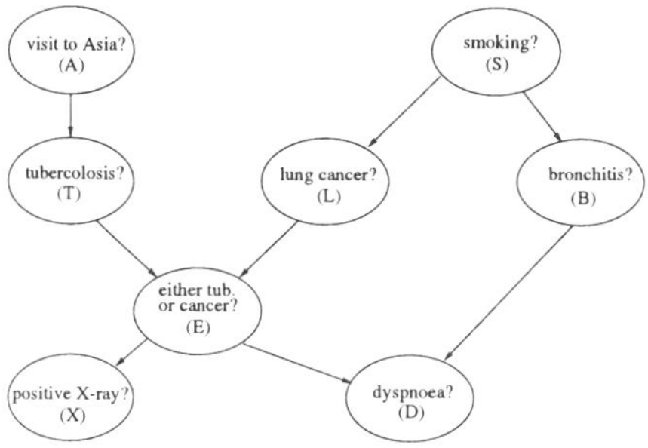


```
asia(A,T,E,S,L,B,X,D) :-
    a(A), s(S), t_a(T,A),
    l_s(L,S), b_s(B,S),
    e_tl(E,T,L), d_eb(D,E,B),
    x_e(X,E).

0.01:a(1).        0.99:a(0).
0.50:s(1).        0.50:s(0).

0.05:t_a(1,1). 0.95:t_a(0,1). 0.1:t_a(1,0). 0.9:t_a(0,0).
e_tl(0,0,0).   e_tl(1,0,1).   e_tl(1,1,0).   e_tl(1,1,1).
% l_s/2, b_s/2, d_eb/2 and x_e/2 definitions omitted
```

Figure 3. *Asia* Bayesian net and its encoding as an SLP.

Bayesian net and its translation to an impure unnormalised SLP. The structure of the Markov net can be completely described with a single clause, and the 6 clique potentials each get their own predicate symbol.

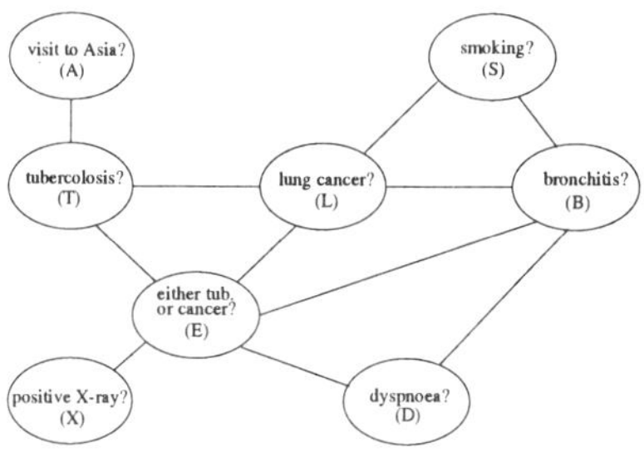


```
asia(A,B,D,E,L,S,T,X) :-
 c6(E,X), c5(E,B,D), c4(L,B,S),
 c3(L,E,B), c2(E,L,T), c1(A,T).

0.0005:c1(1,1). 0.0095:c1(1,0).
0.0099:c1(0,1) 0.9801:c1(0,0).
%Clauses for c2, c3,
%c4, c5, c6  omitted
```

Figure 4. *Asia* Markov net and its encoding as an SLP.

Since SLPs generalise stochastic context-free grammars (SCFGs) it is easy to encode SCFGs as SLPs. Consider the context-free grammar $S \rightarrow aSa \mid bSb \mid aa \mid bb$ which generates palindromes. By placing a probability distribution over the four productions we have an SCFG which defines a distribution over palindromic strings of as and bs. $\mathcal{S}_{\mathrm{PALINDROME}}$ in Figure 5 encodes such an SCFG as an SLP where $p_{(\lambda,\mathcal{S}_{\mathrm{PALINDROME}},\leftarrow s(X,[]))}$ is the distribution over strings. Hidden Markov models, which are essentially stochastic regular grammars, can be dealt with similarly. Figure 6 shows an SLP defining a distribution over the the language $\{a^n b^n c^n\}$ which is *not* context-free.

Consider now the SLP in Figure 7 which represents a simple linear Markov net. A is independent of C given B ($A \perp C|B$). For instance $P(A = foo|B = foo) = P(A = foo|B = foo, C = foo)$. This conditional independence phenomenon

```
0.5:s(A,D) :- A=[a|B], s(B,C), C=[a|D].  0.1:s([a,a|T],T).
0.3:s(A,D) :- A=[b|B], s(B,C), C=[b|D].  0.1:s([b,b|T],T).
```

Figure 5. $\mathcal{S}_{\text{PALINDROME}}$, an SLP representation of an SCFG.

```
anbncn(A) :- build(A-B,B-C,C-[]).

0.3: build(A-A,B-B,C-C).
0.7: build([a|A]-Ap,[b|B]-Bp,[c|C]-Cp) :-
        build(A-Ap,B-Bp,C-Cp).
```

Figure 6. Stochastic non context-free grammar defined with an SLP

is central to probabilistic graphical models such as Markov nets. But note that A is independent of C given *any* value of B. Sometimes such a strong assumption will not be justified. It may be that A is only independent of C given *particular values of* B. This *conditional conditional independence*[2] or *context-specific* independence between A and C crop up often in applications and has been investigated by Boutilier *et al* [Boutilier *et al.*, 1996].

```
linear(A,B,C) :-
      c1(A,B),
      c2(B,C).

0.2 : c1(foo,foo).        0.3 : c2(foo,foo).
0.1 : c1(foo,bar).        0.2 : c2(foo,bar).
0.3 : c1(bar,foo).        0.3 : c2(bar,foo).
0.4 : c1(bar,bar).        0.2 : c2(bar,bar).
```

Figure 7. Linear Markov net SLP

To represent context-sensitive independence, it is necessary to differentiate between these two sorts of values of B. Assume that there are two predicates, `strong/1` and `weak/1` defined to be mutually exclusive which achieve this. The SLP in Figure 8 then defines an appropriate mixture model. A neater alternative might be to use negation to differentiate, using `strong(B)` and `\+strong(B)`[3], but the use of negation in SLPs has yet to be properly investigated, hence the current restriction to definite clauses.

[2] conditional on a variable *and* conditional on values of that variable

[3] \+ is ISO Prolog notation for `not`.

```
mixlin(A,B,C) :-
      strong(B),
      c1(A,B),
      c2(B,C).

mixlin(A,B,C) :-
      weak(B),
      c3(A,B,C).

%labelled defs of c1, c2 and c3 omitted
```

Figure 8. Mixture model SLP defining context-specific independence

2.4 Representing degrees of belief

In an SLP, a labelled rule $l : p(X,Y) \leftarrow q(X,Y), r(Y)$ does not define the probability that some ground atomic formula $p(a,b)$ is *true* (as in KBMC, see Section 3.2), nor does it provide bounds on the probability that $p(a,b)$ is true as in [Shapiro, 1983; Ng and Subrahmanian, 1992]. Instead, there is a binary distribution associated with $p(X,Y)$ which defines the probability of instantiations such as $\{X/a, Y/b\}$. In order to reason about the probability of the truth of atomic formulae, atomic formulae are simply augmented by introducing an extra logical-random variable to represent the truth value of unaugmented atomic formulae, and then this logical-random variable is treated exactly as any other. This is in keeping with a conservative approach—if the truth value of an atomic formula as it varies across different "possible worlds" is of interest—then this variation is modelled in the standard way: with a random variable.

Consider a probabilistic rule from [Koller and Pfeffer, 1997], where the "rule says that when a person's parent has a gene, the person will inherit it with probability 0.5".

```
genotype(P,G) :- (0.5) parent(P,Q), genotype(Q,G).
```

It is possible to encode such "degree of belief" probability in an SLP with a Boolean truth-value variable as in Figure 9. However, this parameterised clause is only guaranteed to define the desired probability if it is the only clause for `genotype/3`.

To find the probability that $genotype(bob, big\_ears)$ is true, the SLP is used to compute the probabilities of $genotype(bob, big\_ears, 1)$ *and* $genotype(bob, big\_ears, 0)$. This amounts to demanding arguments (=proofs) for the truth of $genotype(bob, big\_ears)$ *and* for its falsity. The strength of these proofs are then balanced when deciding on the probability of truth.

```
genotype(P,G,T) :- parent(P,Q),
   genotype(Q,G,1), half(T).

0.5 : half(1).
0.5 : half(0).
```

Figure 9. Representing degree of belief with an extra variable

3 RESEARCH ON LOGIC AND PROBABILITY: A SAMPLE

Here a number of alternative approaches to integrating logic and probability are considered. This is not a comprehensive survey; but, in the spirit of statistical inference, it hopefully constitutes a sufficiently representative sample to allow suitably qualified inferences.

3.1 Probabilities in logic programming

Clark and McCabe

Adding an extra argument to atomic formulae to represent the degree of belief associated with such atomic formulae, as in Section 2.4, has a long history in logic programming, although the mechanisms for propagating probabilities through Prolog rules have often been *ad hoc*.

In [Clark and McCabe, 1982], the extra argument is a probability so that if the query $\leftarrow r(a, b, P)$ renders an answer substitution $P = 0.2$, this means that $r(a, b)$ holds with probability 0.2. Extra literals are added to the clause which compute the probability in the head atomic formula as a function of probabilities of the body atomic formulae: they embody a *combination rule*. There the combination rule is simple multiplication (which represents the independence assumptions in SLPs). Clark and McCabe also propose more complex combination rules where a query matches more than one clause head. Again it is up to the user to devise these rules. With SLPs only combination by mixing is allowed.

Obtaining a degree-of-belief probability (in, say, $r(a, b)$ being true) is both more direct and more flexible in [Clark and McCabe, 1982] than with SLPs. It is more direct, because there is a single extra argument representing degree of belief. With SLPs all refutations of $\leftarrow r(a, b, T)$ are found and the ratio of $T = 0$ to $T = 1$ answer substitutions is computed. (Assume that the SLP has been so written so that other substitutions are not possible.) When implementing an SLP in Prolog, there are two extra arguments, one for the truth-value and one for the clause parameter. In [Clark and McCabe, 1982], it is up to the "expert" to define how probabilities propagate up to the query, whereas SLPs are representations of particular sorts of distributions.

Shapiro's logic programs with uncertainties (LPUs)

Shapiro's approach is close to that of SLPs: the calculation of probabilities is carried out by a meta-interpreter and if all of Shapiro's *certainty factors* are 1, then operational and declarative semantics degenerate to those of standard logic programs.

LPUs involve certainty factors and certainty functions mapping multisets of certainty factors (e.g. the multiset of certainty factors from the body of a clause) to a single certainty factor. This is close to the Prolog implementation of SLPs, except that the SLP 'certainty function' (simple multiplication) is not necessarily monotonic increasing as Shapiro's have to be.

With an LPU, deducing $p(a)$ with certainty α is tantamount to proving that $Pr(p(a)) \geq \alpha$. Shapiro defines a minimal model for a LPU where each atomic formula is associated with the minimal probability with which that atomic formula is true. Consequently, the first clause below is redundant. With SLPs, there is no direct representation of degrees of belief and exact probabilities are computed, rather than bounds on probabilities. Also the probabilities associated with different refutations of a goal add up, so that neither clause would be redundant.

```
< p(a) :- true, 0.2>
< p(a) :- true, 0.3>
```

Shapiro introduces an interpreter which essentially searches for a proof that a query is true with probability above a certain threshold. It is possible to prune this search in many cases—this would be harder with SLPs since it is necessary to add up a large (possibly infinite) number of refutations of a goal to get its probability.

Probabilistic logic programming

Probabilistic Logic Programming (hereafter PLP) is an extension of Logic Programming developed by Ng and Subrahmanian in a series of papers [Ng and Subrahmanian, 1992; Ng and Subrahmanian, 1993; Ng and Subrahmanian, 1994]. In [Ng and Subrahmanian, 1992], the basic ideas of PLP are presented, in [Ng and Subrahmanian, 1993] variables ranging over probabilities are introduced and in [Ng and Subrahmanian, 1994] non-monotonic negation is added. As with Shapiro's LPUs, PLPs are designed for the propagation of constraints on probability values in the form of intervals, rather than the computation of exact probabilities as in SLPs.

In PLP, conjunctions and disjunctions of literals are annotated with probability intervals, allowing clauses such as (4–7).

$$eastbound(train1) : [0.7, 0.9] \leftarrow \tag{4}$$

$$bark(X) : [0.95, 1] \leftarrow dog(X) : [1, 1] \tag{5}$$

$$(bark(X) \wedge dog(X)) : [0.95 * V, 0.95 * V] \leftarrow dog(X) : [V, V] \tag{6}$$

$$not\_dog(X) : [1 - V_2, 1 - V_1] \leftarrow dog(X) : [V_1, V_2] \tag{7}$$

The probabilities here are to be interpreted as possible-world probabilities, and can be translated into Halpern's logic as follows:

$$
\begin{array}{lrcl}
(8) & w(eastbound(train1)) \in [0.7, 0.9] & \leftarrow & \\
(9) & \forall x : w(bark(x)) \in [0.95, 1] & \leftarrow & w(dog(x)) = 1 \\
(10) & \forall x, v : w(bark(x) \wedge dog(x)) = 0.95 * v & \leftarrow & w(dog(x)) = v \\
(11) & \forall x, v : not\_dog(x) \in [1 - v_2, 1 - v_1] & \leftarrow & w(dog(X)) \in [v_1, v_2]
\end{array}
$$

Ng and Subrahmanian introduce probability intervals, since point probability values for conjuncts and disjuncts are not derivable from their constituents. In general only intervals are derivable, as shown in (12) and (13).

$$
\begin{array}{ll}
 & P(\alpha) \in [\rho_\alpha, \delta_\alpha], P(\beta) \in [\rho_\beta, \delta_\beta] \\
(12) & \quad \models \; P(\alpha \wedge \beta) \in [\max\{0, \rho_\alpha + \rho_\beta - 1\}, \min\{\delta_\alpha, \delta_\beta\}] \\
 & P(\alpha) \in [\rho_\alpha, \delta_\alpha], P(\beta) \in [\rho_\beta, \delta_\beta] \\
(13) & \quad \models \; P(\alpha \vee \beta) \in [\max\{\rho_\alpha, \rho_\beta\}, \min\{1, \delta_\alpha + \delta_\beta\}]
\end{array}
$$

It is worth stressing that PLP can not express domain probabilities directly. A PLP statement such as

(14) $bronchitic(X) : [0.4, 0.4] \leftarrow smoker(X) : [1.0, 1.0]$

which translates to

(15) $\forall x : w(bronchitic(x)) = 0.4 \leftarrow w(smoker(x)) = 1$

does *not* state that 40% of smokers are bronchitic, but that for each particular smoker (i.e. someone who is a smoker in all worlds) there is exactly a 40% chance that that individual is bronchitic, a much stronger statement. Hence (15) would be contradicted by a single smoker $\mathbf{a}$ who is known definitely to be or not to be bronchitic, since, respectively, this would translate as $w(bronchitic(\mathbf{a})) = 1$ or $w(bronchitic(\mathbf{a})) = 0$.

3.2 *Logic and Bayes nets*

Knowledge-based model construction

Knowledge-based model construction (KBMC) [Ngo *et al.*, 1995] is described by Koller and Pfeffer [1997]:

> *Knowledge-based model construction (KBMC)* goes a considerable way towards bridging this gap [between logic and probability] by allowing a set of *first-order probabilistic logic (FOPL)* rules (first-order rules with associated probabilistic uncertainty parameters) to be used as a basis for generating Bayesian networks tailored to particular problem instances.

Normally, one constructs a fixed Bayesian network such as the "Asia" net, so that it is 'ready' to process any potential query. In KBMC, a Bayesian network is constructed *for a particular query*, and then that network is used to answer the query.

This query-dependence is built into logic programming with SLD-resolution—for each query there is a particular SLD-tree. Also, Prolog searches for a refutation in the tree, rather than constructing the complete tree and then looking for refutations. The latter is not even an option when the tree is infinite.

In *Knowledge-based model construction (KBMC)* [Ngo and Haddaway, 1997; Koller and Pfeffer, 1997; Haddaway, 1999] first-order rules with associated probabilities are used to generate Bayesian networks for particular queries. As in SLD-resolution queries are matched to the heads of rules, but in KBMC this results in nodes representing ground facts being added to a growing (directed) Bayesian network. A *context* is defined using normal first-order rules, perhaps explicitly as a logic program [Ngo and Haddaway, 1997], which specifies logical conditions for labelled rules to be used. The ground facts are seen as Boolean variables (either *true* or *false*). Once the Bayesian network is built it is then used to compute the probability that the query is true.

In KBMC, as in much of the work connecting logic and probability, parameterised first-order rules $\alpha : c(X) \leftarrow a(X)$ are connected to conditional probability statements such as $p(c(b)|a(b)) = \alpha$. Also the objective is to compute the probability that an atomic formula is true. With SLPs, the focus is more on undirected representations, so that $l : p(X,Y) \leftarrow q(X,Y), r(Y)$ forms part of the definition of a binary distribution associated with $p/2$ defined in terms of distributions associated with $q/2$ and $r/1$. *No* attempt to model causality is made. Despite these differences in approach there are clear similarities between KBMC query-specific Bayes net construction and the query-specific exploration of an SLD-tree by Prolog which deserve further investigation.

Ngo and Haddaway

In [Haddaway, 1999], Haddaway states that "...a Bayesian net is essentially a propositional representation of the domain: each node represents a multi-valued propositional variable. Thus it is not possible to express general relationships among concepts without enumerating all the potential instances in advance". The existence of nodes representing random variables with countably and uncountably infinite domains would seem to contradict this view. Such nodes are commonplace in many applications (see, for example, the documentation accompanying the BUGS system [Spiegelhalter *et al.*, 1996]). In any case, SLPs take a different view of the connection between logical and random variables: the nodes (random variables) in a Bayes net are identified with logical variables, ranging over a possibly infinite domain.

In [Ngo *et al.*, 1995] Haddaway uses a normal logic program as a *context base*. Since random variables are functions (measurable ones from some probability space to some other space), Haddaway takes the natural step of representing them as such. Haddaway has $(n+1)$-ary probabilistic p-predicates which map n-tuples to values. The value of the random variable is constrained to be the last argument of the relevant atomic formula. With each p-predicate p there must be a statement of the form $VAL(p) = \{v_1, v_2, \ldots v_n\}$ where the v_i are constants enumerating the set of possible values of the random variable represented by the p-predicate. If $A = p(t_1, \ldots, t_n, t_{n+1})$ is a p-atomic formula then $obj(A)$ is used to denote the tuple $(t_1, \ldots, t_n)$ and $VAL(A)$ denotes the value t_n. There is thus a demarcation between individuals (such as $obj(A)$) and attributes of those individuals: $VAL(A)$.

The probabilistic base of a knowledge base is a set of probabilistic sentences of the form $(P(A_0|A_1, \ldots, A_n) = \alpha) \leftarrow L_1, \ldots L_n$ where the A_i are p-atomic formulae and the L_i are context literals. For example:

$$P(burglary(X, yes)|nbrhd(X, good)) = .3 \leftarrow in\_CALI(X)$$

To this are added various type declarations, and rules for combining rules. Haddaway distinguishes evidence (ground p-atomic formulae) and contextual information, which is a set of context atomic formulae. This is in contrast to the SLP approach where evidence is supplied by instantiating variables in the top-level query to the SLP, or by specialising the logic program. Contextual information is represented in the same way in both approaches. Queries take the form of $P(Q) =?$ where the last argument of Q is a variable in Q - the query is asking for the posterior distribution of $obj(Q)$

Query answering uses an algorithm (Q-procedure), which builds a "necessary portion of the Bayesian network" where each node of the net corresponds to an $obj(A)$, where A is a *relevant* ground p-atomic formula. The Bayesian net is updated using the evidence and the probabilities associated with the query are computed in the normal fashion.

Space prevents giving the BUILD-NET procedure for constructing the Bayesian net from a query, instead just note that there are similarities with query-answering in SLPs. In BUILD-NET all sentences that have the input atomic formula as consequent are collected and the associated probabilities combined according to a combination rule. This is analogous to the SLP approach of finding all refutations of a goal—except we, unlike Haddaway, only allow one form of combination. Haddaway's dynamic query-driven construction of a Bayes net is analogous to the query-driven construction of an SLD-tree.

Jaeger: first-order Bayesian nets

Jaeger [1997] notes that standard Bayesian nets model a multivariate distribution over a fixed number of random variables. These apparently have to be "attributes" of a single "random event"—the Bayes net can not adequately model relations between different random events.

Jaeger's basic idea is to have Bayesian networks where the nodes are relation symbols, r whose values are possible interpretations of r. So the random variables are set-valued. In such a set-up, normal first-order formulae set up hard-constraints on possible values for the relation symbols in the language.

In contrast, the SLP approach does not make any fundamental distinction between attributes and the individuals (Jaeger's "random events") that possess them. This follows standard first-order logic semantics and also standard approaches in statistics for modelling related multivariate distributions. For example, the domain of a model for the first-order theory

```
dist(london,york,333).
has(james,job,family).
social_unit(family).
father(james,rob).
```

contains elements corresponding to the constants `york,london, 333, james, job, family` and `rob` and there is no distinction between individuals and attributes. Indeed something which (intuitively) appears as an attribute in one formula can represent an individual in another, e.g. `family`.

4 CONCLUSIONS AND POSSIBLE FUTURE DIRECTIONS

Sections 2 and 3 contain a number of approaches to integrating logical and probabilistic reasoning. Generally, they are restricted to computing the probabilities of atomic formulae. With the exception of SLPs, the atomic formula-probabilities are probabilities that the atomic formula is *true*. Although this is a natural extension of logical reasoning, it does not allow the identification of logical and random variables which the SLP does. Such an identification is likely to be of practical importance. Probabilistic inference relying on SLD-resolution is highly inefficient. Inference methods which involve storing and transmitting probabilistic information such as Bayes net algorithms based on join trees are considerably more efficient. Writing a meta-logical interpreter to perform such inference should be able to exploit the identification of logical and random variables using the sort of "level-crossing" tricks contained in Earley deduction interpreters [Pereira and Shieber, 1987].

In a number of cases bounds on probabilities rather than exact probabilities are computed. In Probabilistic Logic Programming (PLP), considerable work has been done on propagating such constraints, and checking for consistency. The SLP approach is more traditional in that there is a fully defined probabilistic (loglinear) model from which it is possible to compute all relevant probabilities. It would be interesting to see to what extent these two approaches could be unified. One possibility would be to extend SLPs to define constraints on probabilities rather than exact values: constraint logic programming over reals (CLP(R)) is an obvious choice here.

On a more fundamental level, there are connections between the 'tree-shaped' distributions which an SLP defines via SLD-resolution and the probability trees that Shafer uses as "a language for talking about the structure of contingency" [Shafer, 1996]. Shafer's aim is to provide a more dynamic basis for probability than the usual static set-theoretic Kolmogorovian axioms of probability theory. Rather than concern ourselves merely with the probability that an event happens, Shafer seeks to connect that probability with a probabilistic history of those events that led up to it. Shafer refers to the final event, as a *Moivrean* event—this is the event of standard probability theory. The events leading up to a Moivrean event— the changes "in some object or circumstance at a particular time" are called *Humean* events by Shafer. SLPs also have this division: there is the Humean event of choosing a particular clause head, and there is the Moivrean event of getting a particular instantiation for the variables in a top-level goal.

ACKNOWLEDGEMENTS

Many thanks to Sara-Jayne Farmer for weeding out various errors and omissions. Thanks to Stephen Muggleton for useful discussions on the role of normalisation in SLPs and to Stefan Riezler for clarifying his method. Thanks also Gillian, Jane and Robert Higgins for putting up with me. Finally thanks to Luc de Raedt for encouraging me to investigate first-order Bayesian nets.

This chapter is an updated version of [Cussens, 1999a].

Department of Computer Science, University of York, UK.

BIBLIOGRAPHY

[Abadi and Halpern, 1994] Martín Abadi and Joseph Y. Halpern. Decidability and expressiveness for first-order logics of probability. *Information and Computation*, 112(1):1–36, 1994.

[Abney, 1997] Steven Abney. Stochastic attribute-value grammars. *Computational Linguistics*, 23(4):597–618, 1997.

[Bacchus *et al.*, 1996] Fahiem Bacchus, Adam Grove, Joseph Y. Halpern, and Daphne Koller. From statistics to belief. *Artificial Intelligence*, 87:75–143, 1996.

[Boutilier *et al.*, 1996] Craig Boutilier, Nir Friedman, Moises Goldszmidt, and Daphne Koller. Context-specific independence in Bayesian networks. In *Proceedings of the Twelfth Annual Conference on Uncertainty in Artificial Intelligence (UAI–96)*, pages 115–123, Portland, Oregon, 1996.

[Clark and McCabe, 1982] K.L. Clark and F.G. McCabe. PROLOG: a language for implementing expert systems. In J.E. Hayes, Donald Michie, and Y-H Pao, editors, *Machine Intelligence*, volume 10, chapter 23, pages 455–470. Ellis Horwood, Chichester, 1982.

[Cussens, 1996] James Cussens. Deduction, induction and probabilistic support. *Synthese*, 108(1):1–10, July 1996.

[Cussens, 1999a] James Cussens. Integrating probabilistic and logical reasoning. *Electronic Transactions on Artificial Intelligence*, 3(B):79–103, 1999. Selected Articles from the Machine Intelligence 16 Workshop.

[Cussens, 1999b] James Cussens. Loglinear models for first-order probabilistic reasoning. In Kathryn B. Laskey and Henri Prade, editors, *Proceedings of the Fifteenth Annual Conference on Uncertainty in Artificial Intelligence (UAI–99)*, pages 126–133, Stockholm, 1999. Morgan Kaufmann.

[Cussens, 2000a] James Cussens. Attribute-value and relational learning: A statistical viewpoint. In Luc De Raedt and Stefan Kramer, editors, *Proceedings of the ICML-2000 Workshop on Attribute-Value and Relational Learning: Crossing the Boundaries*, pages 35–39, 2000.

[Cussens, 2000b] James Cussens. Stochastic logic programs: Sampling, inference and applications. In Craig Boutilier and Moisés Goldszmidt, editors, *Proceedings of the Sixteenth Annual Conference on Uncertainty in Artificial Intelligence (UAI–2000)*, pages 115–122, Stanford, CA, 2000. Morgan Kaufmann.

[Cussens, 2001a] James Cussens. Parameter estimation in stochastic logic programs. *Machine Learning*, 44(3): 245–271, 2001.

[Cussens, 2001b] James Cussens. Statistical aspects of stochastic logic programs. In Tommi Jaakkola and Thomas Richardson, editors, *Artificial Intelligence and Statistics 2001: Proceedings of the Eighth International Workshop*, pages 181–186, Key West, Florida, January 2001. Morgan Kaufmann.

[Della Pietra *et al.*, 1997] S. Della Pietra, V. Della Pietra, and J. Lafferty. Inducing features of random fields. *IEEE Transactions on Pattern Analysis and Machine Intelligence*, 19(4):380–393, April 1997.

[Eisele, 1994] Andreas Eisele. Towards probabilistic extensions of constraint-based grammars. Contribution to DYANA-2 Deliverable R1.2B, DYANA-2 project, 1994.

[Haddaway, 1999] Peter Haddaway. An overview of some recent developments in Bayesian problem solving techniques. *AI Magazine*, Spring 1999.

[Halpern, 1990] Joseph Y. Halpern. An analysis of first-order logics of probability. *Artificial Intelligence*, 46:311–350, 1990.

[Jaeger, 1997] Manfred Jaeger. Relational Bayesian networks. In *Proceedings of the Thirteenth Annual Conference on Uncertainty in Artificial Intelligence (UAI–97)*, pages 266–273, San Francisco, CA, 1997. Morgan Kaufmann Publishers.

[Koller and Pfeffer, 1997] Daphne Koller and Avi Pfeffer. Learning probabilities for noisy first-order rules. In *Proceedings of the Fifteenth International Joint Conference on Artificial Intelligence (IJCAI-97)*, Nagoya, Japan, August 1997.

[Lari and Young, 1990] K. Lari and S.J. Young. The estimation of stochastic context-free grammars using the Inside-Outside algorithm. *Computer Speech and Language*, 4:35–56, 1990.

[McCarthy and Hayes, 1969] J. McCarthy and P. Hayes. Some philosophical problems from the standpoint of artificial intelligence. In B. Meltzer and D. Michie, editors, *Machine Intelligence 4*, pages 463–502. Edinburgh University Press, Edinburgh, 1969.

[Muggleton, 1996] S.H. Muggleton. Stochastic logic programs. In L. de Raedt, editor, *Advances in Inductive Logic Programming*, pages 254–264. IOS Press, 1996.

[Muggleton, 2000] S.H. Muggleton. Semantics and derivation for stochastic logic programs. In Richard Dybowski, editor, *Proceedings of the UAI-2000 Workshop on Fusion of Domain Knowledge with Data for Decision Support*, 2000.

[Ng and Subrahmanian, 1992] Raymond Ng and V.S. Subrahmanian. Probabilistic logic programming. *Information and Computation*, 101(2):150–201, 1992.

[Ng and Subrahmanian, 1993] Raymond Ng and V.S. Subrahmanian. A semantical framework for supporting subjective and conditional probabilities in deductive databases. *Journal of Automated Reasoning*, 10(2):191–235, 1993.

[Ng and Subrahmanian, 1994] Raymond Ng and V.S. Subrahmanian. Stable semantics for probabilistic databases. *Information and Computation*, 110(1):42–83, 1994.

[Ngo and Haddaway, 1997] L. Ngo and Peter Haddaway. Answering queries from context-sensitive probabilistic knowledge bases. *Theoretical Computer Science*, 171:147–171, 1997.

[Ngo *et al.*, 1995] Liem Ngo, Peter Haddawy, and James Helwig. A theorectical framework for context-sensitive temporal probability model construction with application to plan projection. In *Proceedings of the Eleventh Annual Conference on Uncertainty in Artificial Intelligence (UAI–95)*, pages 419–426, Montreal, Quebec, Canada, 1995.

[Pereira and Shieber, 1987] Fernando C. N. Pereira and Stuart M. Shieber. *Prolog and Natural-Language Analysis*. CLSI, Stanford, 1987.

[Popper and Miller, 1987] Karl R. Popper and David Miller. Why probabilistic support is not inductive. *Philosophical Transactions of the Royal Society of London*, 321:569–591, 1987.

[Riezler, 1997] Stefan Riezler. Probabilistic constraint logic programming. Arbeitsberichte des SFB 340 Bericht Nr. 117, Universität Tübingen, 1997.

[Riezler, 1998] Stefan Riezler. *Probabilistic Constraint Logic Programming*. PhD thesis, Universität Tübingen, 1998. AIMS Report 5(1), 1999, IMS, Universität Stuttgart.

[Shafer, 1996] Glenn Shafer. *The Art of Causal Conjecture*. MIT Press, Cambridge, Mass., 1996.

[Shapiro, 1983] Ehud Shapiro. Logic programs with uncertainties: A tool for implementing rule-based systems. In *Proc. IJCAI-83*, pages 529–532, 1983.

[Spiegelhalter *et al.*, 1996] D. Spiegelhalter, A. Thomas, N. Best, and W. Gilks. *BUGS 0.5 Bayesian inference using Gibbs Sampling Manual*. MRC Biostatistics Unit, Cambridge, 1996.

PART III

BAYESIANISM AND DECISION THEORY

RICHARD BRADLEY

RAMSEY AND THE MEASUREMENT OF BELIEF

1 THE PROBLEM OF MEASUREMENT

Bayesian decision theories are formal theories of rational agency that tell us both what the properties of a rational state of mind are and what action it is rational for an agent to perform, given her state of mind. What makes a theory of this kind *Bayesian* is commitment to the following claims:

1. The (only) factors relevant to rational decision making are the options available, the relative subjective desirability of the possible outcomes of choosing one or another option and the subjective likelihood that each outcome will be achieved, given the exercise of a particular option.
2. To act rationally is to choose the option (or one of the options) which has the best expected consequences, given one's partial beliefs and desires.
3. Rational degrees of belief are probabilities.
4. Rational degrees of desire are utilities.

Different versions of Bayesian decision theory, such as those of Savage and Jeffrey, differ with respect to the kinds of options they postulate, with respect to the precise interpretation of the second claim (the expected utility hypothesis) and with respect to further constraints they place on rational desire. But such differences will not affect our discussion here. (Nor will the question of whether Bayesians are committed to conditionalisation as means of revising partial beliefs).

This paper will discuss a particular problem for Bayesian decision theories that derives from the fact that its main variables — an agent's degrees of partial belief and desire — are not (directly) observable, but have to be inferred from what the agent says and does. It is my contention that there is much to be learnt in this regard from a proposal made by Frank Ramsey in his paper 'Truth and Probability' [Ramsey, 1926]. His proposal will be accordingly be examined in considerable detail and then, having established it viability, defended against some influential objections.

1.1 Two Kinds of Problems

Let us start by distinguishing two kinds of problems concerning the status of scientific theories which can be called the problems of justification and of measurement. The first is the problem of validating the claims made by a theory about the objects and relations of which it speaks. The second is the problem of determining,

D. Corfield and J. Williamson (eds.), Foundations of Bayesianism, 263–290.

in a particular context, the values of the theory's variables. Of course, it is often the case that the same or similar observations will be used by scientists both to confirm a theory's claims and to determine the values of its variables. But the two problems are clearly distinct: the task of justifying Newton's first law of mechanics is different to that of measuring the forces, masses and acceleration of physical objects. Indeed the truth of theories may be presupposed in the interpretation of measurement observations, such as when Newton's law is used to calculate forces from masses and accelerations.

For a decision theory the problem of justification concerns both the claims it makes about the properties of consistent partial belief, desire and preference and the claims it makes about the rational way to act in the light of them. The measurement problem on the other hand boils down to the problem of determining an agent's partial beliefs and desires on the assumption that they have the properties postulated by the theory. Banal though the distinction between these problems may be, often enough it is neglected in debates within Bayesian decision theory. The reason is, I suspect, that Bayesian decision theories can be interpreted both as normative and as descriptive theories of agency and that its representation theorems can be construed as providing a solution to either the problem of justification or that of measurement, depending on the interpretation of the theory.

Let me explain in a bit more detail. Decision-theoretic representation theorems show that some set of conditions on an agent's preferences determines the existence of a quantitative representation of her partial beliefs and desires, consistent with her preferences. If the conditions on preference are naturally construed as conditions of rationality, then these theorems can serve to address the problem of justification in the sense that they show why the theory is normatively compelling. For they imply that by accepting that rational preference should satisfy the conditions in question, one is committed to accepting the theory's claims about the properties of rational partial belief and desire. The problem of justification then reduces to a defence of these qualitative rationality conditions on preference — supposedly an easier task.

I simplify considerably, of course. Some of the conditions that such theorems postulate are not rationality conditions but idealisations, designed to make the issue mathematically tractable. The ubiquitous completeness and continuity conditions are cases in point. But the thought is that these idealisations do not distort the main results, so that the relation between, for instance, incomplete preferences and incomplete probabilities and utilities should approximate the relation between complete ones. It would be good to see this rigorously demonstrated, but the claim has considerable plausibility. For why should making up my mind about matters concerning which I previously had no attitude affect, say, whether my beliefs about those that I do are probabilities or not?

Representation theorems also function as demonstrations of the measurability of the main decision theoretic variables — degrees of belief and desire. Or rather they show that if the idealised conditions postulated by the axioms of preference are realised then knowledge of agents' preferences suffice for knowledge of their degrees of belief and desire. In this context the way in which we evaluate the axioms is different from before. When justification was at issue we asked of the axioms of preference whether they were really rationality conditions. When the task is measurement, and our interest descriptive rather than normative, we ask whether the conditions are actually satisfied by agent's preferences or whether they can be made to satisfy them. These questions may be connected in practice. There may be evolutionary grounds, for instance, for supposing that the preferences of actual agents will by and large satisfy rationality conditions. But in principle, one is not required to solve the problem of justification in order to solve the measurement problem. In particular it is not necessary to establish the normative validity of Bayesian decision theory to demonstrate the measurability of the variables it postulates.

1.2 The Evidence Base

We now focus on the concern of this paper: the problem of measurement. To solve it we need to do two things. First we need to identify what sort of evidence can be used to determine agents' states of partial belief and desire. And second, we need show how and to what extent the evidence in question determines a measure of these states. We shall attend to the first question in this section and then devote most of the rest of the paper to the second, taking for granted a particular answer to the first.

Behaviourism

Historically discussion of this issue has been dominated by what I will call Epistemological Behaviourism: a rather puritanical doctrine of empiricist descent according to which the only acceptable evidence for an agent's mental states is intersubjectively observable behaviour. Evidence obtained by introspection, in particular, should not be countenanced as introspected states are not intersubjectively observable (even if 'observable' by the person whose states they are). As a consequence reports by someone on their mental states are to count as evidence only insofar as any perceptible effect of a mental cause can count as evidence. It does not follow from someone reporting that they believe x, that they do so, except if it has already been established through behavioural evidence that such belief reports are reliable.

Epistemological Behaviourism is to be distinguished from the associated metaphysical and semantic doctrines according to which mental states are nothing more than dispositions to behaviour or, more radically, than sets of observable behaviours (e.g. to desire that x is to act in ways which tend to bring it about that x, to prefer x to y is to choose x over y whenever both are available). Both forms of Behaviourism have been influential in the development of decision theory. On the metaphysical side, for instance, De Finetti and to some extent Savage saw themselves as giving behavioural meaning to the concept of probability, while

Samuelson saw his axioms of weak preference as giving behavioural meaning to the concept of preference. Ramsey too gestures in this direction.

It is with the epistemological form of Behaviourism that I am concerned, however, because of its influence on Bayesian decision theorists' understanding of the problem of the measurement. In behaviourist spirit, Bayesian decision theorists typically take a satisfactory solution of this problem to be a demonstration that decision theoretic representations of the states of minds of rational agents are characterisable in terms of sets of intersubjectively observable behaviours, that there exist (so to speak) defining sets of observable behaviours for each possible state of mind. This attitude is exemplified by Savage's claim regarding strengths of belief:

> "If the state of mind in question is not capable of manifesting itself in some sort of extraverbal behaviour, it is extraneous to our main interest. If, on the other hand, it does manifest itself through more material behaviour, that should, at least in principle, imply the possibility of testing whether a person holds one event to be more probable than another, by some behaviour expressing, or giving meaning to, his judgement."[1]

If such a reduction of belief and desire to preference were possible it would, of course, represent a solution of a kind, for it would imply that a set of observations of someone's behaviour with the right kinds of properties would suffice to determine the values of the main decision theoretic variables. But contrary to what seems to be prevailing belief, such a reduction has never been successfully achieved and I doubt very much will ever be so. The argument for this is quite simple. An agent's state of indifference between two options is something that decision theories can (and should) represent. But no behaviour can ever attest in a conclusive manner to someone's indifference between two or more possibilities. So decision-theoretic models are necessarily underdetermined by behavioural evidence (even all possible behavioural evidence, elicited under ideal conditions).

Let me elaborate a little by consideration of the example of Savage's representation theorem, since this is probably the best known.[2] Formally what Savage shows is that a binary relation, $\geq$, on a set of actions (actions on his account being functions from events to outcomes) that satisfies Savage's axioms will determine the existence of a utility function, u, on outcomes (unique up to choice of scale) and a unique probability function, Pr, on events such that for all actions, a_1 and a_2, $a_1 \geq a_2 \iff EU(a_1) \geq EU(a_2)$, where $EU(a_1)$ is the expectation of utility given a_1, relative to u and Pr. If $\geq$ is interpreted as the 'at least as preferred as' relation then it is plausible to construe u and Pr as respectively measures of the agent's degrees of desire and belief. We are now very near to what we want. Having reduced

[1] Savage [1954, p. 27–28]. This passage is a little misleading: Savage's views seem on closer examination to be more like those that I will attribute to Ramsey. But it accurately represents the behaviourist tradition in decision theory.

[2] See Savage [1954].

quantitative mental states — degrees of belief and desire — to a qualitative one — preference — all we require now is a reduction of the 'at least as preferred as' mental relation to some behavioural correlate. And this seems like it should not be too difficult as preferences between actions will be directly manifested in the agent's choice of action under suitable circumstances.

So near and yet so far. Even if we set aside obvious practical difficulties and grant that failure to perform an action indicates the presence of a preferred action rather than agent's ignorance of its availability, Savage's theory cannot be construed in a behaviouristically acceptable manner. For an agent's choices reveal, if anything, her strict preferences between options and not the 'at least as preferred as' relations that are the subject of Savage's rationality conditions. Moreover, strict preferences need not satisfy these conditions. In particular it would be wrong to require that strict preferences be complete, not just because this requires agents to have too much knowledge, but because agents can legitimately be indifferent between options.

Could indifference not be revealed by a failure to choose when confronted by a non-empty set of alternatives? But such behaviour has a number of possible interpretations. It could show that the agent prefers doing nothing to all of these options. Or that two or more of the options are equally preferred. Or even that some of the options are not comparable. Each possibility implies something different for the measurement process: the first that the 'do nothing' option has greater subjective expected utility than the others, the second that option set needs to be refined so as to determine which options it is that have equal subjective expected utility; and the third that the measurement cannot be completed since one of its preconditions (the comparability of the options) is not satisfied. Behavioural evidence may allow elimination of the first possibility, but in principle it could not discriminate between the second and third. One might of course eliminate the problem by forcing a choice in every circumstance, but the evidence so obtained could not then be used for the intended purpose without producing measurement artifacts.

The Use of Verbal Evidence

To my mind the search for a *reduction* of belief and desire to choice is misguided. We should accept that if Savage's theorem (or one like it) does what it says then it successfully reduces quantitative belief and desire to preference. We can also accept that observed choice constrains (but does not determine) attributions of preference to agents. But to attribute determinate preferences to an agent we need to accept evidence of a wider source and in particular the verbal reports of the agent concerning their preferences and (crucially) instances of their indifference between possibilities.

This is not to grant agents any kind of epistemic privilege with regard to their mental states. People certainly can be wrong about such things. But Behaviourism draws too sharp a contrast between the status of behavioural and verbal evidence, falsely identifying the former with what is observable and indubitable and the latter with what is not. In fact both in theory and in practice the distinction is not nearly so neat. What people report clearly *can* be evidence for what they really think and feel, certainly not indubitable and perhaps different in kind from the evidence gleaned by observing their behaviour, but not therefore better or worse. Indeed it is hardly conceivable in practice that we could do without verbal reports. For language has the great advantage of allowing precise formulation of the alternatives with respect to which we wish to determine agents' attitudes. Granted, this can raise the question as to whether the observer and the subject have a shared understanding of what is conveyed by particular linguistic expressions. But these difficulties are hardly more severe than those affecting the interpretation of their behaviour — it is just as easy to misinterpret the meaning of a wave of a hand as the meaning of an utterance.

The objection to introspection based evidence is not much more sustainable in theory than in practice. The supposed indubitability of behavioural evidence rests on the idea that because, perceptual illusions aside, we can *see* what someone is doing, there can be no doubting evidence statements of the 'The agent did such and such' kind. But in reality observations of behaviour (or at very least their linguistic representation) are always 'polluted' by interpretations: we don't see someone making circular motions with a cloth on the table, we see them wiping it clean; we don't see limbs describing particular trajectories, we see someone going for a walk; and so on. These interpretations are contestable and observers with different background knowledge will often interpret what someone is doing in different ways. Intersubjective observability does not mean either certainty or consensus.

Once one gives up on the goal of deriving all knowledge claims from indubitable foundations, we need not take an all or nothing attitude to the results of introspection. Under certain conditions introspection may reliably produce evidence of a particular kind, under others it may be less satisfactory. A decision about what sort of evidence to admit in the determination of agents' mental states must be made and motivated. Inevitably there will be trade-offs between the reliability of the evidence admitted and its richness. And depending on the attitude taken to the admissibility of different types of evidence, there will be different problems of measurement and different methods for solving it that depart from different trade-offs. For this reason, in addition to the usual ones, the solution defended here cannot be regarded as the only one deserving consideration.

Ramsey's Problem

Having granted that there is more than one way in which the problem of measurement can be formulated, let us now concentrate on the version of the problem found in Ramsey's essay "Truth and Probability" [Ramsey, 1926]. Ramsey's main concern in this paper is to explicate the concept of probability, but as he held that notions like this had no precise meaning unless a measurement procedure is specified, much of the paper is devoted to addressing our problem. We need not of course endorse Ramsey's operationalism in order to learn from his measurement methods.

Ramsey's thinking on the question of the measurement of belief seems at first to be very much in the behaviourist mould. He argues, for instance, that the idea that believing something more or less strongly was connected to a perceptible feeling of belief of a certain intensity is:

> " ... observably false, for the beliefs which we hold most strongly are often accompanied by practically no feeling at all; no one feels strongly about things he takes for granted."[3]

And that:

> "when we seek to know what is the difference between believing more firmly and believing less firmly, we can no longer regard it as consisting in having more or less observable feelings; at least I personally cannot recognise any such feelings. The difference seems to me to lie in how far we should act on these beliefs ... "[4]

But Ramsey is not in fact rejecting introspection wholesale, only the particular use of introspection associated with the idea of measuring strength of belief in terms of the sensations or feelings that accompany it. Indeed in the argument just quoted he makes use of introspective evidence: his own failure to perceive a feeling corresponding to his belief. And further on, when he argues that although we may feel that:

> "we know how strongly we believe things and that we can only know this if we can measure our belief by introspection ... our judgement about the strength of our belief is really about how we should act in hypothetical circumstances." [5]

the judgement that he refers to — as to how we would act under hypothetical circumstances — is presumably an introspective one. In fact, as I shall argue in greater detail later on, not much sense can be made of Ramsey's measurement procedure unless introspective evidence of this kind is admitted.

Ramsey takes his arguments to show that although we might be able to introspect whether we do or do not believe something, there is no reliable way of introspecting the degree to which we do. It would appear that this suspicion of

[3]Ramsey [1926, p. 65]
[4]Ibid, p.66
[5]Ibid, p.67

introspection, if not his arguments against it, extends to the possibility of qualitative judgements as to whether one believes one thing more strongly than another, despite the fact that they seem not dissimilar in nature to judgements as to how we would choose or act in particular circumstances. In any case, the upshot is that he admits only evidence as to the choices that an agent does or would make between specified alternatives and not their direct reports on their partial attitudes. In this respect Ramsey's approach is very much in line with the norm in decision theory. While this is not the only reasonable formulation of the problem, it has the advantage of expressing it as it is most commonly understood by Bayesian decision theorists.

1.3 Ramsey's Theory of Measurement

Conditional Prospects

The problem, as I have presented it, is to explain how the evidence relating to an agent's choices or actions determines a decision-theoretic representation of her partial belief and desires. Any approach that admits only evidence of this kind faces what is frequently termed the problem of the simultaneous determination of belief and desire: essentially that of untangling the respective contributions made by an agent's beliefs and desires to her choices. To solve it Ramsey makes use of two important ideas: that of a conditional prospect and that of an ethically neutral proposition.

Conditional prospects are possibilities such as that if its hot today, then it will rain tomorrow and, if not, it will snow. Such prospects have come to be termed 'gambles' or 'actions' in the literature on Ramsey, although he does not use either term. Both are misleading in some important ways, and I will largely avoid them.

Ramsey takes it for granted that the desirability of choosing a particular conditional prospect is related in a precise way to the desirability of the possible states of the world consistent with it. Namely, that the former is a weighted average of the latter, with the weights coming from the agent's degrees of belief. Formally, let $\Pr(P)$ be a measure of the degree to which the agent believes that P and $U(\alpha)$ and $U(\beta)$ be measures of the degrees to which she desires that respectively α and β be the actual state of the world. Then the desirability, $U(\Gamma)$, of the conditional prospect Γ, that α be the case if P is and that β be the case if not, is its expectation of utility:

PROPOSITION 1. $U(\Gamma) = U(\alpha).\Pr(P) + U(\beta).(1 - \Pr(P))$.

Ramsey doesn't attempt to justify this assumption, arguing that although the theory upon which it rests:

> "cannot be made adequate to all the facts, ... [it is] a useful approximation to the truth, particularly in our self-conscious or professional life, and it is presupposed in a great deal of our thought."[6]

[6]Ibid, p.69

It might well be objected that Ramsey's assumption about the way in which expectations determine desires is a good deal more specific than the sort of 'folk-psychology' that arguably much of our thought presupposes. This objection would not, I think, worry Ramsey much as he held that concepts like partial belief or utility are at least partially defined by the procedures for measuring them.[7] And Ramsey quite explicitly ties his method to the measurement of partial attitudes qua bases or causal components of actions, the concept of partial attitude that has come to predominate in the social sciences. Ramsey seems on safe ground here in that the quantities implicitly defined by his measurement scheme turn out to have many of the properties commonly attributed to them by social scientists. This fact alone may suffice to motivate the assumptions that Ramsey makes about the nature of partial belief and desire, but it should be remembered that the motivation is only as strong as the consensus amongst social scientists from which it stems.

But we are straying into the issue of justification, which we have already undertaken to set aside. The important fact is that Proposition 1 expresses (albeit formally) no more than what I previously stipulated as one of the defining contentions of Bayesian decision theory: namely that they desirability of an option is given by the expected benefit of choosing it, given one's beliefs. In attempting to solve a measurement problem it is perfectly appropriate to assume the truth of the theory whose variables require measurement. So there is no requirement here to justify Proposition 1 any further.

Ethically neutral propositions

The second critical innovation of Ramsey's is the postulation of what he calls ethically neutral propositions. An ethically neutral proposition is simply one whose truth or falsity is a matter of indifference to the agent and does not affect their attitude to any other prospects e.g. the prospect of a dust storm on Mars does not influence any of my earthly concerns.

Crucially the probabilities of some ethically neutral propositions can be inferred from an agent's preferences. Suppose, for instance, that an agent is not indifferent between the prospect of sun and that of snow, but indifferent between the prospect that if P is true, then it will be sunny, but if P is not, then it will snow, and the prospect that if P is true, then it will snow, but if P is not, then it will be sunny. Then we can infer that they regard P as likely to be true as not. For were it not, they should prefer one of the conditional prospects over the other (this follows directly from Proposition 1). The proposition that a toss of the coin in my hands will land heads up may be an example of such a proposition.

Ramsey's Method (Informally)

Suppose we want to determine Mary's attitudes to the various kinds of weather that the next day might bring: sun, snow, rain, and so on. Then if

[7] "the degree of belief is like a time interval; it has no precise meaning unless we specify how it is to be measured" ibid, p.63

> "we had the power of the Almighty, and could persuade our subject of our power" [8]

we could offer each kind of weather as an option to be exercised if she so chooses. By getting her to choose between any pair of them we obtain a ranking of all of her prospects and can assign some arbitrary number — say 1 and 0 for simplicity — to the top and bottom ranked ones.

Let us suppose that in Mary's case the results of coin tosses are indeed ethically neutral and of probability one-half and that the top ranked prospect is a sunny day tomorrow and the bottom ranked one is snow tomorrow. Now, Mary's attitude to the prospect that there will be sun tomorrow if the coin lands heads and snow if it lands tails will depend on the degree to which she respectively desires sunny and snowy days and the degree to which she believes that the coin will land heads or tails. But she considers the latter to be equi-probable, so she will regard the conditional prospect of sunny and snowy days in the event of heads or tails, as being midway in desirability between the prospects of a sunny day and that of a snowy one. (Again, this follows directly from Proposition 1). Relative then to our arbitrary choice of values for the top and bottom of the ranking, this means it has a utility of 0.5. And so too will any prospect ranked with this 'gamble'.

Just as we identified the value 0.5 with a particular conditional prospect we can, by compounding conditional prospects, identify any point in the interval from 0 to 1 with some 'gamble' on the top and bottom ranked prospects. For instance, 0.25 might be identified with the prospect that, in the event of the first toss of the coin coming up heads, there will be sun tomorrow if a second toss of the coin lands heads as well and snow if it lands tails, and that, in the event of it coming up tails on the first toss, there will be snow tomorrow. This gives us a scale with which to measure the utility of all prospects, whatever their form.

We can also use Mary's choices amongst her options to determine her degrees of belief. Suppose that she is indifferent between the prospect of it being cloudy tomorrow and that of it being sunny tomorrow if it rains tonight and of it being snowy if it doesn't. Then the utility of the prospect of cloudy weather tomorrow, relative to that of sunny and snowy weather, indicates the degree to which Mary believes that it will rain tonight: the closer the utility of cloudy weather is to that of sunny weather, the stronger must be her belief that it will rain, the closer it is to the utility of snowy weather the weaker must be her belief. In general, if Mary is indifferent between the prospect that α and the prospect that β if P is the case and that γ if P is not, then by rearrangement of Proposition 1:

$$\Pr(P) = \frac{U(\alpha) - U(\gamma)}{U(\beta) - U(\gamma)}.$$

So, on the assumption that the right conditional prospect can always be found, we have a means to determine Mary's degrees of belief in every proposition, whether ethically neutral or not.

[8] Ibid, p. 72. It is not in fact really necessary to have any powers other than those of persuasion.

2 RECONSTRUCTING RAMSEY'S ACCOUNT

2.1 The Basic Framework

Our informal presentation suggests that Ramsey has an simple, elegant and effective method for measuring belief and desire to offer us. Unfortunately, Ramsey does not work all the details of his theory, claiming at one point:

> "this would, I think, be rather like working out to seven places of decimals a result only valid to two"[9].

But it is clearly important to determine whether his demonstration can work in principle. The literature on Ramsey is sadly lacking in this respect. Many decision theorists have drawn inspiration from Ramsey, without engaging with the details of his work or by reconstructing it in terms of ideas drawn from Savage and others.[10] What little expository literature there is tends to remain too distant from the details or too uncritical of Ramsey's claims.[11] This has meant that crucial assumptions have gone unnoticed. The only exception that I am aware of is Sobel [1988], which carefully examines Ramsey's argument, filling in some of the details and making suggestions for amendments. But Sobel doesn't finish the job and, in particular, neither provides proofs that are missing (e.g. of the existence of a utility representation of preference) nor carefully examines all the ones that Ramsey does give (e.g. of the additivity of degrees of belief).

Worlds and propositions

Ramsey makes a distinction between the objects of belief — propositions — and the objects of desire — prospects: possible courses of the world (worlds for short) and conditional prospects or 'gambles'. We denote propositions by italic upper case letters, worlds by lower case Greek letters and arbitrary prospects by upper case Greek letters. In Ramsey's framework conditional prospects are essentially functions from partitions of propositions to worlds, with a function from the partition $\{X_1, X_2, .., X_n\}$ to worlds $\alpha_1, \alpha_2, \ldots, \alpha_n$ being written as (α_1 if X_1)(α_2 if X_2) ... (α_n if X_n). We take for granted that the ordering does not matter: (α_1 if X_1)(α_2 if X_2) is, for instance, the same conditional prospect as (α_2 if X_2)(α_1 if X_1).

As Ramsey speaks of worlds as being compatible with propositions or of including their truth, worlds must be the sorts of entities that can stand in logical relationships to propositions. He defines them as:

> "the different possible totalities of events between which our subject chooses — the ultimate organic unities"[12]

[9] Ibid, p. 76

[10] For instance, Davidson and Suppes [1956], Jeffrey [1983], Bolker [1967]

[11] For instance, N.-E. Sahlin, [1990]

[12] Ramsey [1926, p. 72–72]

From this, and on the basis of what he does with worlds, it might seem that we should think of them as something like maximally specific propositions, so that for all worlds α and propositions X, α implies that X or α implies that not X. (I shall represent the fact of α implying that X in the manner of possible worlds semantics, by writing $\alpha \in X$). In one respect, however, this cannot be exactly right. In his definition of an ethically neutral proposition of probability one half, and elsewhere, Ramsey implicitly assumes the existence of worlds that are compatible with both the truth and falsity of ethically neutral propositions. So either we have to say that worlds are only maximally specific about things that matter to the agent (and, hence, qualify the claims of the previous paragraph) or reformulate the relevant definitions. One way of doing so is to introduce world-like entities with just the right lack of specificity concerning the truth values of ethically neutral propositions. The matter has been thoroughly explored in Sobel [1988] and so I permit myself to ignore here the complications that it gives rise to.

Ethical Neutrality

The concept of an ethically neutral proposition is intuitively clear: it is just a proposition whose truth or falsity does not affect an agent's attitude to any of her prospects. But to express this more formally Ramsey needs to be able to say what it is for two prospects to differ from another only with respect to the truth of the proposition in question. Worlds, however, cannot differ with respect to the truth of a single proposition only. To get around this problem Ramsey assumes the existence of a class of atomic propositions; propositions which are true or false independently of one another and such that no two worlds are exactly alike with regard to the truth of all of them. An atomic proposition P is then defined to be ethically neutral for some agent iff she is indifferent between any two worlds which differ only in respect to the truth of P. Finally, non-atomic propositions are said to be ethically neutral iff all their atomic truth-arguments are. This cost of this formulation is that it ties his account to Wittgenstein's logical atomism to a far greater extent than I imagine he would care.[13]

Are there any ethically neutral propositions? The standard candidates are propositions such as that the next card drawn will be an Ace or the coin will land heads. But the truth of such propositions do affect agents' attitudes to some prospects. Take any proposition X and suppose that the agent is not indifferent to prospect that A. Then the truth of X will be a matter of consequence to her attitude to the prospect that A if X, because in the event that X is true, the prospect that A if X amounts to that of A. So X is not ethically neutral. It follows that there are no propositions that are neutral with respect to all prospects.

On the other hand, Ramsey only assumes that ethically neutral propositions leave the agents' attitudes to *worlds* unaffected. So, as long as conditional prospects

[13] Sobel [1988, p. 237] says that Ramsey is not committed to the claim that every proposition is a truth functional compound of atomic ones. But without this claim his definition of ethical neutrality is not completely general. See Sobel for a discussion of what he calls Ramsey's thin logical atomism.

do not serve to discriminate worlds, the supposition of ethically neutral propositions will not seem unreasonable.[14] And Ramsey's definition of an ethically neutral proposition of probability one-half, one of the linchpins of his method, will be well-founded.

DEFINITION 2. Let P be any ethically neutral proposition and α and β be any worlds consistent with both the truth and the falsity of P and such that the agent prefers α to β. Then P is of **probability one-half** iff the agent is indifferent between the prospect of (α if P)(β if $\neg P$) and that of (β if P)(α if $\neg P$).

2.2 *The Existence of a Utility Measure*

Axiomatising Preference

Under what conditions will the kind of measurement process that we previously described yield a measure of the strength of an agent's desires and to what extent will such a measure be unique? Ramsey answers this question by stating a representation theorem for preference orderings of prospects that establishes the existence of quantitative (utility) representations of the agent's degree of preference (or desire). He does not, however, give a full proof of this theorem. As far as I know nobody else has supplied one either. Indeed, despite the fact that Ramsey gives strong indications of how the proof should go, there seems to be very little recognition of the fact that the basic strategy of the proof differs to a considerable extent from others to be found in the decision theoretic literature. Perhaps the problem is a natural tendency to read his work through the lenses of either von Neumann and Morgenstern or Savage.

I will begin with a statement of Ramsey's representation theorem, before returning to the question of his strategy for proving it. Though I will remain fairly close to the letter of Ramsey's account, I will deviate from it at points in the interests of clarity and ease of understanding. Let a preference relation, $>$, and an indifference relation, $\approx$, be defined over the set of all prospects, Γ: a set of worlds plus all conditional prospects defined with respect to them (relative to a given set of propositions). By $\Phi \geq \Psi$ is meant that $\Phi > \Psi$ or $\Phi \approx \Psi$. Although Ramsey does not explicitly postulate it, he takes it for granted in subsequent discussion that agents' preferences are complete i.e. that $\Phi > \Psi$ or $\Psi > \Phi$ or $\Phi \approx \Psi$.

Let us call the set of prospects equally preferred to Φ, its **value** and denote it by $\bar{\Phi}$, avoiding Ramsey's economical, but often confusing, practice of denoting values as well as worlds by Greek letters. We write $\bar{\Phi} \succeq \bar{\Psi}$ iff $\Phi \geq \Psi$. While Ramsey directly axiomatises the relation on values induced by the preference relation on prospects, I will state the axioms in terms of the latter. His way of doing it obscures some issues of importance to our discussion and is easily recovered from ours. Let

[14] The question left open is: if worlds are not distinguished by conditional prospects, what is the relation between the two? This question, indeed the more general one of the logical status of conditional prospects is never addressed by Ramsey. We return to the problem right at the end of the paper.

Δ be a non-empty set of ethically neutral propositions of probability one-half and suppose that P belongs to Δ. Then Ramsey postulates:

R1 If $Q \in \Delta$ and $(\alpha \text{ if } P)(\beta \text{ if } \neg P) \geq (\gamma \text{ if } P)(\delta \text{ if } \neg P)$, then: $(\alpha \text{ if } Q)(\beta \text{ if } \neg Q) \geq (\gamma \text{ if } Q)(\delta \text{ if } \neg Q)$.

R2 If $(\alpha \text{ if } P)(\delta \text{ if } \neg P) \approx (\beta \text{ if } P)(\gamma \text{ if } \neg P)$ then:
(i) $\alpha > \beta \iff \gamma > \delta$
(ii) $\alpha \approx \beta \iff \gamma \approx \delta$.

R3 If $\Phi \geq \Psi$ and $\Psi \geq \Theta$, then $\Phi \geq \Theta$.

R4 If $(\alpha \text{ if } P)(\delta \text{ if } \neg P) \geq (\beta \text{ if } P)(\gamma \text{ if } \neg P)$ and $(\gamma \text{ if } P)(\zeta \text{ if } \neg P) \geq (\delta \text{ if } P)(\eta \text{ if } \neg P)$, then $(\alpha \text{ if } P)(\zeta \text{ if } \neg P) \geq (\beta \text{ if } P)(\eta \text{ if } \neg P)$

R5 $\forall(\alpha, \beta, \gamma)[\exists(\delta)$: $(\alpha \text{ if } P)(\gamma \text{ if } \neg P) \approx (\delta \text{ if } P)(\beta \text{ if } \neg P)]$

R6 $\forall(\alpha, \beta)[\exists(\delta)$: $(\alpha \text{ if } P)(\beta \text{ if } \neg P) \approx (\delta \text{ if } P)(\delta \text{ if } \neg P)]$

R7 Axiom of Continuity

R8 Archimedean Axiom.

I have slightly strengthened Ramsey's first, third and fourth axiom by stating them in terms of the weak order '$\geq$' rather than the indifference relation '$\approx$'. In the presence of R2, my R5 and R6 are jointly equivalent to his fifth and sixth axioms. Ramsey doesn't say what he intends by R7 and R8, though R7 is presumably about the continuity of preferences and R8 about the values of worlds. In particular, what is required of R8, whatever its precise formulation, is that it allows the derivation of the Archimedean condition referred to in Definition 7 below.

Presumably R7 is meant to ensure that for every conditional prospect $(\alpha \text{ if } X)(\beta \text{ if } \neg X)$ there exists a world γ such that $(\alpha \text{ if } X)(\beta \text{ if } \neg X) \approx \gamma$, so that every value contains a world. Apart from simplifying movement between values and worlds, this assumption plays a crucial role in his subsequent derivation of degrees of beliefs. Furthermore, as Ramsey doesn't assume (as we did in our informal presentation of his method) the existence of compound conditional prospects — prospects of the form $(\Phi \text{ if } X)(\Psi \text{ if } \neg X)$, where Φ and Ψ are conditional prospects — he does need something to ensure, for any two prospects, the existence of a conditional prospect whose value lies midway between theirs. This in turn is required for Ramsey to conclude that "These axioms enable the values to be correlated one-one with real numbers . . . ".[15]

[15] Ramsey [1926, p. 75]

Intervals of Values

Ramsey's next move is to "explain what is meant by the difference in value of α and β being equal to that between γ and δ" by defining it to mean that if P is an ethically neutral proposition of probability one-half then the agent is indifferent between $(\alpha \text{ if } P)(\delta \text{ if } \neg P)$ and $(\beta \text{ if } P)(\gamma \text{ if } \neg P)$. Although Ramsey seems to be speaking here of a relation (of difference in value) between *worlds*, what he really needs for his representation theorem is a definition of a difference relation between *values* of worlds.[16] So let us denote the difference between the values $\bar{\alpha}$ and $\bar{\beta}$ by $\bar{\alpha} - \bar{\beta}$ and define an ordering, $\succeq$, of **differences in values** as follows (the coherence of the definition is guaranteed by R1):

DEFINITION 3. If $P \in \Delta$ then $\bar{\alpha} - \bar{\beta} \succeq \bar{\gamma} - \bar{\delta}$ iff $\forall(\alpha \in \bar{\alpha}, \beta \in \bar{\beta}, \gamma \in \bar{\gamma}, \delta \in \bar{\delta}), (\alpha \text{ if } P)(\delta \text{ if } \neg P) \geq (\beta \text{ if } P)(\gamma \text{ if } \neg P)$.

With the concept of difference of value in hand, it becomes much easier to understand the role played by Ramsey's preference axioms. Essentially they are there to ensure that it be possible to give a numerical representation of not only such facts as the agent preferring one thing to another, but also of the extent to which their preference or desire for one thing exceeds their desire for another. To this end axiom R1 ensures the coherence of the definition of differences in values, axioms R5, R6, R7 and R8 a correspondence between values and real numbers, and R2, R3, and R4 that the difference operation on values functions like the subtraction operation on real numbers.

With regard to the latter, note that Axiom R4, opaque in its original formulation, translates as an axiom of transitivity for differences in values:

R4* If $\bar{\alpha} - \bar{\beta} \succeq \bar{\gamma} - \bar{\delta}$ and $\bar{\gamma} - \bar{\delta} \succeq \bar{\eta} - \bar{\zeta}$, then $\bar{\alpha} - \bar{\beta} \succeq \bar{\eta} - \bar{\zeta}$

while, as we now prove, it follows from the definitions of ethical neutrality and the difference operation that if $\bar{\alpha} - \bar{\beta} \succeq \bar{\gamma} - \bar{\delta}$ then $\bar{\beta} - \bar{\alpha} \succeq \bar{\delta} - \bar{\gamma}$ and $\bar{\alpha} - \bar{\gamma} \succeq \bar{\beta} - \bar{\delta}$.

LEMMA 4. *If* $\bar{\alpha} - \bar{\beta} \succeq \bar{\gamma} - \bar{\delta}$ *then:*
(i) $\bar{\delta} - \bar{\gamma} \succeq \bar{\beta} - \bar{\alpha}$
(ii) $\bar{\alpha} - \bar{\gamma} \succeq \bar{\beta} - \bar{\delta}$

Proof. Suppose that $P \in \Delta$. Omitting explicit quantification over worlds where the meaning is obvious, we note that $\bar{\alpha} - \bar{\beta} \succeq \bar{\gamma} - \bar{\delta} \Leftrightarrow (\alpha \text{ if } P)(\delta \text{ if } \neg P) \geq (\beta \text{ if } P)(\gamma \text{ if } \neg P)$. But by Definition 2, (i) $(\alpha \text{ if } P)(\delta \text{ if } \neg P) \geq (\beta \text{ if } P)(\gamma \text{ if } \neg P) \Leftrightarrow (\delta \text{ if } P)(\alpha \text{ if } \neg P) \geq (\gamma \text{ if } P)(\beta \text{ if } \neg P) \Leftrightarrow \bar{\delta} - \bar{\gamma} \succeq \bar{\beta} - \bar{\alpha}$, and (ii) $(\alpha \text{ if } P)(\delta \text{ if } \neg P) \geq (\beta \text{ if } P)(\gamma \text{ if } \neg P) \Leftrightarrow (\alpha \text{ if } P)(\delta \text{ if } \neg P) \geq (\gamma \text{ if } P)(\beta \text{ if } \neg P) \Leftrightarrow \bar{\alpha} - \bar{\gamma} \succeq \bar{\beta} - \bar{\delta}$. ■

Proving the Representation Theorem

We are now in a position to state Ramsey's theorem establishing the existence of utility measures of agents' desires. Ramsey does not give a uniqueness theorem

[16]This ambiguity is reproduced without comment in most expositions of Ramsey.

for such utility measures, but his subsequent discussion of the measurement of probabilities assumes that they are unique up to affine linear transformation (or choice of scale) i.e. that preferences are interval-scale measurable. We state below the theorem he requires.

THEOREM 5 (Existence). *There exists a utility function, U, on the set of all values, $\bar{\Gamma}$, such that* $\forall(\bar{\alpha}, \bar{\beta}, \bar{\gamma}, \bar{\delta} \in \bar{\Gamma})$, $\bar{\alpha} - \bar{\beta} \succeq \bar{\gamma} - \bar{\delta} \Leftrightarrow U(\bar{\alpha}) - U(\bar{\beta}) \geq U(\bar{\gamma}) - U(\bar{\delta})$

THEOREM 6 (Uniqueness). *If U' is another such a utility function, then there exists real numbers a and b, such that $a > 0$ and $U' = a.U + b$.*

The key to understanding Ramsey's representation theorem is to recognise that it implicitly draws on the theory of measurement deriving from the work of the German mathematician Hölder (with which Ramsey would have been familiar). We begin with a statement of the relevant results in this area, drawing from their presentation in Krantz *et al.* [1971, chapter 4].[17]

DEFINITION 7. Let A be a non-empty set and $\succeq$ a binary relation on $A \times A$. Then $< A \times A, \succeq >$ is an **algebraic difference structure** iff $\forall(a, b, c, d, a', b', c' \in A)$:

1. $\succeq$ is a complete and transitive

2. If $ab \succeq cd$, then $dc \succeq ba$

3. If $ab \succeq a'b'$ and $bc \succeq b'c'$ then $ac \succeq a'c'$

4. If $ab \succeq cd \succeq ba$ then there exists $x, x' \in A$, such that $ax \approx cd \approx x'b$

5. Archimedean condition

THEOREM 8. *If $< A \times A, \succeq >$ is an algebraic difference structure, then there exists a real-valued function, ϕ, on A, such that* $\forall(a, b, c, d \in A)$:

$$ab \succeq cd \Leftrightarrow \phi(a) - \phi(b) \geq \phi(c) - \phi(d)$$

Furthermore, ϕ is unique up to positive linear transformation i.e. if ϕ' is another such a function then $\exists(x, y \in \Re : x > 0, \phi = x.\phi + y)$.

Ramsey's basic strategy for proving his representation theorem is to use preference orderings of prospects to define an algebraic difference structure and then to invoke Theorem 8. We will now reconstruct his proof on that basis.

THEOREM 9. *Let $\succeq$ be the relation on $\bar{\Gamma} \times \bar{\Gamma}$ induced by definition 3. Then $< \bar{\Gamma} \times \bar{\Gamma}, \succeq >$ is a difference algebra.*

[17] The authors point out that Hölder's results can be applied to the problem of measurement of degrees of preference, but (oddly) make no attempt do so directly. Nor is there explicit recognition of the use that Ramsey makes of them.

Proof. We prove that $< \bar{\Gamma} \times \bar{\Gamma}, \succeq >$ satisfies the five conditions given in Definition 7.

1. Follows directly from the completeness and transitivity of $\geq$.

2. By Lemma 4(i).

3. By R4*, if $\bar{\alpha} - \bar{\alpha}' \geq \bar{\beta} - \bar{\beta}'$ and $\bar{\beta} - \bar{\beta}' \geq \bar{\gamma} - \bar{\gamma}'$ then $\bar{\alpha} - \bar{\alpha}' \geq \bar{\gamma} - \bar{\gamma}'$. But by Corollary 4 (ii) $\bar{\alpha} - \bar{\alpha}' \geq \bar{\beta} - \bar{\beta}' \Leftrightarrow \bar{\alpha} - \bar{\beta} \geq \bar{\alpha}' - \bar{\beta}'$, $\bar{\beta} - \bar{\beta}' \geq \bar{\gamma} - \bar{\gamma}' \Leftrightarrow \bar{\beta} - \bar{\gamma} \geq \bar{\beta}' - \bar{\gamma}'$ and $\bar{\alpha} - \bar{\alpha}' \geq \bar{\gamma} - \bar{\gamma}' \Leftrightarrow \bar{\alpha} - \bar{\gamma} \geq \bar{\alpha}' - \bar{\gamma}'$. Hence if $\bar{\alpha} - \bar{\beta} \geq \bar{\alpha}' - \bar{\beta}'$ and $\bar{\beta} - \bar{\gamma} > \bar{\beta}' - \bar{\gamma}'$ then $\bar{\alpha} - \bar{\gamma} \geq \bar{\alpha}' - \bar{\gamma}'$.

4. By R5, $\forall(\alpha, \beta, \gamma, \delta)[\exists(\epsilon, \epsilon')$: $(\alpha \text{ if } P)(\gamma \text{ if } \neg P) \approx (\epsilon \text{ if } P)(\beta \text{ if } \neg P)$ and $(\delta \text{ if } P)(\beta \text{ if } \neg P) \approx (\epsilon' \text{ if } P)(\gamma \text{ if } \neg P)]$. But by Lemma 4(i) $(\delta \text{ if } P)(\beta \text{ if } \neg P) \approx (\epsilon' \text{ if } P)(\gamma \text{ if } \neg P) \Leftrightarrow (\beta \text{ if } P)(\delta \text{ if } \neg P) \approx (\gamma \text{ if } P)(\epsilon' \text{ if } \neg P)$. Hence $\exists(\bar{\epsilon}, \bar{\epsilon}') : \bar{\alpha} - \bar{\epsilon} \approx \bar{\beta} - \bar{\gamma}$ and $\bar{\beta} - \bar{\gamma} \approx \bar{\epsilon}' - \bar{\delta}$.

5. Follows from R8. ■

Theorems 5 and 6 clearly follow directly from Theorems 8 and 9. Ramsey does not seek to explicitly establish that the utility function, U, referred to in these theorems also represents the agent's preference ranking of possibilities in the sense that the utilities of prospects go by their position in the preference order. To establish this we would have to make a further, but unobjectionable, assumption. As it turns out, the assumption is presupposed in Ramsey's subsequent derivation of degrees of belief and so there good reason to make it explicit here.

R9 Let P be any proposition and α and β any worlds. Then:

$$\alpha \geq \beta \Longleftrightarrow \alpha \geq (\alpha \text{ if } P)(\beta \text{ if } \neg P) \geq \beta.$$

COROLLARY 10. $\alpha \approx (\alpha \textit{ if } P)(\alpha \textit{ if} \neg P)$

THEOREM 11. *The utility function, U, on $\bar{\Gamma}$, is such that* $\forall(\bar{\alpha}, \bar{\beta} \in \bar{\Gamma}), \bar{\alpha} \succeq \bar{\beta} \Leftrightarrow U(\bar{\alpha}) \geq U(\bar{\beta})$

Proof. By Corollary 10 and Theorem 5 it follows that $\alpha \geq \beta \Leftrightarrow (\alpha \text{ if } P)(\alpha \text{ if } \neg P) \geq (\beta \text{ if } P)(\beta \text{ if } \neg P) \Leftrightarrow \bar{\alpha} - \bar{\beta} \succeq \bar{\beta} - \bar{\alpha} \Leftrightarrow U(\bar{\alpha}) - U(\bar{\beta}) > U(\bar{\beta}) - U(\bar{\alpha}) \Leftrightarrow U(\bar{\alpha}) \geq U(\bar{\beta})$. ■

2.3 *Measuring Partial Belief*

Defining Degrees of Belief

Recall from our informal presentation of his method that Ramsey's next move is to use the utility measure on worlds to determine the agent's degrees of belief in all propositions, including those that are not ethically neutral. The vehicle for doing so is the following definition.

DEFINITION 12 (*Degrees of Belief*). Suppose that worlds $\alpha \in P$, $\beta \in \neg P$ and ξ are such that $\alpha \not\approx \beta$ and $\xi \approx (\alpha \text{ if } P)(\beta \text{ if } \neg P)$. Then:

$$\Pr(P) =_{\text{defn}} \frac{U(\xi) - U(\beta)}{U(\alpha) - U(\beta)}$$

The existence of the world ε in question is presumably secured by R7. As the Uniqueness Theorem establishes that ratios of utility difference are independent of choice of scale for the utility function, Definition 12 determines a unique measure of the agent's degrees of belief. Ramsey notes that in this definition the proposition P is not assumed to be ethically neutral, but that it is necessary to assume both that this definition is independent of the choice of worlds meeting the antecedent conditions and that

> "there is a world with any assigned value in which P is true, and one in which P is false"[18]

Why the latter assumption is necessary will only become clear once we look at Ramsey's proof that degrees of belief are probabilities. But if it is to be tenable it is patently necessary that P be neither logically true nor logically false. But this means that some separate treatment of such propositions is required e.g. by stipulating that $\Pr(P) = 1$, whenever P is logically true.

As regards the former assumption (of independence), Ramsey does not say how it might be formally expressed as a condition on preference or choice. But it must be possible to do so, as we know from the Uniqueness Theorem that the equality (or otherwise) of ratios of utility differences is determined by the preference ranking. One way to proceed would be to define the conditions under which the difference in values of α and β equals a particular fraction of the difference between the values of γ and δ. The definitions are cumbersome, so I will confine myself to illustrating the case of one-half. Suppose that Q is an ethically neutral proposition of probability one-half, that $\zeta \approx (\varepsilon \text{ if } Q)(\delta \text{ if } \neg Q)$ and that $\eta \approx (\varepsilon \text{ if } Q)(\gamma \text{ if } \neg Q)$. Then we can say that difference in value of α and β equals half the difference between the values of γ and δ iff $(\alpha \text{ if } Q)(\zeta \text{ if } \neg Q) \approx (\beta \text{ if } Q)(\eta \text{ if } \neg Q)$. And so on.

THEOREM 13. *If $\alpha \in P$ and $\beta \in \neg P$, then:*

$$U((\alpha \textit{ if } P)(\beta \textit{ if} \neg P)) = U(\alpha).\Pr(P) + U(\beta).(1 - \Pr(P)).$$

Proof. If $\alpha \approx \beta$, then it follows from axiom R9 that $\alpha \approx (\alpha \text{ if } P)(\beta \text{ if } \neg P) \approx \beta$. So the theorem follows immediately. If $\alpha \not\approx \beta$, then suppose that ξ is such that $\xi \approx (\alpha \text{ if } P)(\beta \text{ if } \neg P)$. Then by the definition of $\Pr(P)$, $U(\xi) = U(\alpha).\Pr(P) - U(\beta).(1 - \Pr(P)) = U((\alpha \text{ if } P)(\beta \text{ if } \neg P))$. ■

[18] Ramsey, ibid, p. 75

DEFINITION 14 (*Conditional Degrees of Belief*). Suppose that (α if Q)(β if $\neg Q$) $\approx$ (γ if PQ)(δ if $\neg PQ$)(β if $\neg Q$). Then the degree of belief in P given Q;

$$\Pr(P \mid Q) =_{\text{defn}} \frac{U(\alpha) - U(\delta)}{U(\gamma) - U(\delta)}$$

As with the definition of degrees of belief it must be supposed (though Ramsey does not explicitly say so) that $\gamma \not\approx \delta$, $\alpha \in Q$, $\beta \in \neg Q$, $\gamma \in PQ$ and $\delta \in \neg PQ$, that the definition is independent of particular choices of worlds satisfying the antecedent conditions and that there is a world with any assigned value in which PQ, $P\neg Q$ and $\neg Q$ are true. It would also appear that the existence of equally ranked conditional prospects of kind referred to in the definition is not guaranteed by Ramsey's assumptions. There are a number of ways of filling in this gap. The most conceptually satisfactory would involve the postulation of compounded conditional prospects and a generalisation of R5. But somewhat more economically, we could simply add the following axiom to Ramsey's.

R10 Let $\{P_1, P_2, \ldots, P_n\}$ be a partition of propositions. Then $\forall(\gamma, \delta, \ldots, \beta)$, $\exists(\alpha : (\gamma \text{ if } P_1)(\delta \text{ if } P_2) \ldots (\beta \text{ if } P_n) \approx (\alpha \text{ if } P_1 \cup P_2) \ldots (\beta \text{ if } P_n))$.

Proving Coherence

Ramsey must still demonstrate that the degree of belief function $\Pr(\cdot)$ is truly a probability function. This is done in Theorems 15 and 17. Ramsey's proof of Theorem 17 requires a further assumption, not made explicit by him, but which is quite reasonable if one accepts his framework.

R11 Suppose that P and Q are inconsistent, $\alpha \in P$, $\beta \in Q$, and $\gamma \in P \cup Q$. If $\alpha \approx \beta \approx \gamma$, then ($\alpha$ if P)(β if Q)(δ if $\neg P \neg Q$) $\approx$ (γ if $P \cup Q$)(δ if $\neg P \neg Q$)

Ramsey's own proof makes no use of R10, for the obvious reason that he does not explicitly postulate it. But we have seen that it is required elsewhere and by making use of it here a much simpler alternative proof of Theorem 17 can be given which does not require R11. Both proofs follow below.

THEOREM 15. *Let P be any proposition. Then:*

(i) $\Pr(P) > 0$

(ii) $\Pr(P) + \Pr(\neg P) = 1$

(iii) $\Pr(P \mid Q) + \Pr(\neg P \mid Q) = 1$

Proof. Suppose that ξ is such that $\xi \approx$ (α if P)(β if $\neg P$). Then:

(i) By R9, $\alpha \geq \beta \iff \alpha \geq \xi \geq \beta$. So $U(\xi) - U(\beta) \leq U(\alpha) - U(\beta)$, and it then follows from the definition of Pr that it never takes negative values.

(ii) By definition, $\Pr(P) + \Pr(\neg P) = \frac{U(\xi)-U(\beta)}{U(\alpha)-U(\beta)} + \frac{U(\xi)-U(\alpha)}{U(\beta)-U(\alpha)} = 1$.

(iii) Suppose that $\gamma \in PQ$ and $\delta \in \neg PQ$ are such that (α if Q)(β if $\neg Q$) $\approx$ (γ if PQ)(δ if $\neg PQ$)(β if $\neg Q$). Then by definition of conditional degrees of belief, $\Pr(P \mid Q) = \frac{U(\alpha)-U(\delta)}{U(\gamma)-U(\delta)}$ and $\Pr(\neg P \mid Q) = \frac{U(\alpha)-U(\gamma)}{U(\delta)-U(\gamma)} = \frac{U(\gamma)-U(\alpha)}{U(\gamma)-U(\delta)}$. So $\Pr(P \mid Q) + \Pr(\neg P \mid Q) = \frac{U(\alpha)-U(\delta)-U(\alpha)+U(\gamma)}{U(\gamma)-U(\delta)} = 1$. ■

LEMMA 16. *Suppose that $\beta \in \neg Q$, $\gamma \in PQ$ and $\delta \in \neg PQ$. Then* $U((\gamma$ *if* $PQ)(\delta$ *if* $\neg PQ)(\beta$ *if* $\neg Q)) = (U(\gamma).\Pr(P \mid Q) + U(\delta)(1 - \Pr(P \mid Q))).\Pr(Q) + U(\beta).\Pr(\neg Q)$

Proof. Let $\alpha \in Q$ be such that (α if Q)(β if $\neg Q$) $\approx$ (γ if PQ)(δ if $\neg PQ$)(β if $\neg Q$). Then by Theorem 13, $U(\gamma$ if $PQ)(\delta$ if $\neg PQ)(\beta$ if $\neg Q) = U(\alpha).\Pr(Q) - U(\beta).(1 - \Pr(Q))$. But by Definition 14, $U(\alpha) = U(\gamma).\Pr(P \mid Q) + U(\delta)(1 - \Pr(P \mid Q))$. So $U((\gamma$ if $PQ)(\delta$ if $\neg PQ)(\beta$ if $\neg Q)) = (U(\gamma).\Pr(P \mid Q) + U(\delta)(1 - \Pr(P \mid Q))).\Pr(Q) + U(\beta).\Pr(\neg Q)$. ■

THEOREM 17. $\Pr(P \mid Q) = \frac{\Pr(PQ)}{\Pr(Q)}$

Proof.[Ramsey's Proof] Let $\Pr(Q) = x$ and $\Pr(P \mid Q) = y$. Then we need to show that $\Pr(PQ) = xy$. Let α and β be any worlds in Q and $\neg Q$ respectively, such that, for some real number t, $U(\alpha) = U(\xi) + (1-x)t$ and $U(\beta) = U(\alpha) - t = U(\xi) - xt$. By assumption such worlds α and β exist. Now $U(\xi) = U(\xi).x + U(\xi).(1 - x) = U(\alpha).x + U(\beta).(1 - x) = U((\alpha$ if $Q)(\beta$ if $\neg Q))$. Then by definition, $x = \frac{U(\xi)-U(\beta)}{U(\alpha)-U(\beta)}$.

Now let worlds $\gamma \in PQ$, $\delta \in \neg PQ$ be such that $U(\gamma) = U(\alpha) + \frac{t}{y} - t$ and $U(\delta) = U(\beta) = U(\alpha) - t$. Again by assumption such worlds γ and δ exist. Then by Lemma 16, $U((\gamma$ if $PQ)(\delta$ if $P\neg Q)(\beta$ if $\neg Q)) = (U(\gamma).y + U(\delta)(1 - y)).x + U(\beta)(1-x) = U(\alpha).x + U(\beta).(1-x) = U((\alpha$ if $Q)(\beta$ if $\neg Q))$. So by Definition 14, $y = \frac{U(\alpha)-U(\delta)}{U(\gamma)-U(\delta)} = \frac{U(\alpha)-U(\beta)}{U(\gamma)-U(\beta)}$. So $xy = \frac{U(\xi)-U(\beta)}{U(\alpha)-U(\beta)} \cdot \frac{U(\alpha)-U(\beta)}{U(\gamma)-U(\beta)} = \frac{U(\xi)-U(\beta)}{U(\gamma)-U(\beta)}$. It also follows from Axiom R11, that $U((\gamma$ if $PQ)(\delta$ if $P\neg Q)(\beta$ if $\neg Q)) = U((\gamma$ if $PQ)(\beta$ if $(\neg P \cup \neg Q)))$. Hence $\xi \approx (\gamma$ if $PQ)(\beta$ if $(\neg P \cup \neg Q))$ But then by definition $\Pr(PQ) = \frac{U(\xi)-U(\beta)}{U(\gamma)-U(\beta)} = xy$. ■

Proof.[Alternative Proof] Let worlds $\delta \in \neg PQ$ and $\beta \in \neg Q$ be such that $\delta \approx \beta$. Now by R10, there exists worlds α and ϵ such that $\xi \approx (\alpha$ if $Q)(\beta$ if $\neg Q) \approx (\gamma$ if $PQ)(\delta$ if $\neg PQ)(\beta$ if $\neg Q) \approx (\gamma$ if $PQ)(\epsilon$ if $(\neg P \cup \neg Q))$. Then by definition, $\Pr(Q) = \frac{U(\xi)-U(\beta)}{U(\alpha)-U(\beta)}$, $\Pr(PQ) = \frac{U(\xi)-U(\beta)}{U(\gamma)-U(\beta)}$ and $\Pr(P \mid Q) = \frac{U(\alpha)-U(\delta)}{U(\gamma)-U(\delta)} = \frac{U(\alpha)-U(\beta)}{U(\gamma)-U(\delta)}$. So $\Pr(PQ) = \Pr(Q).\Pr(P \mid Q)$. ■

COROLLARY 18. $\Pr(PQ) + \Pr(\neg PQ) = \Pr(Q)$.

Proof. By Theorem 17, $\Pr(PQ) = \Pr(P \mid Q).\Pr(Q)$ and $\Pr(\neg PQ) = \Pr(\neg P \mid Q).\Pr(Q)$. Therefore $\Pr(PQ) + \Pr(\neg PQ) = (\Pr(P \mid Q) + \Pr(\neg P|Q)).\Pr(Q) = \Pr(Q)$ by 15(iii). ∎

Conditionalism

The importance of Ramsey's assumption that propositions contain worlds of every utility value should now be clear — it is what allows the assumption of the existence of the worlds α, β, γ and δ referred to in his proof of Theorem 17 and of worlds β and δ referred to in the alternative proof. Ramsey seems to think of this as a purely technical condition. But one might derive it from conditionalist premises, the relevant one in this context being as follows.

PROPOSITION 19. *(Ethical Conditionalism) For any propositions P and Q, there are worlds $\alpha \in P$ and $\beta \in Q$ such that $\alpha \approx \beta$.*

Given the 1–1 correspondence between worlds and real numbers, Ethical Conditionalism implies the assumption of whose necessity Ramsey speaks. Namely, that whatever the range of utility values taken by prospects, for every value in that range and every proposition, there is a world which implies the truth of that proposition and which has the utility value in question.

Conditionalism is to my mind an imminently defensible doctrine. Essentially the conditionalist's claim is that however good (or bad) some possibility might be on average, there are imaginable circumstances in which it is not so. No prospect is good or bad in itself, but is only so relative to the conditions under which it is expected to be realised. Suppose, for instance, that P identifies some good prospect like winning the National Lottery and Q some dreadful one, like the death of a relative. The conditionalist claim is that even winning the lottery can be a bad thing, such as when it alienates one from one's friends or causes one to stop activities that gives one's life a sense of purpose. Likewise even the death of a relative can be a good thing, such as when it pre-empts a period of great suffering for them.

Defensible though it may be, Ethical Conditionalism is not consistent with Ramsey's atomistic framework. For consider worlds α and β such that $\alpha \not\approx \beta$ and the proposition — call it A — that α is the actual world. Then since worlds are (nearly) maximally specific it follows that any world in which A is true is ranked with α. But then there is no world in which A is true which is equally preferred to β.

The only way I can see of blocking this argument, is to deny that worlds imply the existence of propositions stating that they are the actual world. But this would be to argue, in effect, that worlds could not be represented propositionally and that contradicts the requirement that the agent be able to choose amongst them (which, I take it, presupposes that she can distinguish them propositionally). So this is not a response open to Ramsey. And though the issue may in some sense be 'merely technical', some modification to his framework will be required to deal with the problem.

3 THE EVALUATION

3.1 *Ethical Neutrality versus State-Independence*

> "Ramsey's essays, though now much appreciated, seem to have had relatively little influence."[19]

Savage's remark applies equally well today and mainstream decision theory descends from Savage and not Ramsey. There are, I think, two reasons for his lack of influence. One is that Ramsey's style is so elliptical, and his writings so lacking in detail, that decision theorists have been unsure as to what exactly he has or has not achieved.[20] The second is that the distinction between the problem of justifying the claims of decision theory regarding the properties of rational belief and desire and the problem of the measurement of the decision theoretic variables — degrees of belief and desire — has not been properly recognised. Due, as I suggested before, to the different possible roles played by representation theorems with respect to the two problems.

This is important, because from the point of view of the problem of justifying the decision theory he employs, Ramsey's representation theorem is not particularly helpful. For one is very unlikely to accept his axioms as definitive of rational preference for conditional prospects unless one accepts the theory of expected utility that motivates them. This is particularly true of axiom R4, which seems to have no justification other than that it secures the meaningfulness of utility differences. Taken as axioms of measurement, however, they do much better for they specify in a precise way the conditions under which a measure of the agent's degrees of desire, unique up to a choice of scale, is determined by her choices amongst prospects.

With respect to problem of justification, on the other hand, a theory like Savage's is a good deal more impressive. Savage chooses his axioms of preference with an eye to their independent plausibility as rationality conditions. Independent, that is, of the quantitative theory of belief and desire that he will derive from them. Such a claim can justifiably be made for the Sure-Thing principle, for instance. One need not grant much plausibility to expected utility theory to grant that of two actions that yield the same outcomes when C is the case, one should choose the one with the preferred outcome when C is not. But Savage builds a very strong and implausible assumption into the very framework of his decision theory. He assumes that the desirability of any possible outcome of an action is independent of the state of the world in which it is realised.

Let us start by getting a general idea of the problem. It is a banal fact about our attitudes to many things that they depend on all sorts of contextual factors.

[19] Savage [1954, p. 96]

[20] For instance, Fishburn [1981] rejects Ramsey-type theories in favour of Savage-type ones on the grounds of his 'restricted act space'. In fact, however, his set of conditional prospects is roughly equivalent to Savage's set of acts.

Hot chocolate can be delightful on a cold evening, but sickly in the heat of a summer's day. Swimming on the other hand is best reserved for those hot days. I shall say, somewhat barbarically, that the swimming or drinking hot chocolate is desirabilistically dependent on the weather. Many things, on the other hand, are to all practical purposes desirabilistically independent, certainly swimming and the temperature on the moon are for me. Any reasonable theory of rational agency ought to recognise these banal facts.

How does Savage's theory violate them? Savage uses observations of choices amongst actions to determine agents' attitudes. Actions, on his account, are functions from states of the world to possible outcomes: when you choose an action you choose to make it true that if the world is in state s_1 then outcome o_1 will be realised, if it is in state s_2 then outcome o_2 will be realised, and so on. Now if we are to recognise that the desirability of the outcomes of actions depend on the state of affairs in which they are realised, then either the utilities we derive for them must be state-dependent i.e. of the form $U(o_1|s_1)$, or the outcome o_1 must include the fact that s_1 prevails (as outcomes of Ramsey's conditional prospects do). But Savage both assumes that any combination of state and outcome is possible and assigns state-independent utilities to outcomes.

On Ramsey's account, outcomes (worlds) are maximally specific with regard to things that matter to the agent, but not all outcomes are achievable in any given state. So his theory requires no violation of the banal facts concerning the interdependence of our attitudes. Instead of building desirabilistic independence into his framework, he postulates the existence of only a very limited class of possibilities — those represented by ethically neutral propositions — which are desirabilistically independent of all others. One may question whether there are any propositions that are truly ethically neutral, but there are clearly some that are good approximations. The postulation of their existence is not a heavy burden for such an idealised account to bare.

This is not, of course, the end of the matter. There have been numerous attempts to solve the problem of state-dependent utilities (as it has become known) within Savage's framework.[21] Many of the proposed solutions are ingenious, but they always come at the cost of greater complexity and more burdensome assumptions. This is not the appropriate place to review the literature, but anyone who has ploughed through it will have little difficulty in recognising the merits of the elegantly simple method that Ramsey invented. Indeed, despite the problems in the details that we discovered, there is nothing that matches it as an answer to the problem of measurement.

3.2 *Jeffrey's Objection*

In motivating his own method of measuring belief, Ramsey argues that the established method of offering bets with monetary rewards to someone to elicit their

[21] For a summary see Schervish *et al.* [1990]

degrees of belief is 'fundamentally sound' but suffers from being both insufficiently general and necessarily inexact. Inexact partly because the marginal utility of money need not be constant, partly because people may be especially averse (or otherwise) to betting because of the excitement involved and partly because ''the proposal of a bet may ... alter his state of opinion''[22]

Ramsey seems to think that his own theory is not vulnerable to these problems, even though his method is similar in many ways to the one he is criticising. Not everyone would agree. Richard Jeffrey, for instance, has argued that just such a problem plagues Ramsey's own account.[23] In order to measure agents' partial beliefs, Ramsey requires that they treat possibilities like it being α if P and β if not as real prospects i.e. things that can be brought about by choice. But to persuade someone of our power to bring it about at will that it will be sunny tomorrow if the coin lands heads and snowy if it lands tails is to cause them to entertain possibilities which at present they do not. That is, one must modify their beliefs in order that one may better measure them! There is of course no guarantee then that the measurements so effected are not, at least partially, artifacts of the measurement process itself.

It is worth noting that such an objection, if sustainable, can be directed with equal force at Savage. For when Savage invites agents to make choices amongst actions, he supposes that they know exactly what consequences the action has in every possible state of the world and hence what they are committing themselves to. This makes the choice of a Savage-type action rather more like a choice amongst Ramsey-type conditional prospects than amongst the sorts of things we normally think of as actions. Indeed, formally, they are just the same thing: functions from events to outcomes. What is of the essence, in any case, is that agents are invited to choose amongst causal mechanisms of some kind whose effects in each possible circumstance are advertised in advance. And the essence of Jeffrey's objection is that agents may legitimately doubt the efficacy of such mechanisms, and make their choice in the light of these doubts. If they do, their choices will reflect not their evaluation of the advertised outcomes of the chosen prospect, but their evaluations of the outcomes that they actually expect. Even in pure gambling situations agents will factor in such possibilities as the casino closing before paying up.

How might Ramsey respond to this problem? Sobel argues that Ramsey must require that the probability of a proposition P be measured only by means of conditional prospects which are such that P's probability is evidentially and causally independent of the conditional prospect being offered (by, for instance, addition of a further restriction in Definition 12).[24] But there is no obvious way of expressing this independence condition in terms of agents' preferences and so no way of applying it until the probability of P has already been measured.

[22] Ramsey [1926, p. 68]
[23] Jeffrey [1983, chapter 10]
[24] Sobel [1988, p.256]

A natural response to Jeffrey's objection would be to say that Ramsey does not, in fact, require that agents really believe in such fanciful causal possibilities. All that he requires is that they choose amongst gambles *as if* they believed that they would truly yield the advertised consequences under the relevant conditions. To be sure, such a response will not satisfy the behaviourist, for introspection on the part of agents must then play a crucial role in the production of their choices. For when we ask Mary to choose between an prospect which yields sunny weather if Labour wins the next election and rainy weather if they do not, and one which yields rainy weather if Labour wins the next election and sunny weather if they do not, we are in effect asking her to determine *for herself* what she would prefer in the event that such gambles were reliable. But then we may as well just ask Mary what she would prefer and forget about the observation of choices altogether.

And indeed why not? Let us see what such a reconstrual of Ramsey's method would amount to in the context of the experimental determination of a subject's degrees of belief and desire, by comparing the following measurement schemes:

1. *Scheme A*: The subject introspects her degrees of belief and desire and then reports them to the observer.

2. *Scheme B*: The observer presents the subject with a number of options and her choice is recorded. The set of options offered is varied until a ranking over all of them has been constructed from the observations of her choices. This ranking is then used to construct a quantitative representation of her degrees of belief and desire.

3. *Scheme C*: The observer questions the subject as to which of various possibilities she would prefer were the true one. Her answers are then used to construct a ranking of all possibilities and this in turn determines a quantitative representation of her mental attitudes.

Scheme A is the method criticised by behaviourists and Ramsey alike for its naive dependence on introspection. Scheme B summarises the behaviourist's method, Scheme C the alternative interpretation of Ramsey's method. Both are underwritten by the representation theorems of Decision Theory. In Scheme C introspection plays an essential role: to provide answers to the experimenter's questions the subject must reflect upon and judge her own preferences. In Scheme B, on the other hand, though it is conceivable that the subject arrives at a choice via introspection of her preferences, she need not do so. She may simply choose without reflection, indeed without even having the concept of preference. Scheme C is a method intimately tied to the possibility of linguistic communication and the kind of self-consciousness that typically accompanies it; Scheme B is just as applicable to earthworms as to philosophers.

I see no reason why Ramsey should be resistant to this interpretation of his method as a version of Scheme C. Although it requires him to disavow the behaviourist pretension that introspection can be completely eliminated in favour of rich observations of behaviour, it does not commit him to the view instantiated in Scheme A (and which he clearly rejects) that partial beliefs and desires can be directly introspected. In this sense this interpretation does not conflict with anything that he says. And it has the crucial advantage of extricating him from Jeffrey's objection.

3.3 Ramsey à la Jeffrey?

In filling in the details of Ramsey's theory of measurement we have had reason to raise a number of questions and to make a number of supplementary assumptions. But only the incompatibility of Conditionalism with his framework seems to raise a serious problem for Ramsey. In fact, however, this problem is largely a technical one and can be solved by modifications to Ramsey's framework that are not contrary to the 'spirit' of his account. I will content myself with sketching the essentials.

The basic move is to take (non-contradictory) propositions rather than worlds to be the elementary objects of preference. One immediate positive spin-off is that the notion of ethical neutrality can then be formulated in a manner less dependent on the peculiarities of Wittgenstein's theory of atomic propositions.[25]

DEFINITION 20. Suppose P and Q are mutually consistent propositions. Then P is **neutral** with respect to Q iff $PQ \approx Q \approx P\neg Q$.

DEFINITION 21. P is **ethically neutral** iff P is neutral with respect to all propositions Q consistent with P.

Conditional prospects must now be defined as functions from partitions of propositions to (non-contradictory) propositions, with the constraint that for any conditional prospect Φ and proposition X, $\Phi(X)$ implies X. But little else need change, since most of Ramsey's formal argument is carried out at the level of values. Of course, utilities as well as probabilities will now be defined on propositions. Finally the relevant definition of Conditionalism is as follows.

PROPOSITION 22. *For any propositions P and Q there exists propositions P' and Q' such that P' implies P, Q' implies Q and $P' \approx Q'$.*

Proposition 22 can be satisfied only if there are no propositions X such that, for all propositions Y, X implies Y or X implies $\neg Y$. In other words, Conditionalism requires that the domain of the preference relation be atomless.

All of this takes Ramsey's framework quite a bit closer to that underlying Richard Jeffrey's decision theory and Ethan Bolker's representation theorem for it.[26] So too did the contention that Ramsey's work should be interpreted in such a way as to rid it of any dependence on dubious causal devices such as gambles. But I do not to propose to go much further in their direction, because from the perspective of the problem of the measurement of belief, the Jeffrey–Bolker theory suffers

[25]See Bradley [1997] for a more detailed development of these ideas.

[26]See Jeffrey [1983] and Bolker [1967].

from a crucial weakness by comparison to Ramsey's.[27] For Bolker's representation theorem does not establish the existence of a unique measure of an agent's beliefs or a measure of her degrees of desire unique up to a choice of scale. In particular it allows for the possibility that two probability measures of an agent's degrees of belief, P_1 and P_2, both be consistent with her expressed preferences yet differ to the extent that there are propositions A and B such that $P_1(A) > P_1(B)$ but $P_2(B) > P_2(A)$.[28]

The essential difference, in this regard, between Ramsey's theory and that of Jeffrey and Bolker is that the latter make do without any conditional prospects of the kind postulated by Ramsey, working only with agents' attitudes to propositions. The price of this ontological economy would seem to be the underdetermination of agents' degrees of belief and desire by the evidence of their expressed preferences. If the price is too high, we have reason to favour a Ramsey-type theory when addressing the problem of measurement.

But this should not obscure the fact that this discussion raises a difficult question concerning the status of Ramsey's conditional prospects. For if conditional prospects could be given propositional expression then it should be possible to strengthen the Jeffrey's theory by simply adding to it suitably translated versions of Ramsey's postulates concerning preferences for conditional prospects. But the evidence is that this cannot be done without leading to some unpalatable consequences. But if Ramsey's conditional prospects have no adequate propositional correlates, as has already been suggested by his definition of ethical neutral propositions, what exactly is their nature?[29]

Department of Philosophy, Logic and Scientific Method,
London School of Economics, UK

BIBLIOGRAPHY

[Bradley, 1998] R. W. Bradley. A Representation Theorem for a Decision Theory with Conditionals, *Synthese*, **116**, 187–229, 1998

[Bradley, 1997] R. W. Bradley. *The Representation of Beliefs and Desires within Decision Theory*, PhD dissertation, University of Chicago, 1997.

[Bolker, 1967] E. D. Bolker. A Simultaneous Axiomatisation of Utility and Subjective Probability, *Philosophy of Science*, **34**, 292–312, 1967.

[Davidson and Suppes, 1956] D. Davidson and P. Suppes. Finitistic Axiomatisation of Subjective Probability and Utility, *Econometrica*, **24**, 264–275, 1956.

[Fishburn, 1981] P. C. Fishburn. Subjective Expected Utility: A Review of Normative Theories, *Theory and Decision*, **13**, 139–199, 1981.

[Jeffrey, 1983] R. C. Jeffrey. *The Logic of Decision*, 2nd edn, Chicago: University of Chicago Press, 1983.

[Joyce, 1999] J. Joyce. *The Foundations of Causal Decision Theory*, Cambridge University Press, 1999.

[27] On the other hand, with respect to the problem of normative justification the Jeffrey–Bolker theory is much better than Ramsey's.

[28] For further discussion of this problem, see Bradley [1997] and Joyce [1999].

[29] An attempt to give their logical properties is to be found in Bradley [1998] and [1997].

[Krantz *et al.*, 1971] D. H. Krantz, R. Duncan Luce, P. Suppes and A. Tversky. *Foundations of Measurement*, Volume 1, Academic Press, 1971.

[Pfanzagl, 1968] J. Pfanzagl. *Theory of Measurement*, New York: Wiley, 1968.

[Ramsey, 1926] F. P. Ramsey. Truth and Probability. In *Philosophical Papers*, ed. D. H. Mellor, Cambridge: Cambridge University Press, 1990..

[Sahlin, 1990] N.-E. Sahlin. *The Philosophy of F. P. Ramsey*, Cambridge: Cambridge University Press, 1990.

[Savage, 1954] L. J. Savage. *The Foundations of Statistics*, 1954. 2nd edition, New York: Dover, 1972

[Schervish *et al.*, 1990] M. J. Schervish, T. Seidenfeld and J. B. Kadane. State-dependent Utilities, *Journal of the American Mathematical Association*, **85**, 840–847, 1990.

[Sobel, 1988] J. H. Sobel. Ramsey's Foundations Extended to Desirabilities, *Theory and Decision*, **44**, 231–278, 1988.

EDWARD F. MCCLENNEN

BAYESIANISM AND INDEPENDENCE

1 INTRODUCTION

The cornerstone of the Bayesian theory of utility and subjective probability, the independence principle, places a significant restriction on the ordering of options that involve risk or uncertainty (in suitably defined senses of each of these terms). For the case of options involving risk, the principle is typically formulated in the following manner:

> **The Independence Principle (IND)**: Let P, P' and Q be any three risky prospects or gambles, and $0 < \lambda \leq 1$ then
>
> $$P \sim P' \Rightarrow \lambda P + (1 - \lambda)Q \sim \lambda P' + (1 - \lambda)Q,$$

where "$\sim$" denotes indifference.[1] That is, substituting indifferent components for one another preserves indifference.

IND invites particularization and reformulation in a variety of different ways. For the matters to be explored below, perhaps the most important particularization is where the components are not themselves lotteries, but "sure" outcomes (e.g., amounts of money). Since an outcome involving no risk can be viewed as a "gamble" in which one gets that outcome with probability 1, **IND** yields directly the following:

> **Independence for Sure Outcomes (ISO)**: Let O_1, O_2, O_3, be any three sure outcomes (monetary prizes, etc.), and $0 < \lambda \leq 1$: then
>
> $$O_1 \sim O_2 \Rightarrow \lambda O_1 + (1 - \lambda)O_3 \sim \lambda O_2 + (1 - \lambda)O_3.$$

Independence, as formulated above, presupposes that λ is a well defined probability value satisfying, in general, the condition $0 \leq \lambda \leq 0$. But the principle has a natural extension to cases in which the agent faces prospects the likelihoods of whose outcomes may not be well defined. In Savage [1972], independence is formulated without reference to probabilities at all, but only to the notion of mutually exclusive and exhaustive events that condition the consequences of various acts. Formulated somewhat less rigorously than it is in Savage, but in a manner that clarifies its connection with the version defined above, the requirement is:

[1]The notations I shall utilize here are taken from Fishburn and Wakker [1995]. That article contains an extremely helpful and for the most part insightful guide to the history of the utilization of various versions of the independence axiom, as well as a very comprehensive bibliography. For the issues to be discussed here, one cannot do better than start with their account. My only complaints are (1) that they do not give sufficient attention to the dominance interpretation that clearly motivated Savage to embrace a version of independence, and (2) they pass over the manifold criticisms that have been mounted to the acceptance of various versions of this axiom.

D. Corfield and J. Williamson (eds.), Foundations of Bayesianism, 291–307.

Savage Independence (**SI**): Let E and $-E$ be mutually exclusive and exhaustive events conditioning the various risky components of four gambles, R, R', S, S' and let the schedule of consequences be as follows:

$$\begin{array}{ccc} & E & -E \\ R & P & Q \\ R' & P' & Q \\ \\ S & P & Q' \\ S' & P' & Q' \end{array}$$

Then: $R \geq R' \Leftrightarrow S \geq S'$, where "$\geq$" denotes the weak preference ordering relation.

In very general terms, the particular formulation to which appeal is made in a given axiomatic construction typically depends in part on the strength of the other axioms employed and in part on considerations of simplicity and/or formal elegance. But, as we shall see, some version or other of independence or something that implies independence is invariably to be found in the construction.

2 AN ALTERNATIVE FORMULATION: THE SURE-THING PRINCIPLE

The independence axiom is only one of two ways in which the key axiom of expected utility and subjective probability has been formulated. The other formulation, following Savage [1972] came to be known as the "sure-thing" principle (**STP**).[2] It is introduced in Friedman and Savage [1952] in the following manner (once again adjusting its formulation to our present notation):

> [S]uppose a physician now knows that his patient has one of several diseases for each of which the physician would prescribe immediate

[2]The earliest reference to what came to be known as the "sure-thing" principle, as far as I have been able to determine, occurs in a discussion by Savage of a decision situation in which risk is not well-defined—what has come to be know as a case of decision making under conditions of uncertainty. Savage imagines an agent who is interested simply in maximizing *expected* income, but who is faced with a situation in which he cannot appeal to well-defined probabilities. Under such circumstances, Savage argues,

> ...there is one unquestionably appropriate criterion for preferring some act to some others: If for every possible state, the expected income of one act is never less and is in some cases greater than the corresponding income of another, then the former act is preferable to the latter. This obvious principle is widely used in everyday life and in statistics, but only occasionally does it lead to a complete solution of a decision problem. [Savage, 1951]

In neither this article, nor in the one he wrote with Friedman a year later, does Savage characterize the principle in question as the "sure-thing" principle: that term appears to occur for the first time in Savage [1972].

> bed rest. We assert that under this circumstance the physician should, and unless confused, will prescribe immediate bed rest whether he is now, or later, or never, able to make an exact diagnosis.
>
> Much more abstractly, consider a person constrained to choose between a pair of alternatives, R and R' without knowing whether a particular event E does (or will) in fact obtain. Suppose that, depending on his choice and whether E does obtain, he is to receive one of four gambles, according to the following schedule:
>
> | | Event | |
> |---|---|---|
> | | E | $-E$ |
> | Choice | | |
> | R | P | Q |
> | R' | P' | Q' |
>
> The principle in sufficient generality for the present purpose asserts: if the person does not prefer P to P' and does not prefer Q to Q' then he will not prefer the choice R to R'. Further, if the person does not prefer R to R', he will either not prefer P to P' or not prefer Q to Q' (possibly both).
>
> We anticipate that if the reader considers this principle, in the light of the illustration that precedes and such others as he himself may invent, he will concede that the principle is not one he would deliberately violate [Friedman and Savage, 1952, pp. 468–9]

As Savage was to make clear two years later, what is being invoked here is essentially the dominance principle that had been employed by many statisticians as an admissibility criterion.[3] As formulated above, it should be noted that it applies to cases in which the component entities, P, Q, etc., are themselves risky prospects or gambles. It can also, just like **IND**, be particularized to the case where outcomes are not gambles but sure amounts of money or other goods.

3 THE CONNECTION BETWEEN INDEPENDENCE AND DOMINANCE

Some sense of the logical connection between **IND** and **STP** can be gained by considering how one might get from a simple version of dominance to a version of independence. Consider the following version of dominance formulated with respect to two components, and well-defined probabilities:

[3]See Savage [Savage, 1972, p. 114]. The principle can be recast in a form that is applicable to options defined in terms of some partition of n mutually exclusive and exhaustive events, and also applied to cases in which well-defined probabilities can be associated with each event in the partition. In any of its formulations, of course, one must presuppose that the choice of an option does not differentially affect the probabilities of the conditioning events. That is, the conditional probability of E_i given choice of P must be equal to the conditional probability of E_i, given choice of Q.

(**STP***) For $0 < \lambda \leq 1$, and all P, Q, P', Q',
$P > P'$ and $Q \geq Q' \Rightarrow [\lambda P + (1 - \lambda)Q] > [\lambda P' + (1 - \lambda)Q']$.

But since indifference implies weak preference, this yields immediately:

For $0 < \lambda \leq 1$, and all P, Q, P', Q',
$P > P'$ and $Q \sim Q' \Rightarrow \lambda P + (1 - \lambda)Q > \lambda P' + (1 - \lambda)Q'$.

And, on the plausible assumption that any component is indifferent to itself, this in turn implies:

(**IND***) For $0 < \lambda \leq 1$, and all P, Q, P',
$P > P' \Rightarrow \lambda P + (1 - \lambda)Q > \lambda P' + (1 - \lambda)Q$.

Now **IND*** makes the ordering of the two gambles—with fixed probabilities, and one component in common—a simple function of the relative ordering of the other component: this speaks to the notion of independence, and in fact **IND*** is a version of the independence axiom. Principle **STP***, on the other hand, is a version of **STP**—suggesting as it does that, if the components of one gamble are uniformly at least as good as the ones with which they are paired in the other gamble, and in one case better, then this determines the ordering of the two gambles.[4] One can also start with some version of **IND** and derive a version of **STP**.[5]

Many—starting with Friedman and Savage [1952]—have viewed the dominance principle as intuitively the more compelling, and have thus viewed this as an effective way to motivate the independence condition. But the logical connection between the two principles can also be used to undermine the intuitive status of the sure-thing principle. If one had evidence that within some domain the independence condition is suspect, one could conclude that since independence is a logical consequence of the sure-thing principle (together with certain other putatively unproblematic axioms), the sure-thing principle in that context must also be regarded as suspect. In effect, any such argument in favor of independence, via *modus ponens*, could be used to undercut the sure-thing principle, via *modus tollens*.

Since it was Savage himself who originally framed the sure-thing principle, and who, together with Friedman, used it to secure a version of independence, one might have expected that he would introduce it as a postulate in the book he subsequently published, *The Foundations of Statistics*. Interestingly enough, however, there he adopts just the reverse approach. The appeal to "sure-thing" considerations serves to simply *informally* motivate the independence postulate:

> A businessman contemplates buying a certain piece of property. He considers the outcome of the next presidential election relevant to the

[4]Friedman and Savage [1952, p. 469] proceed in this manner, by starting with a version of the sure-thing principle framed in terms of a partition of events rather than probabilities, together with certain other assumptions, and deriving a version of the independence axiom.

[5]To be sure, as Fishburn and Wakker [1995, p. 1137] argue, for a full-blown axiomatization one can only substitute one of these principles for the other in the presence of certain other axioms.

> attractiveness of the purchase. So, to clarify the matter for himself, he asks whether he would buy if he knew that the Republican candidate was going to win, and decides he would do so. Similarly, he considers whether he would buy if he knew the Democratic candidate were going to win, and again finds that he would do so. Seeing that he would buy in either event, he decides he should buy, even though he does not know which event obtains or will obtain, as we would ordinarily say. It is all too seldom that a decision can be arrived at on the basis of the principle used by this businessman, but, except possibly for the assumption of simple ordering, I know of no other extralogical principle governing decisions that finds such ready acceptance. [Savage, 1972, pp. 21–22]

But right on the heels of this informal discussion, it is a version of the independence principle that is introduced as a postulate (and a version of the sure-thing principle is then shown to be derivable as a theorem).

Whether one takes the sure-thing or the independence principle as basic, it is important to note that what is requisite for the expected utility and subjective probability constructions is some version or other that is explicitly taken to hold for gambles (or uncertain prospects) defined over gambles. Specifically, Independence for Sure Outcomes (**ISO**) will not suffice.[6] This point has not always been stated with the clarity that it deserves. In many expositions, independence or dominance is illustrated with reference to a case involving "sure" outcomes (for example, monetary payoffs), but what is employed in the construction is the much stronger generalized version of the principle. Where the issue is the plausibility of the principle in its requisite strength, it does no good to cite weaker versions of the same.

4 INDEPENDENCE AS NON-COMPLEMENTARITY

The introduction of the term "independence" appears to have been motivated by a perceived analogy to the economic concept of independent goods, in which the value of a bundle of various quantities of different goods is an additive function of the value of the quantities of the various separate goods that make up that bundle. It is, of course, a well-known fact that independence with respect to the value of a bundle of commodities can fail. The value of the combination of x amount of one good and y amount of another good may not be equivalent to the sum of the value of x amount of the one good and the value of y amount of the other good. Failure of independence in such cases is said to be due to complementarity or interaction. That is, the value of one good may be enhanced or reduced in virtue of its being combined with some other good, as, for example, in the proverbial case in which white wine is said to complement fish, and red wine to complement beef.

[6]Nor would a version of **STP** framed with regard to sure outcomes.

Starting with von Neumann and Morgenstern [von Neumann and Morgenstern, 1953], however, one finds repeated appeal to the argument that such a problem of complementarity cannot arise in the case of what are disjunctive (as distinct from conjunctive) bundles of goods, i.e., lotteries over goods, and hence that the assumption of independence in this context is warranted. Here is how the argument emerges in their work (adjusting the quote to the notation introduced above):

> By a combination of two events we mean this: Let the two events be denoted by P and Q and use, for the sake of simplicity, the probability 50%–50% . Then the "combination" is the prospect of seeing P occur with probability 50% and (if P does not occur) Q with the (remaining) probability of 50% . We stress that the two alternatives are mutually exclusive, so that no possibility of complementarity and the like exists.

Samuelson [1952] explicitly marks the analogy and, while acknowledging that complementarities can arise in the case of (conjunctive) bundles of goods, insists that the nature of a disjunctive (or stochastic) bundle, in which one is to get just one of the disjuncts, makes it plausible to impose independence as a condition on preferences for gambles.

The argument for non-complementarity in the case of disjunctive bundles is, however, simply not compelling at all. Disjunctive bundles may not be subject to the problem of commodity complementarity, but this does not rule out the possibility of forms of "complementarity" that are special to disjunctive prospects.[7]

By way of illustration, consider the following type of decision situation, first isolated by Allais [1953], in which the conditioning events are the numbered tickets to be drawn in a lottery, with various associated prizes (in units of $ 100,000) attaching to each:

	Lottery ticket number		
	1	2–11	12–100
P	5	5	5
Q	0	25	5
P'	5	5	0
Q'	0	25	0

[7]The possibility of complementarity is discussed in [Manne, 1952; Allais and Hagen, 1979; McClennen, 1983; Loomes, 1984] and [Sen, 1985]. Broome [1991, pp. 37–38, fn. 13] has complained that I do not understand that "complementarity" is a term (as used by economists) that applies just in the context of conjunctive bundles of commodities. I am prepared to yield to Broome with regard to a point of terminology; but terminology, it seems to me, is not the issue. As the examples I am about to present show, I believe, the considerations that undercut the application of the independence condition to certain cases of choice under conditions of risk and/or uncertainty closely parallel the considerations that undercut the application of an analogous independence condition to certain cases of choice among commodity bundles.

Significant proportions of subjects register a preference for P over Q but a preference for Q' over P'.[8] For one who accepts both that the set of options should be weakly ordered, and also accepts a standard reduction principle, this preference pattern leads to a violation of **IND.**

Here, the set of alternatives consisting of $\{P, Q\}$ differs from the set of alternatives consisting of $\{P', Q'\}$ only in respect to the level of the constant prize associated with ticket numbers 12–100. With an appropriate repartitioning of conditioning states—by letting E be the state in which the ticket drawn has a number between 1 and 11, and *not-E* be the state in which the ticket drawn has a number between 12 and 100—one can then directly appeal to **SI**—Savage's version of the independence principle—to show that if P is preferred to Q then P' must be preferred to Q'.

Consider, however, an agent who is concerned in risky choices with the dispersion in the monetary value of the various possible prizes, and who, other things being equal, prefers less to more dispersion.[9] The gamble P has zero dispersion, and this may, despite its lower expected monetary return, make it attractive relative to Q. When P and Q are transformed, respectively, into P' and Q', by reducing the payoff for lottery tickets 12–100 from 5 units to 0, the expected monetary value of both P' and Q' is reduced by the same amount. But this is not the case with regard to the increases in dispersion. The increase in dispersion from P to P' is greater than the increase from Q to Q', regardless of how dispersion is measured. If dispersion considerations are relatively important, then, the fact that in the case of Q' the alteration in payoffs results in less increase in dispersion might, then, tip the balance in favor of Q'.

The way in which dispersion considerations can result in preference patterns that fail to conform to **IND** can be illustrated in an even more striking manner by reference to an example offered by Kahneman and Tversky [1979]. Suppose the agent is to make a selection from each of the following two sets of paired alternatives:

	R (66)	Y (33)	Black (01)
P	\$ 2400	\$ 2400	\$ 2400
Q	\$ 2400	\$ 2500	\$ 0
P'	\$ 0	\$ 2400	\$ 2400
Q'	\$ 0	\$ 2500	\$ 0

In this instance, the options are defined in terms of drawing a colored ball from an urn—Red, Yellow, or Black—where the urn contains only balls of those three

[8] Empirical findings are surveyed in, e.g., [Kahneman and Tversky, 1979; MacCrimmon and Larsson, 1979] and [Shoemaker, 1980]. Savage himself, who was firmly committed to **IND**, admitted that on first consideration these were his preferences. See Savage [1972, p. 103].

[9] For simplicity, I assume here that the agent is concerned simply with monetary payoffs and probabilities (including, of course, in this case the dispersion features of the probability distributions) and that he treats the value of sure amounts of money as a linear function of monetary amount.

colors, and in proportion 66/33/1.

Again, a significant number of persons prefer P to Q but also prefer Q' to P', in violation of **IND.** The preferences in question can arise, however, if the agent ranks the options by appeal to the following dispersion-sensitive rule. Suppose the agent prefers, other things being equal, a higher expected monetary return to a lower one but also prefers, other things being equal, a smaller expected shortfall, where expected shortfall is defined as the expected deviation below the mean (that is, 1/2 the expected deviation from the mean). Assume, for the sake of simplicity, that these are the only relevant factors and that the agent's implicit tradeoff rate between expected return and expected shortfall, and hence his rule for evaluating such gambles, is given by the following linear function:

$$V(P) = E(P) - kS(P),$$

where $V(P)$ is the value, $E(P)$ is the expected return (mean value), $S(P)$ is expected shortfall, of the gamble P, and k is some constant, defined on the open interval $[0, 1]$.[10] For illustration, let $k = 1/2$. Then one gets the following values:

	E	S	$E - .5S$
P	2400	0	2400
Q	2409	30.03	2393.985
P'	816	538.56	546.72
Q'	825	552.75	548.625

If this method of evaluation—based on the specified linear function of expectation and shortfall (dispersion)—is used, P will be preferred to Q but Q' will be preferred to P', in violation of **IND.**[11]

Now, if a person utilized such a evaluation procedure—one that incorporated considerations of expected monetary value and dispersion—he or she could argue that precisely why a violation of **IND** occurs here is because a special kind of complementarity arises in the concatenation of risky prospects. To see this it will prove useful to change the example slightly. In the example just discussed, P is presumed to be preferred to Q. But suppose that P is replaced with P^* which yields the agent a payoff \$ 2393.985 instead of \$ 2400, and the agent employs the rule stated above. Then one obtains the following values:

[10] Again, for the sake of the example, I assume that the agent is concerned simply with monetary payoffs and probabilities (including the dispersion features of the probability distributions) and that he treats the value of sure amounts of money as a linear function of monetary amount. Similar results can be obtained if shortfall, and, more generally, dispersion, is measured in terms of the Gini-coefficient.

[11] For a very interesting, and much more formal, exploration of measures of risk that are sensitive to both mean values and dispersion (specifically variance) see Pollatsek and Tversky [1970]. As they go on to note, this approach to the measurement of risk is incompatible with standard expected-utility theory.

	E	S	$E - .5S$
P^*	2393.985	0	2393.985
Q	2409	30.03	2393.985
P'	803.9549	530.610234	538.649783
Q'	825	552.75	548.625

The gambles P^* and Q are now indifferent to each other, but Q' is still preferred to P'. In this instance, then, when Q is substituted for P^* in the more complex gamble, even though P^* and Q are indifferent to one another, the latter enhances the value of the resultant complex gamble more than does the former: that is, Q' is preferred to P'. Moreover, there is no mystery as to this differential impact on the value of the resultant complex gambles. The rule employed makes the value of the complex gamble—the whole—a function of mean value and shortfall. The expectation factor's impact is "well-behaved". The embedding of P^* in the more complex gamble results in a proportional decrease in expected value that is strictly equal to the proportional decrease that results from embedding Q: that is, $E(P^*)/E(P') = E(Q)/E(Q')$. The differential impact is due to the shortfall factor. The proportional increases in dispersion are not equivalent. Thus, for one who is concerned about dispersion, even though P^* is indifferent to Q, when the two are considered in isolation, there is a better "fit" between Q and [. , 34/100; $ 0, 66/100] than there is between P^* and [. , 34/100; $ 0, 66/100]. Combining Q with the balance results in a smaller proportional increase in dispersion than that which results from combining P^* with the balance.

There are two other much discussed counterexamples to the independence principle, which serves to suggest a distinct type of complementarity that can arise in the concatenation of risky prospects. The examples are both due to Ellsberg [1961] and are directed not at **IND** but at Savage's version, **SI**. In the first example, Ellsberg considers a situation in which the agent is to choose between various gambles based upon drawing a ball at random from an urn that contains Red, Black, and Yellow balls, where one knows that there are 30 Red balls in the urn, and that 60 are either Black or Yellow, but the relative proportion of Black and Yellow is unknown:

	(30) Red	(60) Black	Yellow	Range of EMR
P	$ 100	0	0	33 1/3
Q	0	100	0	0 to 66 2/3
P'	$ 100	0	100	33 1/3 to 1
Q'	0	100	100	66 2/3

Since the probabilities of the conditioning events are only partially defined, one cannot associate with each such option an unambiguous expected monetary return.

But, as the column to the far right serves to indicate, one can still specify the possible range of such values.

Ellsberg notes that many people prefer P to Q, while preferring Q' to P'. He also notes that the following rule of evaluation generates this preference ordering: rank options in terms of increasing minimum expected monetary return. Now note that the pair of options $\{P, Q\}$ differ from the pair of options $\{P', Q'\}$ only in respect to the payoffs in the event that a Yellow ball is drawn. But in each case the amount to be received if a Yellow ball is drawn is constant. Thus, once again with an appropriate repartitioning of the states, **SI** applies, and requires that P be preferred to Q just in case P' is preferred to Q', contrary to the described preferences. Thus, preferences based on the rule in question violate **SI**

Once again, one can interpret what is happening here in terms of the notion of complementarities with respect to the value of disjunctions of outcomes. The shift from a situation in which, under the condition of drawing a yellow ball one receives \$ 0, regardless of which act is chosen, to a situation in which, under the same chance conditions, one receives \$ 100, regardless of which act is chosen, results in "contamination" (to use Samuelson's term for complementarity). And, once again, there is no mystery here as to how this happens. The person who adopts Ellsberg's rule can be characterized as uncertainty (or, as Ellsberg himself terms it, "ambiguity") averse: uncertain prospects (as distinct from those whose associated expected return is well defined) are discounted to their minimum expected return. Although one can think of P' and Q' as resulting from a modification of P and Q, respectively—in each case the addition of \$ 100 to the payoff when a yellow ball is drawn, this proportional increase in payoffs has differential impact with regard to uncertainty or ambiguity. In the case of choice between P and Q it is Q that presents an uncertainty; but given the substitution, it is now the counterpart to P, namely, P', (and not the counterpart to Q, namely Q') that presents the uncertainty.

In Ellsberg's other example one is to suppose that the agent is asked to choose between the following three gambles:

P [\$ 100, E; \$ 0, $-E$]
Q [\$ 0, E; \$ 100, $-E$]
R [\$ 100, H; \$ 0, T],

E is some event whose likelihood is completely unknown and (H, T) are the two possible outcomes of a toss of what is presumed to be a fair coin, so that a given subject can be presumed to assign a subjective probability of $1/2$ to each of H and T. Many persons report that they prefer R to both P and Q but that (predictably) they are indifferent between P and Q. Such a ranking typically expresses itself as a willingness to pay more for R than for either P or Q, but to be willing to pay exactly the same amount for either P or Q. That is, for the same prizes or outcomes, persons typically prefer even-chance odds to ambiguous (i.e., uncertain) odds. Consider now, the following compound gambles, based on the above options (where, once again, 'H' and 'T' refer to the outcomes of the flip of the same fair coin):

R' $[P, \text{H}; Q, \text{T}]$
P' $[P, \text{H}; P, \text{T}]$

By appeal to the standard reduction rule, the agent must rank R' as indifferent to R.[12] Moreover, by analogous reasoning, P' must be ranked indifferent to P. Hence, given the preferences projected above for R and P, and acceptance of the standard weak ordering axiom, the agent must prefer R' to P'. But this violates **IND.** P' and R' differ only in that where R' has an occurrence of Q, P' has an occurrence of P; but, by hypothesis, the agent is indifferent between P and Q, hence by **IND**, P' should be indifferent to R'.

The agent who prefers R to P, however, has a natural rejoinder and one that appeals once again to a special kind of complementarity for disjunctive prospects—a complementarity that occurs when the agent's method of evaluation is sensitive to ambiguity or uncertainty. While the agent is indifferent between $P$and Q, the combination of Q with P in the even-chance lottery R' results in the ambiguity or uncertainty of the odds associated with each gamble taken separately canceling each other out, while the combination of P with P in P' results in no corresponding reduction of ambiguity. This is not due to anything pertaining to P or Q taken separately. The resolution of the ambiguity is a function of the particular way in which the component gambles are combined in P' and in R'.

What has been isolated here is something that is distinct from the complementarity that arises in connection with conjunctive bundles. Here there is no question of some sort of interaction between prizes, both of which are to be received. It arises within the context of a disjunctive concatenation of prizes or goods, and turns on the implication of combining both well-defined and indeterminate odds. But it bears all the marks of being a type of complementarity. The gambles P and Q are clearly indifferent to one another when considered in isolation. When each is disjunctively combined with P to form P' and R', however, there is a "fit" that obtains in case of R' that does not obtain in the case of P'. The fit, moreover, is perfectly explicable: the rules for combining probabilities imply that R' involves no ambiguity with respect to the odds of receiving \$ 100; while those same rules imply that P' is maximally ambiguous. For one who is uncertainty (or ambiguity) averse, then, it does make a difference as to which of the (indifferent) components are combined with which.

The implication of both the Ellsberg and the Allais examples is quite clear. One cannot infer that **IND** is a plausible condition to impose on the ordering of disjunctive bundles simply by appeal to the consideration that complementarities of the type that arise in connection with conjunctive bundles cannot arise in connection with disjunctive bundles. That argument is a non-starter, for it ignores a kind of "complementarity" that is unique to disjunctive bundles, and that forms an intelligible basis for discriminating between prospects, if one is (as in the case of

[12] If one treats the probability of E as an unknown, p, combining P and Q together by tossing a coin to decide which one to choose results in p being cancelled out, so its value is irrelevant. That is, the coin toss results in one's having an ex ante 50–50 chance of getting \$ 100.

the Allais example) concerned with dispersion or (as in the case of the Ellsberg examples) concerned with uncertainty (ambiguity).

5 SURE-THING REASONING AS A BASIS FOR THE INDEPENDENCE PRINCIPLE

How do things fare if we approach the axiomatization by appeal to **STP** rather than **IND**? There is no question that the dominance idea to which **STP** appeals is intuitively very plausible. Recall, once again, the argument presented by Friedman and Savage [1952] in support of **IND**. A version of **IND** follows directly from **STP**, in the presence a standard reduction principle. Now, **STP** mandates preference for P over Q if, no matter what the turn of events, the outcome of choosing P is at least as good as the outcome of choosing Q and, for *some* turn of events, the outcome of choosing P is better than the outcome of choosing Q. And that seems plausible enough. If you strictly prefer the consequences of P to those of Q, given, say, that the event E occurs, and if you would regard the consequences of P as at least as good as those of Q, given that the event *not-E* occurs, then choice of P over Q promises a "sure-thing" with respect to consequences: by choosing P you cannot do worse, and may end up doing better, than if you choose Q. Moreover, if you strictly prefer the consequences of P to those of Q, given E, and if you strictly prefer the consequences of P to those of Q, given *not-E,* then choice of P over Q promises a "sure-thing" in an even stronger sense: it is guaranteed that one will do better having chosen P rather than Q, regardless of the turn of events.

Despite the fact that many have taken this to be a decisive consideration in favor of **STP**, this line of reasoning is also flawed. **STP** is *very* strong. It is framed with respect to the outcomes that can be associated with arbitrarily selected partitions of conditioning states. The principle requires that if there exists *any* event-partition for which the outcomes of P dominate the outcomes of Q, then P must be preferred to Q. In particular, then, the principle is not limited in its application to outcomes that can be characterized in terms of sure or riskless outcomes.

This raises a substantial issue. Consider a variant of the Kahneman and Tversky problem we discussed above, where $p(E) = 33/34$, and $p(F) = 34/100$:

P [\$ 2400, E or $-E$]
Q [\$ 2500, E; \$ 0, $-E$]

P' [P, F; \$ 0, $-F$] = [[\$ 2400, E or $-E$], F; \$ 0, $-F$]
Q' [Q,F; \$ 0, $-F$] = [[2500, E; \$ 0, $-E$], F; \$ 0, $-F$]

Once again, many report they prefer P to Q, but Q' to P'. In the case of P' and Q', then, we have a partition for which the associated outcomes satisfy the conditions for dominance: P preferred to Q, and \$ 0 at least as good as \$ 0. Thus, by **STP** the agent should rank P' over Q'. But what qualifies these outcomes as relevant for the purposes of assessing the choice between P' and Q' from a "sure-thing" perspective? Within the framework of a *finer* partitioning of events—and one that is an *explicit* feature of the problem—it is simply not true that one does at least as well by choosing P' as by choosing Q', regardless of the turn of (all relevant) events. By inspection, the outcome of Q' in the event that both E and F occur is \$ 2500, which is, by hypothesis, strictly preferred to any of the possible outcomes of P'. I do not mean to suggest, of course, that application of **STP** can be undercut in such cases simply by displaying *some* other partition of events such that preferences for the outcomes under that partition fail to satisfy the antecedent condition of **STP.** The issue here concerns the propriety of appealing to a partition under which the antecedent conditions are satisfied even though there exists an explicit *refinement* of that very same partition for which the antecedent conditions are *not* satisfied. If there is such an explicit refinement, then by reference to the consequences under that (refined) description, it is no longer clear what is the force of an appeal to dominance considerations.

Savage himself was well aware of the full scope of **STP**, and explicitly raised the question of whether it might be appropriate to restrict it to cases where the outcomes themselves are not defined in probabilistic terms. Focusing on the case of event-defined gambles over "sure" amounts of money, he rejects this suggestion on the following grounds:

> A cash prize is to a large extent a lottery ticket in that the uncertainty as to what will become of a person if he has a gift of a thousand dollars [for example] is not in principle different from the uncertainty about what will become of him if he holds a lottery ticket.

This amounts to denying that there is anything like a bedrock level of certainty. On this account, *it is risk all the way down.* Suppose, however, we grant this, and hence understand that one cannot distinguish a more restrictive version of **STP.** What makes this an argument for accepting **STP** rather than rejecting it?

The agent can acknowledge, of course, that if he chooses Q' over P', then he moves *through* a state in which a dominance relation obtains. More specifically, if we think of the F-events as occurring first, followed by the E-events, then no matter what the outcome of the F-events, and before the E-events are run, the prospect the agent then faces, if he has chosen Q', is dispreferred to, or no better than, the prospect he would then be facing if he had chosen P'. He could argue, however, that this is something that he suffers only *en passant*, and since he is concerned only with final outcomes and their probabilities, it is of no consequence to him.

Now, Savage's reply, as reported above, is that, in effect, it is always a matter of what we face *en passant,* since it is risk all the way down. This means, however, that any problem involves choice between gambles, and thus that the agent can never be sure he will always do better choosing one way rather than another. But, then, granting Savage's point, why not turn it upside down and regard it as undercutting the whole idea of an appeal to dominance with respect to outcomes?

But perhaps we need not take such a drastic position. Any principle such as **STP** must be interpreted as constraining preferences among alternatives, *under a* *given description* of those alternatives. If the agent has not refined his description of certain component gambles, and treats them as simply possible outcomes over which he has a preference ranking, then it can be argued that it is appropriate to appeal to dominance considerations. Suppose, however, that he has refined his description of those outcomes—recognizing explicitly the nature of the further risks to be encountered. In such a case, since at *that* level of discrimination the principle is revealed not to apply, it is unclear what force there is to an argument that invokes dominance at a coarser level of description.

I conclude, then, that while sure-thing considerations provide a highly plausible basis for a version of **STP** that is framed with respect to riskless outcomes (relative to some base-description), there is little to support the extension of this line of reasoning to the full blown principle **STP**. **STP**, no less than the **IND,** is subject to serious question.

6 RAIFFA'S ARGUMENT

Raiffa [1961; 1968] offers a quite different defense of **IND**, and one that has been extensively cited by others. Consider once again the Kahneman and Tversky example:

	R (66)	Y (33)	Black (01)
P	\$ 2400	\$ 2400	\$ 2400
Q	\$ 2400	\$ 2500	\$ 0
P'	\$ 0	\$ 2400	\$ 2400
Q'	\$ 0	\$ 2500	\$ 0

Raiffa suggests that the reported preference pattern would presumably hold if both preferentially *inferior* options, Q and P' were augmented by some very small amount under each of the conditioning events, say, in the following fashion:

	R (66)	Y (33)	Black (01)
P	\$ 2400	\$ 2400	\$ 2400
Q^*	\$ 2410	\$ 2510	\$ 10
P^*	\$ 10	\$ 2410	\$ 2410
Q'	\$ 0	\$ 2500	\$ 0

Raiffa argues that if $P$is preferred to Q^*, and Q' is preferred to P^*, by some individual, then it is reasonable to suppose that if the individual is offered the opportunity to choose between P and Q^* if a fair coin comes up Heads, and to choose between P^* and Q' if the coin lands Tails, the option [Choose P, if Heads; and Q', if Tails] will be preferred to option [Choose Q^*, if Heads; and P^*, if Tails]. The point is that the first of these more complex options promises him that he will get either one or the other of what he regards as the two superior options; and the second promises him that he will get either one or the other of the two inferior options. Call the first of these options P/Q' and the second Q^*/P^*. If we write out the schedule of payoffs, we get the following:

	R	Y	B
P/Q'	[$ 2400, H; $ 0, T]	[$ 2400, H; $ 2500, T]	[$ 2400, H; $ 0, T]
Q^*/P^*	[$ 2410, H; $ 10, T]	[$ 2410, H; $ 2510, T]	[$ 2410, H; $ 10, T]

But, by inspection, we can see that the second—the inferior plan dominates the first—the superior plan, and does so with respect to sure outcomes! That is, we have here a violation of **DSO.**

But once again the argument is hardly convincing. What Raiffa has assumed is that the agent will rank the option P/Q' over Q^*/P^*. But why? What mandates that ordering of the contingency plans is simply another version of **IND**, according to which if P is preferred to Q^*, and Q' is preferred to P^*, then $[P, 1/2; Q', 1/2]$ must be preferred to $[Q^*, 1/2; P^*, 1/2]$. It is no surprise, then, that the agent who rejects **IND** at the first level, and then invokes **IND** at the second level (the level of more complex options) will get into serious difficulties. But there is no reason to suppose that having rejected **IND** at one level, the agent will accept it at the next level. For the agent under discussion, the very observations that Raiffa makes will suffice for his or her rejection of the "reasonable" assumption that P/Q' will be ranked over Q^*/P^*. So once again, the case for **IND** seems to elude us.

7 DYNAMIC VERSIONS OF RAIFFA'S ARGUMENT

In recent years quite a number variations on Raiffa's argument have been tried out, most of them within the framework of a theory of dynamic choice. The objective has been to show that an agent who rejects **IND or STP** can be subjected to a *sequence* of choices such that, in the end, the agent will end doing less well, as measured, for example, by sure amounts of money, than if they had not rejected one or the other of the principles in question. The original version of this argument is to be found in Raiffa [1968]. The issues here are admittedly rather complicated. The reader will find my own detailed review of this line of reasoning in McClennen [1990].[13] My position, as discussed in McClennen [1990, pp. 173–82], is two-fold. First, arguments of this sort only go through on the assumption that (as in Raiffa's simple argument discussed above) one supposes the agent to reject some version of **IND,** and then accept another version of the same axiom at another stage in the dynamic choice process. Second, within a dynamic framework any problem of violating the simple principle **DSO** can be completely avoided by settling upon

[13]One especially powerful argument of this sort was presented by Seidenfeld [1988a]. I offered a rejoinder in the same issue of the journal in question, McClennen [1988], and Seidenfeld responded, again in the same issue, in Seidenfeld [1988b].

a plan and resolutely carrying it out. For these reasons I fail to see how any of these dynamic variations on Raiffa's original argument provide any real support for either **IND** or **STP.**

8 SUMMARY

My concern here has been to examine a variety of standard arguments that have been offered in support of a key axiom for the Bayesian theory of rational choice. Each of the three arguments examined must be judged as a non-starter. While one can agree with Samuelson that complementarities of the classical sort, namely those that arise in connection with (conjunctive) bundles of commodities, cannot arise in connection with disjunctively defined prospects, still this does not preclude the possibility of complementarity-like effects that are special to disjunctive combinations of goods. At the level of strength required for the constructions, **STP** strikes me as not at all plausible. And, finally, one should not be surprised if someone who both rejects **IND** and then accepts it at another level involving the same gambles as components can find themselves in the embarrassing position of having preferences that violate what is in fact a plausible, weak version of **STP**. I conclude that the Bayesian theory is in great need of something it has yet to find: a convincing defense of the Independence Axiom.

Department of Philosophy, Logic, and Scientific Method,
London School of Economics and Political Science.

BIBLIOGRAPHY

[Allais, 1953] M. Allais. Le comportement de l'homme rationnel devant le risque: Critique des postulats et axiomes de l'ecole americaine. *Econometrica* **21**: 503–546, 1953.

[Allais and Hagen, 1979] M. Allais and O. Hagen, eds. *Expected Utility Hypothesis and the Allais Paradox*. D. Reidel Publishing Company, 1979.

[Broome, 1991] J. Broome. Rationality and the sure-thing principle. In *Thoughtful Economic Man*. g. Meeks, ed. Cambridge University Press, 1991.

[Ellsberg, 1961] D. Ellsberg. Risk, ambiguity, and the Savage axioms. *Quarterly Journal of Economics* **75**: 643–669, 1961.

[Fishburn and Wakker, 1995] P. Fishburn and P. Wakker. The invention of the independence condition for preferences. *Management Science* **41**: 1130–1144, 1995.

[Friedman and Savage, 1952] M. Friedman and L. J. Savage . The expected-utility hypothesis and the measurability of utility. *Journal of Political Economy* **60**: 463–474, 1952.

[Kahneman and Tversky, 1979] D. Kahneman and A. Tversky. Prospect theory: an analysis of decision under risk. *Econometrica* **47**: 263–291, 1979.

[Loomes, 1984] G. Loomes. The importance of what might have been. *Progress in Utility and Risk Theory*. O. Hagen, ed. pp. 219–235. D. Reidel Publishing Company, 1984.

[MacCrimmon and Larsson, 1979] K. R. MacCrimmon and S. Larsson. Utility theory: axioms versus "Paradoxes". In *Expected Utility and the Allais Paradox*. M. Allais and O. Hagen, eds. pp. 333–409. D. Reidel Publishing Company, 1979.

[Manne, 1952] A. S. Manne. The strong independence assumption — gasoline blends and probability mixtures (with discussion). *Econometrica* **20**: 665–669, 1952.

[McClennen, 1983] E. F. McClennen. Sure-thing doubts. In *Foundations of Utility and Risk Theory with Applications*. B. P. Stingum and F. Wenstop, eds. 1983.

[McClennen, 1988] E. F. McClennen. Ordering and independence: a comment on Professor Seidenfeld. *Economics and Philosophy* **4**: 298–308, 1988.

[McClennen, 1990] E. F. McClennen. *Rationality and Dynamic Choice: Foundational Explorations*. Cambridge University Press, 1990.

[Pollatsek and Tversky, 1970] A. Pollatsek and A. Tversky. A theory of risk. *Journal of Mathematical Psychology* **7**: 540–553, 1970.

[Raiffa, 1961] H. Raiffa. Risk, ambiguity, and the Savage axioms: comment. *Quarterly Journal of Economics* **75**: 690–694, 1961.

[Raiffa, 1968] H. Raiffa. *Decision Analysis*. Reading, MA, Addison-Wesley, 1968.

[Samuelson, 1952] P. A. Samuelson. Probability, utility, and the independence axiom. *Econometrica* **20**: 670–678, 1952.

[Savage, 1951] L. J. Savage. *The theory of statistical decision*. Journal of the American Statistical Association, 1951.

[Savage, 1972] L. J. Savage. *The Foundations of Statistics*. New York, Dover, 1972. First published, 1954.

[Seidenfeld, 1988a] T. Seidenfeld. Decision theory without 'independence' or without 'ordering', what is the difference? *Economics and Philosophy* **4**: 267–290, 1988..

[Seidenfeld, 1988b] T. Seidenfeld. Rejoinder. *Economics and Philosophy* **4**: 309–315, 1988.

[Sen, 1985] A. Sen. Rationality and uncertainty. *Theory and Decision*, **18**: 109–27, 1985.

[Shoemaker, 1980] P. J. H. Shoemaker. *Experiments on Decision Under Risk: The Expected-Utility Hypothesis*. Boston, Martinus Nijhoff Publishing, 1980.

[von Neumann and Morgenstern, 1953] J. von Neumann and O. Morgenstern. *Theory of Games and Economic Behavior*, Third Edition. Princeton University Press, Princeton, 1953.

PHILIPPE MONGIN

THE PARADOX OF THE BAYESIAN EXPERTS

1 INTRODUCTION

Suppose that a group of experts are asked to express their preference rankings on a set of uncertain prospects and that all of them satisfy the economist's standard requisite of Bayesian rationality. Suppose also that there is another individual who attempts to summarize the experts' preference judgments into a single ranking. What conditions should the observer's ranking normatively be expected to satisfy? A natural requirement to impose is that it be Paretian, i.e., it should respect unanimously expressed preferences over pairs of prospects. Another condition which appears to be desirable is that the observer's and the experts' rankings should conform to one and the same decision theory, i.e., the observer himself should be Bayesian. The next question is then, are these seemingly compelling normative assumptions compatible with each other?

As a specific application, think of an insurer who considers selling a new insurance policy and consults a panel of experts before deciding which specification of the insurance policy, if any, should be marketed. Suppose that the insurer knows little or nothing about how to elicit the experts' subjective probabilities. The only way in which he could take advantage of the panel's expertise seems to be this: he will require the experts to state their preferences between the logically possible specifications, and then aggregate these data to define his own ranking. If one further assumes that the experts are Bayesian, the question naturally arises of how the insurer could conform to the double consistency requirement just explained. Notice that this question makes perfectly good sense even if one has assumed that the insurer is not familiar with Bayesian elicitation methods. Writers in the tradition of de Finetti (1974–75) have emphasized that to conform to the Bayesian axioms is tantamount to being "coherent" in one's betting behaviour, regardless of whether or not one knows the probability calculus.

Several writers in the field of collective choice or decision-making have investigated aggregative problems that are formally similar to the Bayesian experts problem. Their nearly unexceptional conclusion is that logical difficulties will result from the double imposition of Bayesianism and Paretianism on the observer's preference. In an earlier paper [Mongin, 1995], we provided an up-to-date analysis of these difficulties, using the axiom system which enjoys the highest theoretical status among Bayesians, i.e., Savage's [1972]. Essentially, the imposition of relatively weak Paretian conditions, such as Pareto Indifference or Weak Pareto, leads to impossibility results in a quasi-Arrovian style, i.e., to dictatorial conclusions, whereas the imposition of the Strong Pareto condition involves a sheer logical impossibility, unless the individuals have identical probabilities or utilities. In each case, the inference depends on "technical" assumptions the role and relevance of

D. Corfield and J. Williamson (eds.), Foundations of Bayesianism, 309–338.

which should carefully be ascertained; unexpected possibility results emerge when they are relaxed. Our Savagean conclusions encompass most of the more partial or elementary variants of the Bayesian experts paradox that have been discussed thusfar. We refer the reader to this earlier paper for references and comparisons.

The logical difficulties of "consistent Bayesian aggregation" led some writers to relinquish the first consistency condition — Paretianism — while others abandonned the second — Bayesianism. Either way out of the paradox involves a *diminutio capitis*. We have just suggested that the two requirements seemed equally natural. There are indeed serious arguments in favour of each, which makes the choice of a weaker version of consistency very awkward. In the present paper, we shall explore an altogether different potential solution to the paradox of the Bayesian experts. It consists in retaining the double consistency condition while varying the chosen notion of "Bayesianism". The impossibility results derived from applying Savage's axioms suggest that one should take a fresh look at them. A natural candidate to play the culprit's part is the sure-thing principle; but it is not our intention here to weaken it. If only for the purpose of theoretical experimentation, we want to remain within the confines of Bayesianism. There are also significant and well-recognized problems connected with the use of Savage's divisibility axiom. The present paper will take notice of them, but its primary target is to investigate the role of those axioms which ensure that the utility value of consequences is independent of what state of the world occurs.

Several writers in the Bayesian tradition, such as Karni [1985] and Drèze [1987], have repeatedly emphasized that state-independence is an inappropriate assumption to make in general. The standard example to support this claim involves the partition of states into the events "the agent lives" and "the agent dies". Insurance economics is replete with examples of a less dramatic sort in which the assumption of state-independence appears to be indefensible, both normatively and factually. On the constructive side, Drèze, Karni and others have devised axiom systems which deliver *state-dependent* subjective expected utility representations. They provide the generalization of Bayesianism that we want to put to the test. The general question of this paper is then, does state-dependent utility theory offer a solution to the paradox of the Bayesian experts?

For reasons of tractability rather than of substance, most of the work on state-dependent utility does not employ Savage's framework but the alternative, highly accessible framework introduced by Anscombe and Aumann [1963]. We shall follow the existing literature and rephrase both the paradox and its tentative solutions accordingly. As is well-known, Anscombe and Aumann's (AA) approach to uncertainty involves a loss of generality with respect to Savage's, in that it assumes a lottery structure on the consequence set. On the other hand, it makes it possible to dispense with Savage's divisibility axiom, and thus to deal with finite state sets – a welcome extension of subjective expected utility (SEU) theory. As far as technical derivations are concerned, the analysis of "consistent Bayesian aggregation" *à la* Savage depended on the measure-theoretic properties of his construction, in particular the nonatomicity of his derived subjective probability. The reader should expect the present analysis to revolve around the convexity properties of the *consequence set*, as conveniently assumed by AA.

The paper is organized as follows. Section 2 presents the definitions and axioms from SEU theory that will be used throughout. We shall briefly contrast AA's initial system – which is state-independent – with two later state-dependent variants. It is easy enough to axiomatize a *completely state-dependent* system of SEU. The well-known difficulty with this construction is that it leaves the individual's subjective probability indeterminate. Most of the work by Drèze, Karni and others has actually consisted in defining systems of *intermediary strength*, which allow for a state-dependent utility valuation of consequences but preserve the determination (if perhaps not the uniqueness) of the individual's subjective probability. Among the variants of AA's construction, only these intermediary systems can be claimed by Bayesianism. To accept complete state-dependence is really to take off the edge of the doctrine; this appears to be a well-recognized point. We have selected here an influential intermediary system first introduced by Karni, Schmeidler and Vind [1983].

Section 3 restates the initial paradox by applying AA's own state-independent system. As suggested, the results exactly parallel those reached in the Savage case but are easier to derive. Section 4 makes a start with an easy possibility result: the paradox disappears in the pure state-dependent generalization of AA's system. Then, we proceed to reexamine the paradox in the light of the relevant intermediary system of Karni, Schmeidler and Vind (KSV). The general conclusion of section 4 is that impossibility results can be derived in the KSV framework too, but the required technical assumptions are even stronger than those put to use in the state-independent case. This is why we choose to impose these assumptions only on a subset of the state space, and accordingly obtain only *local* analogues of our earlier dictatorial or logical impossibility theorems. Section 5 discusses a two-individual illustration of our impossibility results and compares them with those of Schervish, Seidenfeld and Kadane (1991), who have also investigated a state-dependent version of the Bayesian experts problem. Section 6 elaborates on the implications of the present results for the theory of collective decision-making. The proofs of all formal statements are in Appendix A.

2 DEFINITIONS AND AXIOMS FROM SUBJECTIVE EXPECTED UTILITY THEORY

As in Anscombe and Aumann [1963], we assume that there is a finite set S of states of the world, to be denoted by $s = 1, \cdots, T$, and there is a set X of final outcomes, to be denoted by $A, B, C, \cdots$ Throughout, we require that there are at least two distinct states and two distinct final outcomes. (A stronger cardinality restriction will be introduced in section 3.) The consequence set is $\mathcal{R} = \Delta_F(X)$, i.e., the set of all simple probabilities on X. The set of uncertain prospects, or (to use Savage's word) acts, is the set H of all functions $S \to \mathcal{R}$, to be denoted by

$f, g, h, \cdots$ Then, $f(s)$ is a simple probability on X; denote by $f(s, A)$ the value it gives to $A \in X$. Since the state set is finite, it is often convenient to denote acts as vectors:

$$f = [R_1, \cdots, R_s, \cdots, R_T],$$

where R_s stands for $f(s)$. (Then, we rewrite $f(s, A)$ as $R_s(A)$.) Finally, consider the set $\mathcal{R}^* = \Delta_F(H)$, i.e., the set of all simple probabilities on H. A typical element of $\mathcal{R}^*$ may be written as:

$$(\lambda_1 f^1, \cdots, \lambda_k f^k) = (\lambda_1 [R_1^1, \cdots, R_T^1], \cdots, \lambda_k [R_1^k, \cdots, R_T^k]),$$

where $\lambda = (\lambda_1, \cdots, \lambda_k)$ is a probability vector and indexes $1, \cdots, k$ refer to particular acts in H.[1] Now, following many writers in AA theory, we can identify this element of $\mathcal{R}^*$ with the following, altogether different mathematical object:

$$\lambda_1 f^1 + \cdots + \lambda_k f^k \in H.$$

As is well-known, to identify these two mathematical entities with each other is equivalent to assuming AA's "reversal of order" axiom. The resulting simplification has a price, because it then becomes impossible to discuss the extension of AA's approach to "moral hazard", as promoted by Drèze [1987].

Granting the identification $\mathcal{R}^* \simeq H$, H becomes the decision-maker's choice set. The preference relation $\succeq$ on H can then be subjected to all or part of the following axioms.

AXIOM 1 (VNM axiom). $\succeq$ satisfies the von Neumann-Morgenstern axioms.

Any axiomatic version of VNM theory will do (see [Fishburn, 1982], for details).

AXIOM 2 (Nontriviality). There are two outcomes A^*, A_* such that

$$[A^*, \cdots, A^*] \succ [A_*, \cdots, A_*].$$

The axiom below relies on the derived concept of a conditional preference. For any $s \in S$, define $\succeq_s$ ("the preference conditional on s") by[2]:

$$f \succeq_s g \text{ iff } [\forall f', g' \in H : f'_{-s} = g'_{-s}, f(s) = f'(s), g(s) = g'(s)]\ f' \succeq g'.$$

Define a state s to be null if its conditional preference $\succeq_s$ is trivial, i.e., $\succeq_s = H \times H$.

[1] We shall adopt the convention of using square brackets for uncertain prospects and curved ones for risky prospects (i.e., prospects with preassigned probabilities).

[2] We use the game-theoretic notation f'_{-s}, g'_{-s} to refer to the subvectors obtained from f', g' by deleting their s-component.

AXIOM 3 (State-Independence). For all non-null states s, t, and all constant $f, g \in H$, $f \succeq_s g$ iff $f \succeq_t g$.

Under state-independence it becomes meaningful to identify the constant act $[R, \cdots, R]$ with its value $R \in \mathcal{R}$, so that $\succeq$ induces a preference relation on $\mathcal{R}$. In this context Axiom 2 states that there are $A^*, A_* \in X$ such that $A^* \succ A_*$. Another version of Axiom 2 will also be used at a later stage in this paper:

AXIOM 2′. There is a non-null state s.

PROPOSITION 1 ([Anscombe and Aumann, 1963]). *If $\succeq$ satisfies Axioms 1, 2 and 3, there exist a nonconstant VNM function u on $\mathcal{R}$ and a probability $p = (p_1, \cdots, p_T)$ on S such that for all $f, g \in H$:*

$$(*) \qquad f \succeq g \text{ iff } \sum_{s \in S} p_s u(f(s)) \geq \sum_{s \in S} p_s u(g(s)).$$

State s is null iff $p_s = 0$. Any other pair (u', p') that satisfies the same properties as (u, p) is such that $p' = p$ and u' is a positive affine transform of u.

PROPOSITION 2 ([Fishburn, 1970]). *If $\succeq$ satisfies Axiom 1, there exist VNM functions $u_1, \cdots, u_T$ on $\mathcal{R}$ such that for all $f, g \in H$:*

$$(**) \qquad f \succeq g \text{ iff } \sum_{s \in S} u_s(f(s)) \geq \sum_{s \in S} u_s(g(s)).$$

The function u_s represents the conditional preference $\succeq_s$, and is constant iff s is null. Any other T-tuple $(u'_1, \cdots, u'_T)$ satisfying the same properties as $(u_1, \cdots, u_T)$ is a positive affine transform of the latter vector.

By a VNM function u on $\mathcal{R}$ we mean a function which has the following, mixture-preserving property: for any $\lambda \in [0, 1]$ and any R, R',

$$u((\lambda R, (1 - \lambda)R')) = \lambda u(R) + (1 - \lambda)u(R'),$$

or equivalently (since we are considering here only simple probabilities R, R'), a function which has the expected-utility form:

$$u(R) = \sum_{A \in X} R(A)u(A).$$

The fact that the utility representations defined on the *consequence set*, i.e., u in Proposition 1 and the u_i in Proposition 2, are VNM is a characteristic feature of the Anscombe-Aumann approach as a whole. Conceptually, this feature is irrelevant to the aim of the construction; technically, it is an ingenious device – actually, comparable with Savage's $(P6)$ — to facilitate the derivation of the SEU formula.[3]

OBSERVATION. If $\succeq$ satisfies not only Axiom 1, but also Axiom 2, there are VNM functions $v_1, \cdots, v_T$ on $\mathcal{R}$ and a probability $p = (p_1, \cdots, p_T)$ on S such that for all $f, g, \in H$:

$$(***) \qquad f \succeq g \text{ iff } \sum_{s \in S} p_s v_s(f(s)) \geq \sum_{s \in S} p_s v_s(g(s)).$$

Any $(T+1)$-tuple $(q, v'_1, \cdots, v'_T)$ such that q is a probability on S and:

$$p_s = 0 \Leftrightarrow q_s = 0,$$
$$v'_s = \frac{p_s}{q_s} v_s \text{ if } q_s \neq 0, \; v'_s \text{ arbitrary otherwise,}$$

can be substituted for $(p, v_1, \cdots, v_T)$ in (***).

NOTATION. We define $U(f) = \sum_{s \in S} p_s u(f(s))$ and $V(f) = \sum_{s \in S} p_s v_s(f(s))$ when the suitable assumptions hold. As a rule, U, V, W will refer to representations of preferences over acts, and u, v, w to representations of preferences over consequences.

The comparison between Proposition 1, Proposition 2, and the ensuing well-known observation, brings out the classic difficulty of state-dependent utility theory. The system consisting of only Axioms 1 and 2 is not rich enough to determine the decision-maker's subjective probability. To add Axiom 3 makes it possible to uniquely determine p if one selects a state-independent representation u on $\mathcal{R}$, as do AA in their seminal article. However, Axiom 3 is too restrictive; it amounts to excluding the relevant complication of state-dependent preferences. Hence a dilemma of determination and relevance; see, among others, [Fishburn, 1970; Drèze, 1987; Karni, 1985; Karni, 1993; Schervish *et al.*, 1990].[4]

Various methods have been put forward to escape from the dilemma just suggested. Most (but not all) of them consist in assuming Axioms 1 and 2, and then introducing further axioms to determine the state-dependent functions $v_1, \cdots, v_T$ that underlie the uninformative representations $u_1, \cdots, u_T$ of Proposition 2. Once

[3] We made another assumption on the consequence set $\mathcal{R}$ which — by contrast to the VNM assumption — is dispensable within the AA approach. To save notation, we assumed a state-independent consequence set $\mathcal{R}$. Some expositions of state-dependent utility theory, such as Fishburn's [1970], and actually the original paper by Karni, Schmeidler and Vind [1983], adopt a more general framework in which not only the evaluations but also the availability of consequences vary from one state to another. As far as we can judge, the results of the present paper can be extended unproblematically to this more general framework. On the issue of state-dependent consequences in expected utility theory, see also [Hammond, 1998].

[4] The last paper usefully emphasizes that Anscombe and Aumann's choice of a state-independent u on $\mathcal{R}$ is to some extent question-begging. Even when Axiom 3 holds, it is trivially possible to replace (*) with infinitely many equivalent *state-dependent* representations, each of which corresponds to one particular subjective probability.

the vector $(v_1, \cdots, v_T)$ is known (up to a positive affine transformation, PAT),[5] it becomes possible to write $u_s = p_s v_s$, where the p_s are well-determined (and ideally unique) probability values. We shall not attempt at covering all the variants of this axiomatization strategy. A representative system will be enough for the purpose of this paper.

As in Karni, Schmeidler and Vind [1983], we introduce an auxiliary binary relation $\tilde{\succeq}$. It is meant to describe the preference that the decision-maker would express between acts *if his subjective probability were some given* $q = (q_1, \cdots, q_T)$. KSV's strategy is to infer the agent's actual state-dependent utilities from the (supposedly meaningful and even observable) hypothetical preference $\tilde{\succeq}$, and then determine his actual, unknown subjective probability p by using this utility-relative information.

Formally, fix a probability $q = (q_1, \cdots, q_T)$ on S with $q_s > 0$ for all s and associate with each $f \in H$ an *hypothetical act* f' defined as follows: f' is on S, and for each s, $f'(s) = f(s)q_s$, that is to say, $f'(s)$ is that function on X which satisfies $f'(s, A) = f(s, A)q_s$ for all $A \in X$. Note that $f'(s)$ is *not* a probability on X, unlike $f(s)$, but f' can be viewed as a probability on $S \times X$ unlike f. (The computation is obvious.) Given the positivity assumption made on q, the set H' of all hypothetical acts is clearly in a one-to-one relationship to H; this makes the notation f, f' unambiguous. The element f' describes the effect of compounding the given probability q on S with each of the lotteries that f assigns to $s = 1, \cdots, T$. Define the *hypothetical preference* ($\tilde{\succeq}$) to be a preference relation on the set of hypothetical acts. This formal construct is meant to capture the modification in the individual's preferences "if his subjective probability were q".[6]

It is consistent to impose the same decision-theoretic constraints on both $\tilde{\succeq}$ and $\succeq$, i.e., to subject hypothetical preference to the VNM axioms. Beyond this, some coordinating condition should relate $\succeq$ to $\tilde{\succeq}$. In effect, KSV impose the ("consistency") axiom that conditional preferences $\tilde{\succeq}_s$ and $\succeq_s$ are the same whenever s is non-null for $\succeq$. The point of this axiom is to ensure that hypothetical preference data deliver usable information on the individuals' state-dependent utilities.[7]

AXIOM 4 (Hypothetical Preference). For all $s \in S$ that are non-null with respect to $\succeq$, and for all $f, g \in H$,

$$f \succeq_s g \Leftrightarrow f' \tilde{\succeq}_s g',$$

where f' and g' are the elements in H' associated with f and g respectively.

[5] Formally, two vectors $(v'_1, \cdots, v'_T)$ and $(v_1, \cdots, v_T)$ are identical up to a PAT if there are a number $\mu > 0$ and a vector $(\nu_1, \cdots, \nu_T)$ such that $(v'_1, \cdots, v'_T) = \mu(v_1, \cdots, v_T) + (\nu_1, \cdots, \nu_T)$.

[6] Notice carefully that although hypothetical acts carry preassigned probabilities with them, they do not reduce to VNM lotteries. States of the world matter in the construction of hypothetical acts.

[7] KSV's exposition is considerably more complex, due to their detailed analysis of null states. Wakker [1987], and Schervish, Seidenfeld and Kadane [1990], provide alternative restatements; we do not follow them here.

PROPOSITION 3 ([Karni *et al.*, 1983]). *Assume that* $\succeq$ *satisfies Axioms 1 and 2′. Take any probability* q *on* S *with* $q_s > 0$ *for all* $s \in S$. *Assume that the induced hypothetical preference* $\tilde{\succeq}$ *also satisfies Axioms 1 and 2′, and that* $\succeq$ *and* $\tilde{\succeq}$ *jointly satisfy Axiom 4. Then, there are VNM functions* $v_1, \cdots, v_T$ *on* $\mathcal{R}$ *and a probability* $p = (p_1, \cdots, p_T)$ *on* S *such that for all* $f, g \in H$:

(i) $f \succeq g$ *iff* $\sum_{s \in S} p_s v_s(f(s)) \geq \sum_{s \in S} p_s v_s(g(s))$

(ii) $f' \tilde{\succeq} g'$ *iff* $\sum_{s \in S} q_s v_s(f(s)) \geq \sum_{s \in S} q_s v_s(g(s))$.

If s *is non-null for* $\succeq$, *then* $p_s > 0$. *Any other* $(T+1)$*-tuple* $(v'_1, \cdots, v'_T, p')$ *that satisfies conditions (i) and (ii) is such that the vector* $(v'_1, \cdots, v'_T)$ *is a PAT of* $(v_1, \cdots, v_T)$ *and* $p'_s/p'_t = p_s/p_t$ *for all* s, t *non-null for* $\succeq$.

Notice that the theorem does not entirely determine the agent's probability on null states: if s is null for $\succeq$ and not for $\tilde{\succeq}$, then comparison of (i) and (ii) leads to the conclusion that $p_s = 0$; if s is null for *both* $\succeq$ and $\tilde{\succeq}$, nothing can be said about p_s (but necessarily, v_s = constant).

Although the conclusions of the theorem are stated atemporally, they might be interpreted in terms of the two step-experiment mentioned at the outset. Irrespective of whether Axiom 4 provides a satisfactory formal rendering, the experiment *itself* raises a conceptual problem: the agent might well attach no sense to the expression of his preferences conditionally on the use of a subjective probability which is not his own.[8]

At least, the KSV procedure has a significant negative argument to recommend itself: to the best of our knowledge, existing alternatives *either* entail only a partial solution to the indeterminacy-of-probability issue, *or* involve the same operational difficulties as the KSV procedure, *or* imply an even more radical departure from standard Bayesian assumptions. Karni and Schmeidler's [1993] state-dependent variant of Savage's axiomatization exemplifies the first problem. In the AA framework, Karni's [1993] assumption of *given* transformations between the $v_s(\cdot)$ illustrates the second problem, while Drèze's [1987] use of a "moral hazard" assumption illustrates the third. For all its shortcomings, KSV's article is a serious representative of the work done in the field of state-dependent utility theory. This is sufficient to make it relevant to a paper which is primarily concerned with theoretical experimentation.

3 IMPOSSIBILITY RESULTS IN THE STATE-INDEPENDENT CASE

The present section will first introduce a multi-individual extension of the AA approach broadly speaking, and then restrict attention to the state-independent case

[8][Drèze, 1987] expresses his critique of the KSV approach differently. He claims that it relies on information obtained from *verbal* behaviour, which he says is unreliable in principle and should be ignored. In essence, Drèze disqualifies KSV's contribution on the grounds that they do not follow the methodology of revealed preference theory. The critical point in the text does not depend on one's adhering to a revealed preference methodology.

with a view of deriving the AA variant of the Bayesian experts paradox.

Let us assume that there are individuals, to be represented by indices $i = 1, \cdots, n$, and an observer, to be represented by index $i = 0$, who express their subjective probabilities indirectly, i.e., by stating their preferences $\succeq^i$ over uncertain prospects. Throughout, we shall require $\succeq^i$ to satisfy some subset of the axioms of section 2, for all $i = 0, 1, \cdots, n$. This requirement reflects the assumption that both the individual experts and the observer are Bayesian; it encompasses one of the two consistency conditions discussed in the introduction. The remaining, Paretian condition can be made precise in terms of one of the following standard requirements: for all $f, g \in H$,

$$f \sim^i g,\ i = 1, \cdots, n \ \Rightarrow\ f \sim^0 g. \qquad \text{(C)}$$

$$f \succeq^i g,\ i = 1, \cdots, n \ \Rightarrow\ f \succeq^0 g. \qquad \text{(C}_1\text{)}$$

$$f \succ^i g,\ i = 1, \cdots, n \ \Rightarrow\ f \succ^0 g. \qquad \text{(C}_2\text{)}$$

$$f \succeq^i g,\ i = 1, \cdots, n \text{ and } \exists j \in \{1, \cdots, n\} : f \succ^j g \ \Rightarrow\ f \succ^0 g. \qquad \text{(C}_3\text{)}$$

In social choice theory, these are the conditions of Pareto-Indifference, Pareto-Weak Preference, Weak Pareto, Strict Pareto, respectively. We also introduce the Strong Pareto condition:

$$(\mathrm{C}^+) \ = \ (\mathrm{C})\ \&\ (\mathrm{C}_3).$$

Obviously, $(\mathrm{C}_1) \Rightarrow (\mathrm{C})$ and $(\mathrm{C}_3) \Rightarrow (\mathrm{C}_2)$. Given the rich structure of the consequence set in the AA approach, more can be said on the logical relations between the Pareto conditions. It will shortly be seen that under a minor restriction on preferences, (C_1), hence (C) are implied by any other condition. Let us introduce the following restriction of Minimum Agreement on Acts:

(MAA) $\qquad \exists f_*, f_{**} \in H,\ \forall i = 1, \cdots, n,\ f_* \succ^i f_{**}.$

Notice the difference with the requirement of Minimum Agreement on Consequences used in Mongin [1995, Section 3]. In the present context the latter would state that:

(MAC) $\qquad \exists R_*, R_{**} \in \mathcal{R},\ \forall i = 1, \cdots, n,\ R_* \succ R_{**}.$

In a pure state-independent context such as that of the earlier article, (MAC) provided an appropriate notion of minimum agreement among the individuals. We want a weaker condition here since it should also be applicable to the state-dependent context of the following sections.

For $i = 0, 1, \cdots, n$ denote by U^i the SEU representation of $\succeq^i$ when this relation satisfies all of the assumptions of Proposition 1, and V^i the more general additive representation of $\succeq^i$ that satisfies the unique assumption of Proposition 2. Then, $U^i(f) = \sum_{s \in S} p^i_s u^i(f(s))$ and $V^i(f) = \sum_{s \in S} u^i_s(f(s))$. For any vector-valued function $(\varphi_1, \cdots, \varphi_k)$, denote its range (i.e., set of values) by $Rge(\varphi_1, \cdots, \varphi_k)$. A basic consequence of imposing *any* of the AA systems of section 2 on the observer's and individual preferences $\succeq^0, \succeq^1, \cdots, \succeq^n$ is that the vector of corresponding utility representations has a convex range. Lemmas 4, 5 and 6 spell out this fact and its important consequences in terms of the V^i representations. The same results obviously apply to the U^i since they are restricted forms of the V^i.

LEMMA 4. *If* $\succeq^0, \succeq^1, \cdots, \succeq^n$ *satisfy the assumption of Proposition 2 (= Axiom 1), then* $Rge(V^0, V^1, \cdots, V^n)$ *is convex.*

De Meyer and Mongin [1995] have investigated the aggregative properties of a real function φ_0 which is related to given real functions $\varphi_1, \cdots, \varphi_n$ by unanimity conditions analogous to (C), $(\mathrm{C}_1), \cdots$ and by the assumption that $(\varphi_0, \varphi_1, \cdots, \varphi_n)$ has convex range. These aggregative results are applicable here because of Lemma 4 and will be used throughout the paper. Here is the first application:[9]

LEMMA 5. *If* $\succeq^0, \succeq^1, \cdots, \succeq^n$ *satisfy Axiom 1, then* (C) *holds if and only if there are real numbers* $a_1, \cdots, a_n, b$ *such that* $V^0 = \sum_{i=1}^n a_i V^i + b$. (C_1) *[resp.* (C^+)*] holds if and only if this equation is satisfied for some choice of non-negative [resp. positive] numbers* $a_1, \cdots, a_n$.

Another consequence of Lemma 4 is the following tightening of the logical implications between unanimity conditions:

LEMMA 6. *If* $\succeq^0, \succeq^1, \cdots, \succeq^n$ *satisfy the assumptions of Proposition 1, and if (MAA) holds, then*

$$(\mathrm{C}_2) \Rightarrow (\mathrm{C}_1) \text{ and } (\mathrm{C}_3) \Leftrightarrow (\mathrm{C}^+).$$

Thus, the list of conditions becomes simplified under (MAA). Returning now to the conclusion of Lemma 5, we know that it can be applied to the state-independent representations U^i. Hence, it seems as if this lemma delivered an aggregative rule of the familiar sort — what social choice theorists call generalized utilitarianism (e.g., [d'Aspremont, 1985]). A simple algebraic argument adapted from Mongin [1995, Section 4]) will demonstrate that this is *not* the case in general. Impossibility results lurk behind the apparently well-behaved affine decomposition $U^0 = \sum_{i=1}^n a_i U^i + b$. Dictatorial rules will emerge from the analysis of the weaker unanimity conditions (C), (C_1), (C_2), while sheer logical impossibility will result from imposing the stronger conditions (C_3) or (C^+).

Given a preference profile $\succeq_0, \succeq_1, \cdots, \succeq_n$ satisfying the assumptions of Proposition 1, hence representable by

$$U^0(\cdot) = \sum_s p_s^0 u^0(\cdot),\ U^1(\cdot) = \sum_s p_s^1 u^1(\cdot), \cdots, U^n(\cdot) = \sum_s p_s^n u^n(\cdot),$$

[9] Lemma 5 is an encompassing version of a famous social aggregation theorem first stated by Harsanyi [1955].

we say that i is a *probability dictator* if $p^0 = p^i$, that i is a *utility dictator* if $u^0 = u^i$ (up to a PAT), and that i is an *overall dictator* if he is both a probability and a utility dictator. We define i to be an *inverse* utility dictator or an *inverse* overall dictator by changing the clause that $u^0 = u^i$ into $u^0 = -u^i$. We shall also say that *probability agreement* prevails if $p^1 = \cdots = p^n$ and that *pairwise utility dependence* (p.u.d.) prevails if for all $i, j \geq 1, u^i = u^j$ (up to an affine transformation of any sign). Probability agreement and p.u.d. are two degenerate cases of individual profiles; in general, both probabilities and utilities should be expected to vary from one individual to another.

How to capture individual diversity in the language of formal choice theories is a difficult problem. As in Coulhon and Mongin [1989], or Mongin [1995], we shall use the convenient shortcut of defining diversity in terms of algebraic independence. Recall that a set of elements $\{\varphi_1, \cdots, \varphi_k\}$ of a vector space is *affinely independent* if for any set of real numbers $a_1, \cdots, a_k, b$,

$$a_1\varphi_1 + \cdots + a_k\varphi_k + b = 0 \Rightarrow a_1 = \cdots = a_k = b = 0.$$

This concept, rather than the weaker one of linear independence, provides the relevant notion of algebraic independence in the case of utility functions. Plainly, affine and linear independence become equivalent in the case of probabilities. A relevant fact to report here is that a set of VNM functions $u^1, \cdots, u^n$ is affinely independent if and only if these functions are "separated" from each other by suitable lotteries. This equivalence can be immediately extended to AA representations:

LEMMA 7. *Suppose that $u^1, \cdots, u^n$ are VNM utility functions on $\mathcal{R}$. They are affinely independent if and only if for every $i = 1, \cdots, n$, there are $R^i_*, R^i_{**} \in \mathcal{R}$ such that:*

$$u^i(R^i_*) > u^i(R^i_{**}) \textit{ and } u^j(R^i_*) = u^j(R^i_{**}) \textit{ for } j \neq i.$$

Similarly, the $V^1, \cdots, V^n$ derived in Proposition 2 are affinely independent if and only if for every $i = 1, \cdots, n$, there are f^i_*, f^i_{**} such that:

$$V^i(f^i_*) > V^i(f^i_{**}) \text{ and } V^j(f^i_*) = V^j(f^i_{**}).$$

If affine independence assumptions formalize individual diversity in an obvious sense, it is also the case that in a VNM context, they imply some form of minimum agreement between individuals. This rather curious consequence deserves emphasis here since it means that in some algebraic contexts (MAA) and (MAC) are given for free:[10]

LEMMA 8. *Suppose that $u^1, \cdots, u^n$ are affinely independent VNM functions. Then, (MAC) holds. Similarly, if the $V^1, \cdots, V^n$ of Proposition 2 are affinely independent, (MAA) holds.*

[10]Compare with the related statements in [Weymark, 1993, Proposition 3] and [Mongin, 1995, Corollary 4.3].

We are now in a position to state the two impossibility theorems which formalize the paradox of the Bayesian experts. In part (*) of both Propositions 9 and 10 we introduce a linear independence restriction on individual probabilities. To ensure that this restriction applies, we shall assume in part (*) that the state space S has cardinality at least n.

PROPOSITION 9. *Assume that* $\succeq^0, \succeq^1, \cdots, \succeq^n$ *satisfy Anscombe and Aumann's axioms of state-independent utility, i.e., the assumptions of Proposition 1. Denoting by* $p^1, \cdots, p^n$ *the probabilities and by* $u^1, \cdots, u^n$ *the utility functions on consequences provided by Proposition 1, assume that either:*

(*) $p^1, \cdots, p^n$ *are linearly independent,*

or

(**) $u^1, \cdots, u^n$ *are affinely independent.*

Then, if (C) holds, there is either a utility or an inverse utility dictator in case (*), and there is a probability dictator in case (**). There is an overall or an inverse overall dictator when both (*) and (**) apply. If either (C_1) *or* (C_2) *holds, the same results follow, except that there is always a utility dictator in case (*).*

When there is an overall dictator, all of the unanimity conditions are obviously satisfied, so that we could have stated part of Proposition 9 in terms of "if and only if" conditions.[11] This observation also implies that the problem of Consistent Bayesian Aggregation does not involve any *logical* impossibility in the case of conditions (C), (C_1) and (C_2). The stronger conditions (C_3) and (C^+) lead to altogether different conclusions.

PROPOSITION 10. *The assumptions are as in Proposition 9. Then, if* (C_3) *or* (C^+) *holds, case (*) implies that pairwise utility dependence prevails and that there is a utility dictator; case (**) implies that probability agreement prevails and that there is a probability dictator.*

Notice that in both Proposition 9 and 10, (MAA) is an inference, not an assumption. A modest strengthening of the first part of Proposition 10 would follow from assuming (MAC). Then, *positive* p.u.d. prevails (i.e., all individual utilities are identical up to a positive scale factor).

Proposition 10 can be restated as follows: under the assumptions of Proposition 9, (MAA) and (C_3), if either the n probabilities are linearly independent and (at least) two utility functions are affinely independent, or the n utility functions are affinely independent and (at least) two probabilities are distinct, then there is no solution to the Bayesian experts problem. This wording makes it clear that under appropriate distinctiveness restrictions, (C_3) is a *logical* impossibility; given these restrictions, *even dictatorship* fails to deliver a solution.

[11] Note also that inverse utility dictatorship is impossible when (C_2) and (MAA) hold. Utility dictatorship and inverse utility dictatorship can coexist with each other under the weaker assumption (C_1), as the following shows: take $n = 2, U_0 = U_1$ and $U_2 = -U_1$.

A word of comparison with the Savagean formulation of the paradox is in order. The main technical step in Mongin [1995] was to derive a version of Lemma 5. Since Savage does not assume anything on the consequence set, this had to be done by a special construction based on his divisibility-of-events axiom ($P6$). Once the affine decomposition of Lemma 5 is obtained, the algebra of impossibility results follows similar paths in the Savage and the Anscombe-Aumann variants.[12]

4 THE STATE-DEPENDENT CASE

Suppose that we just impose Axiom 1 on the preference relations $\succeq^0, \succeq^1, \cdots, \succeq^n$. This is the pure state-dependent case, as characterized by Proposition 2; each $\succeq^i$ is represented by $V^i(f) = \sum_{s\in S} u^i_s(f(s))$. It is easy to check that nontrivial solutions to the aggregation problem now exist, whatever individual preferences might be. To see that, take any profile $\succeq^1, \cdots, \succeq^n$ that satisfies Axiom 1 and consider the added preference relation $\succeq^0$ defined by means of the following representation:

$$(+) \qquad V^0(f) = \sum_{i=1}^{n} a_i (\sum_{s\in S} u^i_s(f(s)),$$

where $a_i > 0$ for all i. Obviously, $\succeq^0$ satisfies the whole list of Pareto conditions (C), $\cdots$, (C$^+$). It is also clear that $\succeq^0$ satisfies Axiom 1 (since a sum of VNM functions is also VNM, and Axiom 1 does not require anything beyond that property). A little more explicitly, (+) can be rearranged as:

$$(++) \qquad V^0(f) = \sum_{s\in S} u^0_s(f(s)),$$

by defining $u^0_s = \sum a_i u^i_s$ for all $s \in S$. This rewriting makes it plain that $\succeq^0$ and the $\succeq^i$ obey the same (weak) decision theory.

Hence, in the pure state-dependent case, the paradox of the Bayesian experts vanishes. This mathematically trivial resolution can strike one as conceptually relevant only if one regards Axiom 1 as a sufficient foundation for Bayesianism. We have already suggested that this is *not* a sensible position to take. Without some restriction on the many subjective probabilities that are compatible with state-dependent utilities, Bayesianism vanishes at the same time as the paradox.

Before we proceed to axiomatic systems of intermediary strength, we should complete the analysis of the first paragraph. What is not so trivial as the "resolution" just sketched is the fact that equation (+) delivers a *necessary* solution to

[12] A two-person version of Propositions 9 and 10 was obtained by [Seidenfeld *et al.*, 1989], using an expected utility framework in the style of Anscombe and Aumann. Schervish, Seidenfeld and Kadane [1991, Theorem 2] state this result more formally. We defer comparison to section 5.

the aggregation problem. This fact follows from Lemma 5 above, when conditions (C), (C_1) or (C^+) hold, and from Lemmas 5 and 6 when (C_2) and (C_3) hold (assuming (MAA)). Let us take stock of the characterization just obtained:

PROPOSITION 11. *Assume that $\succeq^0, \succeq^1, \cdots, \succeq^n$ satisfy the unique assumption of Proposition 2, i.e., axiom 1, and that $V^i(f) = \sum_{s \in S} u^i_s(f(s)), i = 0, 1, \cdots, n$ are the state-dependent representations derived in Proposition 2. Then, (C) holds if and only if there are real numbers $a_1, \cdots, a_n, b$ such that:*

$$\forall s \in S, \quad u^0_s = \sum_{i=1}^{n} a_i u^i_s + b.$$

Similarly, (C_1) [or (C_2), if one assumes (MAA)] holds if and only if there are $a_i \geq 0, i = 1, \cdots, n$, and b such that this equation holds; and assuming (MAA), (C_3) holds if and only if there are $a_i > 0, i = 1, \cdots, n$, and b such that the equation holds.

The remainder of this section investigates a multi-agent application of the KSV approach. We shall assume that hypothetical probabilities q are used to determine the observer's and individuals' state-dependent utilities, following the procedure implicitly described in Proposition 3. More precisely, each of $i = 0, 1, \cdots, n$ is endowed with a preference relation $\succeq^i$, as well as an hypothetical preference relation $\tilde{\succeq}^i$, to be thought of here as i's preference over acts conditionally on some given, strictly positive q^i. We know from section 2 that if $\succeq^i$ and $\tilde{\succeq}^i$ conform to axioms 1, 2′ and 4, for $i = 0, 1, \cdots, n$, there are VNM functions $v^i_1, \cdots, v^i_T$ on $\mathcal{R}$ and subjective probabilities p^i on S such that:

(i) $f \succeq^i g$ iff $\sum_{s \in S} p^i_s v^i_s(f(s)) \geq \sum_{s \in S} p^i_s v^i_s(g(s))$
(ii) $f' \tilde{\succeq}^i g'$ iff $\sum_{s \in S} q^i_s v^i_s(f'(s)) \geq \sum_{s \in S} q^i_s v^i_s(g'(s))$.

These equivalences and the accompanying uniqueness properties will lead to the negative results below. We shall make full use of the flexibility implied by the KSV approach, and take the auxiliary probabilities q^i to be sometimes identical, sometimes different from one individual to another. The upshot of this analysis is that if there is sufficient diversity among the individuals' state-dependent utility functions, a variant of the earlier probability dictatorship and probability agreement theorems holds. Correspondingly, a variant of the earlier utility dictatorship and dependence theorem holds, but as will be explained, the symmetry between probability and utility breaks down in the state-dependent case.

To state these negative results, some further terminology is required. For any $S' \subset S, S' \neq \phi$, we shall say that i is a *probability dictator for* S' if either $p^0(S') = p^i(S') = 0$, or $p^0(S') \neq 0 \neq p^i(S')$ and for all $s \in S'$,

$$p^0(s|S') = p^i(s|S'),$$

and that *probability agreement prevails on* S' if for all $i, j = 1, \cdots, n$, either $p^i(S') = p^j(S') = 0$, or $p^i(S') \neq 0 \neq p^j(S')$ and for all $s \in S'$, $p^i(s|S') = p^j(s|S')$. Similarly, we shall say that i is a *utility dictator on* S' if for all $s \in S', v_s^0 = v_s^i$ (up to a PAT (which might depend on the particular s); and that *pairwise utility dependence* (p.u.d.) *prevails on* S' if for all $s \in S'$ and for all $i, j = 1, \cdots, n, v_s^i = \pm v_s^j$, up to PATs (which might depend on s).

The exposition of impossibility results in this section does not follow the order of last section. We first analyze the probabilistic variant of paradox, and then move to its variant in terms of utility functions.

PROPOSITION 12. *Assume that* $\succeq^0, \cdots, \succeq^n$ *and (for some common q)* $\tilde{\succeq}^0, \cdots, \tilde{\succeq}^n$ *satisfy Axioms 1, 2′ and 4. Denote by* $p^1, \cdots, p^n$ *the individuals' subjective probabilities, and by* $v_1^1, \cdots, v_T^1, \cdots, v_1^n, \cdots, v_T^n$ *the individuals' state-dependent utilities, which are provided by Proposition 3. Assume that (C) applies to both sets of preferences. Then, if* S' *is some nonempty subset of* S *such that for all* $s \in S', v_s^1, \cdots, v_s^n$ *are affinely independent, there is a probability dictator on* S'.

If (C) is replaced by either (C_1), or (C_2) together with (MAA), the same results hold; if (C) is replaced with (C_3) and (MAA), probability agreement prevails, and there is a probability dictator, on S'.

As a particular application of Proposition 12, take S' to be the whole subset of those states which are non-null for at least one $i = 1, \cdots, n$. Then, depending on the Pareto conditions, either the dictator imposes his *absolute* probability, or *absolute* probability agreement prevails, exactly as in the state-independent case. In order to obtain this conclusion, one should resort to the strong assumption that for every relevant state s, the $v_s^1, \cdots, v_s^n$ are affinely independent. As explained in section 3, the significance of this assumption can be appreciated using its equivalent reformulation: for every relevant state, and every individual i, there are lotteries R_*^i, R_{**}^i that "separate" v_s^i from the others' utilities v_s^j. One would hesitate to impose such a strong assumption uniformly across states. To take an example in the style of Savage's, suppose that s' is good weather and s'' bad weather, and that individuals i and j have the following preferences: when s' prevails, i — the adventurous vacationer — prefers rockclimbing to canooing and is indifferent between going to a picnic or taking a swim, while j – the quiet vacationer – is indifferent between the first two lotteries but strictly prefers one of the last two to the other; when s'' prevails, both i and j are indifferent between the four lotteries. Or, to take an economic example, suppose that final outcomes are money amounts and that in some states, widely different amounts are available, whereas in others, only trivial increments around a given money amount are.[13] The "separation" property might well be satisfied in the former case but fail in the latter (since this case might be formalized in terms of linear, hence identical utility functions for money). This discussion suggests that the case in which S' is maximal might be irrelevant. It

[13]Admittedly, this example does not quite fit in the formalism of this paper since it involves not only state-dependent utilities but also state-dependent consequences.

explains why we chose to emphasize local (i.e., event-relative) properties as in Proposition 12.

The next proposition deals with a utility variant of the paradox. It is concerned with the special case of an admixture of state-dependence and state-independence. To deal with this case appropriately, we determine the KSV procedure beyond what was done by these authors. Suppose that there is a subset S' of states — all of which we take to be non-null — having the following property: conditional preferences on constant acts do not vary across states in S, whereas they vary across any two $s \in S, t \notin S'$. Thus, as far as S' is concerned, event-, rather than state-dependence, prevails. Restricting attention to acts taking some fixed value on each $t \notin S'$, it can be seen that the standard Anscombe–Aumann theorem (Proposition 1) applies. Thus, using the AA representation, we have a probability π on S'. The assumptions underlying the KSV procedure in Proposition 3 do not ensure that the conditional of the derived probability p on S' will coincide with π. Since π can be revealed by standard betting techniques, it seems natural to require that the two probabilities be equal. The way of obtaining this result while applying the KSV procedure is to impose that the conditional of the hypothetical probability q on S' be equal to the (independently revealed) π.

Formally (in the notation of section 2):

ASSUMPTION 13. Suppose that there is $S' \subset S, |S'| \geq 2$ such that every $s \in S'$ is non-null, and for every pair of constant acts $f, g \in H$:

$$\forall s, t \in S', f \succeq_s g \text{ iff } f \succeq_t g.$$

Then, we require q in the KSV system to satisfy:

$$\forall s \in S', \frac{q(s)}{q(S')} = \pi(s),$$

where π is the probability on S' derived by applying the assumptions of Proposition 1 to the restriction of $\succeq$ to those acts in H which take some fixed set of values on $S \setminus S'$.

Now, we are ready for the last variant of the paradox.

PROPOSITION 14. *Assume that* $\succeq^0, \cdots, \succeq^n$ *and* $\tilde{\succeq}^0, \cdots, \tilde{\succeq}^n$ *obey the KSV system, i.e., they satisfy Axioms 1, 2' and 4. Suppose that there is* $S' \subset S, |S'| \geq 2$ *such that:*

for all $i = 0, 1, \cdots, n,$ *every* $s \in S'$ *is non-null;* (∗)

for all $i = 0, 1, \cdots, n,$ *all* $s, t \in S',$ (∗∗)
and all pairs of constant acts $f, g \in H, f \succeq_s^i$ *iff* $f \succeq_t^i g$.

Denote by $p^0, \cdots, p^n$ *the subjective probabilities given by Proposition 3 for some set of hypothetical probabilities satisfying Assumption 13, and suppose that:*

$(***)$ $\quad (p_s^1)_{s\in S'}, \cdots, (p_s^n)_{s\in S'}$ *are linearly independent vectors.*

Then, if (C) applies to both sets of preferences, there is a utility dictator on S'.

If (C) is replaced with either (C_1) or (C_2), the same result holds. If (C_3) is used instead, positive pairwise utility dependence prevails, and there is a utility dictator, on S'.[14]

Supposing that for all i and s, s is non-null, Proposition 14 can be applied to $S' = S$ as a particular case, but this would be an uninteresting application. The conclusions would just repeat the utility-relative impossibility results already obtained in Propositions 9 and 10. The point of Proposition 14 is to extend these results slightly by emphasizing event-relative properties. Condition (**) is a limited state-independence assumption. It is compatible with a generally state-dependent framework. Notice also that it does not involve any uniformity from one i to another beyond the mere fact that the preferences of each just depend on S' (rather than on the particular state in S'). Going back to one of the previous examples, take $S' = \{$excellent weather, fair weather$\}$. It can no doubt happen that the two vacationers' and the observer's preferences are non-trivial and uniform across S', as required by (*) and (**), respectively. As far as condition (***) is concerned, it is perhaps no more problematic here than it was in the state-independent framework. Take again the vacationers' example: for (***) to be met, it is enough that they entertain different probabilities of the weather turning fair or excellent. Notice however that there is a rough trade-off in plausibility between (**) and (***): the smaller S' is, the more plausible is (**) but the *less* plausible (***).

5 THE TWO-INDIVIDUAL CASE

In the two-individual case, the impossibility conclusions of sections 3 and 4 can be sharpened, as the following corollaries show.

COROLLARY 15. *Assume that $\succeq^0, \succeq^1, \succeq^2$ satisfy Anscombe and Aumann's axioms of state-independent utility. Assume that $p^1 \neq p^2$ and that u^1, u^2 are not identical up to an affine transformation. Then, (C) holds if and only if there is an overall or an inverse overall dictator; (C_1) or (C_2) holds if and only if there is an overall dictator; and it is impossible for either (C_3) or (C^+) to hold.*

COROLLARY 16. *Assume that $\succeq^0, \succeq^1, \succeq^2$ and $\tilde{\succeq}^0, \tilde{\succeq}^1, \tilde{\succeq}^2$ satisfy Karni, Schmeidler and Vind's system of state-dependent utility as restated in section 2. If (C), (C_1), or (MAA) and (C_2) hold, then for each state $s \in S$, either $p^0(s) = p^1(s)$ or $p^0(s) = p^2(s)$ or $v^0(s,\cdot) = v^1(s,\cdot) = v^2(s,\cdot)$ up to affine transformations. If (MAA) and (C_3) hold, for each state $s \in S$, either $p^0(s) = p^1(s) = p^2(s)$ or $v^0(s,\cdot) = v^1(s,\cdot) = v^2(s,\cdot)$ up to affine transformations.*

---

[14] We included Axiom 2′ among the assumptions just for clarity, since condition (*) makes it redundant.

These two corollaries are closely related to the results of Schervish, Seidenfeld and Kadane [1991, Theorems 2 and 4], who formalize a version of the Bayesian experts paradox in the two-individual case by assuming an AA framework of first state-independent, and then state-dependent theory. In the state-dependent case, they use a special variant of the KSV procedure.[15] Like the main theorem of their paper, our Corollary 16 comes close to predicting that under relevant assumptions, for each state s, either probability agreement or utility-dependence prevails on $\{s\}$. The difference between this wording and their formal statement appears to come mostly from the complication of the null states in the variant they adopt.

It should be clear that the analysis of "consistent Bayesian aggregation" cannot be pursued just in the two-individual case. The conclusion corresponding to Corollary 15 loses its elegant simplicity when $n \geq 3$. The mild requirements that p^1, p^2 should be distinct and that u^1, u^2 should not be essentially identical or opposite functions become the more technical, less interpretable restrictions that $p^1, \cdots, p^n$ are linearly independent, and that $u^1, \cdots, u^n$ are affinely independent. Earlier examples demonstrate that in the state-independent framework, nontrivial solutions to the aggregation problem emerge in the absence of suitable independence assumptions.[16] As far as the state-dependent framework is concerned, Corollary 16 appears to derive a quasi-impossibility theorem without making technical restrictions. Again, the simplicity of this conclusion disappears when $n \geq 3$. We shall give a three-individual example to illustrate how easily nontrivial solutions to the aggregation problem might emerge from the state-dependent case, when algebraic independence restrictions are omitted.

Take $X = \{x_1, x_2, x_3\}$, so that $\Delta(X)$ is S_3^1, i.e., the unit simplex of $\boldsymbol{R}^3$. Denote the elements of $\mathcal{R} = \Delta(X)$ as $R = (R_1, R_2, R_3)$. In the notation used throughout, (C) implies that for all $R \in \mathcal{R}$:

$$(i) \qquad \sum_s p_s^0 v_s^0(R) = \sum_i a_i \sum_s p_s^i v_s^i(R)$$

and (considering now hypothetical instead of actual preferences):

$$(ii) \qquad \sum_s q_s^0 v_s^0(R) = \sum_i b^i \sum_s q_s^i v_s^i(R).$$

It is easy to find specific values such that the KSV assumptions and (C) hold, but for some $S' \subset S$, neither probability dictatorship nor any form of utility dictatorship holds. Take:

[15] See their other paper [Schervish *et al.*, 1990] for a statement of this variant.

[16] See [Goodman, 1988] and [Mongin, 1995, Example 4]. We have belatedly heard of Goodman's contribution to the n-person analysis. Thanks are due to Teddy Seidenfeld for bringing this and other references to our attention.

$$
\begin{aligned}
p^1 &= \begin{pmatrix} \frac{1}{2} \\ \frac{1}{2} \\ 0 \end{pmatrix}, p^2 = \begin{pmatrix} 0 \\ \frac{1}{2} \\ \frac{1}{2} \end{pmatrix}, p^3 = \begin{pmatrix} \frac{1}{2} \\ 0 \\ \frac{1}{2} \end{pmatrix}, p^0 = q = \begin{pmatrix} \frac{1}{3} \\ \frac{1}{3} \\ \frac{1}{3} \end{pmatrix}, \\
u^1_{s_1} &= R_1, u^1_{s_2} = R_3, u^1_{s_3} = 2R_2 + 3R_3, \\
u^2_{s_1} &= R_1 + 3R_2, u^2_{s_2} = R_1, u^2_{s_3} = R_2, \\
u^3_{s_1} &= R_2, u^3_{s_2} = R_3 + 2R_1, u^3_{s_3} = R_3, \\
u^0_{s_1} &= 2R_1 + 4R_2, u^0_{s_2} = 2R_3 + 3R_1, u^0_{s_3} = 3R_2 + 4R_3.
\end{aligned}
$$

The vectors p^0, p^1, p^2, p^3 can be thought of as KSV probabilities. In the particular instance, the common hypothetical probability q is taken to be equal to the observer's. To see that the above data agree with (C), notice that equations (i) and (ii) hold with:

$$a_1 = \frac{4}{3}, a_2 = 2, a_3 = \frac{8}{3}, b_1 = b_2 = b_3 = 1.$$

In contradistinction to the impossibility result stated in Proposition 12, probability dictatorship does not hold on $S' = S$. This fact can be traced to the failure of only one assumption in Proposition 12 — i.e., affine independence. Indeed, the individuals' state-dependent utilities are linearly dependent in each state.

Notice that no form of utility dictatorship prevails either. This frustrates the hope of extending the impossibility conclusion of Corollary 16 without adding suitable technical assumptions.

We close the discussion of Corollary 16 by noting that it is a consequence of Proposition 12 *alone*. That is to say, the probability-relative impossibility result implies a restriction on the observer's utility whenever there are only two individuals. This convenient property is lost in the general case.

6 FINAL COMMENTS: THE *EX POST* SOLUTION TO THE PARADOX

The present paper has offered a comprehensive treatment of the paradox of the Bayesian experts within the framework of the Anscombe-Aumann approach, first by assuming complete state-independence of utility, second by considering the opposite case of complete state-dependence, and third by applying the "intermediary" system of Karni, Schmeidler and Vind [1983] in which utility is state-dependent but the subjective probability is shown to be unique. Propositions 9 and 10 state the paradox in its pure form. They are the AA counterparts of the impossibility theorems recently proved within a state-independent, Savagean framework [Mongin, 1995, Propositions 5 and 7]. Proposition 11 states an easy, but unimpressive possibility result for the pure state-dependent case. By assuming the more informative KSV framework, Propositions 12 and 14 reinstate the paradox, although in a significantly different variant from the initial one.

One might perhaps have conjectured that the paradox would reappear in essentially its original form, once the state-dependence assumption is compounded with a procedure to determine subjective probabilities uniquely. This conjecture fails. The state-by-state analysis uncovers novel and curious situations: combining the assumptions of Propositions 12 and 14 on a sufficiently large state set, one might end up with juxtaposing probability dictatorship or agreement on some events with utility dictatorship or dependence on other events, and a nondescript state of affairs elsewhere. More generally, if "consistent Bayesian aggregation" leads to any paradoxical consequences in the state-dependent framework, these are bound to be state- or at least event-relative. This is the most obvious difference between the negative conclusions delivered by Propositions 9 and 10, on the one hand, and Propositions 12 and 14, on the other.

For more than two experts, technical conditions must be employed to derive dictatorship or uniformity conclusions. We argued that these conditions would be too stringent if they were to apply to each and every subset of the state set. This is why we selected *local* formulations of both our technical conditions and impossibility results. Accordingly, it might be argued that the latter are not really impossibility results, i.e., that the paradox of the Bayesian experts has not been reproduced in the state-dependent framework of this paper. This would be an exaggerate conclusion. It would be tantamount to abstracting from the important differences between a completely unconstrained aggregative rule (Proposition 11) and a relatively constrained one (Propositions 12 and 14). The correct interpretation probably lies half-way between the initial expectation that any sophisticated theory of state-dependent utility would reinstate the original paradox, and the extreme view now under discussion.

Given that we cannot conclude that state-dependent utility theory is the way of escape from the logical difficulties of collective Bayesianism, a more radical alternative must be sought. Within the province of decision theory at large, it remains to investigate suitable relaxations of the sure-thing principle. Within the confines of the present paper, which is restricted to Bayesianism, the remaining logical possibilities are to relax either Paretian consistency or Bayesian consistency. The former solution is illustrated in the field of welfare economics by those writers who reject the *ex ante* formulation of the Pareto principle (i.e., the version which was investigated in this paper), while retaining an *ex post* (i.e., consequence-relative) version.[17] By contrast, the latter solution consists in denying that the aggregate should inherit the individuals' method of decision.

We should like to indicate a (highly qualified) preference for the former over the latter direction of analysis. Even more clearly than some earlier and formally similar cases in welfare economics, the Bayesian experts problem implies that Bayesian consistency should be taken seriously. In the present setting the aggregate does not refer to a collective entity, but to some person acting as an observer.

[17] A leading exponent is Hammond [1982; 1983]. Among the recent applications of the *ex post* point of view, see in particular Zhou's [1996] axiomatization of Bayesian utilitarianism.

To go back to the example of section 1, the aggregate represents the insurer who attempts to summarize the experts' opinions. Given the nature of the observer in this problem situation, it seems natural to subject his preferences to the same choice-theoretic constraints as those prevailing on the (other) individuals' preferences.[18]

Here is a scenario which is compatible with the relaxation of Paretian consistency, whereas Bayesianism is preserved throughout. If, *contrary to our initial assumption*, the insurer understands Bayesian elicitation methods, he will be able to estimate the experts' underlying probability and utility functions. In the state-dependent case, we should then assume that he *himself*, rather than the experimenter, applies the KSV procedure. Once in possession of individual probability and utility data, he will process them *separately* to construct a summary probability and a summary vector of state-dependent utilities. These two items can then be combined unproblematically in the way prescribed by SEU theory. The Pareto principle will be used in the construction of the summary utility vector, but not necessarily in the construction of the summary probability. When applied to the individuals' utilities in each state, it functions as an *ex post* unanimity principle.

The previous paragraph shows that there is one (actually, well-known) way of keeping Bayesian consistency intact while preserving some form of Paretianism. We have just rephrased in terms of our decision-theoretic example the aggregation procedure which has long been recommended by the *ex post* school of welfare economics. In doing so, we have emphasized that there are definitive cognitive assumptions underlying the *ex post* approach — a point which is rarely mentioned in welfare theory. Before arguing that the *ex post* method is a feasible solution to the paradox of the Bayesian experts, one should check whether these assumptions apply. A definitive advantage of the *ex ante* method examined throughout this paper is that it does not require much knowledge on the observer's part.

To claim that the *ex post* approach provides not only a feasible, but a good resolution, a closer examination of the Pareto principle is needed. Implicitly, the defence of the principle trades on a distinction between *factual* and *normative* considerations. The essence of Paretianism is to proclaim that individuals are sovereign in normative matters; this means that their judgments in these matters should never be scrutinized or criticized, but taken for granted. "Normative" here can be diversely understood by reference to values, objectives, or even tastes, as in the "consumer sovereignty" doctrine. These interpretations would correspond to particular statements and defences of the Pareto principle. It is not our task here to list and compare them. The crucial point is that the individuals' sovereignty can be, and has been, argued for in the context of various notions of "normative" judgments, while *there is no concept of factual judgment for which this principle makes sense*. Factual judgements *should* be scrutinized and criticized. Theoretically at

[18]There is a modelling alternative which would make it even clearer that the observer here is just another individual. One could possibly endow him with *two* binary relations, one of which would represent his preferences *qua* ordinary person, the other his preferences *qua* observer. The insurer would then include his own private opinions among those which he tries to amalgamate.

least, they are susceptible of ascertainable truth values; they can be justified or dismissed by logic and evidence. So the Pareto conditions can only hold of a special class of unanimous judgments. Once all this is made clear, it seems as if the *ex post* variant is automatically warranted, and the *ex ante* variant automatically rejected. As it were, the former reaps all the benefits of the normative versus factual distinction.

Let us first clarify the negative part of the argument. Assuming a standard Bayesian framework, comparisons of prospects normally depend on *both* how the agent assesses the values of consequences and how he estimates the likelihood of events. Stochastic dominance, in which only the values of consequences matter, is an exceptional case. Thus, the scope of the *ex ante* Pareto principle exceeds the province of normative judgments. This, in itself, would not make it invalid, just dubious. What makes it invalid is that the excess content of the *ex ante* principle — its encroachment upon the province of factual judgments — leads to spurious recommendations. Here is one: under any state-independent variant of SEU theory, whenever all individuals agree on the strict ranking of two particular consequences, the principle implies that unanimous *probability* judgments should be respected, regardless of the evidence available to each individual. This is a spurious recommendation: evidence should matter to the observer. It can be shown that when conditioning partitions differ from one individual to another, a Bayesian observer who knows what these partitions are will sometimes violate the probabilistic form of the Pareto principle.[19]

Now, consider the positive argument in favour of the *ex post* Pareto principle. It says that the latter is justified because it involves only normative judgments. But in real life, judgments about consequences are infected with factual considerations. A hole in the ozone layer strikes one as an undesirable consequence because of certain scientific facts and laws. To own a large fortune becomes less desirable, or might even become absolutely undesirable, to somebody who knows that he will die tomorrow; and so on. All this suggests that the *ex post* principle could in turn fall a prey to the argument against the *ex ante* principle. By itself, the normative versus factual distinction does not provide the former with a sufficient foundation.[20]

From the above discussion we might conclude that the factual versus normative distinction cuts both ways, and that the foundations of *ex post* reasoning are shaky. But they are at least solider than the foundations of *ex ante* reasoning, which — this paper has attempted to demonstrate — appears to be simply flawed. And in the absence of a third alternative,[21] the *ex post* solution has at least the advantage of providing a feasible way out of the conundrum of collective Bayesianism.

---

[19]We are indebted here to Ed Green and David Schmeidler. Probabilistic unanimity is discussed at grater length in Mongin [1997].

[20]For a similar argument, see [Broome, 1990].

[21]A paper by Levi [1990] appears to sketch a third alternative by defining restrictions on the *ex ante* principle. This is an interesting avenue to explore.

APPENDIX

A PROOFS

Proof [of Lemma 4] Proposition 2 implies that for $i = 0, 1, \cdots, n$, V^i preserves convex combinations of acts. Hence the vector $(V^0, V^1, \cdots, V^n)$ has convex range. ■

Proof [of Lemma 5] See [De Meyer and Mongin, 1995, Proposition 1]. ■

Proof [of Lemma 6] See [De Meyer and Mongin, 1995, Propositions 1 and 2]. ■

Proof [of Lemma 7] The former conclusion is proved in Coulhon and Mongin [Coulhon and Mongin, 1989]. In view of Axiom 1, the latter is an immediate application of the former. ■

Proof [of Lemma 8] Take $R^1_*, R^1_{**}, \cdots, R^n_*, R^n_{**}$ as in the statement of Lemma 7 and construct the following elements of $\mathcal{R}$:

$$P_* = (\frac{1}{n}R^1_*, \cdots, \frac{1}{n}R^n_*) \text{ and } P_{**} = (\frac{1}{n}R^1_{**}, \cdots, \frac{1}{n}R^n_{**}).$$

Then, from the mixture-preserving property of $u^1, \cdots, u^n$:

$$u^i(P_*) > u^i(P_{**}),\ i = 1, \cdots, n,$$

so that (MAC) holds. The case of affinely independent $V_1, \cdots, V_n$ can be dealt with similarly. ■

Proof [of Proposition 9] Suppose that (C) and (*) hold. Lemma 5 implies that there are $a_1, \cdots, a_n, b$ such that

$$(A1) \qquad U^0 = \sum_{i=1}^{n} a_i U^i + b.$$

One of the a_i must be nonzero because of AA's nontriviality assumption (Axiom 2). We may select any $\overline{R} \in \mathcal{R}$ and put $u^0(\overline{R}) = u^1(\overline{R}) = \cdots = u^n(\overline{R}) = 0$; there is no assumption of substance in this normalization. Let us now consider the following class of $f \in H$: there are $s \in S$ and $R \in \mathcal{R}$, such that

$$\begin{cases} f(s) & = R \\ f(t) & = \overline{R} \text{ for all } t \neq s. \end{cases}$$

Applying AA's representation theorem in equation (1), we get:

$$\forall f \in H, \quad \sum_{s \in S} p^0_s u^0(f(s)) = \sum_{i=1}^{n} a_i \sum_{s \in S} p^i_s u^i(f(s)).$$

When we restrict attention to the class of acts just defined, this becomes:

$$\forall s \in S, \forall R \in \mathcal{R}, \quad p_s^0 u^0(R) = \sum_{i=1}^{n} a_i p_s^i u^i(R).$$

From now on in the proof, we shall use functional notation. The last equation becomes:

$$(A2) \qquad p^0 u^0 = \sum_{i=1}^{n} a_i p^i u^i,$$

where the functions on the right- and the left-hand sides are defined on $S \times \mathcal{R}$.

Given that in the state-independent case, constant acts may be identified with consequences, equation (A1) also implies that:

$$(A3) \qquad u^0 = \sum_{i=1}^{n} a_i u^i.$$

Replacing (A3) into (A2) we get:

$$(A4) \qquad \sum_{i=1}^{n} a_i u^i [p^0 - p^i] = 0.$$

If the $p^0 - p^1, \cdots, p^0 - p^n$ were linearly independent, one would have that:

$$a_i u^i = 0 \text{ for all } i = 1, \cdots, n,$$

which is impossible since Axiom 2 implies that the u^i are nonconstant and one a_i must be nonzero. Hence, there is $j \in \{1, \cdots, n\}$ such that for some $b_1, \cdots, b_{j-1}, b_{j+1}, \cdots, b_n$:

$$(A5) \qquad p^0 - p^j = \sum_{i \neq j} b_i (p^0 - p^i).$$

Now, $\sum_{i \neq j} b_i \neq 1$ in view of (*). (Assume that $\sum_{i \neq j} b_i = 1$; then (A5) leads to the absurd equation $p^j = \sum_{i \neq j} b_i p^i$.) We can rewrite (A5) as:

$$p^0 = (1 - \sum_{i \neq j} b_i)^{-1} p^j - \sum_{i \neq j} (1 - \sum_{i \neq j} b_i)^{-1} b_i p^i,$$

which provides a linear decomposition of p^0 in terms of $p^1, \cdots, p^n$. Changing the notation, we have just derived:

$$(A6) \qquad p^0 = \sum_{i=1}^{n} c_i p^i, \text{ for some } c_1, \cdots, c_n \text{ such that } \sum c_i = 1,$$

so that at least one c_i is positive.

Now, replacing (A6) into (A2) leads to:

$$\sum_{i=1}^{n} (c_i u^0 - a_i u^i) p^i = 0.$$

Using (*) again, we conclude that for all $i = 1, \cdots, n$,

$$(A7) \qquad c_i u^0 = a_i u^i.$$

One of the a_i must be non-zero, and for any $i = 1, \cdots, n, a_i \neq 0$ if and only if $c_i \neq 0$ (because $u^0, u^1, \cdots, u^n$ are nonconstant). Hence, there is a utility or inverse utility dictator, as was required to show.

Consider the effect of assuming (C_1) instead of (C), while still assuming that (*) holds. The argument just made remains available, since (C_1) trivially implies (C). But there is now a sign restriction on the a_i (Lemma 5). This restriction, together with the fact that one c_i must be positive, implies that there is a utility dictator.

To deal with (C_2) in case (*) we first note that the latter property implies that:

$$(***) \qquad U^1, \cdots, U^n \text{ are affinely independent.}$$

(Suppose not, and consider the special class of acts at the beginning of the proof; then, for some $j \in \{1, \cdots, n\}$, there are coefficients $d_i, i \neq j$, such that $p^j u^j = \sum_{i \neq j} d_i p^i u^i$, a contradiction.) Then, Lemma 8 says that (MAA) holds, and from Lemma 6 the results reached for (C_1) apply to (C_2).

When (C) and (**) hold, equations (A1) to (A4) remain unchanged. Then, the affine independence property of the u^i implies that for all i,

$$(A8) \qquad a_i(p^0 - p^i) = 0,$$

whence we conclude that there is a probability dictator. This conclusion still holds under either (C_1) or (C_2). The latter case is dealt with by noting that (**) also implies (***). ■

Proof [of Proposition 10] Using Lemmas 6 and 8 as in the previous proof, we see that under either (*) or (**), (C_3) becomes equivalent to (C^+). Hence, from Lemma 5 there is an affine decomposition:

$$U^0 = \sum_{i=1}^{n} a_i U^i + b$$

with positive a_i for $i = 1, \cdots, n$. Now, assuming that (*) is the case, we can reproduce the reasoning of the previous proof and conclude, as in (A7) above, that for all $i = 1, \cdots, n$,

$$c_i u^0 = a_i u^i.$$

Since all of the a_i are positive, and $a_i \neq 0$ if and only if $c_i \neq 0$, we conclude that for $i = 1, \cdots, n$, there are $\alpha_i \neq 0$ such that $u^0 = \alpha_i u^i$, and pairwise utility dependence prevails among $u_1, \cdots, u_n$. Remember that one of the c_i must be positive; this implies that there is a utility dictator. To analyze case (**), we revert to equation (A8) in the proof above:

$$a_i(p^0 - p^i) = 0,$$

and now conclude that probability agreement prevails and that there is a probability dictator. ■

Proof [of Proposition 12] Throughout the proof, we write $v^i(s, R)$ instead of $v^i_s(R)$ and $p^i(s)$ instead of p^i_s, and assume that for some $R_* \in \mathcal{R}$,

$$v^i(s, R_*) = 0 \text{ for all } s \in S \text{ and all } i = 0, 1, \cdots, n.$$

(This normalization is permitted by the uniqueness part of Proposition 3.)

From Proposition 3 we know that if (C) holds, $\succeq^i$ is represented by:

$$V^i(f) = \sum_{s \in S} p^i(s) v^i(s, f(s)),\ i = 0, 1, \cdots, n.$$

Lemma 2 and the chosen normalization imply that:

$$V^0(f) = \sum_{i=1}^{n} a_i V^i(f)$$

for some $a_1, \cdots, a_n$. By identifying the two expressions for $V^0(f)$, and restricting them to those acts which have values R on s and R_* elsewhere, we conclude that:

$$(A9) \qquad p^0(s) v^0(s, R) = \sum_{i=1}^{n} a_i p^i(s) v^i(s, R) \quad \forall s \in S, \forall R \in \mathcal{R}$$

Repeating the argument for the auxiliary preferences $\tilde{\succeq}^0, \cdots, \tilde{\succeq}^n$ and their functional representations leads to:

$$q(s)v^0(s,R) = \sum_{i=1}^{n} b_i q(s) v^i(s,R) \quad \forall s \in S, \forall R \in \mathcal{R},$$

for some $b_1, \cdots, b_n$. Since $q(s) > 0$ for all s,

$$(A10) \qquad v^0(s,R) = \sum_{i=1}^{n} b_i v^i(s,R) \quad \forall s \in S, \forall R \in \mathcal{R}.$$

Replacing (A10) into (A9) we have that:

$$(A11) \qquad \sum_{i=1}^{n} [b_i p^0(s) - a_i p^i(s)] v^i(s,R) = 0 \quad \forall s \in S, \forall R \in \mathcal{R}.$$

Now, consider S' as in the first part of the Proposition. For any fixed $s \in S'$, since the $v^i(s,\cdot)$ are linearly independent, the equation in R:

$$(A12) \qquad \sum_{i=1}^{n} [b_i p^0(s) - a_i p^i(s)] v^i(s,\cdot) = 0$$

implies that:

$$(A13) \qquad b_i p^0(s) = a_i p^i(s), \quad i = 1, \cdots, n.$$

Consider the sets of indexes:

$$I = \{i = 1, \cdots, n | \, a_i \neq 0\} \text{ and } J = \{i = 1, \cdots, n | \, b_i \neq 0\}.$$

From Axiom $2'$, as applied to $\succeq_0$ and $\tilde{\succeq}_0$ respectively, we know that $I \neq \phi \neq J$. Suppose that $I \cap J = \phi$. Then, (A13) implies that:

$$p^0(s) = 0, \text{ and for at least one } i \in I, p^i(s) = 0.$$

If we repeat the reasoning for $s' \in S, s \neq s'$, we find that $p^i(s') = 0$ for the same i. Hence, in the case in which $I \cap J = \phi$, there is i such that $p^0(S') = p^i(S') = 0$, a case of probability dictatorship.

Now, consider the case in which $I \cap J \neq \phi$. There is i such that $a_i \neq 0 \neq b_i$, and:

$$p^0(s) = a_i b_i^{-1} p^i(s), \quad \forall s \in S'.$$

Either $p^i(S') = 0 = p^0(S')$, or $p^i(S') \neq 0$ and

$$\frac{p^0(s)}{p^0(S')} = \frac{p^i(s)}{p^i(S')}, \forall s \in S',$$

which again shows that probability dictatorship prevails. The analysis of the other conditions than (C) makes use of Lemmas 6 and 7, as in the corresponding parts of the proofs of Propositions 10 and 11. Details are left for the reader. ■

Proof [of Proposition 14] We first spell out the implications of Assumption 13 for each KSV representation taken individually. Axiom 1 can be applied to the restriction of $\succeq^i$ to the set $H_{s'}$ of acts having some fixed set of values outside S', and because of (*) and (**), Axiom 2 and a version of Axiom 3 hold for this preference relation (which we also denote by $\succeq^i$). From Proposition 1 there is a state-independent function w^i on $\mathcal{R}$ and a probability π^i on S' such that:

$$\forall f, g \in H_{S'},\ f \succeq^i g \text{ iff } \sum_{s \in S'} \pi_s^i w^i(f(s)) \geq \sum_{s \in S'} \pi_s^i w^i(g(s)).$$

Now, the conclusions of Proposition 3 also apply to the restricted preference. Using Assumption 13, the $(|S'| + 1)$-tuple $(w^i, \cdots, w^i, \pi^i)$ is seen to satisfy conditions (i) and (ii) in Proposition 3, as applied to acts in $H_{S'}$, so that by the uniqueness part of this proposition:

$$(v_s^i)_{s \in S'} = (w^i, \cdots, w^i) \text{ up to a PAT},$$

and:

$$\pi^i(s) = \frac{p^i(s)}{p^i(S')}, \forall s \in S'.$$

Hence, for $i = 0, 1, \cdots, n$, we may replace the initial vector of KSV representations relative to $H_{S'}$ by $(w^i, \cdots, w^i)$, and use (C) and condition (***) to prove impossibility results as if state-independence prevailed. The reader is referred to the relevant parts of the proofs of Propositions 9 and 10. ■

Proof [of Corollary 15] Immediate from Propositions 9 and 10. ■

Proof [of Corollary 16] If $v^1(s, \cdot), v^2(s, \cdot)$ are affinely independent, Proposition 12 implies that either $p^0(s) = p^1(s)$ or $p^0(s) = p^2(s)$ whenever (C), (C_1), or (MAA) and (C_2) hold, a conclusion which is strengthened into $p^0(s) = p^1(s) = p^2(s)$ whenever (MAA) and (C_3) holds. If $v^1(s, \cdot), v^2(s, \cdot)$ are affinely dependent, the conclusion that $v^0(s) = v^1(s) = v^2(s)$, up to relevant affine transformations, follows from inspecting equation (A2) in the proof of Proposition 12. ■

ACKNOWLEDGEMENTS

Earlier versions of this paper were presented in 1995 at the Tokyo Center of Economic Research Conference and the Economics Department, Copenhagen University; and in 1996, at the Economics Departments, Duke University and Princeton University. The author is grateful to the Centre for the Philosophy of the Natural and Social Sciences, The London School of Economics, and the Economics Department, Duke University, for hospitality when he was working on this paper. Special thanks are due to E. Green, P. Hammond, M. Kaneko, E. Karni, H. Moulin, D. Schmeidler, P. Wakker, J. Weymark.

Reprinted from *Journal of Mathematical Economics*, **29**, Philippe Mongin, "The paradox of the Bayesian experts and state-dependent utility theory", pp. 331–361, copyright 1998, with permission from Elsevier Science.

Laboratoire d'économétrie, Centre National de la Recherche Scientifique & Ecole Polytechnique, Paris, France.

BIBLIOGRAPHY

[Anscombe and Aumann, 1963] Anscombe, F.G. and R.J. Aumann, 1963, A definition of subjective probability, Annals of Mathematical Statistics 34, 199-205.

[d'Aspremont, 1985] d'Aspremont, C., 1985, Axioms for social welfare orderings, in: L. Hurwicz, D. Schmeidler, H. Sonnenschein, eds., Social goals and social organization (Cambridge, C.U.P.) 19-76.

[Broome, 1990] Broome, J., 1990, Bolker-Jeffrey expected utility theory and axiomatic utilitarianism, Review of Economic Studies 57, 477-502.

[Coulhon and Mongin, 1989] Coulhon, T. and P. Mongin, 1989, Social choice theory in the case of von Neumann-Morgenstern utilities, Social Choice and Welfare 6, 175-187.

[de Finetti, 1974–75] de Finetti, B., 1974-75, Theory of probability (New York, Wiley, 2 volumes).

[De Meyer and Mongin, 1995] De Meyer, B. and P. Mongin, 1995, A note on affine aggregation, Economics Letters 47, 177-183.

[Drèze, 1987] Drèze, J., 1987, Essays on economic decisions under uncertainty (Cambridge, C.U.P.).

[Fishburn, 1970] Fishburn, P.C., 1970, Utility theory for decision making (New York, Wiley).

[Fishburn, 1982] Fishburn, P.C., 1982, The foundations of expected utility (Dordrecht, Reidel).

[Goodman, 1988] Goodman, J., 1988, Existence of compromises in simple group decisions, unpublished Ph.D. Thesis (Carnegie-Mellon University).

[Hammond, 1982] Hammond, P.J., 1982, Ex-ante and ex-post welfare optimality under uncertainty, Economica 48, 235-250.

[Hammond, 1983] Hammond, P.J., 1983, Ex-post optimality as a dynamically consistent objective for collective choice under uncertainty, in: P.K. Pattanaik and M. Salles, eds., Social choice and welfare (Amsterdam, North Holland).

[Hammond, 1998] Hammond, P.J., 1998, Subjective expected utility theory, in: S. Barberà, P. Hammond, and C. Seidl, Handbook of utility theory (Dordrecht, Kluwer).

[Harsanyi, 1955] Harsanyi, J.C., 1955, Cardinal welfare, individualistic ethics, and interpersonal comparisons of utility, Journal of Political Economy 63, 309-321.

[Karni, 1985] Karni, E., 1985, Decision making under uncertainty: the case of state-dependent preferences (Cambridge, Mass., Harvard University Press).

[Karni, 1993] Karni, E., 1993, A definition of subjective probabilities with state-dependent preferences, Econometrica 61, 187-198.

[Karni and Schmeidler, 1993] Karni, E. and D. Schmeidler, 1993, On the uniqueness of subjective probabilities, Economic Theory 3, 267-277.

[Karni *et al.*, 1983] Karni, E., D. Schmeidler and K. Vind, 1983, On state-dependent preferences and subjective probabilities, Econometrica 51, 1021-1031.

[Levi, 1990] Levi, I., 1990, Pareto unanimity and consensus, Journal of Philosophy 87.

[Mongin, 1995] Mongin, P., 1995, Consistent bayesian aggregation, Journal of Economic Theory 66, 313-351.

[Mongin, 1997] Mongin, P., 1997, Spurious unanimity and the Pareto principle, paper presented at the conference "Utilitarianism Reconsidered" (New Orleans, March 1997).

[Savage, 1972] Savage, L.J., 1972, The foundations of statistics, (New York, Dover, 1st edition, 1954).

[Schervish *et al.*, 1990] Schervish, M.J., T. Seidenfeld and J.B. Kadane, 1990, State-dependent utilities, Journal of the American Statistical Association 85, 840-847.

[Schervish *et al.*, 1991] Schervish, M.J., T. Seidenfeld and J.B. Kadane, 1991, Shared preferences and state-dependent utilities, Management Science 37, 1575-1589.

[Seidenfeld *et al.*, 1989] Seidenfeld, T., J.B. Kadane and M.J. Schervish, 1989, On the shared preferences of two Bayesian decision makers, Journal of Philosophy 86, 225-244.

[Wakker, 1987] Wakker, P., 1987, Subjective probabilities for state-dependent continuous utility, Mathematical Social Sciences 14, 289-298.

[Weymark, 1993] Weymark, J.A., 1993, Harsanyi's social aggregation theorem and the weak Pareto principle, Social Choice and Welfare 10, 209-222.

[Zhou, 1996] Zhou, L., 1996, Bayesian utilitarianism, CORE Discussion Paper 9611 (Universit Catholique de Louvain).

PART V

CRITICISMS OF BAYESIANISM

MAX ALBERT

BAYESIAN LEARNING AND EXPECTATIONS FORMATION: ANYTHING GOES

1 GREAT EXPECTATIONS

When Muth [1961] introduced the rational–expectations hypothesis (REH), his basic idea was that agents form expectations by rationally acquiring and processing information (weak REH). From this, he immediately jumped to a stronger hypothesis (strong REH), which in the following years radically transformed macroeconomic theory and policy. The strong REH is implied by the assumption that agents know the true (statistical) model of their environment. Except for trivial cases, this model comprises a causal model relating endogenous to exogenous variables, and objective probability distributions of the exogenous variables. Agents' expectations are the objective probability distributions of future developments conditional on their current information about past realizations of exogenous and endogenous variables. In terms of the famous Knightian distinction, the strong REH implies risk and not uncertainty. Rationality then requires that agents choose a strategy (i.e., a plan specifying actions for all contingencies) that maximizes the expected utility on the basis of a v. Neumann–Morgenstern (NM) utility function.[1]

However, the optimistic spirit of the "rational expectations revolution" [Begg, 1982] has long since evaporated. Theoretical and empirical weaknesses of the strong REH have become apparent, and there seem to be good arguments for going back to the weak REH. Rationally acquiring and processing information without knowing the true statistical model of the environment (i.e., under conditions of *uncertainty* rather than risk) means rational learning, which is the domain of the subjectively expected utility (SEU) theory a.k.a. *Bayesianism.*[2]

The simplest version of Bayesian learning requires the agent to proceed from a *subjective* joint probability distribution for all conceivable future observations. This distribution reflects personal *degrees of belief*. The optimal strategy maximizes *subjectively* expected utility. The initial subjective distribution, the so–called prior distribution, is again revised by conditioning on observed events, which yields the so–called posterior distributions. The whole revision process, which is equivalent to the use of Bayes' theorem, is also called "updating the prior" because the posterior at one stage of the process serves as the prior on the next.

[1]Cf. Pearl [2000] for a discussion of causality in relation to statistics. Cf. Hacking [1990] on objective vs subjective probabilities. For strict subjectivists, the strong REH makes sense only for a group of agents, where it translates into the Common Prior Assumption (cf. Aumann [1987, 12ff]. See also 2.1 below.

[2]For critical discussions of the strong REH cf. Frydman and Phelps [1983] and Pesaran [1989]; for an overview of the learning literature cf. Kirman and Salmon [1995].

D. Corfield and J. Williamson (eds.), Foundations of Bayesianism, 341–362.

Bayesianism's claim to importance rests on the possible use of a two–stage procedure for deriving the prior. The agent first considers several models (or hypotheses or theories; we use these terms interchangeably) since the true model is unknown. Each model leads to different expectations. Then a prior over the set of models is chosen, which leads to a weighted average of the model–specific expectations. Updating the prior implies a shift in the weights of the models. This two–stage procedure connects Bayesianism with scientific procedures, leading to a unified theory of rationality in economics, statistics and practical decision making.[3]

In a Bayesian context, the strong REH is implied by the assumption that the agent's prior is degenerate and assigns probability 1 to the true model (*true beliefs*). Even without true beliefs, however, rational expectations are possible. Suppose that Adam the Agent tries to predict the outcome of once tossing a fair coin, and that he considers two hypotheses, namely, a probability of head equal to 0.25 or 0.75, respectively. Adam has rational expectations if his prior assigns a probability of 0.5 to each of the two hypotheses, implying that he assigns zero probability to the truth. By definition, rational expectations only require that the subjective probability distribution of the observable variables implied by the prior coincides with the corresponding objective probability distribution [Pesaran, 1989; 1]

Does Bayesian learning converge to rational expectations? Again, early optimism turned into disappointment. Even the beliefs of an ideal Bayesian learner who does *not* dismiss the true model from the outset and who faces *no* costs of gathering further information are not inevitably bound to converge to rational expectations [Blume and Easley 1995: 16–20].

It is not clear, however, what lesson, if any, should be drawn from the possible failure of convergence. A non-convergent learning process is not necessarily an indication of misguided decision making [Kiefer and Nyarko, 1995]. The justification of the Bayesian approach lies not in the convergence properties of Bayesian learning but in the appeal of certain axioms for preferences on the set of strategies. These axioms ensure the existence of an NM utility function and a prior such that the SEU of the strategies reflects the preference ordering.[4]

Therefore, much of the discussion has centered on the Bayesian system of ax-

[3]The classical exposition of Bayesianism is Savage [1954]; see also Kiefer and Nyarko [1995] for a summary and defense emphasizing learning and expectations formation. The approach of Anscombe and Aumann [1963] is appropriate if beliefs refer to objective probability distributions. For a single agent or under the Common Prior Assumption (see 2.1 below), weak and strong REH are formally identical, because the models can be treated like unobservable exogenous variables with a given known distribution (namely, the prior).

[4]Preferences over strategies entail preferences over objective–probability distributions of outcomes (expressed by the NM utility function) and beliefs (expressed by the prior). It is fundamental to Bayesianism that preferences in the narrower sense and beliefs are separable [Binmore, 1993: 207; Aumann 1987: 13 n. 13]. This implies that an agent can adopt an NM utility function independently from her beliefs, or beliefs independently of her NM utility function. On the former case, see also Binmore [1993: 207] on Savage and "massaging the priors". The latter case is illustrated by Savage's own use of the sure-thing principle as a device for (implicitly) adjusting his evalutation of NM utilities in the Allais Paradox (where probabilities are given); cf. Pope [1991].

ioms. Nevertheless, we postpone any comments on the axioms until the very end of the paper. Instead, we focus on the fact that Bayesianism is *empty*. As a positive theory, it implies no "operationally meaningful theorems" (OMTs), i.e., consequences that could potentially be refuted by observations.[5] Any behavior can be rationalized on the basis of some prior, even if the NM utility function is given.

For this reason, Bayesianism is also empty as a normative theory. Assume that Mike the Manager asks Betty the Bayesian for advice. Betty cannot take Mike's current beliefs for granted because it is an open question which beliefs are rational given Mike's previous experiences. So it is natural for Mike to ask whether there are any OMTs: Given what I know about the past, and given my NM utility function, is there anything a rational person would *not* do? Is there a sequence of future choices, including reactions to new information, that can be classified as *irrational*? Since the answer is no, Bayesianism is empty as a normative theory. To a Bayesian, *there is no such thing as irrational behavior.*

Although several such rationalizability results have appeared in recent years, they seem to be not widely known, and their implications are not yet fully appreciated.[6] The result discussed in the present paper assumes that an agent considers a simple chaotic process as explanation for observed phenomena. Such explanations are actually considered in economics and elsewhere; they cannot be dismissed as unreasonable. Chaotic systems have the typical features of very rich sets of hypotheses. Since perfectly rational agents by definition consider such very rich sets, they are necessarily, i.e., *independently from the actual complexity of their environment,* in the situation of a person trying to predict a chaotic system. This implies that their expectations are completely arbitrary. Muth's [1961] conjecture that the weak REH provides a solution to the problem of expectations formation is thus refuted, at least if the weak REH is identified with Bayesianism, as it is usually done in economics.

Section 2 reviews the literature and discusses some *prima facie* arguments against the view that Bayesinism is empty. Section 3 introduces an abstract decision problem, section 4 a set of hypotheses based on a simple chaotic system. Section 5 shows that this set can be used to rationalize any strategy. Section 6 concludes with a consideration of arguments against the position, taken in the present paper, that the emptiness of Bayesianism is a serious flaw.

2 A FOLK THEOREM

Presumably, practitioners tend to believe that there are objectively wrong decisions or mistakes and that decision theory provides the means to avoid them.[7] Theoreti-

[5]Samuelson's [1947: 3] phrase, rather than "empirical content", is used to remind readers of the present paper's close relation to Samuelson's work on revealed preference (see 2.3 below).

[6]The present paper is based on Albert [1996; 1999]. A slightly different result is contained in Nyarko [1997], who refers to an unpublished 1992 paper of J. S. Jordan for yet another version.

[7]See, e.g., Bernstein [1996: 336]. Goldman [1999: 76] also seems to believe that the so–called Dutch Book argument demonstrates that Bayesianism protects against unnecessary losses. However,

cians think differently. It is the folk theorem of decision theory that the notion of rationality employed in economics is "weak".[8] As its counterpart in game theory, the theorem is of unknown origin and implies that (almost) anything goes. Until quite recently, there has been no general proof, but proofs for finite cases are trivial.

There is a difference, however, between the claim that Bayesian rationality is "weak" and that it is completely empty. We therefore discuss three *prima facie* objections to the latter view. 1. It is sometimes suggested that prior beliefs of rational agents are not completely arbitrary. 2. Bayesians have always argued that their definition of rationality implies the rejection of certain other decision rules like the maximin rule; conflict, however, presupposes content. 3. Bayesianism encompasses the theory of demand, which is known to imply OMTs, the so–called axioms of revealed preference.

2.1 Rational Priors

Some Bayesians defend the view that, even though there are no restrictions on priors, all rational agents should hold the same subjective probabilities if they have been exposed to the same experience (e.g., [Aumann 1987: 7, 13f]). This view is known as the Common Prior Assumption (CPA). Aumann [1987] refers to Savage in this context (without giving a reference) and conjectures that Savage would have accepted the CPA. I disagree (cf. [Savage 1962: 11, 13, 14]). However, Savage was convinced that *in practice* experience often leads to convergence of opinion. But this is not a starting point for Bayesianism; it is a fact in need of explanation. For convergence, one needs priors that are not too different. The CPA just begs the question in assuming identical priors.

The CPA makes sense only if there exist canonical or rational priors before any experience. This leads to the classical problem of whether there is an acceptable "principle of insufficient reason" determining probabilities before experience. This idea, going back to Laplace, has been criticized by many authors (cf. [Leamer 1978: 11, 22–39, 61–63, 111, 114; Howson and Urbach 1989: 45–48, 285, 289; Earman 1992: 14–17, 138–141]). It had been revived by Keynes and others in the form of a theory of "logical" probabilities, i.e., uniquely determined *a priori*

---

the argument assumes a situation without any uncertainty concerning gains or losses and, therefore, completely misses the point when used as a defense of Bayesianism.

[8]Hahn [1996: 186] writes that rationality "buys only a small bit of an answer" in an intertemporal context since it has to be supplemented by a theory on agents' beliefs. Blume and Easley [1995: 26] conclude that the content of Bayesian rationality mostly derives from restrictions on the set of beliefs. Bicchieri [1993: 14, esp. n. 9] restricts the predictive usefulness of Bayesian rationality to stable environments and choice situations familiar to the agent, and mentions convergence problems in the case of complicated priors. Arrow [1990: 29] writes that the rationality hypothesis by itself is "weak" and that its force derives from supplementary hypotheses. By varying utility functions for given beliefs, Ledyard [1986] demonstrates that Bayesianism is empty for a quite general game-theoretic setting. However, he is still convinced of its value as a normative theory [Ledyard 1986: 60, 80f]. Bray [1983: 123f] quotes Lucas to the effect that Bayesianism "in many applications" has "little empirical content" but defends it on account of its convergence properties.

probabilities of the logically possible hypotheses. One of the arguments in favor of Bayesianism has been the discovery that such probabilities do not exist.[9] It seems not to be a promising way of further development to revive this idea again. As the history of the subject presents itself, the burden of proof that there is an acceptable "principle of insufficient reason" rests with those in favor of the CPA.

2.2 *The Dominance Principle*

Clashes between Bayesianism on the one hand and decision rules for behavior under uncertainty like the maximin rule on the other hand are due to the fact that the latter violate what we will call the dominance principle.[10] We can use the NM utility function to define a set of *strictly dominated strategies*. A strategy A is *strictly dominated* if and only if for every choice of prior probabilities, there always exists at least one strategy with a higher SEU than A. Bayesianism implies one restriction on behavior for a given NM utility function, namely, that no strictly dominated strategy is chosen. Let us call this restriction the dominance principle.

We illustrate this principle for a case of three strategies leading to different consequences in two mutually exclusive and jointly exhaustive states (figure 1). Given an NM utility function, the three strategies can be represented by their utilities in a two–dimensional coordinate system. Bayesian analysis implies the linearity of the indifference loci in this diagram; the maximin criterion would lead to L–shaped indifference loci. Therefore, the latter criterion allows for choices that are ruled out by Bayesian analysis.

The set of strictly dominated strategies is by definition independent of beliefs. Nevertheless, the dominance principle yields no OMTs. Identifying strictly dominated strategies already requires at least some knowledge about one's environment. If the NM utility function is given, it is always possible to imagine, for each strategy A, a state s such that A yields a higher utility than any other strategy if s holds. The assumption that a strategy is strictly dominated means that logical possibilities are excluded. If Bayesianism is empty as a theory of learning, the dominance principle yields neither predictions nor advice, because before anything has been learned, logical possibilities *cannot* be excluded.

2.3 *The Revealed-Preference Approach*

In economics, it is usually taken for granted that economic theory only provides assumptions about the *general structure* of agents' preferences, and that the details

[9]This appears already to have been a conjecture of Ramsey, the earliest of the modern Bayesians, who made this argument against Keynes; cf. Hacking [1990: 165, 170]. The Keynesian program was taken up later by Carnap; it was intended to provide one of the cornerstones of logical positivism. There is a widespread agreement today that this program foundered in just the way Ramsey conjectured: there are no logical probabilities; cf. Howson and Urbach [1989: 48-56].

[10]This clash has most often been stressed in connection with statistical decision theory, see, e.g., Lindley [1972: 13-15].

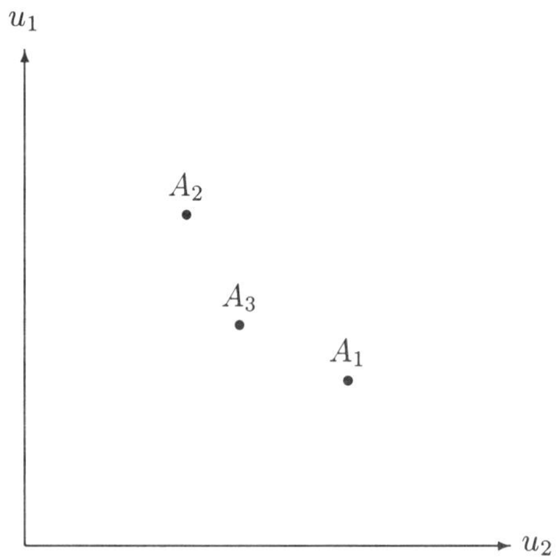

Figure 1. A strictly dominated strategy. The axes measure the NM utilities of three strategies in the case of two mutually exclusive and jointly exhaustive states. The dominated strategy A_3 is selected by the maximin criterion.

necessary for making predictions have to be gathered by observation. According to this view, economic theory provides a framework that allows us to use empirical observations to make predictions, much as Newton's theory allows us to use empirical observations to determine the masses of the planets, which in turn can be used to predict the planets' movements.

An important question in this context is which kind of observations can be used. The economic tradition allows only the use of observations about actual choices. This is consistent with the idea that—presumably as a result of selective pressure—agents act *as if* they were perfectly rational, or at least *as if* deviations from perfectly rational behavior were unsystematic. The as-if approach views rationality as a feature of behavior, not as a feature of the process of deliberation.[11]

This attitude is the basis of the revealed-preference (RP) approach to the theory of demand (cf., e.g., [Varian 1992: 133]). The approach focuses on OMTs—usually called *axioms* of RP since they exhaust the theory—implying that information about past choices can be used to rule out certain future choices if preferences have a certain structure.

The theory of demand implicitly assumes that observer and agent share true beliefs in a single deterministic model. Bayesianism can only profit from the results of this theory if the observer can check whether the agent holds appropriate beliefs. According to the economic tradition, such a check must be based on observations of actual choices and nothing else. Hence, an extension of the RP

[11]Cf. Simon's [1976] characterization of economic rationality as "substantive" rather than "procedural", which seems to be meant as a clarification of Friedman's [1953] as-if approach.

approach to the case of uncertainty is required. Such an extension would provide OMTs demonstrating how, assuming the axioms of Savage [1954] or Anscombe and Aumann [1963], information about past choices can be used to rule out certain future choices.

There are some extensions of the RP approach, but none that cover Bayesian learning.[12] The present paper addresses this issue in a very general way. As a concession to Bayesianism, knowledge of the NM utility function summarizing an agent's behavior under certainty and risk is granted. This leaves only beliefs as missing determinants of behavior. An RP approach to Bayesian learning requires OMTs implying that, given the NM utility function, information about past choices can be used to rule out certain future choices.

3 QUESTIONS

The technical aspects of our problem, and the way to a solution, can be explained with the help of a simple decision problem; generalizations are trivial (see 5.2 below).

Adam and Eve and the Money Spinner. Adam the Agent owns a mysterious black box connected with a screen and a keyboard. The screen displays either 0 or 1; in fixed intervals the screen goes black and then again shows one of the two digits. The tth observation is denoted by x_t. At every point in time $t = 0, 1, \ldots, \infty$, Adam places his bets on the next digit by typing in a number $y_t \in \{0, 1, ..., N\}$. Doing nothing implies $y_t = 0$. So the sequence of events is as follows. At $t = 0$, Adam chooses y_0. At $t = 1$, x_1 appears and Adam chooses y_1. And so on. At time t, Adam chooses y_t. If $x_{t+1} = 0$, the black box produces y_t perfect one–dollar notes. If $x_{t+1} = 1$, it produces $N - y_t$ equally perfect one–dollar notes. Adam cares only for the money he receives; his NM utility function ν: $[0, N] \mapsto \mathbb{R}$ at each point in time is increasing, strictly concave, and finite. Adam's information at time t encompasses all the facts just explained and the history of digits and choices up to time t.

Eve the Economist observes Adam. Her information at time t coincides with Adam's; specifically, she knows his NM utility function. We simplify the problem by assuming that both Adam and Eve know (i.e., rightly believe) that Adam's choices have no influence on the sequence of digits appearing on the screen. ■

Let X be the set of all finite sequences or histories of observations (0s and 1s). These sequences are of varying length; $\ell(\mathbf{x})$ is the length of $\mathbf{x} \in X$. Similarly, let

[12] Border [1992] develops an RP approach to choice among lotteries with monetary rewards. The observer knows only that more money is preferred to less. If observer and observed agree on all (objective) probabilities, any choice behavior that is not stochastically dominated can be rationalized by postulating a suitable utility function. The RP approach of Green and Osband [1991] is based on assumptions that deviate from Savage's [1954] framework in several ways. A direct comparison of results is therefore difficult. Kim [1992] considers choice under uncertainty but excludes learning, i.e., conditionalization on past observations.

Y be the set of all finite sequences of choices (natural numbers from 0 to N), with $\ell(\mathbf{y}) = t$ as the length of $\mathbf{y} \in Y$. In both cases, we include the sequences of zero length ("vacuous" histories). The set XY denotes all pairs $(\mathbf{x}, \mathbf{y})$ from $X \times Y$ with $\ell(\mathbf{x}) = \ell(\mathbf{y})$.

The following three questions define the problem we are interested in by recourse to the situation of Adam and Eve.

Question 1. Can Eve exclude some histories $(\mathbf{x}, \mathbf{y}) \in XY$ as inconsistent with the assumption that Adam is a perfectly rational Bayesian agent?

Question 2. Can Eve, on the basis of finitely many observations $\mathbf{x} \in X$, give good advice to Adam from a Bayesian point of view?

Question 3. Assume that Eve is a Bayesian–minded economist who has observed a sequence of digits and choices $(\mathbf{x}, \mathbf{y}) \in XY$. Is there any restriction on Eve's beliefs concerning Adam's future behavior resulting from the hypothesis that Adam is a perfectly rational Bayesian agent?

Question 1 concerns Bayesianism as a positive theory that should yield predictions of Adam's behavior. Question 2 concerns Bayesianism as a normative theory that could be used by Eve to advise Adam. Question 3 concerns Bayesianism as a methodology used by Eve to analyze Adam's behavior.

Of course, all three questions are strongly interrelated. If Adam can rationalize any choice of strategy, question 1 must be answered in the negative. The same goes for question 2. If any strategy can be rationalized, there is nothing a Bayesian advisor can say except "Do what you want".

The answer to question 3 is slightly more involved. There is a difference between questions 1 and 3. A negative answer to question 1 implies that no OMTs exist. But a Bayesian could still claim that Bayesianism as a methodology allows one to conclude that certain sequences of digits and choices become very improbable if Adam is rational. If that were possible, it would provide a Bayesian argument against the requirement that positive theories should provide OMTs.

Let us shortly summarize the Bayesian analysis of Adam's problem. First of all, Adam should choose a set $\mathcal{H}$ of mutually exclusive hypotheses, each of which implies objective probabilities for all potential future observations. Then, he should choose a subjective probability measure or prior μ on (a σ–algebra of subsets of) $\mathcal{H}$. The prior μ is chosen such that it generates a preference ordering on the set of all strategies. Hence, we know that, for every history $\mathbf{x} \in X$, the pair $(\mathcal{H}, \mu)$ implies conditional probabilities $P_\mu(x_{\ell(\mathbf{x})+1} = i \,|\, \mathcal{H} \wedge \mathbf{x}), i = 0, 1$.[13]

Adam knows that there is no influence from his choice at one point in time to consequences at other points in time. The only connection between choices is

[13]The symbol $\wedge$ in $\mathcal{H} \wedge \mathbf{x}$ denotes the conjunction. Read as statements, $\mathcal{H}$ is a (possibly uncountable) disjunction of hypotheses, and $\mathbf{x}$ is equivalent to the conjunction of statements "At time s, x_s is observed", $s = 1, \ldots, \ell(\mathbf{x})$. If $\ell(\mathbf{x}) = 0$, $\mathcal{H} \wedge \mathbf{x}$ is of course equivalent to $\mathcal{H}$. If Adam observes a sequence $\mathbf{x}$ with subjective probability 0, the conditional probabilities $P_\mu(x_{\ell(\mathbf{x})+1} = i \,|\, \mathcal{H} \wedge \mathbf{x})$ are not defined. He is then free to choose a new prior distribution, which does not improve the chances of predicting his actions. However, we can exclude this case (see below).

learning. Given a "forecast function" [Nyarko 1997: 181] $\rho: X \mapsto [0,1]$ assigning the probability $p = \rho(\mathbf{x})$ to the event $x_{\ell(\mathbf{x})+1} = 0$ (the next digit is 0), the action at $t = \ell(\mathbf{x})$ can be considered separately from other actions. Adam maximizes his SEU, solving the problem

$$(1) \quad \max_y \{p\nu(y) + (1-p)\nu(N-y): p = \rho(\mathbf{x}),\ \ y \in \{0,\dots,N\}\}.$$

His optimal strategy, then, is described by the function $y: X \mapsto \{0,\dots,N\}$ assigning the utility-maximizing choice to each conceivable history:

$$(2) \quad y(\mathbf{x}) \stackrel{\text{def}}{=} \arg\max_y \{p\nu(y) + (1-p)\nu(N-y): p = \rho(\mathbf{x}),\ \ y \in \{0,\dots,N\}\}$$

His actual choice $y_{\ell(\mathbf{x})}$ of course depends on the actual history $\mathbf{x}$.

Bayesian rationality requires that Adam's forecast function reflects some prior:

$$(3) \quad \exists(\mathcal{H},\mu)\forall\mathbf{x} \in X\big[\rho(\mathbf{x}) = P_\mu\big(x_{\ell(\mathbf{x})+1} = 0\,|\,\mathcal{H} \wedge \mathbf{x}\big)\big]$$

If the strategy $y(\mathbf{x})$ is optimal given the set of hypotheses $\mathcal{H}$ and the attached prior μ, the pair $(\mathcal{H},\mu)$ is a rationalization of $y(\mathbf{x})$.

The analysis is simplified by the fact that a strategy can be rationalized, if at all, without assuming any of the conditional probabilities to be 0, because Adam chooses between discrete values. Given discreteness, any decision that is optimal if some event has zero probability will also be optimal if the probability of the respective event is small enough. Of course, all policy variables in real–world decision problems are discrete since the precision of measurement is always finite. Moreover, each possible choice y in (1) is optimal for some values of p. If, therefore, we can find a rationalization for arbitrary forecast functions $\rho: X \mapsto (0,1)$, where $(0,1)$ is the open unit interval, we can rationalize any strategy $y: X \mapsto \{0,\dots,N\}$.

4 THE CHAOTIC CLOCK

The problem in finding rationalizations is that a set of hypotheses might not be rich enough to provide a rationalization for a given strategy. However, there are very trivial sets of hypotheses that are always rich enough.

4.1 The Basic Mechanism

Assume that the evolution of the inner states of Adam's black box follows a deterministic process depending on a starting point. The law of the deterministic process is the baker-map dynamics, which can be graphically illustrated as the output of a chaotic clock (figure 2).

There is one pointer that can point to all real numbers in the intervall $I = [0,1)$, where the vertically upward position is zero and the vertically downward position is $\frac{1}{2}$. Initially, the pointer deviates by an angle $\omega = 2\theta\pi$ from the vertically upward position, thus pointing at the real number θ. At $t = 1,2,\dots,\infty$, the pointer moves

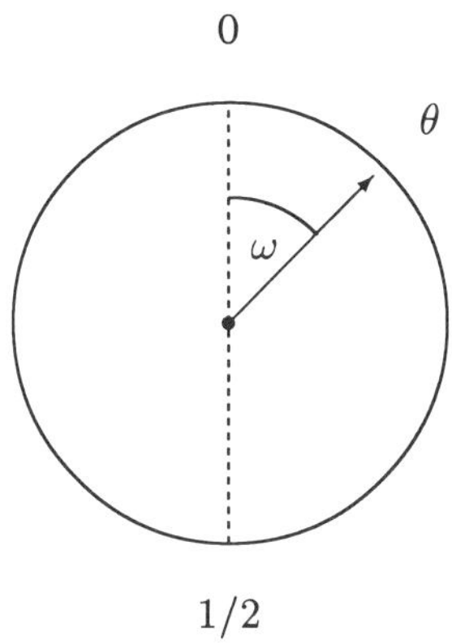

Figure 2. A chaotic clock. At each point in time, the angle ω is doubled. When the pointer is in the first (second) half of the dial, the digit 0 (1) appears on a screen. The resulting sequence of digits is described by the baker-map dynamics.

by doubling the angle ω. If the pointer comes to rest in the first half of the dial and points at a number in $[0, \frac{1}{2})$, the screen of Adam's black box shows 0; otherwise it shows 1.

According to the chaotic-clock hypothesis, the inner states of the black box at $t = 0, 1, \ldots, \infty$ are described by a real variable, the pointer position $z_t \in I$. The inner states evolve deterministically, but it can only be observed whether the pointer position is $z_t \in [0, \frac{1}{2})$ or $z_t \in [\frac{1}{2}, 1)$. These two states result in $x_t = 0$ and $x_t = 1$, respectively. The deterministic law by itself does not allow for a prediction of future observations; an assumption concerning the starting point $z_1 = \theta$ is also necessary. Thus, there is a set of hypotheses, one for each starting point $\theta \in I$. The corresponding dynamical system is the baker-map dynamics:[14]

$$(4)\quad \begin{array}{lrcl} \text{(a)} & x_t & = & g(z_t) \stackrel{\text{def}}{=} 2z_t \text{ div } 1 \\ \text{(b)} & z_{t+1} & = & h(z_t) \stackrel{\text{def}}{=} 2z_t \text{ mod } 1 \\ \text{(c)} & z_1 & = & \theta \end{array}$$

Note that the chaotic clock cannot produce an unbroken infinite sequence of 1s. If the pointer is in the second half of the dial (thus generating a 1 on the screen), doubling the angle ω moves the pointer beyond 0, thus leading to a *smaller* value

[14]"div" denotes integer division; "mod" denotes the indivisible rest of the integer division, i.e., $x \text{ mod } n \stackrel{\text{def}}{=} x - (x \text{ div } n)$. On the baker–map dynamics, see Ford [1983], Devaney [1989: 18 example 3.4, 39, 52] and Schuster [1988: 107f]. The graphical illustration is due to Davies [1987: ch. 4].

of ω. As long as the pointer comes to rest in the second half of the dial, ω falls at every tick of the clock with increasing rates, until, after a finite number of ticks, the pointer is in the first half of the dial, which implies that the screen shows 0.

4.2 Falsification Dynamics

Assume that Adam believes that his black box contains a chaotic clock. In order to analyze the consequences of uncertainty concerning θ, Adam has to know how x_t develops for a given θ. This is very simple in principle. Every $\theta \in I$ can be expressed as

$$(5) \quad \theta = \sum_{n=1}^{\infty} \frac{\theta_n}{2^n} = 0.\theta_1\theta_2\theta_3 \cdots \theta_t \cdots$$

where θ_n is 0 or 1 (dyadic development). In order to enforce uniqueness of the representation (5), we require infinitely many 0s on the right–hand side. Thus, $\frac{1}{2}$ should be represented as $0.1\overline{0} = 0.1$ and not as $0.0\overline{1}$ (where the bar denotes infinite repetition).[15]

The sequence generated by the chaotic clock is just the dyadic development of the starting point, i.e., x_t is equal to θ_t in (5). This follows from two mathematical facts. (i) We have $\theta \geq \frac{1}{2}$ if and only if $\theta_1 = 1$. (ii) Doubling θ shifts the point in the dyadic development one step to the right; the modulo function sets the digit before the point to 0, thus ensuring that the new value is again smaller than 1.

The first digit θ_1 in (5) determines whether θ is smaller than $\frac{1}{2}$ or not. Hence, when Adam observes the first digit $x_1 = \theta_1$, he finds out whether the starting point θ is in the first half ($x_1 = 0$) or in the second half ($x_1 = 1$) of the dial. The second digit places the starting point into one of the four quarters. For example, if $x_1 = 1$ and $x_2 = 0$, the starting point must be in the third quarter $\left[\frac{2}{4}, \frac{3}{4}\right)$. And so on. The sequence of digits corresponds to a sequence of bisections of the set I of potential starting points; with each further digit, the location of the starting point is narrowed down to the upper or lower half of the remaining interval on the dial. At time t, Adam has made t observations revealing one of the 2^t *basic intervals*

$$(6) \quad I_t(m) \stackrel{\text{def}}{=} \left[\frac{m}{2^t}, \frac{m+1}{2^t}\right), \quad m \in \{0, \ldots, 2^t - 1\}$$

as the location of the starting point θ. In infinite time, these bisection steps converge to θ.

In other words, each observation falsifies half of the remaining hypotheses concerning the starting point θ. Hence, Adam's beliefs converge to the truth if he is right in assuming that the black box contains a chaotic clock. This does not mean, however, that his chances of predicting the future improve over time. Obviously, knowing the first t digits of θ's dyadic development is no help in predicting the next digits.

[15]On coin tossing and dyadic development see also Brémaud [1988: 28–31].

4.3 *Souping up the Clock*

The chaotic clock places one restriction on Adam's system of beliefs: it is impossible that the number of 0s is finite. In order to get rid of this restriction, we assume that the angle between the pointer and the zero position is not doubled as before but quadrupled:

$$(7)\quad \begin{array}{lrcl} \text{(a)} & x_t & = & g(z_t) \\ \text{(b)} & z_{t+1} & = & h(2z_t) \\ \text{(c)} & z_1 & = & \theta \end{array}$$

The functions g, h are defined as in (4). For any starting point θ with dyadic development (5), the original system (4) yields the observations $x_t = \theta_t$. The modified system (7) yields $x_t = \theta_{2t-1}$, i.e., every second digit of the the starting point's dyadic development is irrelevant. Therefore, starting points like $\theta = 0.\overline{10}$ or $\theta = 0.\overline{1011}$ generate an unbroken infinite number of 1s although they feature infinitely many 0s as required.

Almost the same analysis as before applies to the modified chaotic clock. Since every second digit of the starting point's dyadic development never appears on the screen, an infinite number of starting points lead to the same sequence of observations. The first t observations determine 2^{t-1} basic intervals of length $\frac{1}{2^{4t-2}}$, one of which contains the starting point. Each basic interval is as rich as the unit interval. It still contains points that can produce any kind of extension to the observed sequence. In other words, the set of hypotheses represented by the set of starting points is so rich that, for any possible future, there are (actually: infinitely many) hypotheses consistent with the past and predicting this future. *No future developments can be excluded on the basis of past observations.*

Adam's beliefs still converge to the truth but not to the full truth: the observations reveal a subset of possible starting points containing the true one. This set corresponds to a subset of observationally equivalent hypotheses, one of which is true. As before, convergence to the truth presupposes that the black box actually contains a modified chaotic clock. If this is not the case, however, the analysis of the learning process is unaffected since no sequence of observations, not even an infinite one, is inconsistent with the assumed mechanism.

However, Adam does not *know* whether the black box actually contains a chaotic clock. One may ask, then, why he should ever consider such a crazy mechanism. There are several reasons.

First of all, Adam is a perfectly rational being and, therefore, logically omniscient (cf. [Earman 1992: 121f]). He is aware not only of the chaotic-clock hypotheses but of many more when he considers the question of choosing a prior. Therefore, the question should rather be why not more hypotheses are considered.

Secondly, if the truth is deterministic, it is observationally equivalent to (7) (with a specific starting point). Thus, (7) can be viewed as a "reduced form", as econometricians would say, of any deterministic theory (plus initial conditions) that is capable of explaining an infinite sequence of the events in question. In this sense, the chaotic clock represents all deterministic theories, and Adam's problem does not become significantly greater when we add further hypotheses to the set of chaotic–clock hypotheses.[16] Hence, considering the chaotic clock is actually a simplification (although an insignificant one) of Adam's problem.

Last but not least, the chaotic–clock model fulfills all the formal requirements of a scientific theory. It assumes a simple mechanism governed by a law of motion that produces different results according to the initial position of the mechanism. While not even the Swiss could actually produce the chaotic clock, processes that lead to chaotic dynamics are not rare, and imperfect observability can produce the kind of irregular behavior characteristic of the chaotic clock. Moreover, although the chaotic–clock model assumes a continuum of starting points (corresponding to more hypotheses than we could ever consider explicitly), it is less complicated than models encountered in physics or economics. It would be difficult to find any acceptable formal requirement that excludes the chaotic clock from consideration.[17]

Let us call "empirically adequate" those hypotheses that are observationally equivalent to the truth. Subsequently, convergence to the truth is taken to mean that, in the limit, the subjective probability of the set of empirically adequate hypotheses is 1. If the truth is deterministic, then, convergence to the truth is ensured.

5 ANSWERS

5.1 Adam's Problem

We turn to the dynamics of probabilistic beliefs generated by the modified chaotic–clock hypotheses described by (7). The hypotheses form a set $\mathcal{H}^*$, where $H_\theta \in \mathcal{H}^*$, $\theta \in I$ denotes the single hypothesis that the starting point of the modified chaotic clock is θ. Since these hypotheses are deterministic, the probabilities $P(x_t = 0 \,|\, H_\theta)$ are either 0 or 1, i.e., H_θ yields certain or point predictions. Uncertainty enters via the uncertainty concerning $\theta \in I$. As a Bayesian, Adam chooses a subjective probability measure on (a σ–algebra of subsets of) $\mathcal{H}^*$, and since each hypothesis in $\mathcal{H}^*$ is represented by a $\theta \in I$, we can consider instead a subjective probability measure on (a σ–algebra of subsets of) I.

Adam needs well-defined conditional probabilities $P_\mu\big(x_{\ell(\mathbf{x})+1} = i \,|\, \mathcal{H}^* \wedge \mathbf{x}\big)$ for any potential sequence of observation $\mathbf{x}$. Hence he must include all the basic intervals $I_t(m)$ in his σ algebra of subsets of I. We can therefore consider the σ algebra generated by the basic intervals (which is just the σ–algebra of the Borel sets). Any probability measure μ on this σ–algebra then determines the conditional

[16] Even adding probabilistic hypotheses would make no difference, as will become obvious below.

[17] The chaotic clock poses a generalized version of Goodman's [1955] "new riddle of induction". The set of hypotheses considered by Goodman is countable and, therefore, too small to lead to the problems discussed in the present paper. Using the chaotic clock for presenting the problem of induction has the advantage that no "gruesome" predicates appear.

probabilities Adam needs to solve his decision problem. Henceforth, we simply speak of a probability measure μ on I or, equivalently, on $\mathcal{H}^*$.

We have already seen in section 3 that the central question is if we can find a rationalization $(\mathcal{H}^*, \mu)$ for arbitrary forecast functions $\rho\colon X \mapsto (0,1)$. This question is answered by the following theorem [Albert 1999: theorem 1].

Theorem (Anything Goes) *Let $\mathcal{H}^*$ be the set of modified chaotic-clock hypotheses. Consider an arbitrary forecast function $\rho\colon X \mapsto (0,1)$. Then, there exist infinitely many probability measures μ on $\mathcal{H}^*$ such that the rationality condition $\rho(\mathbf{x}) = P_\mu\big(x_{\ell(\mathbf{x})+1} = 0 \,|\, \mathcal{H}^* \wedge \mathbf{x}\big)$ holds for all $\mathbf{x} \in X$.*

In interpreting the theorem, we have to remember that Adam as a perfectly rational person is always aware of the implications of all the assumptions he is considering. When choosing the prior μ on $\mathcal{H}^*$, he is aware of the implicit assignments of numerical values to the conditional probabilities. The theorem says that, instead of choosing a prior on $\mathcal{H}^*$, Adam may as well choose arbitrary conditional probabilities.[18] And since these probabilities can generate any contingent choices whatsoever, it is immaterial for Adam whether he asks himself "What should I do?" or "What prior should I choose?". The Bayesian apparatus provides no restrictions and therefore no help in making this choice.

Let us consider two simple cases. Adam might, for instance, choose a constant forecast function $\rho(\mathbf{x}) = \frac{1}{2}$ for all $\mathbf{x} \in X$. This is rationalized by the uniform prior with density $f(\theta) = 1$ on I. This choice implies that, independently from past observations $\mathbf{x}$, the probability of 0 and 1 is always $\frac{1}{2}$. Any other constant forecast function leads to a misleadingly complicated prior distribution that has no density.[19]

Generalizing slightly, Adam might set $\rho(\mathbf{x}) = p[\ell(\mathbf{x})]$ with an arbitrary function $p\colon \mathbb{N} \mapsto (0,1)$, thus fixing his decision at time t independently from and prior to any observations. Adam poses as a Bayesian learner but is actually completely "dogmatic" (and completely unpredictable) in the sense of ignoring any experience.

A Bayesian dogmatic is already unpredictable. A run–of–the–mill Bayesian, who allows experience to influence his behavior whenever it suits him, is all the more unpredictable because he has more options. This answers question 1. Bayesian rationality is empty as a positive theory. Eve cannot exclude any behavior of Adam on the basis of the hypothesis that Adam is rational in the Bayesian sense. Nor can Eve give any advice to Adam, even if she knows his NM utility function, since no strategy is irrational, whatever observations have been made. This answers question 2: Bayesianism is empty as a normative theory.

In relation to the tent-map dynamics (which is mathematically equivalent to (4)), Blume and Easley [1995: 19f, 36f] show that convergence to the truth need not mean that predictions improve: if the prior is continuous at the starting point,

---

[18] Thus, including probabilistic hypotheses in addition to $\mathcal{H}^*$ would not change the results.

[19] Cf. [Brémaud, 1988: 29] who, however, discusses only the uniform distribution.

the posterior distribution converges to the uniform distribution, implying a probability of 0.5 for observing 0. Hence, convergence to the truth does not imply convergence to rational expectations.[20]

This is an interesting point but not the one we are making. Blume and Easley's point provides no argument against Bayesianism (and is not meant to do so), because no procedure can improve on Bayesian learning in these cases. We can get more mileage out of the machinery of chaotic dynamics.

In Blume and Easley's analysis, the *true* law of the process generating the 0s and 1s is chaotic, and this law is *known* to the agent. In our analysis, the true law governing the sequence of 0s and 1s is unimportant. The anything-goes theorem neither refers to some true law nor places restrictions on the sequence of events observed by the decision maker. The problem is not the complexity of the environment but the complexity of a large set of hypotheses. The chaos is in the decision maker's head.

The assumptions of the present analysis are actually quite favorable to Bayesianism since convergence to the truth (in the empirical-adequacy sense) is ensured. Allowing for probabilistic hypotheses would open up the possibility of non-convergence to the truth.[21] As Gillies [2001] shows, a Bayesian learner might then never discover that he had dismissed the truth from the outset, which provides a strong argument in favor of seriously considering a large set of hypotheses.[22]

5.2 *Eve's Problem*

We turn to Eve's prediction problem, i.e., to the problem of an economist trying to predict the behavior of rational agents and using Bayesianism as a methodology. If Eve is a Bayesian herself, she is not overly impressed by the fact that Bayesianism yields no OMTs concerning Adam's behavior. The Bayesian methodology allows a continuum of beliefs between the two categories "ruled out" and "possible" that are considered when we speak of OMTs. However, this is not going to help Eve. Her hypothesis that Adam is rational provides no restrictions concerning Adam's behavior. Therefore, she is in no better position to predict Adam than to predict the digits generated by the black box.

For a formal proof, we have to generalize the previous results to larger spaces of observables. This presents no difficulties.

---

[20]Hence, "merger of opinions" for different persons (cf. [Earman 1992: ch. 6]) has nothing to do with convergence of expectations (opinions concerning the future).

[21]Unfortunately, non-convergence cannot be quite as dramatic as suggested by theorem 2 in Albert [1999], which is incorrect. Under the conditions of theorem 2, convergence to rational expectations is ensured although the probability of convergence to the truth may be 0 (as is typically the case if, e.g., both hypotheses assign a probability of 0.5 to observing 0 in the limit).

[22]A hypothesis is dismissed from the outset iff it is not in the support of the prior. The support is a set with zero–probability complement; sometimes, it is also required that the support's intersection with any open set, if not empty, has positive probability. In our analysis, the support is I since all open sets are measurable and contain basic intervals, which never have probability 0.

Eve observes the digits on the screen and Adam's choices. Moreover, she might observe other things, like Adam's facial expression or his pattern of consumption, that are or are not related to Adam's behavior. Sticking to our premise that, realistically, all observable variables can only range over a finite set of values, we assume that Eve's observable universe can be "digitalized": each state can be described by a binary string of 0s and 1s with maximum length $n \geq 2$. If the number of different states is between 2^{n-1} and 2^n for some n, several strings describe the same state.

Again, we look for a set of hypotheses sufficiently rich to allow Eve to rationalize any forecast. Such a set is again provided by the (modified) chaotic clock. Assume that Eve considers Adam and the money–spinner as a big black box that displays one of 2^n combination of observables at each point in time. The combinations are determined by a chaotic clock that makes n angle–quadrupling ticks at each point in time; the resulting string s_t of n digits is then revealed as a solid block instead of a succession of digits. Formally, we can leave the chaotic clock as it is; we just have to assume that, at each $t = 1, \ldots, \infty$, s_t is observed instead of just one digit x_t:

$$(8) \quad \begin{array}{llll} \text{(a)} & x_\tau = g(z_\tau) & \text{(b)} & z_{\tau+1} = h(2z_\tau) \\ \text{(c)} & z_1 = \theta & \text{(d)} & s_t = \left\{x_{(t-1)n+1}, \ldots, x_{tn}\right\}, t = \tau \text{ div } n \end{array}$$

This dynamical system is identical to (7) except for the fact that at each point in time t, the string s_t produced by the last n ticks of the clock is observed. The sequence generated by the system is the dyadic development of the starting point with every second digit removed.

The chaotic clock is a "theory of everything" for universes the evolution of which can be described by an infinite sequence of binary string of length n. Since the possibilities of assigning probabilities to sequences are not affected by the fact that these sequences are now revealed in a blockwise fashion, the previous results still apply: according to the anything-goes theorem, arbitrary assignments to these probabilities are possible. This answers question 3. Bayesianism as a methodology is completely useless in predicting rational behavior because there are no OMTs covering this behavior. Eve's expectations concerning Adam's behavior are completely arbitrary.

For example, Eve may decide to view Adam as attaching equal probabilities to two arbitrary hypotheses like "The probability of $x_t = 0$ is $\frac{1}{5}$" and "The probability of $x_t = 0$ is $\frac{2}{5}$" while she herself attaches a probability of $\frac{1}{2}$ to $x_t = 0$. She then can find a prior such that exactly the conditional probabilities implied by this view will hold, no matter what happens.

Moreover, the analysis of Eve's problem shows that the restriction of Adam's problem to the prediction of single digits is immaterial. Everything works as before as long as there is a maximum amount of information he can access at each point in time.

6 CONCLUSIONS

Bayesian rationality becomes empty if the decision maker considers a set of hypotheses that is as large as the set described with the help of the chaotic clock. Whatever the actual process generating a sequence of observations, considering a chaotic–clock explanation already implies that any experience can be accommodated without implications for expectations concerning the future. The inclusion of further hypotheses does not add to the complexity of the learning problem.

This is just another version of the problem of induction. Logically, one can never infer the laws governing the world from a finite number of past observation. While many theories may be eliminated over time, it is quite trivial that there always remain enough theories consistent with any kind of future. The so–called pragmatic problem of induction says that learning, if guided by experience and deductive logic alone, yields no restrictions for decision making.[23] The anything-goes theorem shows that Bayesianism, although employing more than just deductive logic, cannot solve the pragmatic problem of induction either.

This result is no surprise once it is clear how many degrees of freedom Bayesianism leaves to the decision maker in setting up the initial beliefs. But it is surprising that a sufficiently rich set of hypotheses can be introduced in such a simple and compact way. If not much sophistication is needed to experience the problem generated by too many logical possibilities, maximum sophistication or perfect rationality will necessarily lead to this problem.

It is a mathematical fact that any strategy can be rationalized even for a given NM utility function. Several arguments might be raised against the position, taken in the present paper, that this fact speaks strongly against adopting Bayesianism as a positive or normative theory. Specifically, four arguments seem to be important. Their discussion will conclude this paper.

6.1 Supplementing Bayesianism

It is not necessarily alarming if a theory is devoid of empirical content. As long as the theory is not analytic, it might still be possible to supplement it by further hypotheses, creating a larger theory the empirical content of which comes neither from the original theory alone nor from the supplement alone. The same goes, mutatis mutandis, for a normative theory. Even if Bayesianism allows one to rationalize any strategy, there might be supplementary rules that distinguish between good and bad rationalizations.[24]

However, such supplementary rules are equivalent to a principle of insufficient reason. This can easily be shown. Prima facie, there seem to be two different options. Supplementary rules or hypotheses providing content can either a) restrict

[23]Cf. Musgrave [1989: section 4] and Miller [1994: 20–23, 38–45], whose solution rests on the assumption that it is possible to reduce the number of acceptable theories drastically.

[24]Nyarko [1997: 176] makes this point but just provides results concerning the implications of different restrictions placed on priors.

prior beliefs *on the basis of* experience or b) restrict prior beliefs *without any basis in* experience.

However, option a) is really not different from option b). Assume that there is a rule R that restricts the prior on the basis of experience. Thus, for every set of data E, the rule R selects a prior. This is exactly what Bayesian updating does on the basis of a still earlier prior chosen before E becomes known. We have seen that a decision maker can choose the prior before experience such that, for every set of data E, an arbitrary predetermined posterior results. Thus, whatever the rule R, a prior can be chosen before any experience that is in conformity with the recommendations of R. It follows that R can be replaced by restrictions on the admissible set of priors before any experience; we are in effect left with option b).

A rule determining a prior, or at least restricting the choice of priors, without any basis in experience is a "principle of insufficient reason". As has been argued before, Bayesianism is the product of a history of failures to provide such a principle. Thus, embracing option b) looks not very promising for normative Bayesianism.

For Bayesianism as a positive theory, of course, it would be irrelevant whether a "principle of insufficient reason" looks reasonable or not as long as the package is successful empirically. However, this kind of success is missing so far, or at least this seems to be the public opinion in economics.

Given the record of Bayesianism as a positive theory of behavior, the advocates of Bayesianism in economics have stressed Bayesianism's methodological virtues as a systematic theory of behavior as compared with the "adhockery" of bounded-rationality approaches.[25]

As we have seen, Bayesianism does not provide a theory of behavior on its own. Without a theory of priors, the actual hypothesis in each application must needs be chosen in an ad hoc fashion by selecting a prior or an admissible set of priors. Thus, Bayesianism as it stands is not methodologically superior to the bounded-rationality approach. It might be true that we can guess in each application what the beliefs of the agents will be. On the other hand, we might equally well guess which of a set of rules of thumb they are likely to use. The degree of adhockery on both sides seems comparable.

6.2 *Bayesianism as Adaptation*

Another defense of Bayesianism is the idea that Bayesian perfect rationality might be an idealization anticipating—prematurely, so to speak—the long–run effects of adaptation and training. This means that selective pressures favor Bayesian perfect rationality in the long–run.

[25]For this argument and the next, cf. Selten's account of a fictitious discussion between exponents of the different approaches to the explanation of behavior, where the Bayesian defends his position by these arguments [Selten 1989: 5, 11, 21]. Selten introduces several counterarguments, which, however, seem to be based on the assumption that Bayesian rationality has content.

Let us compare the hypothetical fate of a perfectly rational with that of a boundedly rational agent. Let us call these hypothetical individuals Priscilla and Brian, respectively. Brian is not logically omniscient; he does not consider a set of hypotheses sufficiently rich to rationalize any behavior. Brian starts with a restricted set of hypotheses, and, for whatever reasons, there are only some priors that appeal to him. Moreover, he is bound to make logical mistakes; thus, even if he tried to maximize his subjectively expected utility, he would often fail to recognize the subjectively optimal actions. On the other hand, Brian might just adopt some rule of thumb for decision making and ignore his own beliefs. Would Brian have any disadvantages as compared with Priscilla? Will the Brians of this world either learn to mimick Priscilla's cleverness or vanish in the long run?

It seems not, at least from a Bayesian point of view. The difference between Brian and Priscilla has nothing to do with the strategies they pick and, consequently, nothing with their success. The difference is just the extent to which their choices can be rationalized in terms of their beliefs. In fact, both might choose the same actions under identical circumstances. It is not as if the rules of Bayesianism offered protection against *mistakes* that could be identified as such by clever Priscilla. To Priscilla, there are actions that are mistaken in the light of one's beliefs but no actions that are mistaken just in light of the known facts. Priscilla is able to rationalize any behavior; even if Brian were unable to do the same, she could do it for him.

Thus, while there might be selective pressure to avoid certain kinds of behavior, there cannot be any selective pressure in favor of perfecting the rationalization of behavior. Nobody stands a better chance in any competition just on account of being a Bayesian. Of course, among Bayesians, there might be selective pressure against certain priors. But this is a completely different point.

The idea that there is selective pressure in favor of perfect rationality is historically connected with the as-if defense of the rationality postulate. The as-if defense has been the war cry of empiricist positive economics: "Never mind how people actually think; when it comes to action, they behave *as if* they were rational." The anything-goes theorem robs the as-if defense of any empiricist appeal, at least if rationality is taken to be Bayesian rationality. The statement "people behave as if they were Bayesians" turns out to be analytic; it boils down to "people do whatever they do." The adaptationist argument has been used to defend the as-if argument: Why should people behave as if they were rational? The adaptationist answer: Because those who do are more successful in the long run than those who do not. Obviously, this argument, if applied to Bayesian rationality, wrongly presupposes that Bayesian rationality helps to avoid mistakes.

6.3 *Logic, Coherence, and the Axioms*

Almost everybody agrees that deductive logic and logical consistency are valuable. However, deductive logic restricts only the structure of beliefs and not the choice of strategies. One might argue, then, that it cannot be held against Bayesianism if the

Bayesian logic of probabilistic beliefs and the corresponding notion of consistency (often called "coherence") display the same weakness.[26]

This argument, however, is insufficient to defend Bayesianism. Logical consistency serves a purpose. Beliefs cannot possibly be true if they are inconsistent. Thus, if one wants truth, logical consistency is necessary. An analogous argument in favor of Bayesianism would have to point out some advantage of coherence unavailable to those relying on non-probabilistic beliefs and deductive logic alone. Such an argument is missing.

It is true that Bayesianism provides a logic of beliefs that rational persons must respect—*if and only if* their beliefs take the form of subjective probabilities. Those who reject this view—Popperians, classical statisticians, and others—can always point out to a Bayesian that their procedures could be rationalized on Bayesian grounds. The argument that the rationalizing prior might be "bad"—e.g., assign a positive probability to hypotheses they do not consider in earnest—will not worry them since for them there are no good priors anyway. In their view, beliefs just do not take the form of a probability distribution.

Of course, it is a theorem that the beliefs of rational agents should take the form of probabilities, not an assumption. The theorem follows from the axioms for preferences on the set of all strategies. If one argues that beliefs need not take the form of probabilities, the real question from a Bayesian point of view is, What's wrong with the axioms?

In my opinion, the axioms are quite reasonable if one is looking for a complete preference order on the set of all strategies. In the case of decision making under certainty or risk, a complete preference order might indeed be helpful.[27] But this is different in the case of choice under uncertainty.

Consider the following thought experiment. Adam has to choose from a menu in a restaurant. He orders chicken because he prefers it most. However, chicken is out. If pork was the second–best choice before, it is the best choice under the new conditions. A complete preference order means that he has decided in advance what to order if items are deleted from the menu.

There is no corresponding thought experiment for choice under uncertainty. Bayesian coherence amounts to a preference order among all conceivable strategies, where each strategy specifies the reactions to any new information. Hence, the reaction to the information that something is not available after all is already part of the chosen strategy. The preference order among the discarded strategies is therefore irrelevant by definition of the term strategy. Bayesianism not only is no help in choosing a strategy; it additionally requires that one chooses an order among the remaining irrelevant strategies. From this point of view, Bayesianism is worse than useless.

[26]Analogies between Bayesianism and deductive logic are stressed by Howson [1997].

[27]Even then, rational choice does not require a complete preference order. Look at a simple example. Which of the following options would you prefer: losing your left hand, your right foot, or $ 10? I find the choice easy although completely ordering the alternatives is beyond me.

ACKNOWLEDGEMENTS

For useful hints, comments, and critical remarks on previous versions, I am grateful to Friedrich Breyer, Donald Gillies, Franco Gori, Daniel Hausman, Frank Heinemann, Ronald A. Heiner, Colin Howson, Karl–Josef Koch, Alberto Loayza–Grbic, Jürgen Meckl, Alan Musgrave, Dieter Schmidtchen, Reinhard Selten, Thusnelda Tivig, Heinrich Ursprung, Hermann Vetter, and Elie Zahar.

Institute of Economics, University of Koblenz–Landau, Germany.

BIBLIOGRAPHY

[Albert, 1996] M. Albert. Bayesian learning and expectations formation: Anything goes, Discussion Papers of the Faculty of Economics and Statistics, No 284, University of Konstanz, 1996.

[Albert, 1999] M. Albert. Bayesian learning when chaos looms large, *Economics Letters* 65, 1–7, 1999.

[Arrow, 1990] K. Arrow. Economic theory and the hypothesis of rationality. In Eatwell *et al.*, pp. 25–37, 1990.

[Anscombe and Aumann, 1963] F. Anscombe and R. J. Aumann. A definition of subjective probability. *Annals of Mathematical Statistics*, 34, 199–205, 1963.

[Aumann, 1987] R. J. Aumann. Correlated equilibrium as an expression of Bayesian rationality, *Econometrica* 55, 1–18, 1987.

[Begg, 1982] D. K. H. Begg. *The Rational Expectations Revolution in Macroeconomics. Theories and Evidence,* Oxford: Phillip Allan, 1982.

[Bernstein, 1996] P. L. Bernstein. *Against the Gods. The Remarkable Story of Risk.* Wiley, New York, 1996.

[Bicchieri, 1993] C. Bicchieri. *Rationality and Coordination,* Cambridge: Cambridge University Press, 1993.

[Bicchieri *et al.*, 1997] C. Bicchieri, R. Jeffrey and B. Skyrms, eds. *The Dynamics of Norms,* Cambridge: Cambridge University Press, 1997.

[Binmore, 1993] K. Binmore. De–Bayesing game theory. In K. Binmore, A. Kirman and P. Tani, eds., *Frontiers of Game Theory,* pp. 321–339. Cambridge/Mass.: MIT Press 1993.

[Binmore, 1994] K. Binmore. *Game Theory and the Social Contract, Vol. I: Playing Fair,* Cambridge/Mass.: MIT Press, 1994.

[Blume and Easley, 1995] L. E. Blume and D. Easley. What has the rational learning literature taught us?. In Kirman and Salmon, pp. 12–39, 1995.

[Border, 1992] K. C. Border. Revealed preference, stochastic dominance and choice of lotteries, *Journal of Economic Theory* 56, 20–42, 1992.

[Bray, 1983] M. Bray. Convergence to rational expectations equilibria. In Frydman and Phelps, pp. 123–132, 1983a.

[Brémaud, 1988] P. Brémaud. *An Introduction to Probabilistic Modelling,* New York: Springer, 1988.

[Davies, 1987] P. Davies. *Cosmic Blueprint,* London: Heinemann, 1987.

[Devaney, 1989] R. L. Devaney. *An Introduction to Chaotic Dynamical Systems,* 2nd ed., Redwood City/Cal.: Addison–Wesley, 1989.

[Earman, 1992] J. Earman. *Bayes or Bust?,* Cambridge/Mass.: MIT Press, 1992.

[Eatwell *et al.*, 1990] J. Eatwell, M. Milgate and P. Newman, eds. *The New Palgrave: Utility and Probability,* New York: Norton, 1990.

[Ford, 1983] J. Ford. How random is a coin toss? *Physics Today,* April 1983.

[Friedman, 1953] M. Friedman. The methodology of positive economics, in: Milton Friedman, *Essays in Positive Economics,* Chicago and London: Chicago University Press 1953, pp. 3–43.

[Frydman and Phelps, 1983] R. Frydman and E. S. Phelps. Introduction. In Frydman and Phelps, pp. 1–30, 1983a.

[Frydman and Phelps, 1983a] R. Frydman and E. S. Phelps. *Individual Forecasting and Aggregate Outcomes,* Cambridge: Cambridge University Press 1983.

[Gillies, 2001] D. A. Gillies. Bayesianism and the Fixity of the Theoretical Framework. In present volume, pp. 373–390, 2001.

[Goldman, 1999] A. I. Goldman. *Knowledge in a Social World,* Oxford: Oxford University Press, 1999.

[Goodman, 1955] N. Goodman. *Fact, Fiction, and Forecast,* Indianapolis: Bobbs–Merrill, 1955.

[Green and Osband, 1991] E. J. Green and K. Osband. A revealed preference theory for expected utility, *Review of Economic Studies* 58, 677–696, 1991.

[Hacking, 1990] I. Hacking. Probability. In Eatwell, *et al.*, pp. 163–177, 1990.

[Hahn, 1996] F. Hahn. Rerum cognoscere causas, *Economics and Philosophy* 12, 183–195, 1996.

[Howson, 1997] C. Howson. Logic and probability, *British Journal for the Philosophy of Science* 48, 517–531, 1997.

[Howson and Urbach, 1989] C. Howson and P. Urbach. *Scientific Reasoning: The Bayesian Approach,* La Salle/Ill.: Open Court. 1989.

[Kiefer and Nyarko, 1995] N. M. Kiefer and Y. Nyarko. Savage–Bayesian models of economics, in: Kirman and Salmon, pp. 40–62, 1995.

[Kim, 1992] T. Kim. The subjective expected utility hypothesis and revealed preference, *Economic Theory* 1, 251–263, 1992.

[Kirman and Salmon, 1995] A. Kirman and M. Salmon, eds. *Learning and Rationality in Economics,* Oxford: Blackwell, 1995.

[Leamer, 1978] E. E. Leamer. *Specification searches,* New York: Wiley, 1978.

[Ledyard, 1986] J. O. Ledyard. The scope of the hypothesis of Bayesian equilibrium, *Journal of Economic Theory* 39, 59–82, 1986.

[Lindley, 1972] D. V. Lindley. *Bayesian Statistics. A Review,* Philadelphia: SIAM, 1972.

[Miller, 1994] D. Miller. *Critical Rationalism. A Restatement and Defense,* Chicago and La Salle/Ill.: Open Court, 1994.

[Musgrave, 1989] A. Musgrave. Saving science from scepticism. In F. D'Agostino and I. C. Jarvie, eds. *Freedom and Rationality,* Dordrecht: Kluwer 1989, pp. 297–323.

[Muth, 1961] J. F. Muth. Rational expectations and the theory of price movements, *Econometrica* 29, 315–335, 1961.

[Nyarko, 1997] Y. Nyarko. Savage–Bayesian agents play a repeated game. In Bicchieri *et al.*, pp. 175–197, 1997.

[Pearl, 2000] J. Pearl. *Causality. Models, Reasoning, and Inference,* Cambridge: Cambridge University Press, 2000.

[Pesaran, 1987] M. H. Pesaran. *The Limits to Rational Expectations,* Oxford: Blackwell, 1987.

[Pope, 1991] R. E. Pope. The delusion of certainty in Savage's sure–thing principle, *Journal of Economic Psychology* 12, 209–241, 1991.

[Samuelson, 1947] P. A. Samuelson. *Foundations of Economic Analysis,* enlarged edition, Cambridge/Mass. and London: Harvard University Press 1983 (first published 1947).

[Savage, 1954] L. J. Savage. *The Foundations of Statistics,* New York: Wiley, 1954.

[Savage *et al.*, 1962] L. J. Savage *et al. The Foundations of Statistical Inference,* London: Methuen/New York: Wiley, 1962.

[Schuster, 1988] H. G. Schuster. *Deterministic Chaos,* 2nd rev. ed., Weinheim: VCH, 1988.

[Selten, 1989] R. Selten. *Evolution, Learning, and Economic Behavior,* 1989 Nancy L. Schwartz Memorial Lecture, J.L. Kellogg Graduate School of Management, Northwestern University.

[Simon, 1989] H. A. Simon. From substantive to procedural rationality. In Spiro J. Latsis, ed., *Method and Appraisal in Economics,* pp. 129–148. Cambridge University Press, 1976,

[Varian, 1992] H. R. Varian. *Microeconomic Analysis,* 3rd rev. ed., Norton, New York, 1992.

DONALD GILLIES

BAYESIANISM AND THE FIXITY OF THE THEORETICAL FRAMEWORK

1 INTRODUCTION. BAYESIANISM VERSUS CLASSICAL STATISTICS

Bayesianism is a powerful current of thought in quite a number of different areas, which include: artificial intelligence, decision theory, economics, philosophy of science, and statistics. In the present paper, I will deal only with Bayesianism in statistics. In fact since the beginning of this century, the principal controversy within statistics has been between Bayesianism and the so-called classical statistics. I will begin therefore by attempting to characterise, in outline at least, these two approaches to statistics.

Let us start with Bayesianism. When this is applied in a particular problem situation, it is usually assumed that there is a given a set of possible statistical hypotheses H_θ where $\theta \varepsilon I$, for some set I, normally an interval of the real line. Some data or evidence e say is collected and the problem is to judge the hypotheses in the light of this evidence. To do this, the parameter θ is given a prior probability distribution $p(\theta)$ say. This represents the degree of belief of the statistician that θ has various values before the evidence e is considered. Given $p(\theta)$, the posterior probability distribution given e, i.e. $p(\theta|e)$ is then calculated using Bayes' Theorem. Our Bayesian statistician now adjusts his or her beliefs from $p(\theta)$ to $p(\theta|e)$, a process known as *Bayesian conditionalisation*. The merits of the various hypotheses H_θ are now judged using $p(\theta|e)$. Statistical inference on this account consists essentially of a change from a set of beliefs represented by p(θ) to another set represented by $p(\theta|e)$, or, to put it another way by a change of beliefs brought about by Bayesian conditionalisation.

While the concept of change of belief lies at the heart of Bayesianism, the corresponding concept for classical statistics is, in my view. that of hypothesis testing. I regard statistical tests as the core of classical statistics. This means that classical statistics, despite being allegedly 'classical', is in reality much more recent than Bayesianism. Bayesianism began with the publication of the famous paper of Bayes and Price in [1763]. It received a powerful mathematical development from Laplace in his [1812] By contrast classical statistics can be dated from 1900 because, in a paper published that year, Karl Pearson introduced the χ^2 test — the first really important and widely used statistical test. Further statistical tests and a theory of statistical testing were subsequently developed by the founders of classical statistics 'Student' (W. S. Gosset), R. A. Fisher, E. S. Pearson, and J. Neyman. The methodology of classical statistics is essentially that of testing. Statistical hypothesis are put forward tentatively to explain observed data, and are then subjected to statistical tests. If they pass these tests, they continue to be held.

D. Corfield and J. Williamson (eds.), Foundations of Bayesianism, 363–379.

If they fail the tests, they have to be abandoned or modified. The method here is that of conjectures and refutations as advocated by Popper in his [1963].

Having stated what I see as the difference between Bayesianism and classical statistics, I will now outline the criticism of Bayesianism which I wish to develop in this paper. It is not a criticism which attempts to show that Bayesianism is wrong in all circumstances. Indeed there are some situations where a Bayesian analysis seems to me quite correct — for just one example see my joint paper with Phil Dawid, [1989]. What the argument seeks to do is to place a limit on the situations in which Bayesianism should be applied. Roughly the thesis is that Bayesianism can be validly applied only if we are in a situation in which there is a fixed and known theoretical framework which it is reasonable to suppose will not be altered in the course of the investigation.[1] I call this the condition of *the fixity of the theoretical framework*. For Bayesianism to be appropriate, the framework of general laws and theories assumed must not be altered during the procedure of belief change in the light of evidence. If this framework were altered at any stage, this could lead to changes in the probabilities which were made not in accordance with Bayes theorem and Bayesian conditionalisation. It follows that, if we are studying a process whose nature is not well known, statistical testing using the methodology of classical statistics is essential.

I will try both to elaborate this criticism and to render it plausible by considering two examples, one in each of the next two sections. The first of these examples is of an investigation carried out by an eminent classical statistician, and the second by an eminent Bayesian. Our eminent classical statistician is Jerzy Neyman, and his investigation was into the distribution of larvae in the plots of an experimental field. I will try to show in section 2 that this investigation was an admirable one, and that its success depended crucially on the use of the methodology of testing employed in classical statistics. Our eminent Bayesian is Bruno de Finetti, and in section 3 I will consider his use of exchangeability in his [1937]. I will argue that this gives reasonable answers if the process under study consists objectively of independent events, but can go drastically wrong if the process is e.g. a Markov chain. This shows that we have to be very sure of the correctness of our theoretical framework (in this case that the process consists of independent events) before applying Bayesianism. After going through these two examples, I will conclude

[1]The phrase 'a fixed theoretical framework' comes from Lakatos [1968, p. 161], although he uses it in a somewhat different sense. Lakatos is criticising Carnap's inductive logic, and points out that Carnap's confirmation function (c-function) depends on the language employed so that it cannot cope with changes in the language. Lakatos puts the argument like this:

> 'Although growth of the evidence *within* a fixed theoretical framework (the language L) leaves the chosen c-function unaltered, growth of the theoretical framework (introduction of a *new* language L^*) may change it radically.' [1968, p. 161]

Lakatos here identifies a theoretical framework with a language. By contrast I am using 'theoretical framework' to refer to the set of theories under consideration. Thus a theoretical framework in my sense changes when a new theory is introduced even though the language does not change. Despite this difference, the general structure of Lakatos' argument is quite similar to that of the argument developed in this paper.

the paper in section 4, by considering some ways in which Bayesianism might be defended against the criticisms presented.

2 AN INVESTIGATION OF NEYMAN'S

Neyman describes his investigation in [Neyman, 1952, pp. 33–7]. His account begins as follows:

> 'Problems of pest control led to studies of the distribution of larvae in small plots. An experimental field planted with some crop is divided into a number of small plots, ... Then all the larvae found in each plot are counted. Naturally the number of larvae varies considerably from one plot to another.' [1952, p. 33]

Neyman wanted to find a mathematical model which would account for this variation. The first such model which suggested itself to him was the Poisson distribution, according to which the probability of there being a number n of larvae in a small plot (p_n) is given by $p_n = \exp(-\lambda)\lambda^n/n!$ for some value of the parameter λ. In a loose sense this corresponds to the assumption that the larvae are distributed randomly throughout the field. It was thus a very plausible hypothesis, and indeed Neyman says explicitly that it was [1952, p. 33] '...one strongly suggested by intuition.' Neyman had moreover used the same hypothesis of a Poisson distribution for a very similar problem concerned with the distribution of bacteria on a Petri-plate, and there it had proved very successful. Despite these favourable a priori indications, Neyman followed the methodology of classical statistics by subjecting the hypothesis of a Poisson distribution to a series of tests, and, rather surprisingly, these showed that the hypothesis was false.

In his 1952, Neyman gives the results of 5 trials of the Poisson distribution hypothesis. In each case this hypothesis was subjected to a χ^2 test. In one case the test resulted in a confirmation with a value of χ^2 of 4.0 with 2 degrees of freedom, corresponding to 13.5%. The remaining four tests, however, were clear refutations with χ^2 values corresponding to around 0.1% or less, resulting in falsifications even at a 1% level of significance. There could be no doubt in the light of these results that the hypothesis of a Poisson distribution was incorrect. As Neyman says:

> 'In all cases, the first theoretical distribution tried was that of Poisson. It will be seen that the general character of the observed distribution is entirely different from that of Poisson. There seems to be no doubt but that a very serious divergence exists between the actual phenomenon of distribution of larvae and the machinery assumed in the mathematical model. When this circumstance was brought to my attention by Dr. Beall, we set out to discover the reasons for the divergence.' [1952, p. 34]

As the last sentence of the quotation shows, Neyman did not consider any hypotheses other than that of the Poisson distribution until after the Poisson distribution hypothesis had been refuted by statistical tests. As so often in science, it was the *falsification* of a hypothesis which stimulated theoretical reasoning. This point will be important when we consider how this case might be analysed from the Bayesian point of view. Let us now see how Neyman continued with the investigation. He describes his next steps as follows:

> '... if we attempt to treat the distribution of larvae from the point of view of Poisson, we would have to assume that each larva is placed on the field independently of the others. This basic assumption was flatly contradicted by the life of larvae as described by Dr. Beall. Larvae develop from eggs laid by moths. It is plausible to assume that, when a moth feels like laying eggs, it does not make any special choice between sections of a field planted with the same crop and reasonably uniform in other respects. Therefore, as far as the spots where a number of moths lay their eggs is concerned, it is plausible that the distribution of spots follows a Poisson Law of frequency, depending on just one parameter, say m, representing the average number of spots per unit area.
>
> However, it appears that the moths do not lay eggs one at a time. In fact, at each "sitting" a moth lays a whole batch of eggs and the number of eggs varies from one cluster to another. Moreover, by the time the counts are made the number of larvae is subject to another source of variation, due to mortality.
>
> After hatching in a particular spot, the larvae begin to look for food and crawl around. Since the speed of their movements is only moderate, it is obvious that for a larva to be found within a plot, the birthplace of this larva must be fairly close to this plot. If one larva is found, then it is likely that the plot will contain more than one from the same cluster.' [1952, pp. 34–5]

It is worth noting here that in his attempt to find a new better hypothesis to describe the distribution of the larvae, Neyman made use of background knowledge about the larvae obtained from the domain expert, Dr. Beall. This led him to suppose that the larvae would be distributed in clusters round points where batches of eggs had been laid. The points where the eggs were laid would follow a Poisson distribution, but not the larvae themselves. Neyman produced a mathematical model of this situation which led to the conclusion that the larvae would be distributed in what he called a 'Type A distribution' depending on two parameters. Using the same data as before, Neyman again applied the χ^2 test in the 5 cases, and this time all the tests confirmed the hypothesis. Neyman had clearly succeeded in explaining a surprising experimental finding, and his successful investigation shows the merits of classical statistics, or, what is the same thing, Popper's methodology of conjectures and refutations applied using statistical tests to obtain the refutations.

Neyman himself observes [1952, p. 37]: 'In this example, in order to have agreement between the observed and predicted frequencies, it was imperative to adjust the mathematical model.' Moreover far from being dogmatic about his new Type A distribution, he is anxious to point out that it, like the Poisson distribution, has its limitations. Indeed he says:

> '...there are organisms (e.g., scales) whose distribution on units of area of their habitat does not conform with type A. An investigation revealed that the processes governing the distribution of these organisms were much more complex than that described and therefore, if a statistical treatment is desired, a fresh effort to construct an appropriate mathematical model is necessary.' [1952, p. 37]

That concludes my account of Neyman's investigation of the distribution of larvae in a field, and I now turn to the question of whether a Bayesian statistician could have carried out this investigation as successfully as the classical statistician Neyman. I do not see how this could have been possible. A Bayesian would start in the same way by formulating a set of possible hypotheses H_λ where $0 < \lambda < \infty$. Here H_λ is just the Poisson distribution with parameter λ. The next step would have been to set up a prior probability distribution $p(\lambda)$ representing the Bayesian statistician's prior degree of belief in the various hypotheses. This would have been changed in the light of the evidence e to a posterior distribution $p(\lambda|e)$. Yet it is difficult to see how all these changes in degrees of belief by Bayesian conditionalisation could have produced the solution to the problem, namely a Type A distribution. The Bayesian mechanism seems capable of doing no more than change the statistician's degree of belief in particular values of λ. This illustrates very nicely my thesis that Bayesianism requires the fixity of the theoretical framework. The theoretical framework at the beginning of the investigation was the assumption of a Poisson distribution. If this framework had been adequate, as it was in the example of bacteria on a Petri-plate, then Bayesianism would have dealt with the problem satisfactorily. However the theoretical framework was not adequate for the example of larvae in a field. It had to be changed from the assumption of a Poisson distribution to that of a Type A distribution, and the procedure of Bayesian conditionalisation cannot cope with such a change in belief.

To this it might be objected by a Bayesian that the initial set of possible hypotheses should have included both Poisson distributions and Type A distributions. If this had been done, then Bayesian conditionalisation would have dealt with the problem in a perfectly satisfactory manner. However, the difficulty with this proposal is that, as already pointed out, Neyman only thought of his Type A distribution after the assumption of a Poisson distribution had been refuted by a series of χ^2 tests. Neyman certainly did not consider Type A distributions as an a priori possibility at the beginning of the investigation. Indeed Type A distributions did not exist in the literature of probability and statistics at the beginning of Neyman's investigation. It was his analysis of the particular problem with the help of the domain expert Dr. Beall, which caused Neyman to introduce Type A distributions for the first time. Moreover it was only the stimulus provided by the falsification of his initial hypothesis which led Neyman to carry out the subtle analysis which led him to formulate the Type A distribution.

A persistent defender of Bayesianism might still argue that a proper analysis of the problem at the beginning of the investigation could have led to the introduction of the Type A distribution at that stage. I rather doubt whether this is a serious possibility, but let us suppose for the moment that it is. The methodology corresponding to this approach would be for the Bayesian statistician to begin with a lengthy analysis of the problem, consulting domain experts, and introducing all the various distributions which might be relevant. While the views of Dr. Beall suggested the Type A distributions, the views of other domain experts, since domain experts often disagree, might have suggested further possible distributions, say distributions of types B, C, and D. Moreover distributions other than Type A are sometimes necessary for problems of this kind, as Neyman's discussion of the distribution of scales, quoted earlier, shows. The Bayesian could then formulate his prior belief distribution over all these hypotheses, and proceed from there. Unfortunately such an approach could very often prove a complete waste of time. Suppose a Bayesian statistician had tried such an approach on the example of bacteria on a Petri-plate. By the time he or she had formulated the first one or two of the hypothetical new distributions which might be possible, Neyman would already have confirmed by a series of χ^2 tests that the simple Poisson distribution was quite adequate in this case. This shows how easy and straightforward is the methodology of classical statistics. It allows us to start with a simple conjecture such as the Poisson distribution, provided only we obey the golden rule of testing our conjecture severely. If the conjecture passes our tests, then it can be accepted provisionally until some further investigations suggest the need for a modification. In the interim we have found a workable hypothesis without the need for elaborating a whole series of possible alternatives. Since Bayesianism depends on the fixity of the theoretical framework, Bayesian statisticians are faced with an awkward choice. Either they must, at the very beginning of the investigation, consider a whole series of arcane possible hypotheses, or they must risk never subsequently arriving at the hypothesis which constitutes the solution of the problem. Their difficulty here arises from the very essence of Baysianism, namely its limitation of changes of belief to those produced by Bayesian conditionalisation.

There are some further ways in which Bayesianism might be defended in the context of this particular example, but it will be convenient to postpone their consideration until section 4, and proceed in the next section to give my second example. In the first example, I have tried to show the merits of the methodology of classical statistics when applied by a leading classical statistician. In the second example I will move in the opposite direction by giving an analysis by a leading Bayesian, namely de Finetti's use of exchangeability, and then trying to show that this analysis only give satisfactory results if no changes are needed in the theoretical framework which is implicitly assumed.

3 DE FINETTI ON EXCHANGEABILITY

In Chapter III of his [1937], de Finetti poses the question (p. 118): 'Why are we obliged in the majority of problems to evaluate a probability according to the observation of a frequency?', commenting that this question (p. 119): 'includes in reality the problem of reasoning by induction.' He continues:

> 'In order to fix our ideas better, let us imagine a concrete example, or rather a concrete interpretation of the problem, which does not restrict its generality at all. Let us suppose that the game of heads or tails is played with a coin of irregular appearance.' [1937, p. 119]

We will now explain how de Finetti analyses this example of the biassed coin from his subjective Bayesian point of view. It will emerge that this concrete example does, in a significant respect, fail to represent the full generality of the problem of reasoning by induction.

De Finetti's first step is to consider a sequence of tosses of the coin which we suppose gives results: $E_1, \ldots, E_n, \ldots$, where each E_i is either heads (H_i) or tails (T_i). So, in particular, H_{n+1} = Heads occurs on the n + 1th toss. Further let e be a complete specification of the results of the first n tosses, that is a sequence n places long, at the ith place of which we have either H_i or T_i. Suppose that heads occurs r times on the first n tosses. The subjective Bayesian's method is to calculate $P(H_{n+1}|e)$, and to show that under some general conditions which will be specified later $P(H_{n+1}|e)$ tends to r/n for large n. This shows that whatever value is assigned to the prior probability $P(H_{n+1})$, the posterior probability $P(H_{n+1}|e)$ will tend to the observed frequency for large n. Thus different individuals who may hold widely differing opinions initially will, if they change their probabilities by Bayesian conditionalisation, come to agree on their posterior probabilities. Such is the argument. Let us now give, in our simple case, the mathematical proof which underpins it.

Suppose that $P(E_i) \neq 0$ for all i, so that also $P(e) \neq 0$. We then have by the definition of conditional probability

$$(1) \quad P(H_{n+1}|e) = \frac{P(H_{n+1}\&e)}{P(e)}$$

To proceed further we introduce the condition of *exchangeability*. Suppose Mr B is making an a priori bet that a particular n-tuple of results ($E_{i1}E_{i2}\ldots E_{in}$ say) occurs. Suppose further that heads occurs r times in this n-tuple. Mr B's betting quotients are said to be *exchangeable* if he assigns the same betting quotient to any other particular n-tuple of results in which heads occurs r times, where both n and r can be chosen to have any finite integral non-negative values with $r \leq n$. Let us write his prior probability (or betting quotient) that there will be r heads in n tosses as $\omega_r^{(n)}$. There are ${}^nC_r$ different ways in which r heads can occur in n tosses, where, as usual, ${}^nC_r = \frac{n!}{(n-r)!r!} = \frac{n(n-1)\dots(n-r+1)}{r(r-1)\dots1}$. Each of the corresponding n-tuples must, by exchangeability, be assigned the same probability, which is therefore $\omega_r^{(n)}/{}^nC_r$. Thus

$$(2) \quad P(E_{i_1}E_{i_2}\dots E_{i_n}) = \frac{\omega_r^{(n)}}{{}^nC_r}$$

Now e, by definition, is just a particular n-tuple of results in which heads occurs r times. Thus, by exchangeability,

$$(3) \quad P(e) = P(E_1E_2\dots E_n) = \frac{\omega_r^{(n)}}{{}^nC_r}$$

Now $H_{n+1}\&e$ is an $(n+1)$-tuple of results in which heads occurs $r+1$ times. Thus, by the same argument,

$$(4) \quad P(H_{n+1}\&e) = \frac{\omega_{r+1}^{(n+1)}}{{}^{n+1}C_{r+1}}$$

And so, substituting in (1), we get

$$\begin{aligned}(5) \quad P(H_{n+1}|e) &= \frac{{}^nC_r}{{}^{n+1}C_{r+1}} \frac{\omega_{r+1}^{(n+1)}}{\omega_r^{(n)}} \\ &= \frac{n!}{(n-r)!r!} \frac{(r+1)!(n-r)!}{(n+1)!} \frac{\omega_{r+1}^{(n+1)}}{\omega_r^{(n)}} \\ P(H_{n+1}|e) &= \frac{r+1}{n+1} \frac{\omega_{r+1}^{(n+1)}}{\omega_r^{(n)}}\end{aligned}$$

Formula (5) (which is de Finetti's formula (6), [1937, p. 122], with a slightly different notation) gives us the result we want. Provided only $\frac{\omega_{r+1}^{(n+1)}}{\omega_r^{(n)}} \rightarrow 1$ as $n \rightarrow \infty$ (a very plausible requirement), we may choose our prior probabilities $\omega_r^{(n)}$ in any way we please, and still get that, as $n \rightarrow \infty$, $P(H_{n+1}|e) \rightarrow r/n$ (the observed frequency), as required.

We can, however, obtain an even simpler result if we choose the prior probabilities in a particular way. In n tosses, we can have either 0, 1, 2,…, or n heads. So, by coherence,

$$(6) \quad \omega_o^{(n)} + \omega_1^{(n)} + \omega_2^{(n)} + \dots + \omega_r^{(n)} + \dots + \omega_n^{(n)} = 1$$

In the subjective theory, we can choose the $\omega_r^{(n)}$ (the prior probabilities) in any way we choose subject only to (6). However we can also, though this is not compulsory, make the 'principle of indifference' choice of making them all equal so that

$$(7) \quad \omega_o^{(n)} = \omega_1^{(n)} = \omega_2^{(n)} = \ldots = \omega_r^{(n)} = \ldots = \omega_n^{(n)} = 1/(n+1)$$

Substituting this in (5), we get

$$(8) \quad P(H_{n+1}|e) = \frac{r+1}{n+2}$$

This is a classical result — Laplace's Rule of Succession, which de Finetti derives in the above way [de Finetti, 1937, p. 144].

In the above calculations, de Finetti appears to show that subjective Bayesians will be led by the process of Bayesian conditionalisation to choose posterior probabilities which approximate to the observed frequency. He thus appears to have provided a foundation for reasoning by induction. I next want to argue that these calculations, despite their seeming generality, are only appropriate within a specific theoretical framework, and can lead us astray if used when that framework does not hold in reality. In order to identify this framework, I will now give some further results from de Finetti's [1937]. These relate the concept of *exchangeability*, which de Finetti himself had introduced, to the older concept of *independence*. De Finetti's ideas on the relationship between exchangeability and independence are discussed in [Galavotti, 2001].

De Finetti proved a general theorem showing exchangeability and independence are linked, I will now state his result. Let us first define exchangeability for a sequence of random variables (or random quantities as de Finetti prefers to call them) $X_1, \ldots, X_n, \ldots$. These are exchangeable if, for any fixed $n, X_{i_1}, X_{i_2}, \ldots, X_{i_n}$ have the same joint distribution no matter how $i_1, \ldots, i_n$ are chosen. Now let Y_n be the average of any n of the random quantities X_i i.e. $Y_n = (1/n)(X_{i_1} + X_{i_2} + \ldots + X_{i_n})$, since we are dealing with exchangeable random quantities it does not matter which $i_1, \ldots, i_n$ are chosen. de Finetti first shows [1937, p. 126] that the distribution $\Phi_n(\xi) = P(Y_n \leq \xi)$ tends to a limit $\Phi(\xi)$ as $n \to \infty$, except perhaps for points of discontinuity. He goes on to say:

> 'Indeed, let $P_\xi(E)$ be the probability attributed to the generic event E when the events $E_1, E_2, \ldots, E_n, \ldots$ are considered independent and equally probable with probability ξ; the probability $P(E)$ of the same generic event, the E_i being exchangeable events with the limiting distribution $\Phi(\xi)$, is
>
> $$P(E) = \int_0^1 P_\xi(E) d\Phi(\xi).$$

> This fact can be expressed by saying that the probability distributions P corresponding to the case of exchangeable events are linear combinations of the distributions P_ξ corresponding to the case of independent equiprobable events, the weights in the linear combination being expressed by $\Phi(\xi)$.' [1937, pp. 128–9]

This general result can be illustrated by taking a couple of special cases. Suppose that we are dealing with a coin tossing example and the generic event E is that heads occurs r times in n tosses. Then

$$P_\xi(E) = {}^nC_r \xi^r (1-\xi)^{n-r}$$

So

$$P(E) = \omega_r^{(n)} = {}^nC_r \int_0^1 \xi^r (1-\xi)^{n-r} d\Phi(\xi)$$

If, in particular, $\Phi(\xi)$ is the uniform distribution, we have

$$\begin{aligned} \omega_r^{(n)} &= {}^nC_r \textstyle\int_0^1 \xi^r (1-\xi)^{n-r} d\xi \\ &= {}^nC_r B(r+1, n-r+1), \text{ where } B \text{ is the Beta function} \\ &= 1/(n+1) \text{ (cf. formula 7 above).} \end{aligned}$$

Comparing these results with our earlier calculations involving exchangeability, we can see how exchangeability and independence are related.

Roughly speaking we can say that the situation which an objectivist would describe as one of independent events in which particular outcomes have fixed but unknown probabilities corresponds to what de Finetti would describe as one of exchangeable events. Of course de Finetti would not have liked this formulation, since he regarded the 'unknown probabilities' postulated by objectivists and classical statisticians as metaphysical and meaningless. Thus he says:

> 'If ... one plays heads or tails with a coin of irregular appearance, ..., one does not have the right to consider as distinct hypotheses the suppositions that this imperfection has a more or less noticeable influence on the "unknown probability", for this"unknown probability" cannot be defined, and the hypotheses that one would like to introduce in this way have no objectivemeaning.' [1937, pp. 141–1]

de Finetti therefore concludes:

> '... the nebulous and unsatisfactory definition of "independent events with fixed but unknown probability" should be replaced by that of "exchangeable events".' [1937, p. 142]

Naturally I cannot agree with de Finetti's attempt to eliminate the concept of unknown probability. To postulate such probabilities, as is done in classical statistics, is neither meaningless nor metaphysical. Conjectures about such unknown probabilities can be tested using statistical tests, and either confirmed or refuted, and this shows that such conjectures are scientific rather than metaphysical. It is thus both meaningful and scientific to postulate that a particular process consists of independent events with fixed but unknown probability. My thesis is that this postulate gives the theoretical framework within which de Finetti's calculations using exchangeability lead to sensible results. If we try to use these calculations in situations where this theoretical framework does not hold objectively, they are liable to give absurd and quite inappropriate conclusions. This can be easily shown by seeing what happens when we apply the exchangeability calculations to a situation which is not one of independent events but of dependent events.

To illustrate my argument, it would be possible to use any one of a wide variety of sequences of events which are dependent rather than independent. To be concrete, I have first selected the simplest type of dependent sequence, namely a Markov chain, and then chosen one very simple and at the same time striking example of a Markov chain. This is the game of 'Red or Blue'.[2] At each go of the game there is a number s which is determined by the previous results. A fair coin is tossed. If the result is heads, we change s to $s' = s + 1$, and if the result is tails, we change s to $s' = s - 1$. If $s' \geq 0$, the result of the go is said to be blue, while if $s' < 0$, the result of the go is said to be red. So, although the game is based on coin tossing, the results are a sequence of red and blue instead of a sequence of heads and tails. Moreover, while the sequence of heads and tails is independent, the sequence of red and blue is highly dependent. We would expect much longer runs which are all blue, than runs in coin tossing which are all heads. If we start the game with $s = 0$, then there is a slight bias in favour of blue which is the initial position. However, it is easy to eliminate this by deciding the initial value of s by a coin toss. If the toss gives heads we set the initial value of s at 0, and if the toss gives tails we set it at -1. This makes red and blue exactly symmetrical, so that the limiting frequency of blue must equal that of red and be 1/2. It is therefore surprising that over even an enormously large number of repetitions of the game, there is high probability of one of the colours appearing much more often than the other. Feller [1950, pp. 82–3] gives a number of examples of these curious features of the game. Suppose for example that the game is played once a second for a year, i.e. repeated 31,536,000 times. There is a probability of 70% that the more frequent colour will appear for a total of 265.35 days, or about 73% of the time, while the less frequent colour will appear for only 99.65 days, or about 27% of the time.

Let us next suppose that a subjective Bayesian (Mr B) is asked to analyse a sequence of events, each member of which can have one of two values. Unknown to them this sequence is in fact generated by the game of red or blue. Possibly the

[2]The game of 'Red or Blue' is described in Feller, [1950, pp. 67–95], which contains an interesting mathematical analysis of its curious properties. Popper read of the game in Feller, and had the idea of using it to argue against various theories of induction. Popper uses the game to criticise what he calls 'the simple inductive rule' in his [1957, pp. 358–60] (reprinted in his [1983, pp. 301–5]. I have adapted this argument of Popper's to produce the critique of de Finetti's use of exchangeability given here.

sequence might be produced by a man-made device which flashes either 0 (corresponding to red) or 1 (corresponding to blue) onto a screen at regular intervals. However, it is not impossible that the sequence might be one occurring in the world of nature. Consider for example a sequence of days, each of which is classified as 'rainy' if some rain falls, or dry otherwise. In a study of rainfall at Tel Aviv during the rainy season of December, January, and February, it was found that the sequence of days could be modelled successfully as a Markov chain. In fact the probabilities found empirically were: probability of a dry day given that the previous day was dry = 0.75, and probability of a rainy day given that the previous day was rainy = 0.66. (For further details see [Cox and Miller, 1965, pp. 78–9].) It is clear that this kind of dependence will give longer runs of either rainy or dry days than would be expected on the assumption of independence. It is thus not impossible that the sequence of rainy and dry days at some place and season might be represented quite well by the game of red or blue.

Let us now return to our subjective Bayesian Mr B, who has been asked to deal with a process which is really governed, unknown to Mr B, by the game of 'Red or Blue'. Being an admirer of de Finetti's, Mr B will naturally make an assumption of exchangeability. Let us also assume that he gives a uniform distribution a priori to the $\omega_r^{(n)}$ (see formula 7 above) so that Laplace's rule of succession holds (formula 8). This is just for convenience of calculation. The counter-intuitive results would appear for any other coherent choice of the $\omega_r^{(n)}$. Suppose that we have a run of 700 blues, followed by 2 reds. Mr B would calculate the probability of getting blue on the next go using formula 8 with $n = 702$, and $r = 700$. This gives the probability of blue as 701/704 = 0.996 to 3 significant figures. Knowing the mechanism of the game, we can calculate the true probability of blue on the next go, which is very different. Go 700 gave blue, and go 701 gave red. This is only possible if s on go 700 was 0, the result of the toss was tails, and s became -1 on go 701. The next toss must also have yielded tails or there would have been blue again on go 702. Thus s at the start of go 703 must be -2, and this implies that the probability of blue on that go is zero. Then again let us consider one of Feller's massive sessions of 31,536,000 goes. Suppose the result is that the most frequently occurring colour appears 73% of the time (as pointed out above there is a probability of 70% of this result which is thus not an unlikely outcome). Mr B will naturally be estimating the probability of this colour at about 0.73 and so much higher than that of the other colour. Yet in the real underlying game, the two colours are exactly symmetrical.

We see that Mr B's calculations using exchangeability will give results at complete variance with the true situation. The reason for this is clear. By making the assumption of exchangeability, Mr B is implicitly assuming that the process he is considering consists of independent events with a fixed but unknown probability. As long as this theoretical framework holds, his Bayesian calculations will give him reasonable results, but if the theoretical framework does not hold in a particular case, then the same Bayesian calculations will give him completely inappropriate results. My conclusion is, once again, that Bayesianism only works if the condition of the fixity of the theoretical framework is satisfied.

Our situation involving the game of 'Red or Blue' does not pose the same problems for a classical statistician. Suppose such a statistician (Ms C say) is confronted with a sequence of events which, unknown to her, is really governed by the game of 'Red or Blue'. It would be perfectly reasonable for Ms C to begin by making the simplest and most familiar conjecture, namely that the events are independent. Thus Ms C starts tackling the problem in much the same was as Mr B. However, being, unlike Mr B, a good Popperian, Ms C will test her conjecture rigorously with a series of statistical tests for independence. It will not be long before she has rejected her initial conjecture, and she will then start exploring other hypotheses involving various kinds of dependence among the events. If she is a talented scientist, she may soon hit on the red or blue mechanism, and be able to confirm that it is correct by another series of statistical tests. In this case the classical statistician seems better equipped to deal with the problem than the Bayesian. However there are some replies to this argument which could be made from the Bayesian point of view, and I will consider them in the final section of the paper (section 4).

4 POSSIBLE DEFENCES OF BAYESIANISM

de Finetti himself does say one or two things which are relevant to the problem. Having shown that exchangeable events are the subjective equivalent of the objectivist's independent and equiprobable events, he observes that one could introduce subjective equivalents of various forms of dependent events, and, in particular, of Markov chains. As he says:

> 'One could in the first place consider the case of classes of events which can be grouped into Markov "chains" of order $1, 2, \ldots, m, \ldots$, in the same way in which classes of exchangeable events can be related to classes of equiprobable and independent events.' [1937, Footnote 4, p. 146]

We could call such classes of events *Markov-exchangeable*. De Finetti argues that they would constitute a complication and extension of his theory without causing any fundamental problem:

> 'One cannot exclude completely *a priori* the influence of theorder of events ... . There would then be a number of degrees of freedom and much more complication, but nothing would be changed in the setting up and the conception of the problem ... , before we restricted our demonstration to the case ofexchangeable events ...' [1937, p. 145]

Perhaps de Finetti has in mind something like the following. Instead of just assuming exchangeability, we consider not just exchangeability but various forms of Markov-exchangeability. To each of these possibilities we give a prior probability.

No doubt exchangeability will have the highest prior probability. If the case is a standard one, like the biased coin, this high prior probability will be reinforced, and the result will come out more or less like that obtained by just assuming exchangeability. If, however, the case is an unusual one, then the posterior probability of exchangeability will gradually decline, and that of one of the other possibilities will increase until it becomes much more probable than exchangeability.

This approach to the problem is basically the same as that we attributed to the Bayesian in our discussion of Neyman's investigation in section 2, and it is liable to the same difficulties which we noted there. If a Bayesian is to adopt this approach seriously, he or she must begin every investigation by considering all possible hypotheses which might be encountered in the course of the investigation. This is scarcely possible, and, even if it were possible, it would often be a waste of time. There are many situations in which the most obvious and straightforward hypothesis actually works so that a consideration of a large number of arcane alternatives would be useless toil. The classical statisticians do not need to indulge in such toil. They can begin with any assumption (or conjecture) they like, provided only they obey the golden rule of testing it severely. If the assumption passes such tests, it can be provisionally adopted. If it fails, some other better assumption must be sought. Thus the classical statistician proceeds, so to speak, one step at a time, and there is never any need to engage in the hopeless and time-wasting task of surveying all possible hypotheses which might apply to the problem in hand.

There are moreover, as Albert has shown in his contribution to the present volume, further difficulties in this defence of Bayesianism. To see what these are, let us go back to the formulation of Bayesianism given at the beginning of the paper. As I said there, it is usually assumed in a Bayesian statistical analysis that there is a given a set of possible statistical hypotheses H_θ where $\theta \in I$, for some set I, normally an interval of the real line. The problem we are now considering is how the set H_θ where $\theta \in I$ should be chosen. If we select a rather narrow set H_θ we may leave out the hypothesis which would provide the required solution. If we try to make H_θ broad and inclusive, we set ourselves a very difficult task which may well prove a waste of time in a case in which the most simple and obvious solution actually works in practice. What Albert shows in his [2001] is that the second strategy of searching for a broad and inclusive set H_θ is liable to a further difficulty.

Albert considers the possibility of extending the set H_θ by including hypotheses involving chaos theory. Specifically he defines in section 4 of his paper what he calls a 'Chaotic Clock'. In a simple case in which we are considering a sequence of 0's and 1's generated by some unknown process, Albert formulates a set H_θ of hypotheses based on a mechanism involving a chaotic clock. He then gives in section 5.1 of his paper a remarkable result called the *Anything Goes Theorem*. Suppose Mr B adopts any learning strategy whatever, i.e. he chooses his conditional probabilities given evidence in any arbitrary way. There then exists a prior probability distribution p over the set H_θ of hypotheses based on the chaotic clock such that Mr B's probabilities are produced by Bayesian conditioning of p.

Albert's result is very striking indeed. His chaotic clock hypotheses are by no means absurd. After all chaos theory is used in both physics and in economics. Indeed hypotheses involving chaos are quite plausible as a means of explaining, for example, stock market fluctuations. If Mr B were really faced with a bizarre sequence of 0's and 1's, why should he not consider a hypothesis based on chaos theory? Yet if Mr B is allowed to consider the chaotic clock set of hypotheses, then any learning strategy he adopts becomes a Bayesian strategy for a suitable choice of priors. In effect Bayesianism has become empty.

It follows that a Bayesian (Mr B say) is caught on the horns of a dilemma. Mr B may adopt a rather limited set of hypotheses to perform his Bayesian conditionalisation, but then, as the example of the game of Red or Blue shows, if his set excludes the true hypothesis, his Bayesian learning strategy may never bring him close to grasping what the real situation is. This is the first, or 'Red or Blue', horn of the dilemma. If Mr B responds by saying he is prepared to consider a wide and comprehensive set of hypotheses, these will surely include hypotheses from chaos theory and thus anything he does will become Bayesian, making the whole approach empty. This is the second, or 'Chaotic Clock', horn of the dilemma.

The Bayesian is faced with quite severe difficulties here, but there is one further way out which is sometimes suggested, and I will conclude the paper by giving it a brief consideration. The suggestion is that in we should start with a reasonably specific set of initial hypotheses H_θ but add to this set a 'catch all' hypothesis K, which simply says that some hypothesis other than the H_θ is correct. We then give our prior distribution over the H_θ and K. If it is a standard case, then one of the H_θ will emerge as the most probable hypothesis given the evidence. If, however, we are dealing with a non-standard case, then K will gain in probability while the probability of each of the H_θ becomes very small. In such a situation, we will divide up K into some specific set J_θ say, and a new catch all K', and repeat the process. In this way we should, even in a problematic situation, be led to the correct hypothesis.

While such a procedure sounds very reasonable when stated in outline, any attempt actually to implement it in detail brings to light a whole host of difficulties and complexities, and it is not surprising that there is no instance to my knowledge of such a plan being actually carried out in detail by a Bayesian. Let us begin by considering how the prior probabilities should be divided between the set H_θ and the catch all K. Surely K should have a very large prior probability since our background knowledge concerning the development of science would suggest that most hypotheses considered at a particular time are eventually shown to be inadequate to some degree or in some respects. Yet if K is given a large prior probability, this may prevent any of the H_θ ever acquiring a large probability, even in a straightforward case.

Suppose this initial difficulty is overcome, we are the faced with another. Let us take one of the problematic cases in which we assume to begin with one set of hypotheses H_θ say, and another set J_θ are in fact correct. H_θ could be Poisson distributions and J_θ could be Type A distributions, or H_θ could be the hypothesis of independent events with fixed probability θ and J_θ could be hypotheses of a Markov chain of some type. In this case we have got to show how the probability of the catch all K changes from its prior value $p(K)$ say to a posterior value $p(K|e)$ in the light of evidence. How is such a calculation to be carried out? It is no easy matter, and it must be done in such a way that $p(K|e)$ increases to such a value that we decide to abandon the H_θ and subdivide K into J_θ and the new catch all K'. I really think such a calculation is scarcely possible. Of course a Bayesian could show that I am wrong by carrying out such a calculation in one of the cases dealt with in this paper, but the result would undoubtedly be very complicated. At this point one can reasonably ask why the Bayesian wants to get involved in such complexities rather than to adopt the methods of classical statistics which, as I have shown, deal with the problem in an extremely simple and straightforward way, using the method of conjectures and refutations.

My conclusion is that Bayesianism should only be applied if we are in a situation in which there is a fixed and known theoretical framework which it is reasonable to suppose will not be altered in the course of the investigation, that is to say if the condition of the fixity of the theoretical framework is satisfied. As regards many processes whose nature is not exactly known, statistical testing using the methodology of classical statistics is essential.

Department of Philosophy, King's College London, UK.

BIBLIOGRAPHY

[Albert, 2001] M. Albert. Bayesian learning and expectations formation: anything goes. In this volume, pp. 347–368.

[Bayes and Price, 1763] T. Bayes and R. Price. An Essay towards Solving a Problem in the Doctrine of Chances, reprinted in E.S.Pearson and M.G.Kendall (eds.) *Studies in the History of Statistics and Probability*, Griffin, 1970, pp. 134–53. Originally published 1763.

[Cox and Miller, 1965] D. R. Cox and H. D. Miller. *The Theory of Stochastic Processes*. Methuen, 1965.

[Dawid and Gillies, 1989] P. Dawid and D. A. Gillies. A Bayesian Analysis of Hume's Argument concerning Miracles, *The Philosophical Quarterly*, **39**, 57–65, 1989.

[de Finetti, 1937] B. de Finetti. Foresight: Its Logical Laws, Its Subjective Sources. English translation in H.E. Kyburg and H.E. Smokler (eds.), *Studies in Subjective Probability*, pp. 93–158. Wiley, 1964.

[Feller, 1950] W. Feller. *Introduction to Probability Theory and Its Applications*. Third edition, 1971, Wiley.

[Galavotti, 2001] M. C. Galavotti. Subjectivism and Objectivity in Bruno de Finetti's Bayesianism, in present volume, pp. 171–184.

[Lakatos, 1968] I. Lakatos. Changes in the Problem of Inductive Logic. Reprinted in John Worrall and Gregory Currie (eds.), *Imre Lakatos, Philosophical Papers, Volume 2*, pp. 128–200, Cambridge University Press, 1968.

[Laplace, 1812] P. S. Laplace. *Théorie analytique des probabilités*. Reprinted as vol. 5 of *Oeuvres complètes de Laplace*, 14 vols, Gauthier-Villars, 1878–1912.

[Neyman, 1952] J. Neyman. *Lectures and Conferences on Mathematical Statistics and Probability*. 2nd edition, revised and enlarged, Washington: Graduate School of U.S. Department of Agriculture, 1952.

[Pearson, 1900] K. Pearson. On the Criterion that a given System of Deviations from the probable in the case of a Correlated System of Variables is such that it can be reasonably be supposed to have arisen from Random Sampling, reprinted in *Karl Pearson's Early Statistical Papers*, Cambridge University Press, pp. 339–57, 1956.

[Popper, 1957] K. R. Popper. Probability Magic or Knowledge out of Ignorance, *Dialectica*, **11**, 354–74, 1957.

[Popper, 1963] K. R. Popper. *Conjectures and Refutations*. Routledge and Kegan Paul, 1963.

[Popper, 1983] K. R. Popper. *Realism and the Aim of Science*. Hutchinson, 1983.

D. G. MAYO AND M. KRUSE

PRINCIPLES OF INFERENCE AND THEIR CONSEQUENCES

> The likelihood principle emphasized in Bayesian statistics implies, among other things, that the rules governing when data collection stops are irrelevant to data interpretation. It is entirely appropriate to collect data until a point has been proved or disproven ... [Edwards *et al.*, 1963, p. 193].

1 INTRODUCTION

What do data tell us about hypotheses or claims? When do data provide good evidence for or a good test of a hypothesis? These are key questions for a philosophical account of evidence and inference, and in answering them, philosophers of science have often appealed to formal accounts of probabilistic and statistical inference. In so doing, it is obvious that the answer will depend on the principles of inference embodied in one or another statistical account. If inference is by way of Bayes' theorem, then two data sets license different inferences only by registering differently in the Bayesian algorithm. If inference is by way of error statistical methods (e.g., Neyman and Pearson methods), as are commonly used in applications of statistics in science, then two data sets license different inferences or hypotheses if they register differences in the error probabilistic properties of the methods.

The principles embodied in Bayesian as opposed to error statistical methods lead to conflicting appraisals of the evidential import of data, and it is this conflict that is the pivot point around which the main disputes in the philosophy of statistics revolve. The differences between the consequences of these conflicting principles, we propose, are sufficiently serious as to justify supposing that one "cannot be just a little bit Bayesian" [Mayo, 1996], at least when it comes to a philosophical account of inference, but rather must choose between fundamentally incompatible packages of evidence, inference, and testing. In the remainder of this section we will sketch the set of issues that seems to us to serve most powerfully to bring out this incompatibility.

EXAMPLE 1 (ESP Cards). The conflict shows up most clearly with respect to the features of the *data generation process* that are regarded as relevant for assessing evidence. To jump right into the crux of the matter, we can consider a familiar type of example: To test a subject's ability, say, to predict draws from a deck of five ESP cards, he must demonstrate a success rate that would be very improbable if he were merely guessing. Supposing that after a long series of trials, our subject

D. Corfield and J. Williamson (eds.), Foundations of Bayesianism, 381–403.

attains a "statistically significant" result, the question arises: Would it be relevant to your evaluation of the evidence if you learned that he had planned all along to keep running trials *until* reaching such an improbable result? Would you find it relevant to learn that, having failed to score a sufficiently high success rate after 10 trials, he went on to 20 trials, and on and on until finally, say on trial number 1,030, he attained a result that would apparently occur only 5% of the time by chance?

A plan for when to stop an experiment is called a *stopping rule*. So our question is whether you would find knowledge of the subject's stopping rule relevant in assessing the evidence for his ESP ability. If your answer is *yes*, then you are in sync with principles from standard error statistics (e.g., significance testing and confidence interval estimation). Interestingly enough, however, this intuition conflicts with the principles of inference espoused by other popular philosophies of inference, i.e., the Bayesian and Likelihoodist accounts. In particular, it conflicts with the *likelihood principle* (LP). According to the LP, the fact that our subject planned to persist until he got the desired success rate, the fact that he *tried and tried again*, can make *no* difference to the evidential import of the data: the data should be interpreted in just the same way as if he had decided from the start that the experiment would consist of exactly 1,030 trials.

This challenge to the widely held supposition that stopping rules alter the import of data was L. J. Savage's central message to a forum of statisticians in 1959:

> The persistent experimenter can arrive at data that nominally reject any null hypothesis at any significance level, when the null hypothesis is in fact true....
>
> These truths are usually misinterpreted to suggest that the data of such a persistent experimenter are worthless or at least need special interpretation ...The likelihood principle, however, affirms that the experimenter's intention to persist does not change the import of his experience [Savage, 1962, p. 18].

Savage rightly took this conflict as having very serious consequences for the foundations of statistics:

> In view of the likelihood principle, all of [the] classical statistical ideas come under new scrutiny, and must, I believe, be abandoned or seriously modified [Savage, 1962, pp. 17–18].

This conflict corresponds to two contrasting principles on what is required by an account of inference, "*evidential-relationship*" (E-R) principles and "*testing*".

2 EVIDENTIAL-RELATIONSHIP VS. (ERROR STATISTICAL) TESTING ACCOUNTS

In what we are calling E-R accounts, the evidential bearing of data on hypotheses is determined by a measure of support, probability, confirmation, credibility or the like to hypotheses given data. Testing approaches, in contrast, do not seek to assign measures of support or probability to hypotheses, but rather to specify methods by which data can be used to test hypotheses. Probabilistic considerations arise to characterize the probativeness, reliability, or severity of given tests, and specific inferences they license.

The difference between E-R and testing approaches is most dramatically revealed by the fact that two data sets $\boldsymbol{x}$ and $\boldsymbol{y}$ may have exactly the same evidential relationship to hypothesis H, on a given E-R measure, yet warrant very different inferences on testing accounts because $\boldsymbol{x}$ and $\boldsymbol{y}$ arose from tests with different characteristics. In particular, the two tests may differ in the frequency with which they would lead to erroneous inferences (e.g., passing a false or failing a true hypothesis). That is, the tests may have different *error probabilities*. We will refer to the testing philosophy as the *error-statistical approach.*

In *statistical* approaches to evidence, the main E-R measure is given by the probability conferred on $\boldsymbol{x}$ under the assumption that H is correct, $P(\boldsymbol{x}; H)$, i.e., the *likelihood* of H with respect to $\boldsymbol{x}$.[1] The LP, informally speaking, asserts that the evidential import of $\boldsymbol{x}$ on any two hypotheses, H and H', is given by the ratio of the likelihoods of H and H' with respect to $\boldsymbol{x}$.

To get a quick handle on the connection between the LP and stopping rules, suppose $\boldsymbol{x}$ arose from a procedure where it was decided in advance to take just n observations (i.e., n was *predesignated*), and $\boldsymbol{y}$ arose from our ESP subject's 'try and try again' procedure, which just happened to stop at trial n (*sequential* sampling). If, for every hypothesis $H, P(\boldsymbol{x}; H) = P(\boldsymbol{y}; H)$, then according to the LP it can make no difference to the inference which procedure was used. So, the fact that the subject stopped along the way to see if his success rate was sufficiently far from what is expected under chance makes no difference to "what the data are saying" about the hypotheses. This sentiment is quite clear in a seminal paper by Edwards, Lindman and Savage:

> In general, suppose that you collect data of any kind whatsoever — not necessarily Bernoullian, nor identically distributed, nor independent of each other ...— stopping only when the data thus far collected satisfy some criterion of a sort that is sure to be satisfied sooner or later, then the import of the sequence of n data actually observed will be exactly the same as it would be had you planned to take exactly n observations in the first place [Edwards *et al.*, 1963, pp. 238–239].

[1]Note that $P(\boldsymbol{x}; H)$ is not a conditional probability usually written as $P(\boldsymbol{x}|H)$ because that would involve assigning prior probabilities to H — something outside the standard error statistical approach. The way to read $P(\boldsymbol{x}; H)$ is "The probability that X takes value $\boldsymbol{x}$ *according to* statistical hypothesis H." Any statistical hypothesis H must assign probabilities to the different experimental outcomes.

This is called the *irrelevance of the stopping rule* or the *Stopping Rule Principle* (SRP), and is an implication of the LP.[2]

To the holder of the LP, the intuition is that the stopping rule is irrelevant, and it is a virtue of the LP that it accords with this intuition. To the error statistician the situation is exactly the reverse. For her, the stopping rule is relevant because the persistent experimenter is more likely to find data in favor of H, even if H is false, than one who fixed the sample size in advance. Peter Armitage, in his comments to Savage at the 1959 forum, put it thus:

> I think it is quite clear that likelihood ratios, and therefore posterior probabilities, do not depend on a stopping rule. Professor Savage, Dr Cox and Mr Lindley take this necessarily as a point in favour of the use of Bayesian methods. My own feeling goes the other way. *I feel that if a man deliberately stopped an investigation when he had departed sufficiently far from his particular hypothesis, then 'Thou shalt be misled if thou dost not know that'.* If so, prior probability methods seem to appear in a less attractive light than frequency methods, where one can take into account the method of sampling [Armitage, 1962, p. 72], (emphasis added).

It is easy enough to dismiss long-run frequencies as irrelevant to interpreting given evidence, and thereby deny Armitage's concern, but we think that would miss the real epistemological rationale underlying Armitage's argument.[3] Granting that textbooks on "frequency methods" do not adequately supply the rationale, we propose to remedy this situation. Holders of the error statistical philosophy, as we see it, insist that data only provide genuine or reliable evidence for H if H survives a *severe test*. The severity of the test, as probe of H — e.g., the hypothesis that our ESP subject does better than chance — depends upon the test's ability to find that H is false when it is (i.e., when null hypothesis H_0 is true). H is *not* being put to a stringent test when a researcher allows trying and trying again until the data are far enough from H_0 to reject it in favor of H. This conception of tests provides the link between a test's error probabilities and what is required for a warranted inference based on the test. It lets us understand Armitage as saying that one would be misled if one could not take into account that two plans for generating data correspond to tests with different abilities to uncover errors of concern.

In the 40 years since this forum, the conflict between Bayesian and "classical" or error statistics has remained, and the problems it poses for evidence and inference are unresolved. Indeed, in the past decade, as Bayesian statistics has grown in acceptance among philosophers, the crux of this debate seems to have been largely forgotten. We think it needs to be revived.

[2]There are certain exceptions (the stopping rule may be "informative"), but Bayesians do not regard the examples we consider as falling under this qualification. See section 6.1.

[3]This dismissal is the basis of Howson and Urbach's response to Gillies' [1990] criticism of them.

3 THE LIKELIHOOD PRINCIPLE (LP)

The LP is typically stated with reference to two experiments considering the same set of statistical hypotheses H_i about a particular parameter, μ, such as the probability of success (on a Bernoulli trial) or the mean value of some characteristic.

> According to Bayes' theorem, $P(\boldsymbol{x};\mu)\ldots$ constitutes the entire evidence of the experiment, that is, it tells all that the experiment has to tell. More fully and more precisely, if $\boldsymbol{y}$ is the datum of some other experiment, and if it happens that $P(\boldsymbol{x};\mu)$ and $P(\boldsymbol{y};\mu)$ are proportional functions of μ (that is, constant multiples of each other), then each of the two data $\boldsymbol{x}$ and $\boldsymbol{y}$ have exactly the same thing to say about the values of $\mu\ldots$ I, and others, call this important principle the likelihood principle. The function $P(\boldsymbol{x};\mu)$ — rather this function together with all others that result from it by multiplication by a positive constant — is called the likelihood [Savage, 1962, p. 17]. (We substitute his $\Pr(x|\lambda)$ with $P(\boldsymbol{x};\mu)$.)

The *likelihood function* gives the probability (or density) of a *given* observed value of the sample under the different values of the unknown parameter(s) such as μ. More explicitly, writing the n-fold sample $(X_1, X_2, \ldots, X_n)$ as $\boldsymbol{X}$, the likelihood function is defined as the probability (or density) of $\{\boldsymbol{x} = (x_1, x_2, \ldots, x_n)\}$ — arising from the *joint distribution* of the random variables making up the sample $\boldsymbol{X}$ — under the different values of the parameter(s) μ.

Even granting that two experiments may have different error probabilities over a series of applications, for a holder of the LP, once the data are in hand, only the actual likelihoods matter:

> The Likelihood Principle. In making inferences or decisions about μ after $\boldsymbol{x}$ is observed, all relevant experimental information is contained in the likelihood function for the observed $\boldsymbol{x}$. Furthermore, two likelihood functions contain the same information about μ if they are proportional to each other (as functions of μ) [Berger, 1985, p. 28].

That is, the LP asserts that:

> If two data sets $\boldsymbol{x}$ and $\boldsymbol{y}$ have likelihood functions which are (a) functions of the same parameter(s) μ and (b) proportional to each other, then $\boldsymbol{x}$ and $\boldsymbol{y}$ contain the same experimental information about μ.[4]

---

[4]We think this captures the generally agreed upon meaning of the LP although statements may be found that seem stronger. For example, Pratt, Raiffa, and Schlaifer characterize the LP in the following way:

> If, in a given situation, two random variables are observable, and if the value $\boldsymbol{x}$ of the first and the value $\boldsymbol{y}$ of the second give rise to the same likelihood function, then observing the value $\boldsymbol{x}$ of the first and observing the value $\boldsymbol{y}$ of the second are equivalent in the sense that *they should give the same inference, analysis, conclusion, decision, action, or anything else* ([Pratt *et al.*, 1995, p. 542]; emphasis added).

4 STATISTICAL SIGNIFICANCE LEVELS: TESTING PARAMETERS OF BERNOULLI TRIALS

The error statistical approach is *not* consistent with the LP because the error statistical calculations upon which its inferences are based depend on more than the likelihood function. This can be seen by considering Neyman–Pearson statistical significance testing.

Significance testing requires identifying a statistical hypothesis H_0 that will constitute the *test* or *null hypothesis* and an alternative set of hypotheses reflecting the discrepancy from H_0 being probed. A canonical example is where $\boldsymbol{X} := (X_1, X_2, \ldots, X_n)$ is a random sample from the Bernoulli distribution with parameter μ, the probability of success at each trial. In a familiar "coin tossing" situation, we test $H_0 : \mu = 0.5$, (the coin is "fair") against the claim that $J : \mu > 0.5$.

Once a null hypothesis is selected, we define a test statistic, i.e., a characteristic of the sample $\boldsymbol{X} = (X_1, X_2, \ldots, X_n)$ that we are interested in such as $\bar{X}$, the proportion of successes in n Bernoulli trials. Then, we define a measure of *fit* or *distance* between the test statistic, and the value of the test statistic expected under H_0 (in the direction of some alternative hypothesis J). For example, in testing hypothesis $H_0 : \mu = 0.5$, a sensible distance measure $d(\boldsymbol{X}; H_0)$ is the (positive) difference between $\bar{X}$ and the expected proportion of successes under H_0, 0.5, in standard deviation units:

$$d\,(\mathbf{X}; H_0) = \frac{(\bar{X} - 0.5)}{\sigma_{\bar{X}}}.$$

Our distance measure may also be set out in terms of the likelihoods of H_0 as against different alternatives in J. A result $\boldsymbol{x}$ is further from H_0 to the extent that H_0 is less likely than members of J, given $\boldsymbol{x}$. This distance measure, which we may write as $d'(\boldsymbol{X}, H)$ gives us a likelihood ratio (LR). That is:

$$d'\,(\mathbf{X}, H_0) = LR = \frac{P(x; H_0)}{P(x; J)}$$

In the case of composite hypotheses we take the maximum value of the likelihood.[5]

No matter which distance measure is used, the key feature of the test is based on considering not just the one value of d that happened to occur, but *all* of the possible values. That is, d is itself a statistic that can take on different values in repeated trials of the experimental procedure generating the data. This probability distribution is called the *sampling* distribution of the distance statistic, and is what allows calculating error probabilities, one of which is the *statistical significance level* (SL):

[5] Otherwise, one would need to have prior probability assignments for each hypothesis within the composite alternative. Some strict likelihoodists, who do not use prior probabilities, regard this likelihood as undefined (e.g., [Edwards, 1992; Royall, 1997]).

> *Statistical Significance Level of the observed difference* $d(\boldsymbol{x})$ (in testing H_0) = the probability of a difference as large as or larger than $d(\boldsymbol{x})$, under H_0.

In calculating the statistical significance level, one sums up the probabilities of outcomes as far as or further from H_0 as $\boldsymbol{x}$ is. The smaller the significance level, the further $\boldsymbol{x}$ is from what is expected under H_0: if it is very small, say 0.05 or 0.01, then the outcome is said to be statistically significant at these levels.

To highlight how an analysis by way of significance levels violates the LP, the most common example alludes to two different ways one could generate a series of n independent (Bernoulli) coin-tossing trials, with μ the probability of "heads" on each trial.

EXAMPLE 2 (Case 1: The Binomial Distribution). In the first case, it is decided in advance to carry out n flips, stop, and record the number of successes, which we can represent as random variable Z. Here, Z is a Binomial variable with parameters μ and n and the probability distribution of Z takes the form:

$$(1)\quad P_1(Z = z; \mu) = \binom{n}{z} \mu^z (1-\mu)^{n-z}$$

Suppose it is decided to observe $n = 12$ trials and the observed result is $Z = 9$ heads. The probability of this result, under the assumption that $\mu = \mu_0$ is:

$$P_1(Z = 9; H_0 : \mu = \mu_0) = \binom{12}{9} \mu_0^9 (1-\mu_0)^3$$

EXAMPLE 2 (Case 2: The Negative Binomial Distribution — A Case of Sequential Sampling). In case 2, by contrast, we are to consider that the experimenter was interested in the number of heads observed, Z, before obtaining r tails, for some fixed value r. In this sampling scheme the random variable Z follows the Negative Binomial distribution:

$$(2)\quad P_2(Z = z; \mu) = \binom{z+r-1}{z} \mu^z (1-\mu)^r.$$

This experiment can be viewed as conducting Bernoulli trials with the following stopping rule: Stop as soon as you get a total of r tails. We are next to imagine that r had been set in advance to 3, and it happens that 9 heads were observed before the third tail, thereby allowing the trials to terminate. We then have:

$$P_2(Z = 9, r = 3; H_0 : \mu = \mu_0) = \binom{11}{9} (\mu_0)^9 (1-\mu_0)^3.$$

In each of the two cases above, the data set consists of 9 heads and 3 tails. We see immediately that (1) and (2) differ only by a constant. So, a set of z heads and r tails in n=z+r Bernoulli trials defines the *same* likelihood whether by Binomial sampling (n fixed) or Negative Binomial sampling (r fixed). In both cases, the likelihood of μ given $z = \mu^z(1-\mu)^{n-z}$. According to the LP, then, this difference between the two cases makes no difference to what the outcomes tell us about the various values of μ:

> If a Bernoulli process results in z successes in n trials, it has the likelihood function $\mu^z(1-\mu)^{n-z}$ and as far as inferences about μ are concerned, it is irrelevant whether either n or r was predetermined ([Pratt *et al.*, 1995, p. 542]. We replace their p with μ for consistency of notation).

Nevertheless, as the holder of the LP goes on to show, the *significance level* attained in case 1 differs from that of case 2, thereby showing that significance levels violate the LP. In particular, we have

(i) The statistical significance level for the Binomial (n fixed at 12)=

$$P_1(Z \geq 9; H_0 : \mu = 0.5) = P_1(z = 9 \text{ or } 10 \text{ or } 11 \text{ or } 12; \mu = 0.5) \approx 0.075$$

whereas

(ii) The significance level for the Negative Binomial (r fixed at 3)=

$$P_2(z = 9 \text{ or } 10 \text{ or } \ldots; \mu = 0.5) \approx 0.0325.$$

Thus, if the level of significance before rejecting H_0 were fixed at 0.05, we would reject H_0 if the observations were the result of Binomial trials, but we would *not* reject it if *those same observations* were the result of *Negative* Binomial trials.

5 THE OPTIONAL STOPPING EFFECT

Although the contrasting analysis demanded by the error statistician in considering the Binomial vs. the Negative Binomial (Example 2) was not very pronounced, the example we used at the opening of our paper points to much more extreme contrasts.

An example that has received considerable attention is of the type raised by Armitage at the "1959 Savage Forum." The null hypothesis H_0 is an assertion about a population parameter μ, the mean value of some quantity, say, a measure of the effectiveness of some medical treatment. The experiment involves taking a random sample of size n, $X_1, \ldots X_n$ and calculating its mean. Let $\bar{X}_n$ be the sample mean of the n observations of the X_is, where we know that each X_i is distributed Normally with *unknown* mean μ and *known* variance 1, i.e., $X_i \sim$ Normal$(\mu, 1)$. The null hypothesis asserts the treatment has no effect:

$$H_0 : \mu = 0.$$

The alternative hypothesis H_1 is the complex hypothesis consisting of all values of μ other than 0:

$$H_1 : \mu \neq 0.$$

As before, we are to consider two different stopping rules:

EXAMPLE 3 (Case 1: Test T-1 (fixed sample size)). In this case we take n samples, evaluate the distance between the observed mean, $\bar{x}_n$, and the mean hypothesized in H_0, namely 0, and then calculate the SL of this difference. For example, if $\bar{x}_n$ is 2 standard deviation units from 0, then the SL is approximately 0.05, regardless of the true value of the mean.[6] This is the *nominal (or computed) SL.*

A familiar test rule is to reject H_0 whenever the SL reaches some level, say 0.05. This test rule can be described as follows:

$$\text{Test T-1: Reject } H_0 \text{ at } SL = 0.05 \text{ iff } |\bar{X}_n| \geq 2/\sqrt{n}.$$

The standard deviation of $\bar{X}_n$ in this example is $1/\sqrt{n}$. We have

$$P(\text{Test T-1 rejects } H_0; H_0) = 0.05.$$

Since rejecting H_0 when H_0 is true is called the *type I error* of the test, we can also say

$$P(\text{Test T-1 commits a type I error}) = 0.05.$$

EXAMPLE 3 (Case 2: Test T-2 (Sequential testing)). In the second case sample size n is *not* fixed in advance. The stopping rule is:

(T-2) Keep sampling until $\bar{X}_n$ is 2 standard deviations away from 0 (the hypothesized value of μ in H_0) in either direction.

So we have

(T-2) Keep sampling until $|\bar{X}_n| \geq 2/\sqrt{n}$.

The difference between the two cases is that in T-2 the tests are applied *sequentially*. If we have not reached a 2 standard deviation difference after, say, the first 10 trials, we are to go on to take another 10 trials, and so on, as in the "try and try again" procedure of Example 1. The more generalized stopping rule T-2 for the Armitage example is:

$$\text{Keep sampling until } |\bar{X}_n| \geq k_\alpha/\sqrt{n}$$

where k_α is the number of standard deviations away from 0 that corresponds to a (nominal) SL of α. The probability that this rule will stop in a finite number oftrials is *one*, no matter what the true value of μ is; it is what is called a *proper* stopping rule.

[6] That is because we here have a two-sided test.

Table 1. The Effect of Repeated Significance Tests (the "Try and Try Again" Method)

Number of trials n	**Probability of rejecting H_0 with a result nominally significant at the 0.05 level at or before n trials, given H_0 is true**
1	0.05
2	0.083
10	0.193
20	0.238
30	0.280
40	0.303
50	0.320
60	0.334
80	0.357
100	0.375
200	0.425
500	0.487
750	0.512
1000	0.531
Infinity	1.000

Nominal SL vs. Actual SL

The probability that Test T-2 rejects H_0 even though H_0 is true — the probability it commits a type I error — *changes* according to how many sequential tests are run before we are allowed to stop. Because of this, there is a change in the *actual* significance level.

Suppose it takes 1000 trials to reach the 2-standard deviation difference. The SL for a 2-standard deviation difference, in Case 1, where n was fixed, would be 0.05, the *computed* or *nominal significance level.* But the actual probability of rejecting H_0 when it is true increases as n does, and so to calculate the actual SL, we need to calculate:

$$P(\text{Test T-2 stops and rejects } H_0 \text{ at or before } n{=}1000; H_0 \text{ is true}).$$

That is, the actual or overall significance level is the probability of finding a 0.05 *nominally* statistically significant difference from a fixed null hypothesis *at some stopping point or other* up to the point at which one is actually found. In other words, in sequential testing, the *actual* significance level accumulates, a fact reflected in Table 1.

While the *nominal* SL is 0.05, the *actual* SL for Case 2 is about 0.53: 53% of the time H_0 would be rejected even though it is true. More generally, applying stopping rule T-2 would lead to an actual significance level that would differ from, and be greater than, α (unless it stopped at the first trial). If allowed to go on long enough, the probability of such an erroneous rejection is one![7]

By contrast, as Berger and Wolpert note:

> The SRP would imply, [in the Armitage example], that if the observation in Case 2 happened to have $n = k$, then the evidentiary content of the data would be the same as if the data had arisen from the fixed [k] sample size experiment in Case 1 [Berger and Wolpert, 1988, p. 76].

So, in particular, if $n = 1000$, there would be no difference in "the evidentiary content of the data" from the two experiments.

Now holders of the LP do not deny that the actual significance levels differ dramatically, nor do error statisticians deny that alternative hypothesis $\mu = \bar{x}$ is more likely than the null hypothesis $\mu = 0$. Where the disputants *disagree* is with respect to what these facts *mean* for the evidential import of the data. Specifically, the error statistician's concern for the actual and not the nominal significance level in such cases leads her to infer that the stopping rule matters.

In contrast, the fact that the likelihood ratio is unaffected leads the proponent of the LP to infer that there is no difference in the evidential import, notwithstanding the difference in significance levels. Thus, according to the intuitions behind the LP, it is a virtue of a statistical account that it reflects this.

> This irrelevance of stopping rules to statistical inference restores a simplicity and freedom to experimental design that had been lost by classical emphasis on significance levels (in the sense of Neyman and Pearson) ... [Edwards *et al.*, 1963, p. 239].

We can grant some simplicity is lost, but that is because the error probability assurances are lost if one is allowed to change the experiment as one goes along, without reporting the altered significance level. Repeated tests of significance (or sequential trials) are permitted — are even *desirable* — in many situations. However, the error statistician requires that the interpretation of the resulting data reflect the fact that the error characteristics of a sequential test are different from those of a fixed-sample test. In effect, a *penalty* must be paid for perseverance. Before-trial planning stipulates how to select a small enough *nominal* significance level to compute at each trial so that the *actual* significance level is still low.[8] By contrast, since data $\boldsymbol{x}$ enter the Bayesian computation by means of the likelihood function, identical likelihood functions yield identical assignments of posterior probability or density — so no alteration is required with the two stopping rules, according to the LP.

---

[7]Feller [1940] is the first to show this explicitly.

[8]Medical trials, especially, are often deliberately designed as sequential. See [Armitage, 1975].

This leads to the question whether Bayesians are not thereby led into a situation analogous to the one that error statisticians would face were they to ignore the stopping rule.

EXAMPLE 3 (continued). Armitage continued his earlier remarks to Savage at the 1959 forum as follows:

> [Savage] remarked that, using conventional significance tests, if you go on long enough you can be sure of achieving any level of significance; does not the same sort of result happen with Bayesian methods? The departure of the mean by two standard errors corresponds to the ordinary five per cent level. It also corresponds to the null hypothesis being at the five per cent point of the posterior distribution. *Does it not follow that by going on sufficiently long one can be sure of getting the null value arbitrarily far into the tail of the posterior distribution?* ([Armitage, 1962, p. 72]; (emphasis added).

That is, if we consider in Armitage's example the "uninformative" prior distribution of μ, uniform over $(-\infty, +\infty)$ and given that $\sigma^2 = 1$, then the posterior distribution for μ will be:

$$\textit{Normal}\ (\bar{x}_n, 1/n).$$

The methods that Bayesians use to draw inferences about μ all depend on this posterior distribution in one way or another.[9] One common method of Bayesian inference involves using $\bar{x}$ to form an interval of μ values with highest posterior density, the "highest posterior density" (HPD) interval. In this case, the (approximate) 0.95 HPD interval will be $C_n(\bar{x}) = (\bar{x} - 2/\sqrt{n}, \bar{x} + 2/\sqrt{n})$. The Armitage stopping rule allows us to stop only when $|\bar{X}_n| > 2/\sqrt{n}$, and so that stopping rule *insures* that $\mu = 0$ is excluded from the HPD, even if $\mu = 0$ is true.

As even some advocates of the LP note, this looks very troubling for the Bayesian:

> The paradoxical feature of this example is that ... the experimenter can ensure that $C_n(\bar{x})$ *does not contain zero*; thus, as a classical confidence procedure, $\{C_n(\bar{x})\}$ will have zero coverage probability at $[\mu = 0]$ .... It thus seems that the experimenter can, through sneaky choice of the stopping rule, "fool" the Bayesian into believing that $[\mu]$ is not zero [Berger, 1985, p. 507].

That is, (using the non-informative prior density) the use of the stopping rule in T-2 ensures the Bayesian will accord a high posterior probability to an interval that *excludes* the true value of μ. Rather than use the HPD intervals, the analogous

[9]That this uninformative prior results in posteriors that match the values calculated as error probabilities is often touted by Bayesians as a point in *their favor*. For example, where the most an error statistician can say is that a confidence interval estimator contains the true value of μ 95% of the time, the Bayesian, with his uniform prior, can assign .95 posterior probability to the specific interval obtained.

point can be made in reference to Bayesian hypothesis testing.[10] Nor can one just dismiss the issue by noting the obvious fact that the probability for any value of the continuous parameter is zero. Bayesians supply many procedures for inferences about continuous parameters, and the issue at hand arises for each of them. One procedure Bayesians supply is to calculate the posterior probability of a small interval around the null, $(0\text{-}\varepsilon, 0\text{+}\varepsilon)$. With ε small enough, the likelihood is constant in a neighborhood of 0, so the posterior probability obtained from the Armitage stopping rule (T-2) will be very low for $(0\text{-}\varepsilon, 0\text{+}\varepsilon)$, even if $\mu = 0$. And since T-2 is a proper stopping rule, such a low posterior for a true interval around 0 is assured.

In discussions of Armitage's example, most of the focus has been on ways to avoid this very extreme consequence–the guarantee (with probability 1) of arriving at an HPD interval that excludes the true value, $\mu = 0$, or a low posterior density to a true null hypothesis. For example, because the extreme consequence turns on using the (improper) uniform prior, some Bayesians have taken pains to show that this may be avoided with countably additive priors (e.g., [Kadane *et al.*, 1999]).[11]

Nevertheless, the most important consequence of the Armitage example is not so much the extreme cases (where one is guaranteed of strong evidence against the true null) but rather the fact that ignoring stopping rules can lead to a high probability of error, and that this high error probability is not reflected in the interpretation of data according to the LP. Even allowing that the Bayesians have ways to avoid the extreme cases, therefore, these gambits fail to show how to adhere to the LP and avoid a high probabilities of strong evidence against a true null.

To underscore this point, consider a modified version of T-2: the experimenter will make at most 1000 trials, but will stop before then if $\bar{X}_n$ falls more than 2 standard deviations from zero. This modified rule (while also proper) does not assure that when one stops one has $|\bar{X}_n| \geq 2/\sqrt{n}$. Nevertheless, were our experimenter to stop at the 1000th trial, the error probability is high enough (over 0.5) to be disturbing for an error statistician. (See Table 1.) So the error statistician would be troubled by any interpretation of the data that was not altered by dint of this high error probability (due to the stopping rule). Followers of the LP do not regard this stopping rule as altering the interpretation of the data — whatever final form of evidential appraisal or inference they favor. None of the discussions of the Armitage example address this consequence of the less extreme cases.

---

[10]HPDs are not invariant under one-one transformations of the parameter space [Berger, 1985, p. 144]. Some Bayesians find this a compelling reason to avoid HPDs altogether, but this method nevertheless is commonly used.

[11]One might propose that after the first observation, one could use the result to arrive at a new countably additive prior. But this altering of the prior so that the so-called "foregone conclusion" is avoided is not the Armitage example anymore, and so does not cut against that example which concerns an after-trial analysis of the data once one stops.

6 REACTIONS TO THE CONSEQUENCES OF THE LP

For the most part, holders of the LP have not shirked from but have *applauded* the fact that the inferential consequences of the LP conflict with those of error statistical principles. Indeed, those who promote Bayesianism over error statistical approaches often tout the fact that stopping rules (and other aspects of the data generation procedure) do not alter the Bayesian's inference.

At the same time, however, many Bayesians and other holders of the LP are plainly uncomfortable with the fact that the LP can lead to high error probabilities and attempt to deny or mitigate this consequence. We do not think that any existing attempts succeed. Before explaining why, we should emphasize that the consequences of the Armitage-style stopping rule example are not the only ways that adherence to the LP conflicts with the control of error probabilities. Because of this conflict, many have rejected the LP — including some who at first were most sympathetic, most notably Allan Birnbaum, who concluded that

> It seems that the likelihood concept cannot be construed so as to allow useful appraisal, and thereby possible control, of probabilities of erroneous interpretations [Birnbaum, 1969, p. 128].[12]

Therefore, in our view, a strategy to block high probabilities of erroneous interpretations as a result of stopping rules will not do unless it can be demonstrated that:

1. It is part of a complete account that blocks high probabilities of erroneous inferences (whatever the form of inference or evidential appraisal the account licenses.)

2. It is not merely *ad hoc*. There must be a general rationale for the strategy that is also consistent with the LP.

6.1 Can the Stopping Rule Alter the Likelihood Function?

Upon first hearing of the Armitage example, one might assume that the stopping rule T-2 *must* make some kind of difference to the likelihood function. This is especially so for those inclined to dabble informally with likelihoods or Bayes' Theorem apart from any explicit mathematical definition of the likelihood function. We know of no formal statistical treatment of the Armitage example that has seriously claimed that the two stopping rules imply different likelihood functions. (Other types of strategies are proposed, which we will consider.) But these informal intuitions are important, especially for philosophers seeking an adequate account of statistical inference.

[12]See Birnbaum [1961; 1962; 1972], Giere [1977], as well as citations in [Barnard and Godambe, 1982] and [Bjornstad, 1992].

To begin with, it is worth noting that there are *other* kinds of situations in which stopping rules *will* imply different likelihood functions. These are known as *informative stopping rules*, an example of which is given by Edwards, Lindman, and Savage:

> A man who wanted to know how frequently lions watered at a certain pool was chased away by lions before he actually saw any of them watering there; in trying to conclude how many lions do water there he should remember why his observation was interrupted when it was [Edwards *et al.*, 1963, p. 239].

Although a more realistic example might seem more satisfactory, in fact, it is apparently very difficult to find a *realistic* stopping rule that is genuinely informative. (For a discussion, see [Berger and Wolpert, 1988, pp. 88–90].) As Edwards, *et al.*, then add: "We would not give a facetious example had we been able to think of a serious one." In any event, this issue is irrelevant for the Armitage-type example because T-2 is not an informative stopping rule. Although the probability of deciding to take more observations at each stage depends on $\boldsymbol{x}$, it does *not* depend on the parameter μ under test.[13]

Nevertheless, we are willing to address those who assume that an informal or subjectivist construal of probabilities gives them a legitimate way to alter the likelihood based on the stopping rule T-2. But to address them we need more than their intuitive hunch, they need to tell us in general how we are to calculate the likelihoods that will be needed, whether the account is purely likelihoodist (e.g., Royall [1992; 1997]) or Bayesian. Are we to substitute error probabilities in for likelihoods? Which ones? And how will this escape the Bayesian incoherence to which error probabilities such as significance levels are shown to lead?

To see that any such suggested alteration of likelihoods runs afoul of the LP, it must be remembered that the likelihood is a function of the observed $\boldsymbol{x}$:

> The philosophical incompatibility of the LP and the frequentist viewpoint is clear, since the LP deals only with the observed $\boldsymbol{x}$, while frequentist analyses involve averages over possible observations. ... enough direct conflicts have been ... seen to justify viewing the LP as revolutionary from a frequentist perspective [Berger and Wolpert, 1988, pp. 65–66].

Once the data $\boldsymbol{x}$ are in hand, the holder of the LP insists on the "irrelevance of the sample space" — the irrelevance of the other outcomes that *could* have occurred but did not when drawing inferences from $\boldsymbol{x}$ (e.g., [Royall, 1997]). This is often

[13]As Berger and Wolpert [1988, p. 90] observe, the mere fact that the likelihood function depends on N, the number of observations until stopping, does *not* imply that the stopping rule is informative: "Very often N will carry information about [the parameter], but to be informative a stopping rule must carry information about [the parameter] additional to that available in [the sample $\boldsymbol{X}$], and this last will be rare in practice" (*ibid.*, 90). For further discussion of informative stopping rules, see [Roberts, 1967].

expressed by saying the holder of the LP is a *conditionalist*: for them inferences are always conditional on the *actual* value $\boldsymbol{x}$.

With respect to stopping rules, the conditionalist asks: Why should our interpretation of the data in front of us, $\boldsymbol{x}$, depend upon what would have happened *if* the trials were stopped earlier than they *actually* were stopped?

> Those who do not accept the likelihood principle believe that the probabilities of sequences that might have occurred, but did not, somehow affect the import of the sequence that did occur [Edwards *et al.*, 1963, p. 238].

But altering the likelihood because of the stopping rule *is* to take into account the stopping plan, e.g., that if he hadn't gotten a significant result at 10 trials, he *would have* continued, and so on, thereby violating the LP. So anyone who thinks a subjectivist or informal construal of likelihoods gives them a legitimate way out, must be aware of this conflict with the conditionalist principle. Certainly this would put them at odds with leading subjective Bayesians who condemn error statisticians for just such a conflict:

> A significance test inference, therefore, depends not only on the outcome that a trial produced, but also on the outcomes that it could have produced but did not. And the latter are determined by certain private intentions of the experimenter, embodying his stopping rule. It seems to us that this fact precludes a significance test delivering any kind of judgment about empirical support ... For scientists would not normally regard such personal intentions as proper influences on the support which data give to a hypothesis [Howson and Urbach, 1993, p. 212].

Thus, the intuition that the stopping rule should somehow alter the likelihood is at odds with the most well-entrenched subjective Bayesian position and constitutes a shift toward the error statistical (or "frequentist") camp and away from the central philosophy of evidence behind the LP. According to the LP philosophy:

> [I]t seems very strange that a frequentist could not analyze a given set of data, such as $(x_1,\ldots, x_n)$ [in Armitage's example] if the stopping rule is not given. ... data should be able to speak for itself [Berger and Wolpert, 1988, p. 78].

We say the shift is to the error statistical camp because it reflects agreement with the error statistician's position that one cannot properly 'hear' what the data are saying without knowing how they were generated — whenever that information alters the capabilities of the test to probe errors of interest, as in the case of stopping rules. It is precisely in order to have a place to record such information that Neyman and Pearson were led to go beyond the likelihood ratio (LR):

> If we accept the criterion suggested by the method of likelihood it is still necessary to determine its sampling distribution in order to control the error involved in rejecting a true hypothesis, because a knowledge of [the LR] alone is not adequate to insure control of this error [Pearson and Neyman, 1930, p. 106].

When test T-2 stops, it is true that the LR (in favor of H_0) is small. However, to the error statistician, we cannot thereby infer we should be *justified* in rejecting the hypothesis H_0, because:

> In order to fix a limit between 'small' and 'large' values of [LR] we must know how often such values appear when we deal with a true hypothesis. That is to say we must have knowledge of ... the chance of obtaining [LR as small or smaller than the one observed] in the case where [H_0] is true (*ibid*, p. 106).

Accordingly, without the error probability assessment, Pearson and Neyman are saying we cannot determine if there really is any warranted evidence against H_0.[14] Stopping rules give crucial information for such an error statistical calculation.

It is no surprise, then, that the error statistician regards examples like Armitage's as grounds for rejecting the LP. To those who share the error statistical intuitions, our question is: on what grounds can they then defend the LP?

6.2 *Can Stopping Rules Alter the Prior?*

In order to avoid assigning the high posterior to a false non-null hypothesis, as Berger and Wolpert (1988) point out, "the Bayesian might ... assign some positive prior probability, λ, to μ being equal to zero" (p. 81) perhaps to reflect a suspicion that the agent is using stopping rule T-2 because he thinks the null hypothesis is true.[15] Assume, for example, that one assigns a prior probability mass of 0.50 to the null hypothesis and distributes the rest Normally over the remaining values of

[14]It should be emphasized that it is not that the N-P inference consists merely of a report of the significance level (or other error probabilities), at least not if the tests are being used for inference or evidence. It is rather that determining the warranted inference depends on the actual significance level and other error probabilities of tests. Granted, the onus is on the error-statistician to defend a philosophy of inference that uses and depends on controlling error probabilities (though this is not our concern here). See Note 22.

[15]A positive prior probability, λ, can be assigned to $\mu = 0$ and the rest, $1 - \lambda$, distributed over the remaining values of μ. (This amounts to giving $\mu = 0$ a non-zero mass, and every other hypothesis zero mass.) When $1 - \lambda$ is distributed Normally over the remaining hypotheses with mean 0 and variance ρ^2, the posterior probability distribution will be:

$$P\left(\mu = 0/\bar{x}_n = \frac{K}{\sqrt{n}}\right) = \left[1 + \left(\frac{1}{\lambda} - 1\right)\frac{1}{\sqrt{(1+n\rho^2)}}e^{\frac{K^2 n\rho^2}{2(1+n\rho^2)}}\right]^{-1}$$

where K is the number of standard deviations stipulated in the stopping rule and n is the number of observations needed to stop [Berger and Wolpert, 1988, p. 81].

See also [Berger and Berry, 1987], [Smith, 1961, p. 36–37].

μ. If it takes $n = 1000$ trials to stop, the posterior probability assignment to $\mu = 0$ is no longer low, but rather, around 0.37. A virtue of such a prior, as Berger and Wolpert note, is that it results in an increasing posterior probability assignment to $\mu = 0$ as the number of trials before stopping increases. For example, with this prior and $n = 10,000$, the posterior for the null is about 0.65.

Granted, in a case where one *had* this prior, the low posterior assignment to the null hypothesis is avoided, but this does nothing to mitigate the problem as it arises with the *uniform* prior — a prior the Bayesian often advocates. Perhaps the Bayesian would wish to suggest that *whenever* one is confronted with an experiment with stopping rule T-2, one should reject the uniform prior in favor of one that appears to avoid the problem. But why should a Bayesian alter the prior upon learning of the stopping rule?

There is the motivation suggested by Berger and Wolpert [1988], that if you *suspected* that the person generating the observations was using stopping-rule T-2 *for the purpose of misleading you*, you would raise your prior probability assignment to $\mu = 0$. Does this not violate the LP? Perhaps one could retain the LP on the grounds that one is only allowing the stopping-rule to affect the *prior* rather than the likelihoods (and hence not "*what the data say*").[16]

Nevertheless, a Bayesian should have serious objections to this response to the stopping rule problem. Why, after all, should we think that the experimenter is using T-2 to *deceive* you? Why not regard his determination to demonstrate evidence against the null hypothesis as a sign that the null is *false*? Perhaps he is using T-2 only because he *knows* that $\mu \neq 0$ and he is trying to *convince* you of the truth!

Surely it would be unfair to suppose that those who, like Savage, touted the irrelevance of the stopping rule were sanctioning *deception* when they asserted: "Many experimenters would like to feel free to collect data until they have either conclusively proved their point, [or] conclusively disproved it" [Edwards *et al.*, 1963, p. 239]. Plainly, what they *meant* to be saying is that there is no reason to interpret the data differently because they arose from optional stopping. Equating optional stopping (with rule T-2) with deception runs counter to Savage's insistence that, because "optional stopping is no sin," any measure that *is* altered by the stopping rule, such as the significance level, is thereby inappropriate for assessing evidence [Savage, 1964, p. 185].[17]

Those who advocate the above move, then, should ask why the sensitivity of significance levels to stopping rules violates the LP — and thus is a *bad* thing — but the *same* kind of sensitivity of priors is acceptable. The LP, after all, asserts that *all* the information contained in the data that is relevant to comparisons between different parameter values is given in the likelihood function. But what else could

[16]This 'solution' demands that the agent know not only the *stopping*-rule used, but *why* the experimenter chose that particular stopping-rule, since knowing he wanted to *deceive* rather than to *help* you could make all the difference in the prior you use. Yet Bayesians have delighted in the fact that the LP renders irrelevant the *intentions* of experimenters to the import of the experiment.

[17]There is nothing in the LP to prevent Bayesians from deciding in advance to prohibit certain kinds of stopping rules, but again, one would like to know why.

it mean to say that one's choice of *priors* depends on the stopping-rule other than that the stopping-rule contains information relevant to comparisons between values of μ? It is little wonder that many Bayesians have balked at allowing the stopping rule to alter one's prior probability: "Why should one's knowledge, or ignorance, of a quantity depend on the experiment being used to determine it" [Lindley, 1972, p. 71]. Why indeed?[18]

Finally, even if we put aside the question of stopping rules leading to problematic final posterior probabilities, as long as the Bayesian conceives of *likelihood* as determining "what the data have to say", it is *still* the case that the data from T-2 are regarded as much stronger support for the non-null than the null, according to the Bayesian criterion of support.[19]

6.3 *Does the LP Provide Bounds on Being Misled?*

A third kind of response *grants* that the stopping rule makes no difference at all to either the likelihood function or the priors, and instead attempts to argue that, nonetheless, one who holds the LP can avoid having a high probability of being misled by the data. This argument is sound only for tests that differ in essential ways from the type leading to the Armitage result. Nevertheless, this response is important, if only because it is the one first put forward by Savage in responding to Armitage (Savage 1962).[20]

"Let us examine first a simple case" Savage proposes, where we are testing a simple or point null hypothesis H_0 against a point alternative H_1: that is H_0 asserts $\mu = \mu_0$, and the alternative H_1 asserts $\mu = \mu_1$. Call this a *point against point* test. In that case, if one is intent on sampling until the likelihood ratio (*LR*) in favor of H_1 exceeds r (for any value of $r > 1$), it can be shown that if H_0 is true, the probability is only $1/r$ that one will succeed in stopping the trials. This response turns on the fact that when we have a (true) simple null H_0 against a simple alternative H_1, then there is an upper bound to the probability of obtaining a result that makes $H_1 r$ times more likely than H_0, namely, $1/r$, i.e. $\mathrm{P}(LR > r; H_0) \leq 1/r$.

[18]Lindley is referring to Bayesians like Jeffreys [1961] and Rosenkrantz [1977] who determine 'objective' or 'non-subjective' priors by appealing to formal information-theoretic criteria. They would, for example, recommend different priors in the Binomial vs. the Negative Binomial case [Box and Tiao, 1973]. Doing so apparently violates the LP, and has led many Bayesians to be suspicious of such priors [Hill, 1987; Seidenfeld, 1979], or even to declare that "no theory which incorporates non-subjective priors can truly be called Bayesian, and no amount of wishful thinking can alter this reality" (Dawid, in [Bernardo, 1997, p. 179]). For related discussions contrasting subjective and objective priors, see also [Akaike, 1982; Barnett, 1982; Bernardo, 1979; Bernardo, 1997].

[19]This point does not rely on the technical Bayesian definition of "support" as an increase in the posterior, but holds for any conception based on the likelihood, *e.g.*, weight of evidence [Good, 1983]. Bayesians who reject all such notions of Bayesian support need to tell us what notion of support or evidence they condone.

[20]It is also the first one mentioned by many defenders of the LP, e.g., [Berger and Wolpert, 1988; Oakes, 1986; Royall, 1997].

This impressively small upper bound, however, does nothing to ameliorate the consequences of the Armitage optional stopping example because that example is not a case of a point against point test.[21]

6.4 *Extrapolating From Our Intuitions in Simple Cases*

The simple case of testing "point against point" hypotheses has encouraged some to suppose that the LP offers such protection in *all* cases — yet it does not. Perhaps the tendency to turn to the point against point test when confronted with stopping rule problems explains why the Armitage-type consequence has not received more attention by Bayesians. But there seems to be a different kind of strategy often at work in alluding to the point against point test in defending the LP, and we may regard this as a distinct response to the stopping rule problem.

In appraising the LP, say some, we should trust our intuitions about its plausibility when we focus on certain *simple* kinds of situations, such as testing point against point hypotheses, "rather than in extremely complex situations such as [Armitage's example]" [Berger and Wolpert, 1988, p. 83]. Since looking at just the likelihood ratio (and ignoring the stopping rule) seems intuitively plausible in point against point testing, they urge, it stands to reason that the LP must be adhered to in the more 'complex situation' — even if its consequences in the latter case seem unpalatable. Regarded as an argument for deflecting the Armitage example it is clearly unsound. Bracketing a whole class of counterexamples simply on the basis that they are "extremely complicated" is *ad hoc* — preventing the LP from being subject to the relevant kind of test here. Moreover, such sequential tests are hardly exotic, being standard in medicine and elsewhere.

But perhaps it is only intended as a kind of pragmatic appeal to what is imagined to be the lesser of two evils: their reasoning seems to be that even if the LP leads to unintuitive consequences in the complex (optional stopping) case, its rejection would be so unappealing in the simple cases that it is better to uphold the LP and instead discount our intuitions in the complex cases. By contrast, some have gone the route of George Barnard — the statistician credited with first articulating the LP [Barnard, 1949] — who confessed at the 1959 Savage Forum that the Armitage-type example led him to conclude that whereas the LP is fine for the simple cases it must be abandoned in the more complex ones (see [Barnard, 1962]). The LP adherent owes us an argument as to why Barnard's move should not be preferred.

---

[21]The existence of an upper bound less than 1 can also be shown in more general cases such as when deling with k simple hypotheses, though as k increases, the upper bound is no longer impressively small. The general result, stated in [Kerridge, 1963] is that with $k+1$ simple hypothses where H_0 is true and $H_1, \ldots, H_k$ are false and $Pr(H_i) = (k+1)^{-1}$ for $i = 0, 1, \ldots, k$:

$$P(P(H_0/X_n) \leq p) \leq \frac{kp}{(1-p)}.$$

Moreover, such bounds depend on having countably addititve probability, while the uniform prior in Armitage's example imposes finite additivity.

7 CONCLUDING REMARKS

Philosophers who appeal to principles and methods from statistical theory in tackling problems in philosophy of science need to recognize the *consequences* of the statistical theory they endorse. Nowhere is this more crucial than in the on-going debate between Bayesian and non-Bayesian approaches to scientific reasoning. Since Bayesianism — which is committed to the LP — has emerged as the dominant view of scientific inference among philosophers of science, it becomes all the more important to be aware of the LP's many implications regarding evidence, inference and methodology.

Some of the most important of these implications concern the LP's effect on our ability to control error and thereby the reliability and severity of our inferences and tests — generally regarded as important goals of science. A consequence of our discussion is that there is no obvious way in which approaches consistent with the LP can deliver these goods. In giving the spotlight to the kind of unreliability that can result from ignoring *stopping rules*, our goal is really to highlight some of the consequences for reliability of accepting the LP, not to argue that examples such as Armitage's are common. At the same time, however, it should be realized that examining the effect of stopping rules is just *one* of the ways that facts about how the data are generated can affect error probabilities. Embracing the LP is at odds with the goal of distinguishing the import of data on grounds of the error statistical characteristics of the procedure that generated them.

Now Bayesians and likelihoodists may deny that this appeal to error probabilities is what matters in assessing data for inference. They often deny, for example, that the error statistician's concern with the behavior of a test in a series of repetitions is relevant for inference.[22] Strict adherence to this position would lead one to expect that they would be unfazed by the Armitage result. In reality, however, existing Bayesian and Likelihoodist reactions to Armitage-type examples are strikingly and surprisingly equivocal, and the Bayesian attempts to deflect the Armitage result have been unclear e.g. see [Johnstone *et al.*, 1986]. Sometimes they say "It's not a problem, we do not care about error rates", while at other times the claim is "Even though we don't care about error rates, we can still satisfy one who does." The former response is consistent for a holdr of the LP, but it demands renouncing error probabilities, as we understand that notion. The latter attitude demands an argument showing how to resolve the apparent tension with the LP. We have tried to locate the most coherent and consistent arguments, and found that

[22]The long-standing challenge of how to interpret error statistical tests "evidentially" cannot be delved into here, but we can see the directions in which such an interpretation (or reinterpretation) might take us, by extending what we said about why the error statistician regards the stopping rule as relevant. The error statistician regards data as evidence for a hypothesis H to the extent that H has passed a reliable or severe test of H, and this requires not just that H fit x but also that test T would very probably not have resulted in so good a fit with H were H false or specifiably in error. See [Mayo, 2000], [Mayo and Spanos, 2000]. By contrast, the Armitage stopping rule makes it maximally probable that $\mathbf{x}$ fits a false H, so H passes a test with *minimal* severity.

they failed to live up to this demand. We invite anyone who can further clarify the Bayesian and Likelihoodist position on the Armitage example to do so.

ACKNOWLEDGEMENTS

We are indebted to Aris Spanos for numerous, highly important statistical insights regarding the Armitage case. We thank Teddy Seidenfeld, and the participants of D. Mayo's 1999 National Endowment for the Humanities Summer Seminar, for a number of challenging questions, criticisms, and suggestions regarding earlier drafts. D. Mayo gratefully acknowledges support for this research from the National Science Foundation, grant no. SBR-9731505.

Virginia Tech, USA.

BIBLIOGRAPHY

[Akaike, 1982] H. Akaike. On the fallacy of the likelihood principle. *Statistics and Probability Letters* **1**, 75–78, 1982.

[Armitage, 1962] P. Armitage. Contribution to discussion in L. Savage, ed. 1962.

[Armitage, 1975] P. Armitage. *Sequential Medical Trials*. Oxford: Blackwell, 1975.

[Barnard, 1949] G. A. Barnard. Statistical inference. *Journal of the Royal Statistical Society*, Series B (Methodologial), **11**, 115–149, 1949.

[Barnard, 1962] G. A. Barnard. Contribution to discussion in L. Savage, ed. 1962.

[Barnard and Godambe, 1982] G. A. Barnard and V. P. Godambe. Memorial article: Allan Birnbaum 1923–1976. *The Annals of Statistics* **10**, 1033–1039, 1982.

[Barnett, 1982] V. Barnett. *Comparative Statistical Inference*, 2nd edition. John Wiley, New York 1982.

[Berger, 1985] J. O. Berger. *Statistical Decision Theory and Bayesian Analysis*. 2nd edition. Springer-Verlag, New York, 1985.

[Berger and Berry, 1987] J. O. Berger and D. A. Berry. The relevance of stopping rules in statistical inference. In *Statistical Decision Theory and Related Topics IV*, vol. 1, S. S. Gupta and J. Berger, eds. Springer-Verlag, 1987.

[Berger and Wolpert, 1988] J. O. Berger and R. L. Wolpert. *The Likelihood Principle*, 2nd edition. Institute of Mathematical Statistics, Hayward, CA, 1988.

[Bernardo, 1979] J. M. Bernardo. Reference posterior distributions for Bayesian inference (with discussion). *Journal of the Royal Statistical Society*, series B:41, 113–147, 1979.

[Bernardo, 1997] J. M. Bernardo. Noninformative priors do not exist: A discussion with José M. Bernardo (with discussion). *Journal of Statistical Planning and Inference* **65**, 159–189, 1997.

[Birnbaum, 1961] A. Birnbaum. On the foundations of statistical inference: binary experiments. *Annals of Mathematical Statistics* **32**, 414–435, 1961.

[Birnbaum, 1962] A. Birnbaum. On the foundations of statistical inference. *Journal of the American Statistical Association*, **57**, 269–306, 1962.

[Birnbaum, 1969] A. Birnbaum. Concepts of statistical evidence. In *Essays in Honor of Ernest Nagel*, Sidney Morgenbesser, Patrick Suppes and Morton White, eds. St. Martin's Press, 1969.

[Birnbaum, 1972] A. Birnbaum. More on concepts of statistical evidence. *Journal of the American Statistical Association*. **67**, 858–861, 1972.

[Bjornstad, 1992] J. F. Bjornstad. Birnbaum (1962) on the foundations of statistical inference. In *Breakthroughs in Statistics*, vol. 1, 461–477. Samuel Kotz and Norman L. Johnson, eds. Springer-Verlag, New York, 1992.

[Box and Tiao, 1973] G. Box and G. Tiao. *Bayesian Inference in Statistical Analysis*. Addison-Wesley, Reading, MA, 1973.

[Edwards, 1992] A. W. F. Edwards. *Likelihood* (2nd edition). Cambridge University Press, 1992.

[Edwards *et al.*, 1963] W. Edwards, H. Lindman and L. J. Savage. Bayesian statistical infe psychological research. *Psychological Review* **70**, 450–499, 1963.

[Feller, 1940] W. K. Feller. Statistical aspects of ESP. *Journal of Parapsychology* **4**, 271–29

[Giere, 1977] R. N. Giere. Alan Birnbaum's conception of statistical evidence. *Synthese*, **36**, 5–13, 1977.

[Gillies, 1990] D. A. Gillies. Bayesianism versus falsificationism. *Ratio*, **3**, 82–98, 1990.

[Good, 1983] I. J. Good. *Good Thinking*. University of Minnesota Press, Minneapolis, MN, 1983.

[Hill, 1987] B. M. Hill. The validity of the likelihood principle. *The American Statistician*, **47**, 95–100, 1987.

[Howson and Urbach, 1993] C. Howson and P. Urbach. *Scientific Reasoning: The Bayesian Approach*, second edition. Open Court, Chicago, 1993.

[Jeffreys, 1961] H. Jeffreys. *Theory of Probability*, 3rd edition. Clarendon Press, Oxford, 1961.

[Johnstone *et al.*, 1986] D. J. Johnstone, G. A. Barnard and D. V. Lindley. Tests of significance in theory and practice. *The American Statistician*, **35**, 491–504, 1986.

[Kadane *et al.*, 1999] J. B. Kadane, M.J. Schervish and T. Seidenfeld. *Rethinking the Foundations of Statistics*. Cambridge University Press, Cambridge, 1999.

[Kerridge, 1963] D. Kerridge. Bounds for the frequency of misleading Bayes' inferences. *Annals of Mathematical Statistics* **34**, 1109–1110, 1963.

[Lindley, 1972] D. V. Lindley. *Bayesian Statistics — A Review*. J. W. Arrowsmith, Bristol, 1972.

[Mayo, 1996] D. Mayo. *Error and the Growth of Experimental Knowledge*. University of Chicago Press, Chicago, 1996.

[Mayo, 2000] D. Mayo. experimental practice and an error statistical account of evidence. *Philosophy of Science*, **67**, (Proceedings), S193–S207, 2000.

[Mayo and Spanos, 2000] D. Mayo and A. Spanos. A Post-data Interpretation of Neyman-Pearson Methods Based on a Conception of Severe Testing. *Measurements in Physics and Economics Discussion Paper Series*, DP MEAS 8/00. Centre for Philosophy of Natural & Social Science, London School of Economics, 2000.

[Oakes, 1986] M. Oakes. *Statistical Inference*, Wiley, 1986.

[Pearson and Neyman, 1930] E. S. Pearson and J. Neyman. On the problem of two samples. *Bull. Acad. Pol. Sci.*, 73–96, 1930. Reprinted in J. Neyman and E. S. Pearson, *Joint Statistical Papers*. pp. 81–106 University of California Press, Berkeley, 1967.

[Pratt *et al.*, 1995] J. W. Pratt, H. Raffia and R. Schlaifer. *Introduction to Statistical Decision Theory*. The MIT Press, Cambridge, MA, 1995.

[Roberts, 1967] H. V. Roberts. Informative stopping rules and inferences about population size. *Journal of the American Statistical Association*. **62**, 763–775, 1967.

[Rosenkrantz, 1977] R. D. Rosenkrantz. *Inference, Method, and Decision: Towards a Bayesian Philosophy of Science*. Boston: Reidel, 1977.

[Royall, 1992] R. Royall. The elusive concept of statistical evidence (with discussion). In Bayesian Statistics 4, J.M. Bernardo, J.O. Berger, A.P. Dawid and A.F.M. Smith, eds. pp. 405–418. Oxford University Press, Oxford, 1992.

[Royall, 1997] R. Royall. *Statistical Evidence: A Likelihood Paradigm*. Chapman & Hall, London, 1997

[Savage, 1962] L. J. Savage. *The Foundations of Statistical Inference: A Discussion*. Methuen, London, 1962.

[Savage, 1964] L. J. Savage. The foundations of statistics reconsidered. In *Studies in Subjective Probability*, H. Kyberg and H. Smokler, eds. John Wiley, New York, 1964.

[Seidenfeld, 1979] T. Seidenfeld. Why I am not an objective Bayesian. *Theory and Decision* **11**, 413–440, 1979.

[Smith, 1961] C. A. B. Smith. Consistency in statistical inference and decision (with discussion). *Journal of the Royal Statistical Society*, (B), Vol. 23, No. 1, 1–37, 1961.

INDEX

χ^2 test, 363, 365, 367

abstract structures, 77
accidentally correlated, 79
ACE, 53–55
action, 31, 32
actual signicance, 391
actual SL, 390
Aczel, P., 150
Adams principle, 156
Adams, E. W., 150, 156
adding-arrows, 106
adhockery, 358
affinely independent, 319
Akaike, H., 399
Albert, M., 12, 157, 343, 354, 355, 376, 378
Allais Paradox, 342
Allais, M., 296, 301, 306
ambiguity, 62, 65–68, 72
analogy, 192
analysis of variance, 43
ancestrally, 101
Anscombe, F. G., 310, 311, 324, 342, 347
anything goes theorem, 354–357, 359, 376
Armitage, P., 384, 388, 391–393, 399
Arrow, K., 344
Ars Conjectandi, 137
artificial intelligence, 117
as-if, 359
as-if approach, 346
atomic states, 107
Aumann, R. J., 310, 311, 324, 341, 342, 344, 347
average causal effect (ACE), 44
avoid mistakes, 359
axioms, 342, 347, 359, 360
  of revealed preference, 344
axioms of RP, 346
axioms, Bayesian, 342

background knowledge, 97
baker-map dynamics, 349, 350
Barnard, G., 400
Barnett, V., 399
Bayes' Memoir, 137
Bayes' theorem, 1, 63, 381, 385
Bayes, T., 1, 363, 378
Bayesian, 47, 60, 381, 384
Bayesian account, 382
Bayesian conditionalisation, 1, 363, 364, 367–369, 371, 377
Bayesian decision analysis, 47
Bayesian decision theory, 68
Bayesian learning, 347, 354, 355
Bayesian network, 7, 11, 76
Bayesian Networks Maximise Entropy, 100
Bayesian rationality, 359
Bayesian theory of utility and subjective probability, 291, 306
Beall, G., 365, 366, 368
Begg, D. K. H., 341
behaviourism, 265
belief formation, 203
Berger, J. O., 385, 391, 392, 395–400
Bernardo, J. M., 399
Bernoulli, 29
Bernoulli trials, 386
Bernoulli's logarithmic function, 142
Bernoulli, D., 142
Bernoulli, J., 137
Bernstein, P. L., 343

betting quotient, 140, 369
Bicchieri, C., 344
Binmore, K., 342
binomial distribution, 387
Birnbaum, A., 394
blind source separation, 131
Blume, L. E., 342, 344, 354, 355
Bolzano, B., 138
Border, K. C., 347
bounded-rationality, 358, 359
Box, G., 399
Bradley, R., 11
Bray, M., 344
Broome, J., 296, 306
Brémaud, P., 351, 354

calibration, 1, 7
Carnap's probability$_1$, 138
Carnap's probability$_2$, 138
Carnap, R., 138, 145, 153, 345
catch all, 377
causal
  analysis, 24
  assumptions, 22, 23, 25
  effect, 34, 43
  extension, 79
  graph, 32
  inference, 37, 55, 58, 68–70
  irrelevance, 99
  judgment, 31
  Markov condition, 7, 75, 76
  mechanisms, 29
  model, 6, 31, 66
  notation, 23
  restriction, 81
  sentences, 31
  vocabulary, 31
causal and statistical concepts, 20
causality reducible to probabilites, 27
causality-definition, 27, 28
causes of effects, 40, 59–61, 64, 66, 68, 71
chaos, 355
chaotic, 343, 355
chaotic clock, 349–353, 356, 357, 376
Church–Turing Thesis, 149
classical statistics, 363–366, 368, 372, 378
coherence, 157, 359, 360
coherence versus strict coherence, 153
Colhon, T., 319
collapsibility, 21
commodity complementarity, 296
common prior assumption (CPA), 341, 342, 344
common sense, 203, 205
common sense principles, 203
compatibility, 57
complementary, 42, 47, 56, 63, 66
complementary outcomes, 64
complementary potential responses, 68
complementary quantities, 71
complete, 34
completeness, 152
composition, 34
computed significance level, 390
concavity, 205
concept definition, 29
concomitant, 60–64, 66, 67, 69, 70, 72
conditional independence, 62–66, 71
conditional mutual information, 110
conditional prospects, 270
conditionalism, 283
conditionalist, 396
Condorcet, 142
confidence interval estimation, 382
confounding, 24
conjecture and refutation, 364, 366, 378
consequence set, 311
consistency, 34, 143, 360
constrained network, 110
context, 69–72
contingency tables, 22
continuity principle, 207, 212
convergence, 342, 344, 351–355

Corfield, D., 10
correlation restriction, 81
countable additivity, 152
counterfactual, 6, 7, 28, 29, 31, 33, 37, 38, 55, 58, 59, 63, 66, 70, 71
  approach, 51, 53
  inferences, 68
  model, 58
  models, 68
  query, 70
  questions, 71
  universe, 69–71
  variables, 30
Courtney, P., 208
covariant, 51, 52
covariate, 42, 65
Cox's theorem, 178
Cox, D. R., 374, 378
CPA, 344, 345
cross entropy, 101, 106, 107
Cussens, J., 11

DAG, 28, 30, 76
Davies, P., 350
Dawid, P., 7, 364, 378, 399
decision analysis, 50
decision making, 342, 357, 360
decision theory, 11, 263, 343, 344
decision tree, 30, 47
decision-analytic approach, 47, 51, 53
deductive logic, 157, 357, 359, 360
degrees of belief, 203
dependence, 93
determining concomitant, 65–68
determinism, 65, 67, 70
deterministic model, 66, 67, 69
deterministic structure, 68
Devaney, R. L., 350
diagrams, 24
direct and indirect effects, 24
direct method, 76
directed acyclic graph (DAG), 28
disjunctive complementarity, 301, 302, 306
disturbance, 25
divisibility axiom, 310
do-calculus, 26–28
domain expert, 366, 368
dominance principle, 345
dominated strategy, 346
Drèze, J., 310–312
Drèze, K., 316
Dutch Book argument, 1, 9, 13, 140, 178, 343

Earman, J., 188, 344, 352, 355
Easley, D., 342, 344, 354, 355
economic rationality, 346
economics, 342–345, 353, 358
Edwards, A. W. F., 386
Edwards, W., 381, 383, 391, 395, 396, 398
effect of action, 33
effectiveness, 33
effects of causes, 40, 41, 60, 71
Ellsberg's counterexamples to indpendence, 299–302
Ellsberg, D., 299–302, 306
empirical Bayesianism, 2
empirical content, 343, 344, 357
empirically adequate, 353, 355
emptiness, 343
empty, 343–345, 354, 357
endogenous, 32
ensemble learning, 125
equivalence principle, 206, 210
error probabilities, 383
error statistical methods, 381, 383, 384
ethically neutral proposition, 270
evidential-relationship (E-R) principles, 382, 383
exchangeability, 2, 22, 42, 52, 58, 364, 369, 371–374
existence, 78
exogenous, 25, 31

expectation, 342, 343, 356, 357
expectations formation, 342
extended inference processes, 209
extended knowledge base, 209
extended maximum entropy process, 219
extra-causal constraint, 79
(extrinsic) confounding, 56

factor-analysis, 63
Fagin, R., 151
Fallis, D., 183
falsification, 366, 368
falsification dynamics, 351
falsifies, 351
fatalism, 40, 49, 50
fatalistic, 54, 65
Feller, W. K., 373, 378, 391
Finetti, B. de, 1, 9, 137, 179, 309, 364, 368, 370–375, 378
Fishburn, P., 291, 306
Fisher, , 54
fixity of the theoretical framework, 364, 367, 368, 375, 378
folk theorem, 343, 344
Ford, J., 350
forecast function, 349, 354
free will, 70
Friedman, M., 292, 294, 306, 346
Frydman, R., 341
functional model, 58
functional relationships, 68

Gabbay, D. M., 11, 150
Gaifman, H., 151
Galavotti, M. C., 9, 371, 378
game theory, 344
Gaussian process, 126
generative model, 131
generative topographic mapping, 130
geostatistics, 126
Gillies, D. A., 13, 14, 355, 363, 378
goat, 53–55, 58
Goldman, A. I., 343
Good, I. J., 399
Goodman, N., 353
Gosset, W. S., 363
graphical model, 58, 68
Green, E. J., 347

Hacking, I., 150, 180, 341, 345
Hagen, O., 296, 306
Hahn, F., 344
Halmos, P., 144
Halpern, J. Y., 151
Harvard Medical School Test, 157
Heifetz, A., 151
Hellman, G., 141
hidden variables, 57
Hill, B. M., 399
Hodges, W., 148
homogeneity, 42, 50
Howson, C., 9, 10, 13, 138, 144, 156, 344, 345, 360, 396
HPD, 393
Hume, D., 156, 182
hypothesis testing, 363
hypothetical act, 315
hypothetical preference, 315

ICE, 53
imperfect treatment compliance, 58
(improper) uniform prior, 393
improvement, 106
Improvement of Adding Arrows, 106
independence and sure-thing principles, logical connections between, 293–295
independence for sure outcomes (ISO), 291
independence principle (IND), 12, 207, 212, 291, 297–299, 306
independent component analysis, 131
individual causal effect (ICE), 43, 44, 59, 66
inductive inference, 71, 156
inductive logic programming, 11
inference process, 203, 205

informative stopping rules, 395
instrumental, 55–57
instrumental variables, 21
interpreted, 77
interval of ambiguity, 62, 63
interventions, 26, 31
intrinsic aliasing, 45
intrinsically confounded, 56
introspection, 268
invariance, 26
irrational, 343, 354
irrational behaviour, 343
irrelevant information principle, 206, 209

Jaynes, E., 176, 177, 186
Jeffrey's Law, 45, 53, 68
Jeffrey, R., 286
joint probabilites of counterfactuals, 35
Jordan, J. S., 343

Kac, M., 139
Kadane, J. B., 311, 326, 393
Kahneman, D., 297, 302, 304, 306
Karni, E., 310, 311, 316, 327
Keynes, J. M., 138, 344, 345
Kiefer, N. M., 342
Kim, T., 347
Kirman, A., 341
knowledge elicitation problem, 77
Kolmogorov, A. N., 151
Krauss, P., 151
Kruse, M., 14

labelled deductive systems, 11
Lakatos, I., 158, 378
language invariance, 206, 210
Laplace's Rule of Succession, 196, 371, 374
Laplace, P. S., 1, 344, 363, 378
Larsson, S., 297
Latin square, 54, 56, 57
Leamer, E. E., 344
learning, 341, 342, 345, 347, 349, 352, 357
Ledyard, J. O., 344
Leibniz, 137
Lewis, D., 156
likelihood, 383
likelihood function, 385
likelihood principle, 14, 381–387, 393, 398
likelihood ratio (LR), 396
likelihoodist account, 382
Lindley, D. V., 345, 399
Lindman, H., 381, 383, 391, 395, 396, 398
line, 220
linear knowledge bases, 203
linear probabilistic constraints, 203
literal, 76
logical Bayesianism, 2
logical consistency, 359, 360
logical positivism, 345
logically omniscient, 10, 177, 352, 359
Loomes, G., 296, 306
Lucas, R. E., 344

MacCrimmon, K. R., 297, 306
machine learning, 8
macroeconomic, 341
Manne, H., 296, 306
Markov chain, 364, 373–375, 378
Markov-exchangeable, 375
mathematics, 175–178, 180–182, 184, 188–193, 195, 198, 200
Max-Weight Approximation, 110
maximin criterion, 345, 346
maximin rule, 344, 345
maximum entropy, 9, 10, 203, 205
maximum entropy principle, 2, 97
Maxwell–Boltzmann vs. Bose–Einstein statistics, 139
Mayo, D., 14, 381
McClennen, E. F., 291, 296, 305, 307
measurement, 263

mechanisms, 28, 31
merger of opinions, 355
metaphysical, 49, 56, 65, 69, 71
  array, 42, 48, 55, 57
  hypothesis, 55
  model, 40, 43
  null hypothesis, 54
  probability model, 63
methodological, 358
methodology, 348, 355, 356
Meyer, B. de, 318
Miller, D., 357
Miller, H. D., 374, 378
Milne, P. M., 156
minimum agreement on consequences, 317
minumum agreement on acts, 317
mission-oriented Bayesianism, 31
model verification, 29
model-building, 55
modifiable structural equations, 33
modified chaotic clock, 352–354, 356
modularity, 26
Mongin, P., 12, 151, 318, 319
monotonicity, 46, 47
Monstrous Moonshine, 197
Monte Carlo method, 125
Morgenstern, O., 296, 307
Musgrave, A., 357
Muth, J. F., 341, 343

N-solutions, 216
negative binomial, 387, 388
Neptune, prediction of, 186
neural computation, 117
neural network, 8, 118
Neyman, J., 54, 363–365, 367, 376, 378, 381, 391, 396, 397
Neyman–Pearson, 386
nominal (or computed) SL, 389
nominal significance, 390, 391
nominal SL, 390
non-compliance, 49
non-deterministic, 63, 72
non-linear constraints, 203
non-recursive models, 31
null hypothesis, 386
Nyarko, Y., 342, 343, 357

Oakes, M., 399
objective Bayesianism, 2
objective priors, 9
objectivity, 1
obstinacy principle, 207, 212
odds, 140
OMTs, 343–348, 355
open-mindedness, 238
operationally meaningful theorems (MTs), 343
optional stopping effect, 388
Osband, K., 347
overfitting, 121

Pólya, G., 10, 175–178, 181, 183, 186–190, 192, 193, 195–197, 200
parents, 32
Paretian, 309
Paretianism, 309, 310, 329
Pareto principle, 328–330
Pareto-Indifference, 309, 317
Pareto-Weak Preference, 317
Paris, J., 10, 143
partial belief, 279
path diagrams, 25, 26
pattern recognition, 117
Pearl, J., 6, 7, 177, 189, 341
Pearson, E. S., 363, 381, 391, 396, 397
Pearson, K., 363, 379
perfect rationality, 343, 352, 354, 357–359
Pesaran, M. H., 341, 342
Phelps, E. S., 341
physical array, 42, 48, 57
physical model, 40, 44
pluralist, 1
Poincaré, H., 185

Poisson distribution, 365–368, 377
Poisson, S.-D., 138
Pollatsek, A., 298, 307
Pope, R. E., 342
Popper, K., 364, 366, 373, 379
positive solution, 223
posterior distribution, 67
potential response, 33, 42, 65
potential-outcome approach, 30
pragmatic problem of induction, 357
Pratt, J. W., 385
predesignated, 383
predictive distributions, 48, 51
predictive inferences, 48
preference, 342, 345, 346, 360
Price, R., 363, 378
principle of indifference, 2, 9, 97, 138, 218, 371
principle of insufficient reason, 344, 345, 357, 358
principle of the common cause, 78
probabilistic causal model, 34
probabilistic-cause, 29
probability
  dictator, 322
  function, 203
  kinematics, 156
  of causation, 24
  of counterfactual, 34
problem of induction, 353, 357
proof strategy, 194
propagation algorithms, 76
proper stopping rule, 389
protection against mistakes, 359
pseudo-determinism, 68
pseudo-structural nested distribution models, 57

quantum physics, 42
quantum theory, 57

Raiffa, H., 304, 305, 307, 385
Ramsey, F., 1
Ramsey, F. P., 11, 137, 263, 345
random variables, 371
randomization, 21
rational expectation, 341, 342, 355
rational line, 220
rational–expectations hypothesis (REH), 341
rationalised on, 343
rationality, 152, 341, 342, 344, 346, 349, 354, 357, 358
rationality hypothesis, 344
rationalizability, 343
rationalization, 349, 354
reasoning by induction, 369, 371
red or blue, 373–375, 377
reference class problem, 2
regularisation, 121
REH, 341–343
relativisation principle, 206, 211
renaming principle, 206, 210
repeated significance tests, 390
representation theorems, 264
restriction, 81
revealed preference (RP), 343, 345, 346
Riemann Hypothesis, 192
risk factors, 21
risk ratio, 21
Roberts, H. V., 395
Rosenkrantz, R., 187
Royall, R., 386, 395, 399
RP, 346, 347

Salmon, M., 341
sampling distribution, 386
Samuelson, P. A., 296, 307, 343
Savage Forum, 1959, 388, 392, 400
Savage Independence, 292, 298
Savage, L. J., 19, 137, 141, 291–295, 297–299, 302, 303, 306, 310, 316, 342, 344, 347, 381–385, 391, 392, 395, 396, 398, 399
Schervish, M. J., 311, 326, 393
Schlaifer, R., 385

Schmeidler, D., 311, 316, 327
Schuster, H. G., 350
Scott, D., 151
screening, 78
Seidenfeld, T., 305, 307, 311, 326, 393, 399
selective pressure, 346, 358, 359
Selten, R., 358
SEM, 26
Sen, A., 296, 307
seqential sampling, 383
sequential testing, 389
SEU, 342, 345, 349
severe test, 384
severity, 383
sheep, 53, 55
Shimony, A., 153
Shoemaker, P. J. H., 297, 307
significance level, 388
significance testing, 382
Simon, H. A., 346
Simpson's paradox, 22
singly-connected, 109
Smullyan, R., 144
social choice theory, 317
space complexity, 109
SRP, 391
St Petersburg Problem, 142
stable unit-treatment value assumption (SUTVA), 50, 57
stage one, 97
stage two, 97
stastistics, 342
state, 76
state independence, 313
state-dependent utility, 12
state-dependent utility theory, 310, 328
state-independence, 310
state-independent utility, 12
statistical analysis, 19
statistical decision theory, 345
statistical significance, 382, 386, 388
statistical tests, 363, 366, 373, 375
statistical uncertainty, 62
statistics, 341
stochastic logic programs, 11
stochastic processes, 20
stopping rule, 14, 382
stopping rule principle, 384
stopping rules, 383
strategy, 341, 345
Strict Pareto, 317
strict subjectivist, 1
strictly dominated, 345, 346
strictly dominated strategies, 345
strong compatibility, 57
Strong Pareto, 309, 317
structural equation modelling (SEM), 26
structural equations, 24
Structural Equations Model, 24
Student, 363
subjective expected utility, 310
Submodel, 32
sufficient concomitant, 62, 63, 65
Support Vector Machine (SVM), 127
Sure-Thing Principle, 12, 292–295, 302, 310, 342
SUTVA, 50
symmetry, 48
symmetry modelling, 56

t-test, 54, 55
Tarskian truth-definition, 143
Teller, P., 155
tent-map dynamics, 354
testing, 382
Tiao, G., 399
time complexity, 109
Todhunter, I., 142
treatment-unit additivity (TUA), 46, 49, 51
triviality theorem, 156
truncated expression, 27
TUA, 47, 53, 54, 59, 66
Tversky, A., 297, 298, 302, 304, 306
two-stage methodology, 97
Type A distribution, 367, 377

type I error, 389

Ulam, S., 139
ultimate causes, 71
uncertainty, 65
uniform prior, 398
uniformity, 46, 47, 65
uniqueness principle, 221
universal approximator, 119
unsharp probabilities, 153
unsupervised learning, 129
Urbach, P., 138, 344, 345, 396
utilities, 346
utility function, 341–345, 347
utility-maximizing, 349

value-assignment process, 25
Varian, H. R., 346
Vencovská, A., 10
Vind, K., 311, 327
von Neumann, J., 296, 307
von Neumann–Morgenstern axioms, 312

Wakker, P., 291, 306
Watts Assumption, 203
weak compatibility, 57
Weak Pareto, 309, 317
weight, 110
weight of evidence, 399
Williams, P., 8
Williamson, J., 7, 11, 150
Wolpert, R. L., 391, 395–400

APPLIED LOGIC SERIES

1. D. Walton: *Fallacies Arising from Ambiguity.* 1996 ISBN 0-7923-4100-7
2. H. Wansing (ed.): *Proof Theory of Modal Logic.* 1996 ISBN 0-7923-4120-1
3. F. Baader and K.U. Schulz (eds.): *Frontiers of Combining Systems.* First International Workshop, Munich, March 1996. 1996 ISBN 0-7923-4271-2
4. M. Marx and Y. Venema: *Multi-Dimensional Modal Logic.* 1996 ISBN 0-7923-4345-X
5. S. Akama (ed.): *Logic, Language and Computation.* 1997 ISBN 0-7923-4376-X
6. J. Goubault-Larrecq and I. Mackie: *Proof Theory and Automated Deduction.* 1997 ISBN 0-7923-4593-2
7. M. de Rijke (ed.): *Advances in Intensional Logic.* 1997 ISBN 0-7923-4711-0
8. W. Bibel and P.H. Schmitt (eds.): *Automated Deduction - A Basis for Applications.* Volume I. Foundations - Calculi and Methods. 1998 ISBN 0-7923-5129-0
9. W. Bibel and P.H. Schmitt (eds.): *Automated Deduction - A Basis for Applications.* Volume II. Systems and Implementation Techniques. 1998 ISBN 0-7923-5130-4
10. W. Bibel and P.H. Schmitt (eds.): *Automated Deduction - A Basis for Applications.* Volume III. Applications. 1998 ISBN 0-7923-5131-2
    (Set vols. I-III: ISBN 0-7923-5132-0)
11. S.O. Hansson: *A Textbook of Belief Dynamics.* Theory Change and Database Updating. 1999 Hb: ISBN 0-7923-5324-2; Pb: ISBN 0-7923-5327-7
    Solutions to exercises. 1999. Pb: ISBN 0-7923-5328-5
    Set: (Hb): ISBN 0-7923-5326-9
    Set: (Pb): ISBN 0-7923-5329-3
12. R. Pareschi and B. Fronhöfer (eds.): *Dynamic Worlds from the Frame Problem to Knowledge Management.* 1999 ISBN 0-7923-5535-0
13. D.M. Gabbay and H. Wansing (eds.): *What is Negation?* 1999 ISBN 0-7923-5569-5
14. M. Wooldridge and A. Rao (eds.): *Foundations of Rational Agency.* 1999 ISBN 0-7923-5601-2
15. D. Dubois, H. Prade and E.P. Klement (eds.): *Fuzzy Sets, Logics and Reasoning about Knowledge.* 1999 ISBN 0-7923-5911-1
16. H. Barringer, M. Fisher, D. Gabbay and G. Gough (eds.): *Advances in Temporal Logic.* 2000 ISBN 0-7923-6149-0
17. D. Basin, M.D. Agostino, D.M. Gabbay, S. Matthews and L. Viganò (eds.): *Labelled Deduction.* 2000 ISBN 0-7923-6237-3
18. P.A. Flach and A.C. Kakas (eds.): *Abduction and Induction.* Essays on their Relation and Integration. 2000 ISBN 0-7923-6250-0

19. S. Hölldobler (ed.): *Intellectics and Computational Logic*. Papers in Honor of Wolfgang Bibel. 2000 ISBN 0-7923-6261-6
20. P. Bonzon, M. Cavalcanti and Rolf Nossum (eds.): *Formal Aspects of Context*. 2000 ISBN 0-7923-6350-7
21. D.M. Gabbay and N. Olivetti: *Goal-Directed Proof Theory*. 2000 ISBN 0-7923-6473-2
22. M.-A. Williams and H. Rott (eds.): *Frontiers in Belief Revision*. 2001 ISBN 0-7923-7021-X
23. E. Morscher and A. Hieke (eds.): *New Essays in Free Logic*. In Honour of Karel Lambert. 2001 ISBN 1-4020-0216-5
24. D. Corfield and J. Williamson (eds.): *Foundations of Bayesianism*. 2001 ISBN 1-4020-0223-8

KLUWER ACADEMIC PUBLISHERS – DORDRECHT / BOSTON / LONDON